VOLUME SEVENTY

# ADVANCES IN
# MICROBIAL PHYSIOLOGY
## Microbiology of Metal Ions

VOLUME SEVENTY

# ADVANCES IN MICROBIAL PHYSIOLOGY

## Microbiology of Metal Ions


Edited by

**ROBERT K. POOLE**

*West Riding Professor of Microbiology*
*Department of Molecular Biology and Biotechnology*
*The University of Sheffield*
*Firth Court, Western Bank*
*Sheffield, UK*



ELSEVIER

ACADEMIC PRESS

An imprint of Elsevier

Academic Press is an imprint of Elsevier
The Boulevard, Langford Lane, Kidlington, Oxford OX5 1GB, United Kingdom
125 London Wall, London, EC2Y 5AS, United Kingdom
50 Hampshire Street, 5th Floor, Cambridge, MA 02139, United States
525 B Street, Suite 1800, San Diego, CA 92101-4495, United States

First edition 2017



ISBN: 978-0-12-812386-7
ISSN: 0065-2911

For information on all Academic Press publications
visit our website at https://www.elsevier.com/books-and-journals



*Publisher:* Zoe Kruze
*Acquisition Editor:* Alex White
*Editorial Project Manager:* Helene Kabes
*Production Project Manager:* Vignesh Tamil
*Cover Designer:* Greg Harris

Typeset by SPi Global, India

# CONTENTS

# CONTRIBUTORS

**Sankalp Arya**
School of Biosciences, the University of Nottingham, Sutton Bonington Campus, Sutton Bonington, Leicestershire, United Kingdom

**Karishma Asiani**
School of Biosciences, the University of Nottingham, Sutton Bonington Campus, Sutton Bonington, Leicestershire, United Kingdom

**Simon V. Avery**
School of Life Sciences, University of Nottingham University Park, Nottingham, United Kingdom

**Bezalel Bacon**
Stony Brook University, Stony Brook, NY, United States

**Anna Barwinska-Sendra**
Institute for Cell & Molecular Biosciences, Faculty of Medical Sciences, Newcastle University, Newcastle upon Tyne, United Kingdom

**Elizabeth Boon**
Stony Brook University, Stony Brook, NY, United States

**Aaron Crawford**
Aberdeen Fungal Group, MRC Centre for Medical Mycology, School of Medicine, Medical Sciences and Nutrition, University of Aberdeen, Institute of Medical Sciences, Foresterhill, United Kingdom

**Cameron L. Crawford**
The University of Alabama at Birmingham, Birmingham, AL, United States

**Alex G. Dalecki**
The University of Alabama at Birmingham, Birmingham, AL, United States

**Karrera Y. Djoko**
School of Chemistry and Molecular Biosciences and Australian Infectious Diseases Research Centre, The University of Queensland, Brisbane, QLD, Australia

**Colleen M. Hansel**
Woods Hole Oceanographic Institution, Woods Hole, MA, United States

**Jon L. Hobman**
School of Biosciences, the University of Nottingham, Sutton Bonington Campus, Sutton Bonington, Leicestershire, United Kingdom

**D.G. Joakim Larsson**
Institute of Biomedicine, Sahlgrenska Academy, University of Gothenburg; Centre for Antibiotic Resistance Research (CARe) at University of Gothenburg, Gothenburg, Sweden

**Dhara Malavia**
Aberdeen Fungal Group, MRC Centre for Medical Mycology, School of Medicine, Medical Sciences and Nutrition, University of Aberdeen, Institute of Medical Sciences, Foresterhill, United Kingdom

**Alastair G. McEwan**
School of Chemistry and Molecular Biosciences and Australian Infectious Diseases Research Centre, The University of Queensland, Brisbane, QLD, Australia

**Lisa-Marie Nisbett**
Stony Brook University, Stony Brook, NY, United States

**Cheryl-lynn Y. Ong**
School of Chemistry and Molecular Biosciences and Australian Infectious Diseases Research Centre, The University of Queensland, Brisbane, QLD, Australia

**Chandan Pal**
Institute of Biomedicine, Sahlgrenska Academy, University of Gothenburg; Centre for Antibiotic Resistance Research (CARe) at University of Gothenburg, Gothenburg, Sweden

**Christopher Rensing**
College of Resources and Environment, Fujian Agriculture and Forestry University, Fuzhou, Fujian, China; J. Craig Venter Institute, La Jolla, CA, United States

**Dov J. Stekel**
School of Biosciences, the University of Nottingham, Sutton Bonington Campus, Sutton Bonington, Leicestershire, United Kingdom

**Andrew G. Turner**
School of Chemistry and Molecular Biosciences and Australian Infectious Diseases Research Centre, The University of Queensland, Brisbane, QLD, Australia

**Cindy Vallières**
School of Life Sciences, University of Nottingham University Park, Nottingham, United Kingdom

**Kevin J. Waldron**
Institute for Cell & Molecular Biosciences, Faculty of Medical Sciences, Newcastle University, Newcastle upon Tyne, United Kingdom

**Mark J. Walker**
School of Chemistry and Molecular Biosciences and Australian Infectious Diseases Research Centre, The University of Queensland, Brisbane, QLD, Australia

**Duncan Wilson**
Aberdeen Fungal Group, MRC Centre for Medical Mycology, School of Medicine, Medical Sciences and Nutrition, University of Aberdeen, Institute of Medical Sciences, Foresterhill, United Kingdom

**Frank Wolschendorf**
The University of Alabama at Birmingham, Birmingham, AL, United States

# PREFACE

Metal ions and metal biochemistry occupy key roles in all biology and nowhere is the importance of metals for life more apparent than in microbiology. Nearly half of all enzymes are thought to require metals for assembly and/or function. Those elements required for biology and microbiology have been the subject of detailed monographs (for example, Frausto da Silva & Williams, 2001; Hughes & Poole, 1989). Although specific metal ions are needed for proper function, most metalloproteins may bind to one or more incorrect metals (including metal pollutants) sometimes more tightly than the correct ones. Thus, the metal handling systems of cells are vital to sustain adequate metal–protein speciation in vivo.

Cells must practise metal homeostasis, ensuring that the abundance of metals in each protein is carefully regulated. The biological challenge is that the metal composition of cells is markedly different from that in the biosphere. Thus cells must import, export, chelate and detect metal ions for correct structure and function. The ways in which cells achieve these functions are well illustrated by many chapters in this special edition of *Advances in Microbial Physiology*.

Metal ions are not only essential in microbial life but are also potentially toxic. Cells must resist the import of toxic metals, detoxify inappropriate metals or export them. There is a lively metal economy across the borders of the microbial cell. Pertinent also is the fact that many potent antimicrobial compounds are metal-containing. To understand how such antimicrobials work, or may be better designed, the fundamental mechanisms of metal homeostasis must be understood.

Finally, microbes play key roles in the biogeochemical cycling of elements, both mobilising and immobilising natural reserves with implications for nutrition, environmental protection and biotechnology. All these aspects are touched upon in this volume.

There is ample evidence for a persistent research interest in metal biology. For example, a long-standing and regular series of Gordon Research Conferences is held on this topic, and in the United Kingdom the Biotechnology and Biological Sciences Research Council (BBSRC), with support from the Engineering and Physical Sciences Research Council (EPSRC), has funded a lively 'Metals in Biology in Industrial Biotechnology and Bioenergy' Network. Its aim is to foster collaborations between academia, industry, policy makers and nongovernmental organisations to identify new approaches to tackle scientific challenges, translate research and deliver key benefits in industrial biotechnology and bioenergy.

Our ability to understand and manipulate metals in biology is critical to harness the potential of biological resources for producing and processing materials, biopharmaceuticals, chemicals and energy.

Robert K. Poole
March 2017

## REFERENCES

Frausto da Silva, J. J. R., & Williams, R. J. P. (2001). *The biological chemistry of the elements: The inorganic chemistry of life* (2nd ed.). Oxford: Oxford University Press.

Hughes, M. N., & Poole, R. K. (1989). *Metals and micro-organisms*. London: Chapman & Hall.

CHAPTER ONE

# Bacterial Haemoprotein Sensors of NO: H-NOX and NosP

**Bezalel Bacon[1], Lisa-Marie Nisbett[1], Elizabeth Boon[2]**
Stony Brook University, Stony Brook, NY, United States
[2]Corresponding author: e-mail address: elizabeth.boon@stonybrook.edu

## Contents



## Abstract

Low concentrations of nitric oxide (NO) modulate varied behaviours in bacteria including biofilm dispersal and quorum sensing-dependent light production. H-NOX (haem-nitric oxide/oxygen binding) is a haem-bound protein domain that has been shown to be involved in mediating these bacterial responses to NO in several organisms. However, many bacteria that respond to nanomolar concentrations of NO do not contain an annotated H-NOX domain. Nitric oxide sensing protein (NosP), a newly discovered bacterial NO-sensing haemoprotein, may fill this role. The focus of this review is to discuss structure, ligand binding, and activation of H-NOX proteins, as well as to discuss the early evidence for NO sensing and regulation by NosP domains.

[1] Authors contributed equally to this work.

*Advances in Microbial Physiology*, Volume 70
ISSN 0065-2911
http://dx.doi.org/10.1016/bs.ampbs.2017.01.004

Further, these findings are connected to the regulation of bacterial biofilm phenotypes and symbiotic relationships.

## ABBREVIATIONS

**ATP** adenosine triphosphate
***Cb*** *Clostridium botulinum*
**CO** carbon monoxide
***Cs*** *Caldanaerobacter subterraneus*
**Cyclic-di-GMP** bis-(3′-5′)-cyclic dimeric guanosine monophosphate
**FIST** F-box intracellular signal transduction protein
**FNR** fumarate and nitrate regulatory proteins
**GMP** guanosine 5′ monophosphate
**GTP** guanosine triphosphate
**HaCE** H-NOX-associated cyclic-di-GMP processing enzymes
**HaHK** H-NOX-associated histidine kinase
**HNOB** haem-nitric oxide binding domain
**H-NOX** haem-nitric oxide/oxygen-binding domain
**Hpt** histidine-containing phosphotransfer protein
**HqsK** H-NOX-associated quorum sensing histidine kinase
***Lpg*** *Legionella pneumophila*
**MCP** methyl-accepting chemotaxis protein
**NaHK** NosP-associated histidine kinase
**NCSD** normal-coordinate structural decomposition
**NMR** nuclear magnetic resonance
**NO** nitric oxide
**NorR** regulator of NO reductase
**NosP** nitric oxide sensing protein
***Ns*** *Nostoc* sp.
**$O_2$** dioxygen
***Pa*** *Pseudomonas aeruginosa*
**pGpG** 5′-phosphoguanylyl-(3′-5′)-guanosine
**rR** resonance Raman
**sGC** soluble guanylate cyclase
***Sili*** *Silicibacter* sp.
***So*** *Shewanella oneidensis*
***Sw*** *Shewanella woodyi*
***Tt*** *Thermoanaerobacter tengcongensis*
***Vc*** *Vibrio cholerae*
***Vf*** *Vibrio fischeri*

## 1. PHYSIOLOGICAL FUNCTIONS OF NITRIC OXIDE

Nitric oxide (NO) is a diatomic, free radical gas molecule. It can readily diffuse into cells (Barraud, Kelso, Rice, & Kjelleberg, 2015) due to its lipophilicity (Shaw & Vosper, 1977) and its high diffusion coefficient of 3300 $Mm^2 s^{-1}$ (Malinski et al., 1993). As such it is capable of rapid intercellular signalling. In eukaryotes, NO signalling plays many important and well-studied physiological roles. For example, NO engages in paracrine signalling (Thomas, Liu, Kantrow, & Lancaster, 2001). It is produced in a generator cell by an NO synthase from L-arginine (Denninger & Marletta, 1999), after which it subsequently diffuses across the lipid bilayer of an adjacent cell and activates the mammalian NO sensor protein sGC (Cary, Winger, Derbyshire, & Marletta, 2006).

NO also plays many roles in bacterial physiology. For example, when bacteria are engaged in infecting eukaryotic cells, immune cells produce high concentrations of NO upon stimulation by pathogen-associated molecular patterns, such as flagellin and lipopolysaccharides (Bellota-Anton et al., 2011). Additionally, bacteria may encounter NO endogenously produced as an intermediate in respiratory denitrification (Bellota-Anton et al., 2011). At elevated concentrations, NO causes nitrosative stress that hinders normal bacterial cell function and leads to cell death. Consequently, to ensure their survival, bacteria have evolved various mechanisms to detect and detoxify NO. Such mechanisms include the detection of NO by various transcriptional regulators including FNR-like transcription factors (fumarate and nitrate regulatory proteins) (Cruz-Ramos et al., 2002), the NO-responsive transcriptional activator NorR (regulator of NO reductase) (D'Autreaux, Tucker, Dixon, & Spiro, 2005), and the NO-sensitive repressor NsrR (repressor of nitrosative stress) (Bodenmiller & Spiro, 2006). Bacteria detoxify NO using NO-binding enzymes such as flavohaemoglobins, flavorubredoxin NO reductases, respiratory NO reductases, and cytochrome *c* nitrite reductases, each of which converts NO into less toxic molecules such as ammonia, nitrate, and nitrous oxide (Gardner, Helmick, & Gardner, 2002; Mills, Rowley, Spiro, Hinton, & Richardson, 2008; Poole & Hughes, 2000; Stevanin, Moir, & Read, 2005; Stevanin, Poole, Demoncheaux, & Read, 2002). While NO detoxification has generally been documented to occur at micromolar concentrations of NO (Gardner et al., 2002; Mills et al., 2008; Poole & Hughes, 2000; Stevanin et al., 2005, 2002), in some bacteria NO detoxification pathways are known to be stimulated by nanomolar amounts of NO (Hassan, Bergaust, Molstad, de Vries, & Bakken, 2016).

Recently, many studies have also demonstrated that bacteria, like eukaryotes, can also detect NO at low concentrations as a sensitive and specific signalling molecule (Arora, Hossain, Xu, & Boon, 2015). In particular, low concentrations of NO are frequently implicated in regulating bacterial biofilm formation (Barraud et al., 2015; McDougald, Rice, Barraud, Steinberg, & Kjelleberg, 2011). Biofilms occur when bacteria accumulate on a surface in a moist environment within a self-secreted exopolysaccharide matrix (Bhinu, 2005). Biofilms are problematic, as when in this state, bacteria not only exhibit up to 1000-fold higher tolerance to immune responses and antibiotics (Hoiby, Bjarnsholt, Givskov, Molin, & Ciofu, 2010; Stewart & Costerton, 2001), but have also been demonstrated to be the leading cause of nosocomial infections (Lindsay & von Holy, 2006; Wenzel, 2007).

Currently, there are two known bacterial haem-based sensors of NO used in this type of signal transduction: the relatively well-studied H-NOX (haem-nitric oxide/oxygen-binding domain) family and the newly discovered NosP (nitric oxide sensing protein) family. Bacterial H-NOX domains are members of the same superfamily of haem sensors that includes soluble guanylate cyclase (sGC), the eukaryotic sensor of NO (Iyer, Anantharaman, & Aravind, 2003). In fact, the large majority of H-NOX proteins are animal proteins. NosP, on the other hand, is almost exclusively coded for in bacterial genomes. This paper focuses on reviewing what is currently known about NO sensing by H-NOX and NosP in bacteria.

## 2. sGC: THE ANIMAL NO SENSING PROTEIN

Bacterial H-NOX proteins were discovered based on homology to sGCs. Therefore, a brief overview of sGC is presented here. sGC is a haem-based sensor. It is a heterodimeric enzyme composed of $\alpha 1$ (mass ranging from 73 to 88 kDa) and $\beta 1$ (mass of ~80 kDa) subunits that bind one protoporphyrin IX group per dimer (Hobbs & Ignarro, 1996). The haem-binding domain of sGC has been localized to the first 194 amino acids of the $\beta 1$ subunit (Karow et al., 2005) with histidine 105 as the proximal ligand (Wedel et al., 1994). In its resting state, the haem iron of sGC is in the reduced, unligated ferrous state [Fe(II)-unligated], and is available for NO detection. sGC binds picomolar NO at the haem iron, forming the ferrous-NO [Fe(II)-NO] complex, which activates the enzyme to convert GTP (guanosine triphosphate) to cGMP (cyclic guanosine monophosphate) (Stone & Marletta, 1996), ultimately mediating downstream events such as vasodilation, platelet aggregation, neurotransmission, and myocardial function via calcium channels, phosphodiesterases, and kinases (Garthwaite, 2008; Lucas et al., 2000).

On a molecular level, NO ligation to the haem iron causes cleavage of the Fe–His bond, which is believed to result in a conformational change that ultimately leads to the activation of the catalytic domain (Cary, Winger, & Marletta, 2005; Fernhoff, Derbyshire, & Marletta, 2009; Russwurm & Koesling, 2004). Much of what is known about the activation of sGC comes from studies of bacterial H-NOX proteins, however, which will be detailed later.

## 3. DISCOVERY OF A BACTERIAL NO SENSING PROTEIN: H-NOX

In 2003, a family of haem proteins in bacteria with 15%–40% identity to the haem domain from rat sGC was identified (Iyer et al., 2003) (Fig. 1). The family includes the eukaryotic sGCs as well as several hundred predicted open-reading frames (ORFs) from bacterial genomes. Based on their high homology with sGC, these domains were initially named HNOBs for haem-nitric oxide-binding domain.

These bacterial haem domains are all about 190 amino acids long. At the time of discovery, the structure of this domain was unknown, but it would later become clear that these bacterial proteins have particularly high identity to sGC in the haem-binding region (Schmidt, Schramm, Schroder, Wunder, & Stasch, 2004). In particular, several amino acids are absolutely conserved and can be considered signatures of the family. Besides the histidine haem ligand (H105 in rat sGC β1), a YxS/TxR motif is conserved (residues 135, 137, and 139, respectively, in rat β1), as is a proline residue

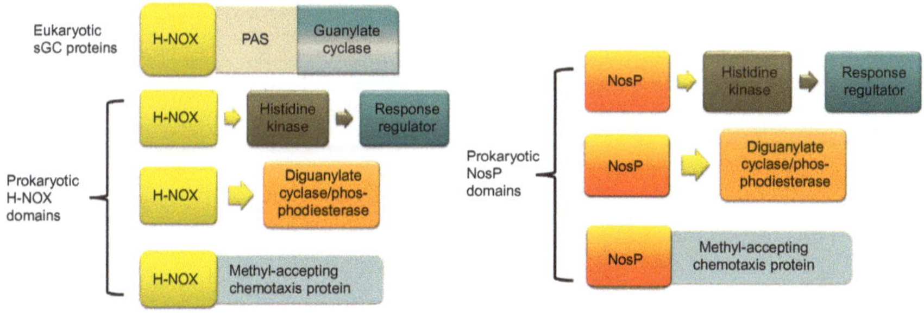


**Fig. 1** H-NOX, the NO sensing domain from the animal NO sensor sGC, is conserved as a haemoprotein sensor in bacteria. In bacteria, it is found in operons with histidine kinases, cyclic-di-GMP synthases and/or phosphodiesterases, and methyl-accepting chemotaxis proteins (*left*). NosP (NO sensing protein), a newly discovered bacterial NO sensing protein has operon arrangements similar to those of H-NOX (*right*).

about 10 amino acids towards the C-terminus from the haem-binding histidine residue (residue 118 in rat β1) (Fig. 2). The YxS/TxR has now been identified as important for coordinating the haem propionate side chains (Pellicena, Karow, Boon, Marletta, & Kuriyan, 2004) (Fig. 3). This motif, along with the proline residue, is now known to be important for maintaining a unique haem structure in this family (see Section 5.2).

In 2004, the Marletta group recombinantly expressed and purified two HNOB domains, one cloned from *Vibrio cholerae* (Vc_0270) and one from *Caldanaerobacter subterraneus* (at the time known as *Thermoanaerobacter*

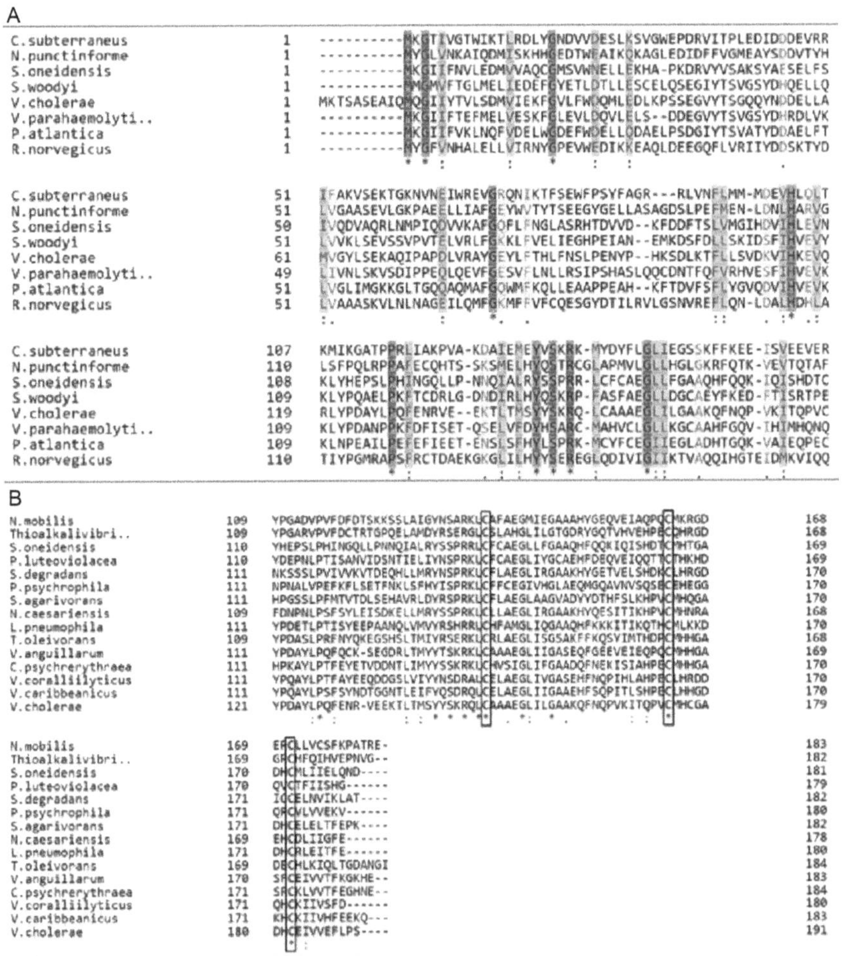


**Fig. 2** (A) Sequence alignment of H-NOX proteins discussed in this review. Of note are the conserved histidine haem ligand (His104 by rat β1 sGC numbering), the YxSxR motif (Tyr135, Ser137, Arg139) that coordinates the haem propionate groups, and the Pro118 residue that maintains a distorted haem cofactor. (B) Multiple sequence alignments of selected gammproteobacteria in which the zinc-binding cysteine residues are conserved.

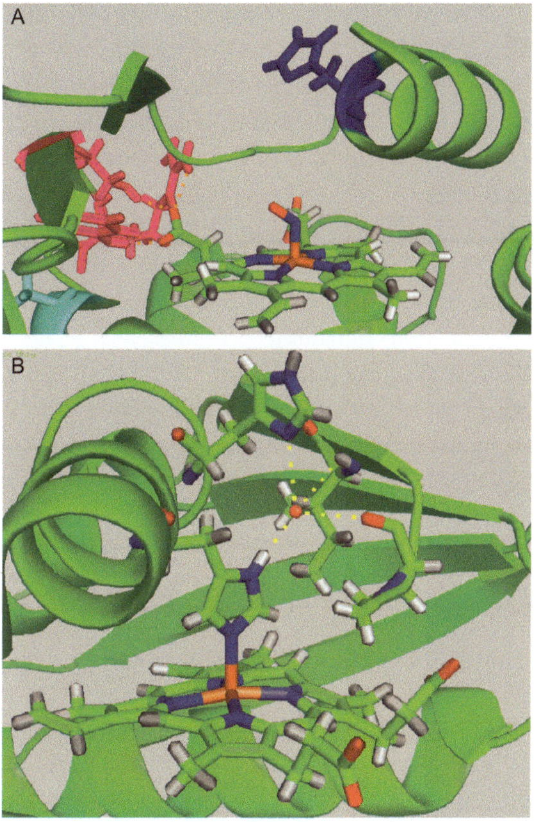


**Fig. 3** (A) The structure of Fe(II)-NO *So*-H-NOX. In *So*-H-NOX, upon NO (*orange*) binding, the histidine proximal ligand (*blue*) dissociates. In fact, NO is bound to the proximal face of haem, a condition possibly brought about by the high NO concentrations that the protein is exposed to during crystallization (Herzik, Jonnalagadda, Kuriyan, & Marletta, 2014). The YxSxR (131, 133, and 135, respectively) motif (*magenta*) is shown coordinating haem propionate groups. The haem in this structure is more relaxed than the haem in the Fe(II)-unliganded state. These data support hypotheses that NO binding triggers Fe–His bond cleavage and haem relaxation that ultimately resulting in signal transduction. Generated in pymol from pdb file 4U9B. (B) The structure of Fe(II)-unligated *So*-H-NOX. Notable features include Pro116-induced (*red*) haem distortion and the bound water molecule coordinating residues His99, Ile118, Pro116, and the proximal haem ligand, His102.

*tengcongensis*; TTE0680 basepairs 1–564) (Karow et al., 2004). They found that, as expected, these domains are histidine-ligated protoporphyrin IX haemoproteins that bind gaseous ligands at the ferrous iron atom of the haem cofactor. It was also discovered that the ferrous protein from *C. subterraneus*, in addition to binding NO, could stabilize a complex with molecular oxygen [Fe(II)–$O_2$]. Based on this discovery, this family of haem sensors was renamed H-NOX, for haem-nitric oxide and/or oxygen-binding domain.

In facultative aerobic bacteria, in general, most H-NOX domains are encoded in a histidine kinase-containing operon; some are encoded in a cyclic-di-GMP synthase and/or phosphodiesterase-containing operon (Iyer et al., 2003) (Fig. 1). H-NOX domains from obligate anaerobes are encoded as the N- or C-terminal domain of methyl-accepting chemotaxis proteins (MCPs) (Cary et al., 2006; Iyer et al., 2003). Based on homology to sGC and genomic association with signalling proteins, it was hypothesized that bacterial H-NOX proteins would also serve as sensors of NO (or possibly $O_2$ in the case of obligate aerobic organisms), regulating signal transduction of the associated signalling proteins. Many subsequent studies have confirmed the role of the interaction of H-NOX with NO in modulating the activities of these proteins (Arora & Boon, 2012; Henares, Higgins, & Boon, 2012; Henares, Xu, & Boon, 2013; Liu et al., 2012; Muralidharan & Boon, 2012); these studies will be described later in this review (see Section 6).

## 4. LIGAND-BINDING PROPERTIES OF H-NOX DOMAINS

Many bacterial H-NOX domains have now been cloned, expressed, purified, and spectroscopically characterized (Arora & Boon, 2012; Boon et al., 2006; Dai, Farquhar, Arora, & Boon, 2012; Henares et al., 2012; Karow et al., 2004; Liu et al., 2012; Ma, Sayed, Beuve, & van den Akker, 2007; Mukhopadyay, Sudasinghe, Schaub, & Yukl, 2016; Price, Chao, & Marletta, 2007; Tsai, Berka, Martin, & Olson, 2012; Tsai et al., 2010; Wang et al., 2010; Wu, Liu, Berka, & Tsai, 2015). UV/vis spectroscopy, paired with resonance Raman (rR) spectroscopy, has demonstrated that ferrous H-NOX domains from facultative aerobes form high-spin 5-coordinate Fe(II)-unligated complexes (Boon et al., 2006; Karow et al., 2004). Upon addition of carbon monoxide (CO), these proteins form low-spin 6-coordinate Fe(II)–CO complexes, while upon ligation to NO the proteins form high-spin 5-coordinate Fe(II)-NO complexes (Karow et al., 2004). Thus, H-NOX domains from facultative aerobes, like sGC, form 5-coordinate NO complexes and rigorously exclude $O_2$ as a ligand (Karow et al., 2004; Price et al., 2007). The ferrous H-NOX domains from obligate anaerobes, however, are similar to the globins, forming stable low-spin 6-coordinate complexes with $O_2$, NO, and CO (Cary et al., 2006; Karow et al., 2004; Tran, Boon, Marletta, & Mathies, 2009).

To date, all purified bacterial H-NOX domains, regardless of species of origin or coordination number of the Fe(II)-NO complex, display slow NO dissociation kinetics (e.g., $15.2 \times 10^{-4}\ s^{-1}$ for *Sw*-H-NOX, $21 \times 10^{-4}\ s^{-1}$ for *Vf*-H-NOX, $3.6 \times 10^{-4}\ s^{-1}$ for rat sGC, $0.05\ s^{-1}$ for *Ns*-H-NOX, $0.3\ s^{-1}$ for *Vc*-H-NOX) (Liu et al., 2012; Stone & Marletta, 1994, 1996; Tsai et al., 2010; Wang et al., 2010; Wu, Liu, Berka, & Tsai, 2013; Zhao, Brandish, Ballou, & Marletta, 1999) with an assumed diffusion-limited association rate of $\sim 10^8\ M^{-1}\ s^{-1}$. Consequently, these domains are thought to possess nanomolar to picomolar affinity for NO, underscoring a role for them in NO sensing and signalling in bacteria (see Section 6).

# 5. STRUCTURE AND THE MOLECULAR BASIS FOR FUNCTION IN H-NOX DOMAINS

Thus far, H-NOX domains from *C. subterraneus*, *Nostoc* sp., and *Shewanella oneidensis* have been characterized with molecular resolution. In this section of the review, we will discuss how key observations from these structural studies provide possible insight into the molecular mechanisms by which H-NOX proteins initiate downstream signal transduction events, most notably haem flattening and Fe–His bond cleavage.

The first H-NOX domain to be crystallized was from the thermophilic anaerobe *C. subterraneus*. The structure of the Fe(II)–$O_2$ complex of the *Cs*-H-NOX domain (at the time known as *Tt*-H-NOX) was solved to 1.7 Å resolution (Pellicena et al., 2004), revealing that the H-NOX family has a novel protein fold that consists of seven α-helices and a four-stranded antiparallel β-sheet. The N-terminal region consists of five helices, αA-αD and αG, which are situated on the distal side of the haem. The C-terminal region, however, is on the proximal side of the haem and consists of the β-sheet, the αF helix (the signalling helix where the histidine proximal ligand is located), and a one-turn helix, αE.

Further analysis of the ferrous-oxy *Cs*-H-NOX structure reveals that a distal pocket hydrogen-bonding network between the iron-bound $O_2$ and three distal pocket amino acids (Trp9, Asn74, and Tyr140 in *Cs*-H-NOX) is present (Pellicena et al., 2004) (Fig. 4). This distal pocket H-bonding network has been shown to play a role in $O_2$ binding by this H-NOX (Boon, Huang, & Marletta, 2005; Hespen, Bruegger, Phillips-Piro, & Marletta, 2016) and is hypothesized to contribute to NO/$O_2$ ligand discrimination in the H-NOX family (Boon et al., 2005; Boon & Marletta, 2005a, 2005b), as will be further discussed in Section 5.2.

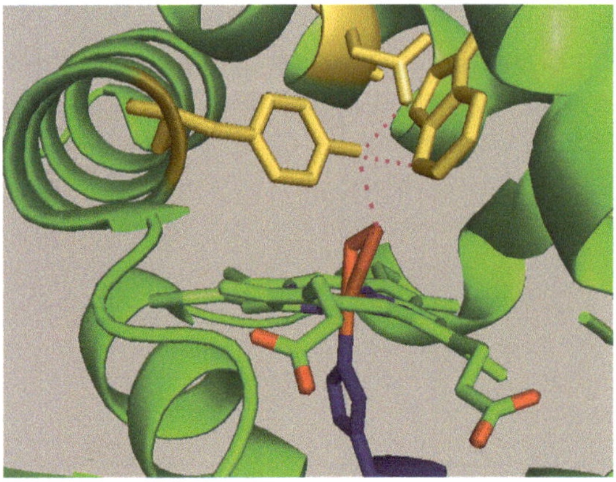

**Fig. 4** The structure of Fe(II)-$O_2$ *Cs*-H-NOX. Oxygen (*red*) binds haem in a bent configuration and is coordinated by Tyr140 and its hydrogen-bonding (*yellow*) network, consisting of Trp9 and Asn74. His104, the proximal iron ligand (*blue*), remains bound upon $O_2$ binding, forming a 6-coordinate complex. The high level of haem distortion can be observed. Generated in PYMOL using PDB file 1U55.

The structure of ferrous-oxy *Cs*-H-NOX also led to the identification of a unique structural feature of the H-NOX family: a highly distorted haem cofactor, as well as the suggestion that a proline residue [Pro115 in *Cs*-H-NOX; this proline is absolutely conserved in the H-NOX family (Karow et al., 2004)] is vitally important in maintaining this haem distortion (Dai & Boon, 2011; Erbil, Price, Wemmer, & Marletta, 2009; Olea, Boon, Pellicena, Kuriyan, & Marletta, 2008; Olea, Kuriyan, & Marletta, 2010; Sun et al., 2016). To assess the significance of Pro115 in haem distortion in *Cs*-H-NOX, the proline to alanine mutant was crystallized and solved to 2.1 Å (Olea et al., 2008). This structure demonstrated that, in addition to haem flattening in the absence of steric crowding from Pro115, this change in haem structure is accompanied by an N-terminal subdomain shift. The connection between the overall protein structure and haem flattening suggests that haem flattening may play a vital role in the H-NOX signalling process. Subsequent structures of *Cs*-H-NOX, as well as other H-NOX domains in a variety of ligand-bound complexes have provided support for this hypothesis and will be discussed further in Section 5.2.

An additional interesting feature revealed by this initial structure of *Cs*-H-NOX, is that the haem is buried in the protein fold (Pellicena et al., 2004). This has raised the question of gas ligand migration within the H-NOX fold. Computational models predict that gas migration to the haem site is guided by molecular tunnels (Zhang, Lu, Cheng, & Li, 2010). Confirming this prediction, *Cs*-H-NOX and *Ns*-H-NOX (H-NOX from *Nostoc* sp.) were crystallized under pressure with xenon gas (Winter, Herzik, Kuriyan, & Marletta, 2011). In comparison to *Cs*-H-NOX (which is not predicted to contain gas tunnel networks), clear continuous tunnels containing xenon were observed in *Ns*-H-NOX.

In order to gain insight into the structural basis of signal transduction, H-NOX constructs representative of pre- and post-NO binding have been obtained and studied. To this end, high-resolution structures of *S. oneidensis* H-NOX were solved by both nuclear magnetic resonance (NMR) and crystallography. *So*-H-NOX in 5-coordinate Fe(II)-NO, 6-coordinate Mn(II)-NO, and Fe(II)-unligated states have been solved to 1.65, 2.45, and 2.00 Å resolution, respectively (Herzik et al., 2014). While *So*-H-NOX usually forms a 5-coordinate complex with NO (due to Fe–His bond cleavage), a mimic of the 6-coordinate intermediate (with metal–His bond intact) was generated by substituting manganese protoporphyrin IX for haem. NMR solution structures of *So*-H-NOX with and without an intact Fe–His bond have also been obtained (Erbil et al., 2009) by comparing the Fe(II)-CO complexes of wild-type *So*-H-NOX and the haem ligand mutant H103G. These studies provide molecular insight to the role of histidine–iron cleavage in H-NOX signalling transduction. Further analysis of the Fe–His interaction and its relation to signalling is discussed in Section 5.3.

## 5.1 Ligand Discrimination in H-NOX Domains

sGC and the H-NOX domains from facultative anaerobes have been demonstrated to exclusively bind NO and CO, but not molecular $O_2$ (Karow et al., 2004; Liu et al., 2012; Price et al., 2007; Stone & Marletta, 1994; Tsai et al., 2010; Wang et al., 2010). The ability to exclude $O_2$ is crucial to these proteins functioning as sensitive NO sensors in an aerobic environment. Consequently, the ability to rigorously exclude $O_2$ by these H-NOX proteins has long been a topic of interest. The discovery of H-NOX domains that bind $O_2$ has offered researchers an opportunity to understand this molecular phenomenon: understanding why *Cs*-H-NOX can stabilize $O_2$ as a ligand should translate into further understanding why sGC and facultative anaerobic H-NOX family members exclude $O_2$.

The high-resolution structure of the *Cs*-H-NOX Fe(II)-$O_2$ complex (Pellicena et al., 2004) revealed that the dioxygen ligand is bent with respect to the plane of the porphyrin, as expected from studies of the globin family.

Furthermore, the $O_2$ molecule was found to interact with a distal pocket hydrogen-bonding network. A primary H-bond is present between the iron-bound $O_2$ and the distal pocket tyrosine 140. Tyr140 is further part of a distal pocket hydrogen-bonding network with tryptophan 9 and asparagine 74. It appears that Tyr140 is held in the optimal position for H-bonding with iron ligands through stabilizing H-bonds from Trp9 and Asn74 (Fig. 4). Sequence alignments reveal that this hydrogen-bonding network is conserved only in H-NOX proteins from obligate anaerobes (Pellicena et al., 2004) (i.e., they bind $O_2$, NO, and CO), leading to the hypothesis that this network is required for $O_2$ binding and its absence may result in ligand discrimination in H-NOX domains.

Support for this hypothesis was initially demonstrated by mutational analysis experiments in *Cs*-H-NOX. Mutation of the distal pocket Tyr140 to leucine in *Cs*-H-NOX (Boon et al., 2005) was found to increase the $O_2$ dissociation rate constant by 20-fold, while the NO dissociation rate constant remained unchanged. Furthermore, introduction of a mutant tyrosine residue into the hydrophobic distal pockets of the rat sGC H-NOX domain (β1 1–194) or the *Legionella pneumophila L2*-H-NOX domain enabled these domains to gain $O_2$-binding function (Boon et al., 2005). These observations suggest a compelling molecular basis for ligand selectivity in the H-NOX family: a distal pocket H-bonding residue is requisite for $O_2$ binding and is used to kinetically distinguish between NO and $O_2$. In the absence of this tyrosine, the $O_2$ dissociation rate is so fast that the $O_2$ complex is never formed, while the rate of NO dissociation remains unchanged, thus providing discrimination (Boon et al., 2005).

Studies of the Pro115 mutant of *Cs*-H-NOX (haem-flattened mutant) have bolstered this hypothesis (Dai & Boon, 2011; Olea et al., 2008; Sun et al., 2016). The *Cs*-H-NOX P115A mutant was shown to have a significantly slower $O_2$ dissociation rate constant in comparison to wild-type *Cs*-H-NOX, but the $O_2$ association rate constant remained unchanged (Olea et al., 2008). As a result, the authors reasoned that the absence of the proline residue causes an increase in the affinity for dioxygen, leading to a more stable Fe(II)-$O_2$ complex. This was elaborated in vibrational correlation spectroscopy experiments (Sun et al., 2016). Photolysis of the Fe–$O_2$ bond in wild-type and P115A *Cs*-H-NOX constructs revealed that the wild-type Fe–$O_2$ bond is more resistant to photolysis and has a longer $O_2$-rebinding time in comparison to the P115A mutant. Consequently, these data suggest that there is an overall stronger iron–$O_2$ bond in the distorted haem of wild type, but a greater degree of ligand trapping in the nondistorted mutant, resulting from a more tightly packed distal pocket and a stronger H-bond between Tyr140 and $O_2$.

Mutational studies introducing a Tyr into the distal pocket of full-length sGC, however, seem to contradict the notion that the distal pocket tyrosine (and by extension the hydrogen-bonding network) is important for facilitating $O_2$ binding to H-NOX domains. When a distal tyrosine (I145Y) was introduced into the distal pocket of full-length sGC, as opposed to the haem-binding domain of sGC only (β1 1–194) (Boon et al., 2005), researchers did not find evidence for $O_2$-binding (Martin, Berka, Bogatenkova, Murad, & Tsai, 2006). Moreover, characterization of the H-NOX domain from the obligate anaerobic bacterium *Clostridium botulinum* revealed that, although *Cb*-H-NOX contains a distal pocket tyrosine, it has no measurable affinity for $O_2$ (Nioche et al., 2004). Consequently, these findings suggest that oxygen binding to H-NOX domains is not solely mediated by the presence of the distal pocket tyrosine residue, but may also be dependent on other structural features.

Indeed, an alternate hypothesis for the role of the hydrogen-bonding network in H-NOX proteins has been proposed, called the sliding-scale hypothesis (Tsai, Berka, et al., 2012). This hypothesis is based on the observation that it is consistently true that gas-binding haem proteins bind NO with about ~1000-fold greater affinity than CO, and CO with ~1000-fold greater affinity than $O_2$, even when the affinity for any one of these ligands varies by more than six orders of magnitude from one protein to another. Notable exceptions to this rule include proteins whose primary function is to bind $O_2$ (Tsai, Berka, et al., 2012). According to the sliding-scale hypothesis, in order to compensate for the much lower affinity haem iron would normally have for $O_2$ than for the other gas ligands, selective structures in the protein are necessary to ensure that oxygen is the preferred binding partner. Therefore, hydrogen-bonding residues in the distal pocket are used to increase the affinity for $O_2$ to near the affinity for CO. This hypothesis is supported by data that show that when the distal pocket H-bonding resides in $O_2$-binding proteins, such as the globins, are mutated to nonpolar residues, the sliding-scale ratios are restored (Tsai, Berka, et al., 2012). This hypothesis predicts, therefore, that the H-bonding network in H-NOX from anaerobic bacteria functions to favour $O_2$-binding with respect to NO- and CO-binding, by lowering the binding rate constants and distorting the sliding scale, while H-NOX from aerobic bacteria conform to the sliding scale (Wu et al., 2013). It follows from this hypothesis that sGC also conforms to the sliding-scale rule and simply does not have the affinity to bind

$O_2$ under physiologically relevant conditions, and that the introduction of a single tyrosine residue is insufficient to lower the $O_2$-binding affinity to measurable levels.

## 5.2 Haem Distortion and Its Role in Signal Transduction

As mentioned earlier, a particularly striking feature of the H-NOX fold is the severely distorted nature of the haem cofactor. Normal-coordinate structural decomposition (NCSD) analysis (Senge et al., 1997) revealed that the *Cs*-H-NOX haem has displacements of −1.0 Å saddling and −1.2 Å ruffling. It is not uncommon for haem proteins to cause distortion of the haem cofactor, but the degree of distortion in H-NOX proteins is significantly higher than that of other known haem proteins. For example, cytochrome *c3* and nitrophorin (protein data bank codes 2CDV and 1ERX), both of which are considered to have severely distorted haem groups, exhibit about 1 and 0.4 Å saddling and −0.4 and −0.8 Å ruffling, respectively (Pellicena et al., 2004).

This large degree of nonplanarity can largely be attributed to two factors specific to H-NOX. First, the haem in H-NOX is buried within the protein, where the propionate groups are constrained by interactions with the highly conserved YxS/TxR motif (Pellicena et al., 2004) (Fig. 3). The arginine has been shown to contribute H-bonds to the haem propionate groups on both pyrrole A and D, while the tyrosine and serine have both been demonstrated to contribute to hydrogen bonding with only the propionate on pyrrole A. Secondly, Van der Waals interactions between the absolutely conserved H-NOX proline residue and one of the pyrroles of the haem causes the pyrrole to shift out of plane (Olea et al., 2008; Pellicena et al., 2004).

Interestingly, the structure of the *Cs*-H-NOX Fe(II)-$O_2$ complex was solved in two different crystal forms, each with two monomers per asymmetric unit, which yielded four independent views of the protein structure (Pellicena et al., 2004). Variation in the structure of the haem group between these structures suggests that there is conformational flexibility in the H-NOX haem structure. This observation has led to the hypothesis that alterations in the degree of haem distortion could be coupled to surface changes in H-NOX, which could ultimately lead to changes in intermolecular interactions with cocistronic partner proteins and subsequent changes in signal transduction downstream.

Crystal structures of H-NOX domains bound to various ligands support this hypothesis. Structures of the H-NOX domain from *Nostoc* sp. as the

Fe(II)-NO, Fe(II)-CO, and Fe(II)-unligated complexes, in comparison with the $O_2$-bound *Cs*-H-NOX structure, reveals that ligand binding causes haem pivoting and bending to occur concomitantly with a shift in the N-terminal helices of the protein (Ma et al., 2007).

The structure of *Cs*-H-NOX in the absence of $O_2$ was recently solved to 2.3 Å and provides further evidence for this theory (Hespen et al., 2016). In this structure, the researchers found that in the absence of $O_2$, the W9/N74/Y140 hydrogen-bonding network is displaced from the face of haem, leading to an overall shift of the helix containing Trp9. This helix also contains isoleucine 5, which is a major contributor to haem distortion (Olea, Herzik, Kuriyan, & Marletta, 2010; Weinert, Phillips-Piro, Tran, Mathies, & Marletta, 2011). Consequently, movement of this residue, and overall shifts in the protein that displace Pro115, results in haem relaxation compared to the ferrous-oxy complex. Another helix that undergoes a major shift upon oxygen binding contains the H-bonding network member Asn74 as well as glycine 71. Gly71 is an absolutely conserved H-NOX family residue that is thought to be vital in helix mobility.

## 5.3 Histidine Dislocation and Its Role in Signal Transduction

Spectroscopic characterization of sGC and facultative anaerobic bacterial H-NOX family members indicates that the haem–iron bond to the histidine ligand (H105 in rat sGC) is broken upon NO ligation to H-NOX (Zhao et al., 1999). Thus it has long been hypothesized that this event is an important molecular step in downstream signal transduction. The crystal structures of H-NOX proteins have solidified the theory that changes in the Fe–His bond are important in signalling, although as discussed later, the details of these structural changes are not yet agreed upon.

For many years, a structure of the Fe(II)-unligated complex of an H-NOX domain was unavailable, making it difficult to compare the structural changes that might take place upon NO ligation and Fe–His bond cleavage. Thus, in order to evaluate the role of the Fe–His bond, *So*-H-NOX Fe(II)-NO, Mn(II)-unligated, and Mn(II)-NO structures were compared (Herzik et al., 2014). The Mn(II)-NO complex remains 6-coordinate, making it possible to evaluate any structural changes that take place upon NO binding independently from those that take place due to histidine dissociation. In the Mn structures, the Mn atom moves towards the distal side of the porphyrin in Mn(II)-NO compared to the Mn(II)-unligated state. No other appreciable structural changes resulted from

NO ligation, however, which supports the hypothesis that cleavage of the histidine bond is indeed an important effector of NO-based signalling.

NMR studies have also contributed to our understanding of the role of Fe–His bond breakage, as well as haem structure, in H-NOX function. NMR structures of *S. oneidensis* H-NOX in the CO-bound state for both wild type and the H103G mutant (the latter protein being expressed and purified in the presence of imidazole) have been solved (Erbil et al., 2009). The H103G mutant was used to mimic a post-NO bound protein with a cleaved His–Fe bond. CO complexes were used to avoid paramagnetic line broadening from the Fe(II)-NO complexes. These structures revealed that the distorted haem cofactor relaxes upon Fe–His bond cleavage, which further supports the earlier findings about haem relaxation and signal transduction mechanism of H-NOX (Erbil et al., 2009). Moreover, a crystal (solved to 2.0 Å) of the homologous proximal histidine mutant of *Cs*-H-NOX (H102G) has corroborated the NMR studies (Olea, Herzik, et al., 2010). As expected, one of the molecular features evident in this crystal structure is loss of haem distortion as a result of His–Fe cleavage. The protein conformational changes associated with haem relaxation are also present. Therefore, it appears that Fe–His bond cleavage may lead to haem relaxation upon NO binding, leading ultimately to structural changes on the surface of H-NOX and signal propagation.

On the other hand, however, X-ray absorption studies on several H-NOX proteins have indicated that Fe–His bond cleavage may not be a significant molecular event. Extended X-ray absorption fine structure (EXAFS) measurements of H-NOX from several species known to form 5- or 6-coordinate complexes with NO were determined (Dai et al., 2012). *Shewanella woodyi* and *Pseudoalteromonas atlantica* H-NOX proteins are known to form 5-coordinates complexes with NO (Arora & Boon, 2012; Liu et al., 2012), which is consistent with dissociation of the proximal histidine. Yet, the EXAFS structure was consistent with a residual weak interaction with the proximal histidine. *Cs*-H-NOX, on the other hand, known to form a 6-coordinate complex with NO, also had appreciable Fe–His bond lengthening (Dai et al., 2012). In short, the Fe–His bond lengths of these three H-NOX structures were not significantly different, which is consistent with a hypothesis that Fe–His bond disruption is not the primary factor leading to signal transduction.

The most definitive evidence that the Fe–His bond is broken upon NO binding comes from the crystal structure of the Fe(II)-NO complex of *So*-H-NOX (Herzik et al., 2014) (Fig. 3B). This structure demonstrates that upon NO binding, the proximal histidine dissociates, resulting in a 5-coordinate complex with NO bound to the proximal side of haem. The authors acknowledge, however, that this might be an artefact of the crystallization conditions that required extremely high NO concentrations. Nonetheless, NO bound to the proximal side is consistent with proposals from previous studies of H-NOX domains from *Nostoc* sp. and *V. cholerae*. Here, multiple rate constants were apparent in NO-binding reactions in the presence of a large excess of NO, but not at stoichiometric NO concentrations (Tsai et al., 2010; Tsai, Berka, et al., 2012; Wu et al., 2013). The authors explain the results by comparing them to studies performed on cytochrome *c′* (Tsai, Martin, Berka, & Olson, 2012). In cytochrome *c′*, weak binding of NO to the distal side of haem weakens the Fe–His bond, allowing for a second NO molecule to displace the proximal histidine ligand as well as the distal pocket NO, leading to NO bound in the proximal haem pocket (Tsai, Berka, et al., 2012; Tsai, Martin, et al., 2012). It is important to note that, if NO is bound to the proximal pocket of H-NOX, it is likely a key contributing factor in $NO/O_2$ ligand discrimination (see Section 5.1), as $O_2$ would not be able to sufficiently weaken the Fe–His bond to allow for a second binding and displacement reaction to occur (Tsai et al., 2010; Tsai, Berka, et al., 2012; Tsai, Martin, et al., 2012; Wu et al., 2013).

There is also evidence against the proximal pocket NO-binding model, however. Interestingly, the H-NOX domains from *Nostoc punctiforme* (*Np*-H-NOX) (Boon et al., 2006), *L. pneumophila* (*L2*-H-NOX) (Boon et al., 2006), and *Vibrio parahaemolyticus* (*Vp*-H-NOX) (Ueno, Fischer, & Boon, unpublished data) form mixtures of 5- and 6-coordinate complexes with NO at room temperature. With each of these H-NOX domains, it has been observed that as temperatures were reduced close to 0°C, the 6-coordinate NO-bound forms were enriched, while as temperatures increased, the percentage of the 5-coordinate NO-bound form increased. This equilibrium is fully reversible with temperature in the absence of free NO. This observation is hypothesized due to a thermally labile proximal Fe(II)–His bond and suggests that in both the 5- and 6-coordinate Fe(II)-NO complexes, NO is bound in the distal haem pocket of the H-NOX fold.

Finally, two additional studies speak to an important role for the proximal histidine ligand in H-NOX structure and function: a molecular modelling study of sGC and X-ray spectroscopy studies of *Cs*-H-NOX. From structural studies, it has been shown that a bound water molecule makes several interactions with the protein backbone and side chains, as well as with the proximal histidine ligand (Fig. 3A). Modelling of sGC (based on the H-NOX crystal structure from *Nostoc* sp.) (Baskaran, Heckler, van den Akker, & Beuve, 2011) has shown that various mutations of residues in this water-network resulted in loss of haem binding and/or loss of sGC activation. The water molecule and its coordinating residues were therefore concluded to be needed for accomplishing the protein conformational changes necessary for signalling. In support of this, the *So*-H-NOX structure showed that this water is lost upon dissociation of the proximal histidine ligand, as the contacts between the histidine and other portions of the protein are disrupted (Herzik et al., 2014).

*Cs*-H-NOX has also been studied by X-ray absorption near edge spectroscopy (XANES) (Dai & Boon, 2011). The excitation energy of the edge of the iron K-shell, the innermost electron shell, is useful for understanding the coordination, oxidation, and ligand-binding properties of haem iron. An interesting observation from this study was that, while it is normally assumed that forming an iron–ligand complex will lower the K-shell excitation energy, the opposite is true in H-NOX due to a commensurate increase in shielding as the remaining valence electrons get pulled to the iron core. By comparing the K-shell energies of wild-type and P115A complexes of *Cs*-H-NOX, it was concluded that haem distortion is important for the energy of ligand binding (as discussed in Section 5.2). Another H-NOX structural detail extracted from these XANES measurements is that the haem iron is displaced, with respect to the plane of the porphyrin nitrogen atoms, in the Fe(II)-unligated state, while upon ligand binding, iron moves away from the histidine towards the distal side of haem. This corresponds well to the NO-dependent conformational changes that are dependent on loss of histidine coordination.

## 6. BIOCHEMICAL FUNCTIONS OF H-NOX PROTEINS

First documented in *Nitrosomonas europaea* (Schmidt, Steenbakkers, op den Camp, Schmidt, & Jetten, 2004) and the cystic fibrosis-associated pathogen *Pseudomonas aeruginosa* (Barraud et al., 2006, 2009), regulation of biofilm formation by nanomolar levels of NO has now been documented in numerous bacteria (Arora et al., 2015; Barraud et al., 2015, 2009; McDougald et al., 2011). In several bacteria, NO/H-NOX signalling through cyclic-di-GMP- and kinase-mediated pathways has been shown to be responsible for these biofilm phenotypes, as well as other bacterial group behaviours such as quorum sensing and symbiosis (Arora & Boon,

2012; Carlson, Vance, & Marletta, 2010; Henares et al., 2012, 2013; Liu et al., 2012; Plate & Marletta, 2012; Rao, Smith, & Marletta, 2015; Wang et al., 2010).

As previously noted, in the genomes of facultative anaerobic bacteria, stand-alone H-NOX domains are commonly encoded in the same operon with two-component signalling histidine kinases and cyclic-di-GMP processing enzymes (Fig. 1). It is noteworthy that the histidine kinase and cyclic-di-GMP enzymes in these H-NOX operons lack sensory domains. Consequently, it is hypothesized that H-NOX functions as an NO sensor to regulate the activities of these enzymes *in trans*. In this section of the review, the effect of H-NOX on the activity of cocistronic partner proteins is briefly discussed. For more extensive details on the molecular mechanisms by which H-NOX proteins explicitly regulate their associated signalling pathways, the following reviews are available (Arora et al., 2015; Nisbett & Boon, 2016; Plate & Marletta, 2013).

## 6.1 H-NOX and HaCE Signalling

Although they are in the minority, some facultative aerobic bacteria code for an H-NOX domain in the same operon as a cyclic-di-GMP synthase and/or phosphodiesterase. We have recently termed these enzymes, collectively, HaCEs for H-NOX-associated cyclic-di-GMP processing enzymes. This category of H-NOX domains is of particular interest as they directly implicate NO/H-NOX signalling in the regulation of cyclic-di-GMP, a secondary messenger molecule in bacteria that regulates biofilm formation (Ross et al., 1987).

HaCE enzymes generally have one or both of two enzymatic domains that directly participate in cyclic-di-GMP regulation: a diguanylate cyclase domain that synthesizes cyclic-di-GMP from two molecules of GTP and/or a cyclic-di-GMP phosphodiesterase domain that hydrolyzes cyclic-di-GMP into pGpG (5′-phosphoguanylyl-(3′-5′)-guanosine) or GMP (guanosine-5′-monophosphate). H-NOX/HaCE complexes have been shown to link NO detection with regulation of HaCE activity, leading to intracellular changes in the concentration of cyclic-di-GMP and biofilm formation.

H-NOX regulation of HaCE activity has been biochemically observed in the bacteria *L. pneumophila* (Carlson et al., 2010), *S. woodyi* (Liu et al., 2012), and more recently, *Agrobacterium vitis* (Nesbitt et al., unpublished data) (Fig. 5). In *L. pneumophila*, *Lpg*-HaCE has only cyclic-di-GMP synthase activity (although it contains a phosphodiesterase domain, it is

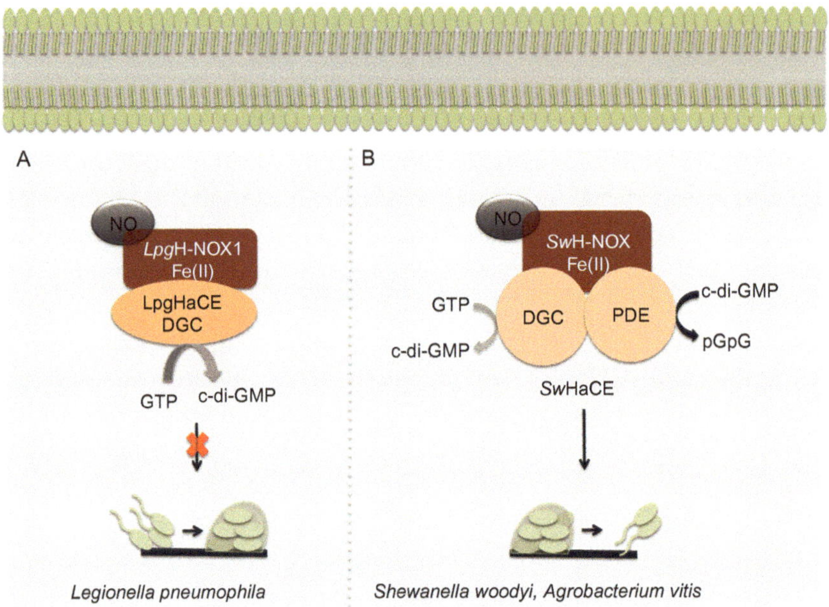


**Fig. 5** NO regulates HaCE activity through ligation to H-NOX. (A) NO bound H-NOX only affects cyclic-di-GMP production in *Legionella pneumophila* as *Lpg*-HaCE is only functional as a cyclic-di-GMP synthase. (B) Fe(II)-NO bound H-NOX directly influences both the production and hydrolysis of cyclic-di-GMP in *Shewanella woodyi* and *Agrobacterium vitis*.

inactive) (Carlson et al., 2010). In both *S. woodyi* and *A. vitis*, however, *Sw*-HaCE (Liu et al., 2012) and *Av*-HaCE (Nesbitt et al., unpublished data) proteins were found to exhibit both cyclic-di-GMP synthase and phosphodiesterase activities in vitro. In *L. pneumophila*, *Lpg*-HaCE cyclic-di-GMP synthase activity is unaffected in the presence of Fe(II)-unligated *L1*-H-NOX (Carlson et al., 2010), while in *S. woodyi*, *Sw*-HaCE cyclic-di-GMP synthase activity is upregulated (by ~10-fold) in the presence of Fe(II)-unligated *Sw*-H-NOX and phosphodiesterase activity remains unchanged (Liu et al., 2012). Thus, it appears that NO-free H-NOX variably affects HaCE activity; NO-bound H-NOX, however, universally results in changes in HaCE activity. In *L. pneumophila*, NO-bound *L1*-H-NOX causes a decrease in *Lpg*-HaCE cyclic-di-GMP synthase activity, leading to a decrease in cyclic-di-GMP concentration in vitro (Carlson et al., 2010). In *S. woodyi*, *Sw*-HaCE cyclic-di-GMP production is also downregulated in the presence of Fe(II)-NO bound *Sw*-H-NOX, but by an slightly different mechanism. Here, in addition to decreasing HaCE cyclic-di-GMP synthase activity, the phosphodiesterase activity of *Sw*-HaCE is increased (Liu et al., 2012), leading to a dramatic decrease in c-di-GMP concentration.

Finally, in light of the *S. oneidensis* H-NOX NMR solution structures and the *C. subterraneus* crystal structures revealing that NO-haem binding may induce haem flattening and subsequent protein conformational changes (Erbil et al., 2009; Olea et al., 2008; Olea, Kuriyan, et al., 2010), it has been hypothesized that changes in both haem and protein conformation may directly translate into changes in downstream signalling events in H-NOX signalling pathways (as discussed in Section 5). Since there was no direct evidence in support of this hypothesis, however, our lab investigated the role of the H-NOX haem structure in the H-NOX/HaCE signalling pathway from *S. woodyi*. In this study, as expected, the relaxed haem proline mutant of *Sw*-H-NOX led to upregulation of the phosphodiesterase activity of *Sw*-HaCE (Muralidharan & Boon, 2012), which is the very same effect that NO-bound H-NOX has on HaCE activity (Liu et al., 2012). This study, therefore, provided the first direct evidence for the role of haem relaxation in H-NOX signal transduction.

## 6.2 H-NOX and Two-Component Signalling

H-NOX genes are cocistronic with two-component signalling histidine kinase (HaHK; H-NOX-associated histidine kinase) genes instead of HaCE genes in most facultative aerobic bacteria (Fig. 1). Two-component signalling networks are signalling systems that bacteria use to sense and respond to various environmental stimuli including nutrient availability, pH, osmolarity, and host factors (Beier & Gross, 2006). Interestingly, thus far, most H-NOX two-component signalling systems have been implicated in c-di-GMP metabolism and biofilm formation in many bacteria, which is functionally consistent with H-NOX/HaCE signalling systems (Henares et al., 2013; Plate & Marletta, 2012; Rao et al., 2015).

Simple two-component signalling networks are comprised of a sensor histidine kinase and a response regulator protein. The sensor histidine kinase detects the environmental stimulus via its sensory domain, and responds by utilizing adenosine triphosphate (ATP) as a phosphodonor to catalyse the autophosphorylation of a conserved histidine residue in the protein's kinase domain. Phosphotransfer from the histidine residue to a conserved aspartic acid residue in the receiver domain of a cognate response regulator protein then results in activation of the appropriate cellular response (Laub,

Biondi, & Skerker, 2007). A common variation in these signalling systems, sometimes called three-component signalling (Elsen, Duche, & Colbeau, 2003; Ortiz de Orue Lucana & Groves, 2009; Szurmant, Bu, Brooks, & Hoch, 2008), is when the histidine kinase sensory domain is replaced by an accessory sensory protein that acts to directly detect the environmental stimuli and, via protein–protein interaction, conveys this information to a histidine kinase (Elsen et al., 2003; Szurmant et al., 2008).

Histidine kinases (HaHKs) that are cocistronic with H-NOX genes do not possess sensory domains and thus H-NOX functions as the sensory domain to regulate these three-component systems in an NO-dependent fashion. To date, these simple H-NOX/HaHK signalling systems have been characterized in *S. oneidensis* (Plate & Marletta, 2012), *V. cholerae* (Mukhopadyay et al., 2016; Plate & Marletta, 2012), and *P. atlantica* (Arora & Boon, 2012) (Fig. 6). In all three bacteria, in the absence of H-NOX, the HaHK proteins exhibit ATP-dependent autophosphorylation activity (Arora & Boon, 2012; Plate & Marletta, 2012; Price et al., 2007) that is strongly inhibited in the presence of NO-bound H-NOX.

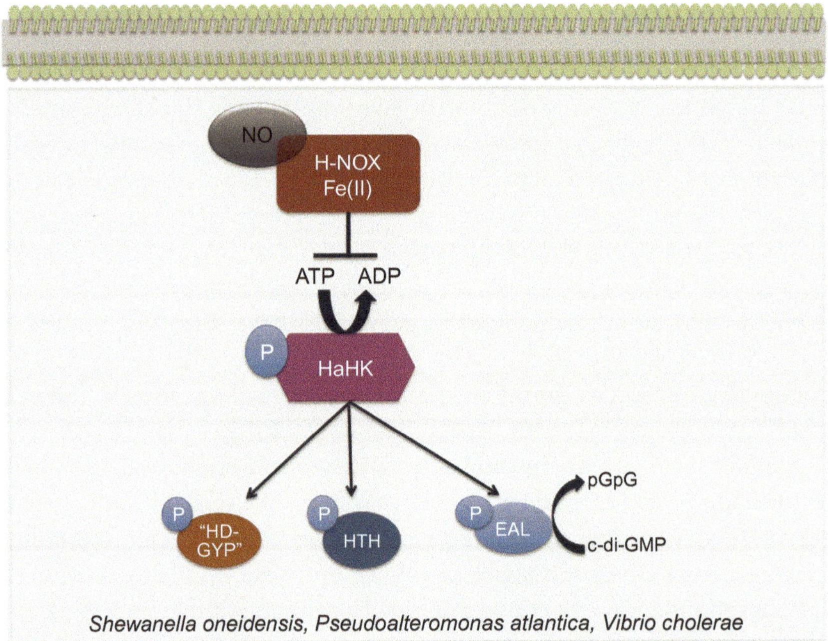


**Fig. 6** NO/H-NOX regulates HaHK autophosphorylation activity and phosphate flow downstream of HaHK. *Shewanella oneidensis*, *Pseudoalteromonas atlantica*, and *Vibrio cholerae* all have similar H-NOX signalling pathways, but *V. cholerae* does not encode a homologous HTH domain-containing response regulator protein.

Another common deviation from simple two-component histidine kinase-response regulator architecture is found in hybrid histidine kinase signalling. Here the histidine kinase has a receiver domain within the same polypeptide (Laub et al., 2007). These hybrid kinase proteins (in response to a stimulus) autophosphorylate their conserved histidine residues and then transfer the phosphate intramolecularly to a conserved aspartic acid residue contained within their receiver domains. The phosphate is then subsequently transferred to a conserved histidine residue in a Hpt (histidine-containing phosphotransfer protein) which then further engages in phosphotransfer with an appropriate response regulator protein (Laub et al., 2007). H-NOX proteins have also been shown to regulate the activity of hybrid histidine kinases and their associated phosphorelay signalling pathways, specifically, those that have been implicated in quorum sensing and symbioses with eukaryotes.

Two H-NOX-associated signalling pathways, those from *Vibrio harveyi* and *V. parahaemolyticus*, have thus far been implicated in quorum sensing (Henares et al., 2012; Ueno et al., unpublished data) (Fig. 7). Quorum

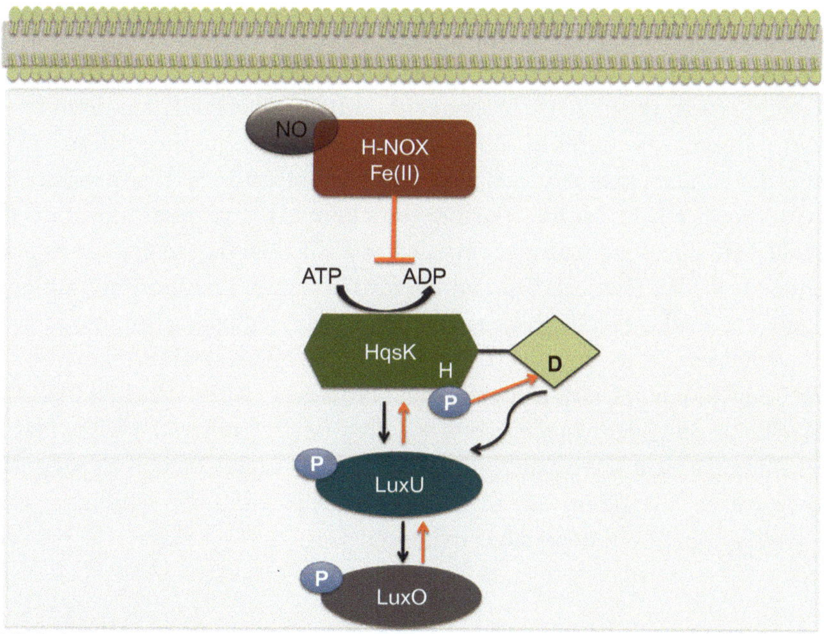

*Vibrio harveyi* and *Vibrio parahaemolyticus*

**Fig. 7** NO/H-NOX regulates HqsK autophosphorylation activity and subsequently influences quorum sensing activity in *Vibrio harveyi* and *Vibrio parahaemolyticus*. In both *V. harveyi* and *V. parahaemolyticus*, Fe(II)-NO bound H-NOX inhibits *Vh*-HqsK autophosphorylation activity which leads to a change in phosphate flow to LuxU/LuxO and this ultimately influences quorum sensing in both bacteria.

sensing is a process that governs population-wide bacterial group behaviours, including virulence gene production, biofilm formation, and bioluminescence, in response to the production, secretion, and detection of autoinducers in a cell density-dependent manner (Papenfort & Bassler, 2016). In both *V. harveyi* and *V. parahaemolyticus*, the HqsK proteins (H-NOX-associated quorum sensing histidine kinase), as expected, exhibit ATP-dependent autophosphorylation in a time-dependent manner that is inhibited in the presence of the Fe(II)-NO complex of H-NOX (Henares et al., 2012; Ueno et al., unpublished data). What makes these systems most interesting is that HqsK signalling merges with the kinase pathways regulated by quorum sensing autoinducers. Thus, NO/H-NOX participates in quorum sensing signalling pathways, with NO acting analogously to an autoinducer in these bacteria.

In addition to regulating quorum sensing, NO/H-NOX regulated hybrid–histidine kinases have also been implicated in regulating microbial symbiotic relationships with various hosts including insects, nematodes, and marine invertebrates (Wang & Ruby, 2011) (Fig. 8). Specifically, *Vibrio fischeri* and its partner the Hawaiian bobtail squid *Euprymna scolopes*, as well as *Silicibacter* sp. strain *Trich*4B and *Trichodesmium erythraeum*, have been demonstrated to be involved in symbiotic relationships that are dependent on H-NOX (Rao et al., 2015; Wang et al., 2010). *Sili*-HaHK has definitively been shown to exhibit ATP-dependent autophosphorylation in a time-dependent manner, activity that is inhibited in the presence of NO-bound *Sili*-H-NOX. Further biochemical characterization of the *Sili*-HaHK signalling pathway revealed that *Sili*-HaHK can engage in phosphorelay with a Hpt, *Sili*-Hpt, which further engages in phosphotransfer to regulate a cyclic-di-GMP synthase, ultimately regulating symbiosis with *T. erythraeum* (Rao et al., 2015). The *Vf*-H-NOX/HaHK pathway has not been biochemically characterized. However, since the *Vf*-H-NOX/HaHK signalling pathway is architecturally similar to other H-NOX/HaHK signalling pathways, we can hypothesize that *Vf*-H-NOX will also regulate downstream signalling, in this case to help establish symbiosis with *E. scolopes*, in a NO-dependent manner.

## 6.3 H-NOX and Methyl Accepting Chemotaxis Signalling

For obligate anaerobes, $O_2$ is toxic. In these bacteria, H-NOX domains are found fused to MCPs (Fig. 1). MCPs typically mediate bacterial movement to either avoid toxins or seek nutrients (Dahl, Boos, & Manson, 1989).

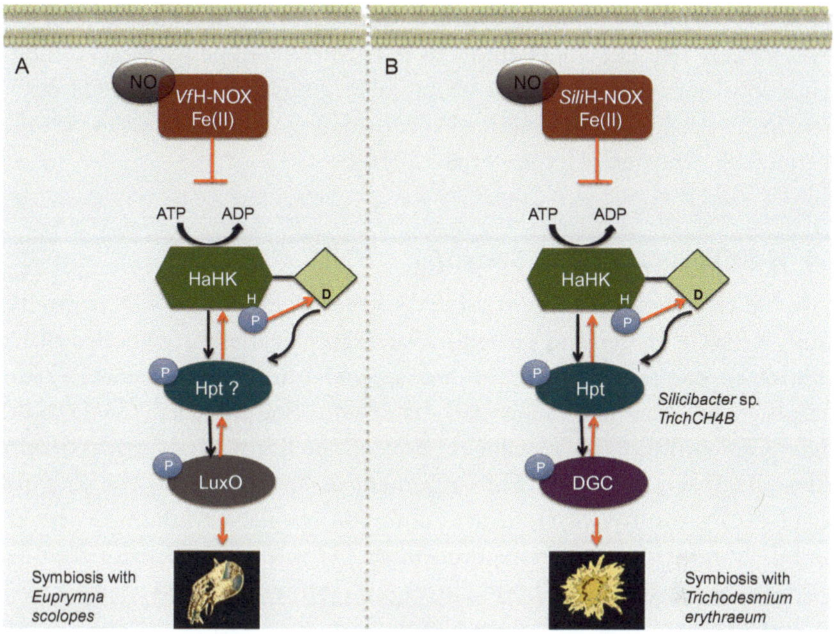


**Fig. 8** NO/H-NOX regulates *Vf*-HaHK and *Sili*-HaHK autophosphorylation activity and their symbiotic relationships with their respective eukaryotic hosts. (A) In *V. fischerii*, Fe(II)-NO bound H-NOX may inhibit *Vf*-HaHK autophosphorylation activity which leads to a change in phosphate flow and symbiosis with *Euprymna scolopes*, the Hawaiian bobtail squid. (B) In *Silicibacter* sp. *TrichCH4B*, NO bound H-NOX has been shown to decrease *Sili*-HaHK autophosphorylation activity, decrease phosphate flow within the *Sili*-HaHK signalling pathway, and influence its symbiosis with *Trichodesmium erythraeum*, an algal symbiont.

MCP receptors are usually transmembrane proteins that regulate a histidine kinase-signalling pathway based on the methylation state of the receptor (Roberts, Papachristodoulou, & Armitage, 2010; Williams & Stewart, 1999).

To date, there has been limited success in measuring the H-NOX-mediated output of MCPs because of the difficulty of purifying active transmembrane proteins. Thus, in order to determine whether the function of H-NOX in *C. subterraneus* is to sense NO or $O_2$, and thus regulate a biochemical output, an orthogonal assay was developed. Here, *Cs*-H-NOX regulation of an orthogonal HaHK cloned from *V. cholerae* (Hespen et al., 2016) was determined. This study showed that Fe(II)-unligated *Cs*-H-NOX-inhibited HaHK autophosphorylation, and the $O_2$-bound, but not the CO- or NO-bound, complex was able to relieve

*Cs*-H-NOX inhibition of *Vc*-HaHK. By extension, the data indicate that $O_2$ is a likely regulator of chemotaxis in many anaerobic bacteria, a finding that is consistent with the earlier structural findings that this class of H-NOX proteins undergoes ligand induced conformational changes in the presence of molecular oxygen (Hespen et al., 2016).

## 6.4 H-NOX as a Redox Sensor

It should be noted that there is some disagreement that H-NOX is strictly a family of $NO/O_2$ sensing proteins. For example, there is evidence that in addition to, or alternatively to, NO sensing, H-NOX proteins may be redox sensors. This was first suggested based on findings that *Ns*-H-NOX undergoes conformational changes between the ferrous and ferric oxidation states (Tsai et al., 2010). This argument is supported by the fact that *Ns*-H-NOX forms a 6-coordinate complex upon NO binding, with the Fe–His bond, therefore, remaining intact. However, as discussed earlier, it is not clear that Fe–His bond cleavage is absolutely necessary to activate downstream signalling. It has been difficult to conclusively test the hypothesis that *Ns*-H-NOX is a redox sensor because downstream partners have not been identified for *Ns*-H-NOX, making a biochemical assay of downstream signalling prohibitive.

Stronger evidence of redox sensing has manifested in gammaproteobacteria, however. Several cysteine residues are conserved among roughly half of H-NOX domains from gammaproteobacteria (Fig. 2). The crystal structure of *So*-H-NOX revealed that these residues function to coordinate a zinc atom (Herzik et al., 2014) (Fig. 9). Furthermore, it was shown that *Vc*-H-NOX, which contains the conserved cysteines, binds stoichiometric zinc, while *Cs*-H-NOX, which does not contain the conserved cysteines, does not bind zinc (Herzik et al., 2014). When the predicted zinc-binding residues were mutated in *So*-H-NOX, the protein was insoluble, so it was originally concluded that the role of zinc is to maintain structural integrity.

More recently, however, it has been suggested that these cysteine residues may serve as redox-dependent mediators of signalling upon the formation and breakage of disulfide bonds (Mukhopadyay et al., 2016). Using *Vc*-H-NOX and *Vc*-HaHK, it was demonstrated that *Vc*-HaHK activity, in addition to being inhibited by the Fe(II)-NO complex of *Vc*-H-NOX, is also inhibited by apo- (haem-free) *Vc*-H-NOX, but only when the conserved cysteines are oxidized. This redox inhibition of *Vc*-HaHK activity

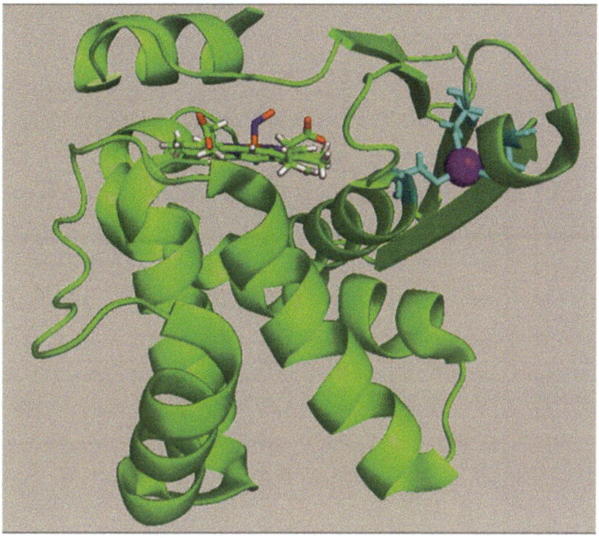

**Fig. 9** *So*-H-NOX structure with zinc (*purple*) bound by cysteine residues (Cys139, Cys164, Cys172; *cyan*). Generated in pymol from pdb file 4U9B.

is attributed to disulfide bonds crosslinking *Vc*-H-NOX and *Vc*-HaHK. In this model, the purpose of zinc is to stabilize the thiolate groups until oxidizing conditions allow these to form disulfide bonds with *Vc*-HaHK.

Therefore, unlike other characterized simple H-NOX/HaHK signalling pathways (*S. oneidensis* and *P. atlantica*), *Vc*-H-NOX appears to have dual functionality and is able regulate to *Vc*-HaHK in both a haem-dependent (upon NO binding) and a haem-independent (upon cysteine oxidation) manner (Mukhopadyay et al., 2016). Future studies need to be conducted, however, to determine if this dual functionality of *Vc*-H-NOX is physiologically relevant in regulating biofilm formation and pathogenicity in *V. cholerae*.

## 7. A NOVEL NO SENSING PROTEIN IN BACTERIA: NosP

Although H-NOX has been shown to the primary NO sensor in many bacteria, there are many more bacteria that lack an H-NOX domain but are still able to respond to low concentrations of NO to regulate processes like quorum sensing and biofilm formation, phenotypes that are reminiscent of NO/H-NOX-mediated pathways. For example, NO regulation of biofilm dispersal in *P. aeruginosa*, a principal pathogen in cystic fibrosis

(Driscoll, Brody, & Kollef, 2007) and hospital-acquired infections (Castiglione et al., 2011), is well documented (Barraud et al., 2006, 2009), but the primary NO sensor is unknown; *P. aeruginosa* does not encode an H-NOX domain. Thus our laboratory has hypothesized the existence of a bacterial NO sensor alternative to H-NOX.

We recently discovered a novel NO-binding protein, a domain we have named NosP. NosP domains belong to a family of uncharacterized proteins currently annotated as FIST (F-box intracellular signal transduction protein) domains due to their predicted secondary structure (Borziak & Zhulin, 2007). NosP domains, like H-NOX domains, are predicted to be cocistronic or fused to MCPs, two-component signalling histidine kinases, and cyclic-di-GMP synthase and/or phosphodiesterase enzymes (Fig. 1). Interestingly, like the H-NOX-associated enzymes discussed earlier, these enzymes do not encode for sensory domains, suggesting NosP domains may function as such *in trans*.

In *P. aeruginosa*, we have demonstrated that a mutant strain lacking components of the NosP pathway loses the ability to disperse biofilms in response to NO (Hossain et al., unpublished data), confirming NosP as a bacterial NO sensor. *Pa*-NosP is cocistronic with a hybrid histidine kinase (that we have named NosP-associated histidine kinase; NaHK) that has been previously implicated in biofilm formation in this bacterial species (Hsu, Chen, Peng, & Chang, 2008). Using purified proteins, we have shown that *Pa*-NosP is a haemoprotein that binds to NO and CO, but does not form a stable ferrous-oxy complex (Hossain et al., unpublished data), consistent with the ligand-binding properties expected of a dedicated NO sensor. Additionally, preliminary results from rR spectroscopy of ferrous NosP from *P. aeruginosa* demonstrate that it is a histidine-ligated 6-coordinate low-spin complex, which upon binding NO, forms a 5-coordinate high-spin complex (Bacon et al., unpublished data). Furthermore, although the NO association rate constant of this protein has not yet been measured, the NO dissociation rate constant has been measured and it is $1.8 \times 10^{-4}$ $s^{-1}$ (Hossain et al., unpublished data). Moreover, we found that NO-bound NosP is able to regulate the phosphorelay activity of NaHK in *P. aeruginosa*. These data collectively suggest *Pa*-NosP is a NO-sensitive haemoprotein (Hossain et al., unpublished data).

Recently, our lab has also characterized a NosP signalling pathway in *S. oneidensis*. Like *Pa*-NosP, *So*-NosP is also a haemoprotein that binds NO and CO but does not form a stable complex with $O_2$ (Binnenkade and Nisbett et al., unpublished data). Additionally, *So*-NosP is predicted to be cocistronic with a histidine kinase known to be involved in the *S. oneidensis* NO-regulated c-di-GMP signalling network (Plate & Marletta, 2012). Using biofilm analysis of NosP and H-NOX mutants of *S. oneidensis*, we show that NosP regulates biofilm formation upstream of H-NOX. Consequently, these preliminary findings indicate that *So*-NosP may directly participate in regulating biofilm formation in *S. oneidensis* by way of regulating the activity of its cocistronic histidine kinase.

## 8. PERSPECTIVES AND CONCLUSIONS

The field of NO signalling in bacteria, including the identification and characterization of the targets of NO in bacteria, is rapidly expanding. In this review, we have presented a summary of the structural information available on select H-NOX domains, the ligand-binding properties of H-NOX domains, how NO/H-NOX domain interactions biochemically result in the regulation of downstream H-NOX-associated signalling pathways, and the discovery of novel NO-sensing mechanisms in bacteria that lack H-NOX domains.

It is now clear that H-NOX proteins, irrespective of whether they are from facultative or obligate anaerobes, are predicted to have approximately picomolar affinity for NO, and H-NOX-associated signalling pathways are governed at nanomolar concentrations of NO (Arora et al., 2015; Nisbett & Boon, 2016; Plate & Marletta, 2013). Several key structural features have been identified in NO-activation of H-NOX, including the dissociation of a proximal histidine ligand (Dai et al., 2012; Erbil et al., 2009; Herzik et al., 2014; Olea, Herzik, et al., 2010) and flattening of a haem cofactor (Dai & Boon, 2011; Erbil et al., 2009; Olea et al., 2008; Olea, Kuriyan et al., 2010; Sun et al., 2016) that is otherwise exceedingly distorted. Features specific to some H-NOX proteins have also been distinguished, including a hydrogen-bonding network that stabilizes $O_2$ binding in H-NOX from obligate anaerobes (Boon et al., 2005; Hespen et al., 2016) and a zinc-binding feature that may be connected to redox sensing in H-NOX domains from some gammaproteobacteria (Mukhopadyay et al., 2016). This detailed understanding of the structure of H-NOX has, in fact, allowed for rational engineering of H-NOX domains to bind specific ligands (Dai & Boon, 2010; Olea, Kuriyan, et al., 2010; Weinert, Phillips-Piro, & Marletta, 2013) and for development of optimized drug agonists for sGC (Kumar et al., 2013; Martin et al., 2010; von Wantoch Rekowski et al., 2013).

Finally, although two NO-sensitive bacterial protein families have been identified, H-NOX and NosP, there are still many bacteria, many Gram-positive bacteria in particular, that are NO-responsive but do not code for either H-NOX or NosP domains in their genomes. Consequently, the identity of the putative primary NO sensor and NO-responsive signalling pathway(s), as well as the molecular mechanism of NO regulation in these bacteria, are all currently unknown.

## REFERENCES

Arora, D. P., & Boon, E. M. (2012). Nitric oxide regulated two-component signaling in Pseudoalteromonas atlantica. *Biochemical and Biophysical Research Communications*, *421*(3), 521–526. http://dx.doi.org/10.1016/j.bbrc.2012.04.037.

Arora, D. P., Hossain, S., Xu, Y., & Boon, E. M. (2015). Nitric oxide regulation of bacterial biofilms. *Biochemistry*, *54*(24), 3717–3728. http://dx.doi.org/10.1021/bi501476n.

Barraud, N., Hassett, D. J., Hwang, S. H., Rice, S. A., Kjelleberg, S., & Webb, J. S. (2006). Involvement of nitric oxide in biofilm dispersal of Pseudomonas aeruginosa. *Journal of Bacteriology*, *188*(21), 7344–7353. http://dx.doi.org/10.1128/JB.00779-06.

Barraud, N., Kelso, M. J., Rice, S. A., & Kjelleberg, S. (2015). Nitric oxide: A key mediator of biofilm dispersal with applications in infectious diseases. *Current Pharmaceutical Design*, *21*(1), 31–42.

Barraud, N., Schleheck, D., Klebensberger, J., Webb, J. S., Hassett, D. J., Rice, S. A., et al. (2009). Nitric oxide signaling in Pseudomonas aeruginosa biofilms mediates phosphodiesterase activity, decreased cyclic di-GMP levels, and enhanced dispersal. *Journal of Bacteriology*, *191*(23), 7333–7342. http://dx.doi.org/10.1128/JB.00975-09.

Baskaran, P., Heckler, E. J., van den Akker, F., & Beuve, A. (2011). Aspartate 102 in the heme domain of soluble guanylyl cyclase has a key role in NO activation. *Biochemistry*, *50*(20), 4291–4297. http://dx.doi.org/10.1021/bi2004087.

Beier, D., & Gross, R. (2006). Regulation of bacterial virulence by two-component systems. *Current Opinion in Microbiology*, *9*(2), 143–152. http://dx.doi.org/10.1016/j.mib.2006.01.005.

Bellota-Anton, C., Munnoch, J., Robb, K., Adamczyk, K., Candelaresi, M., Parker, A. W., et al. (2011). Spectroscopic analysis of protein Fe-NO complexes. *Biochemical Society Transactions*, *39*(5), 1293–1298. http://dx.doi.org/10.1042/BST0391293.

Bhinu, V. S. (2005). Insight into biofilm-associated microbial life. *Journal of Molecular Microbiology and Biotechnology*, *10*(1), 15–21. http://dx.doi.org/10.1159/000090344.

Bodenmiller, D. M., & Spiro, S. (2006). The yjeB (nsrR) gene of Escherichia coli encodes a nitric oxide-sensitive transcriptional regulator. *Journal of Bacteriology*, *188*(3), 874–881. http://dx.doi.org/10.1128/JB.188.3.874-881.2006.

Boon, E. M., Davis, J. H., Tran, R., Karow, D. S., Huang, S. H., Pan, D., et al. (2006). Nitric oxide binding to prokaryotic homologs of the soluble guanylate cyclase beta1 H-NOX domain. *The Journal of Biological Chemistry*, *281*(31), 21892–21902. http://dx.doi.org/10.1074/jbc.M600557200.

Boon, E. M., Huang, S. H., & Marletta, M. A. (2005). A molecular basis for NO selectivity in soluble guanylate cyclase. *Nature Chemical Biology*, *1*(1), 53–59. http://dx.doi.org/10.1038/nchembio704.

Boon, E. M., & Marletta, M. A. (2005a). Ligand discrimination in soluble guanylate cyclase and the H-NOX family of heme sensor proteins. *Current Opinion in Chemical Biology*, *9*(5), 441–446. http://dx.doi.org/10.1016/j.cbpa.2005.08.015.

Boon, E. M., & Marletta, M. A. (2005b). Ligand specificity of H-NOX domains: From sGC to bacterial NO sensors. *Journal of Inorganic Biochemistry*, *99*(4), 892–902. http://dx.doi.org/10.1016/j.jinorgbio.2004.12.016.

Borziak, K., & Zhulin, I. B. (2007). FIST: A sensory domain for diverse signal transduction pathways in prokaryotes and ubiquitin signaling in eukaryotes. *Bioinformatics*, *23*(19), 2518–2521. http://dx.doi.org/10.1093/bioinformatics/btm384.

Carlson, H. K., Vance, R. E., & Marletta, M. A. (2010). H-NOX regulation of c-di-GMP metabolism and biofilm formation in Legionella pneumophila. *Molecular Microbiology*, 77(4), 930–942. http://dx.doi.org/10.1111/j.1365-2958.2010.07259.x.

Cary, S. P., Winger, J. A., Derbyshire, E. R., & Marletta, M. A. (2006). Nitric oxide signaling: No longer simply on or off. *Trends in Biochemical Sciences*, *31*(4), 231–239. http://dx.doi.org/10.1016/j.tibs.2006.02.003.

Cary, S. P., Winger, J. A., & Marletta, M. A. (2005). Tonic and acute nitric oxide signaling through soluble guanylate cyclase is mediated by nonheme nitric oxide, ATP, and GTP. *Proceedings of the National Academy of Sciences of the United States of America*, *102*(37), 13064–13069. http://dx.doi.org/10.1073/pnas.0506289102.

Castiglione, N., Stelitano, V., Rinaldo, S., Giardina, G., Caruso, M., & Cutruzzola, F. (2011). Metabolism of cyclic-di-GMP in bacterial biofilms: From a general overview to biotechnological applications. *Indian Journal of Biotechnology*, *10*(4), 423–431.

Cruz-Ramos, H., Crack, J., Wu, G., Hughes, M. N., Scott, C., Thomson, A. J., et al. (2002). NO sensing by FNR: Regulation of the Escherichia coli NO-detoxifying flavohaemoglobin, Hmp. *The EMBO Journal*, *21*(13), 3235–3244. http://dx.doi.org/10.1093/emboj/cdf339.

Dahl, M. K., Boos, W., & Manson, M. D. (1989). Evolution of chemotactic-signal transducers in enteric bacteria. *Journal of Bacteriology*, *171*(5), 2361–2371.

Dai, Z., & Boon, E. M. (2010). Engineering of the heme pocket of an H-NOX domain for direct cyanide detection and quantification. *Journal of the American Chemical Society*, *132*(33), 11496–11503. http://dx.doi.org/10.1021/ja101674z.

Dai, Z., & Boon, E. M. (2011). Probing the local electronic and geometric properties of the heme iron center in a H-NOX domain. *Journal of Inorganic Biochemistry*, *105*(6), 784–792. http://dx.doi.org/10.1016/j.jinorgbio.2011.03.002.

Dai, Z., Farquhar, E. R., Arora, D. P., & Boon, E. M. (2012). Is histidine dissociation a critical component of the NO/H-NOX signaling mechanism? Insights from X-ray absorption spectroscopy. *Dalton Transactions*, *41*(26), 7984–7993. http://dx.doi.org/10.1039/c2dt30147d.

D'Autreaux, B., Tucker, N. P., Dixon, R., & Spiro, S. (2005). A non-haem iron centre in the transcription factor NorR senses nitric oxide. *Nature*, *437*(7059), 769–772. http://dx.doi.org/10.1038/nature03953.

Denninger, J. W., & Marletta, M. A. (1999). Guanylate cyclase and the .NO/cGMP signaling pathway. *Biochimica et Biophysica Acta*, *1411*(2–3), 334–350.

Driscoll, J. A., Brody, S. L., & Kollef, M. H. (2007). The epidemiology, pathogenesis and treatment of Pseudomonas aeruginosa infections. *Drugs*, *67*(3), 351–368.

Elsen, S., Duche, O., & Colbeau, A. (2003). Interaction between the H2 sensor HupUV and the histidine kinase HupT controls HupSL hydrogenase synthesis in Rhodobacter capsulatus. *Journal of Bacteriology*, *185*(24), 7111–7119.

Erbil, W. K., Price, M. S., Wemmer, D. E., & Marletta, M. A. (2009). A structural basis for H-NOX signaling in Shewanella oneidensis by trapping a histidine kinase inhibitory conformation. *Proceedings of the National Academy of Sciences of the United States of America*, *106*(47), 19753–19760. http://dx.doi.org/10.1073/pnas.0911645106.

Fernhoff, N. B., Derbyshire, E. R., & Marletta, M. A. (2009). A nitric oxide/cysteine interaction mediates the activation of soluble guanylate cyclase. *Proceedings of the National Academy of Sciences of the United States of America*, *106*(51), 21602–21607. http://dx.doi.org/10.1073/pnas.0911083106.

Gardner, A. M., Helmick, R. A., & Gardner, P. R. (2002). Flavorubredoxin, an inducible catalyst for nitric oxide reduction and detoxification in Escherichia coli. *The Journal of Biological Chemistry*, *277*(10), 8172–8177. http://dx.doi.org/10.1074/jbc.M110471200.

Garthwaite, J. (2008). Concepts of neural nitric oxide-mediated transmission. *European Journal of Neuroscience*, *27*(11), 2783–2802. http://dx.doi.org/10.1111/j.1460-9568.2008.06285.x.

Hassan, J., Bergaust, L. L., Molstad, L., de Vries, S., & Bakken, L. R. (2016). Homeostatic control of nitric oxide (NO) at nanomolar concentrations in denitrifying bacteria—Modelling and experimental determination of NO reductase kinetics in vivo in Paracoccus denitrificans. *Environmental Microbiology*, *18*(9), 2964–2978. http://dx.doi.org/10.1111/1462-2920.13129.

Henares, B. M., Higgins, K. E., & Boon, E. M. (2012). Discovery of a nitric oxide responsive quorum sensing circuit in Vibrio harveyi. *ACS Chemical Biology*, 7(8), 1331–1336. http://dx.doi.org/10.1021/cb300215t.

Henares, B. M., Xu, Y., & Boon, E. M. (2013). A nitric oxide-responsive quorum sensing circuit in Vibrio harveyi regulates flagella production and biofilm formation. *International Journal of Molecular Sciences*, *14*(8), 16473–16484. http://dx.doi.org/10.3390/ijms140816473.

Herzik, M. A., Jr., Jonnalagadda, R., Kuriyan, J., & Marletta, M. A. (2014). Structural insights into the role of iron-histidine bond cleavage in nitric oxide-induced activation of H-NOX gas sensor proteins. *Proceedings of the National Academy of Sciences of the United States of America*, *111*(40), E4156–E4164. http://dx.doi.org/10.1073/pnas.1416936111.

Hespen, C. W., Bruegger, J. J., Phillips-Piro, C. M., & Marletta, M. A. (2016). Structural and functional evidence indicates selective oxygen signaling in Caldanaerobacter subterraneus H-NOX. *ACS Chemical Biology*, *11*(8), 2337–2346. http://dx.doi.org/10.1021/acschembio.6b00431.

Hobbs, A. J., & Ignarro, L. J. (1996). Nitric oxide-cyclic GMP signal transduction system. *Methods in Enzymology*, *269*, 134–148.

Hoiby, N., Bjarnsholt, T., Givskov, M., Molin, S., & Ciofu, O. (2010). Antibiotic resistance of bacterial biofilms. *International Journal of Antimicrobial Agents*, *35*(4), 322–332. http://dx.doi.org/10.1016/j.ijantimicag.2009.12.011.

Hsu, J. L., Chen, H. C., Peng, H. L., & Chang, H. Y. (2008). Characterization of the histidine-containing phosphotransfer protein B-mediated multistep phosphorelay system in Pseudomonas aeruginosa PAO1. *The Journal of Biological Chemistry*, *283*(15), 9933–9944. http://dx.doi.org/10.1074/jbc.M708836200.

Iyer, L. M., Anantharaman, V., & Aravind, L. (2003). Ancient conserved domains shared by animal soluble guanylyl cyclases and bacterial signaling proteins. *BMC Genomics*, *4*(1), 5.

Karow, D. S., Pan, D., Davis, J. H., Behrends, S., Mathies, R. A., & Marletta, M. A. (2005). Characterization of functional heme domains from soluble guanylate cyclase. *Biochemistry*, *44*(49), 16266–16274. http://dx.doi.org/10.1021/bi051601b.

Karow, D. S., Pan, D., Tran, R., Pellicena, P., Presley, A., Mathies, R. A., et al. (2004). Spectroscopic characterization of the soluble guanylate cyclase-like heme domains from Vibrio cholerae and Thermoanaerobacter tengcongensis. *Biochemistry*, *43*(31), 10203–10211. http://dx.doi.org/10.1021/bi049374l.

Kumar, V., Martin, F., Hahn, M. G., Schaefer, M., Stamler, J. S., Stasch, J. P., et al. (2013). Insights into BAY 60-2770 activation and S-nitrosylation-dependent desensitization of soluble guanylyl cyclase via crystal structures of homologous nostoc H-NOX domain complexes. *Biochemistry*, *52*(20), 3601–3608. http://dx.doi.org/10.1021/bi301657w.

Laub, M. T., Biondi, E. G., & Skerker, J. M. (2007). Phosphotransfer profiling: Systematic mapping of two-component signal transduction pathways and phosphorelays. *Methods in Enzymology*, *423*, 531–548. http://dx.doi.org/10.1016/S0076-6879(07)23026-5.

Lindsay, D., & von Holy, A. (2006). Bacterial biofilms within the clinical setting: What healthcare professionals should know. *Journal of Hospital Infection*, *64*(4), 313–325. http://dx.doi.org/10.1016/j.jhin.2006.06.028.

Liu, N., Xu, Y., Hossain, S., Huang, N., Coursolle, D., Gralnick, J. A., et al. (2012). Nitric oxide regulation of cyclic di-GMP synthesis and hydrolysis in Shewanella woodyi. *Biochemistry*, *51*(10), 2087–2099. http://dx.doi.org/10.1021/bi201753f.

Lucas, K. A., Pitari, G. M., Kazerounian, S., Ruiz-Stewart, I., Park, J., Schulz, S., et al. (2000). Guanylyl cyclases and signaling by cyclic GMP. *Pharmacological Reviews*, *52*(3), 375–414.

Ma, X., Sayed, N., Beuve, A., & van den Akker, F. (2007). NO and CO differentially activate soluble guanylyl cyclase via a heme pivot-bend mechanism. *The EMBO Journal*, *26*(2), 578–588. http://dx.doi.org/10.1038/sj.emboj.7601521.

Malinski, T., Taha, Z., Grunfeld, S., Patton, S., Kapturczak, M., & Tomboulian, P. (1993). Diffusion of nitric oxide in the aorta wall monitored in situ by porphyrinic microsensors. *Biochemical and Biophysical Research Communications*, *193*(3), 1076–1082. http://dx.doi.org/10.1006/bbrc.1993.1735.

Martin, F., Baskaran, P., Ma, X., Dunten, P. W., Schaefer, M., Stasch, J. P., et al. (2010). Structure of cinaciguat (BAY 58-2667) bound to Nostoc H-NOX domain reveals insights into heme-mimetic activation of the soluble guanylyl cyclase. *The Journal of Biological Chemistry*, *285*(29), 22651–22657. http://dx.doi.org/10.1074/jbc.M110.111559.

Martin, E., Berka, V., Bogatenkova, E., Murad, F., & Tsai, A. L. (2006). Ligand selectivity of soluble guanylyl cyclase: Effect of the hydrogen-bonding tyrosine in the distal heme pocket on binding of oxygen, nitric oxide, and carbon monoxide. *The Journal of Biological Chemistry*, *281*(38), 27836–27845. http://dx.doi.org/10.1074/jbc.M601078200.

McDougald, D., Rice, S. A., Barraud, N., Steinberg, P. D., & Kjelleberg, S. (2011). Should we stay or should we go: Mechanisms and ecological consequences for biofilm dispersal. *Nature Reviews Microbiology*, *10*(1), 39–50. http://dx.doi.org/10.1038/nrmicro2695.

Mills, P. C., Rowley, G., Spiro, S., Hinton, J. C. D., & Richardson, D. J. (2008). A combination of cytochrome c nitrite reductase (NrfA) and flavorubredoxin (NorV) protects Salmonella enterica serovar Typhimurium against killing by NO in anoxic environments. *Microbiology (Reading, England)*, *154*, 1218–1228. http://dx.doi.org/10.1099/mic.0.2007/014290-0.

Mukhopadyay, R., Sudasinghe, N., Schaub, T., & Yukl, E. T. (2016). Heme-independent redox sensing by the heme-nitric oxide/oxygen-binding protein (H-NOX) from Vibrio cholerae. *The Journal of Biological Chemistry*, *291*(34), 17547–17556. http://dx.doi.org/10.1074/jbc.M116.733337.

Muralidharan, S., & Boon, E. M. (2012). Heme flattening is sufficient for signal transduction in the H-NOX family. *Journal of the American Chemical Society*, *134*(4), 2044–2046. http://dx.doi.org/10.1021/ja211576b.

Nioche, P., Berka, V., Vipond, J., Minton, N., Tsai, A. L., & Raman, C. S. (2004). Femtomolar sensitivity of a NO sensor from Clostridium botulinum. *Science*, *306*(5701), 1550–1553. http://dx.doi.org/10.1126/science.1103596.

Nisbett, L. M., & Boon, E. M. (2016). Nitric oxide regulation of H-NOX signaling pathways in bacteria. *Biochemistry*, *55*(35), 4873–4884. http://dx.doi.org/10.1021/acs.biochem.6b00754.

Olea, C., Boon, E. M., Pellicena, P., Kuriyan, J., & Marletta, M. A. (2008). Probing the function of heme distortion in the H-NOX family. *ACS Chemical Biology*, *3*(11), 703–710. http://dx.doi.org/10.1021/cb800185h.

Olea, C., Jr., Herzik, M. A., Jr., Kuriyan, J., & Marletta, M. A. (2010). Structural insights into the molecular mechanism of H-NOX activation. *Protein Science*, *19*(4), 881–887. http://dx.doi.org/10.1002/pro.357.

Olea, C., Jr., Kuriyan, J., & Marletta, M. A. (2010). Modulating heme redox potential through protein-induced porphyrin distortion. *Journal of the American Chemical Society*, *132*(37), 12794–12795. http://dx.doi.org/10.1021/ja106252b.

Ortiz de Orue Lucana, D., & Groves, M. R. (2009). The three-component signalling system HbpS-SenS-SenR as an example of a redox sensing pathway in bacteria. *Amino Acids*, *37*(3), 479–486. http://dx.doi.org/10.1007/s00726-009-0260-9.

Papenfort, K., & Bassler, B. L. (2016). Quorum sensing signal-response systems in Gram-negative bacteria. *Nature Reviews Microbiology*, *14*(9), 576–588. http://dx.doi.org/10.1038/nrmicro.2016.89.

Pellicena, P., Karow, D. S., Boon, E. M., Marletta, M. A., & Kuriyan, J. (2004). Crystal structure of an oxygen-binding heme domain related to soluble guanylate cyclases. *Proceedings of the National Academy of Sciences of the United States of America*, *101*(35), 12854–12859. http://dx.doi.org/10.1073/pnas.0405188101.

Plate, L., & Marletta, M. A. (2012). Nitric oxide modulates bacterial biofilm formation through a multicomponent cyclic-di-GMP signaling network. *Molecular Cell*, *46*(4), 449–460. http://dx.doi.org/10.1016/j.molcel.2012.03.023.

Plate, L., & Marletta, M. A. (2013). Nitric oxide-sensing H-NOX proteins govern bacterial communal behavior. *Trends in Biochemical Sciences*, *38*(11), 566–575. http://dx.doi.org/10.1016/j.tibs.2013.08.008.

Poole, R. K., & Hughes, M. N. (2000). New functions for the ancient globin family: Bacterial responses to nitric oxide and nitrosative stress. *Molecular Microbiology*, *36*(4), 775–783.

Price, M. S., Chao, L. Y., & Marletta, M. A. (2007). Shewanella oneidensis MR-1 H-NOX regulation of a histidine kinase by nitric oxide. *Biochemistry*, *46*(48), 13677–13683. http://dx.doi.org/10.1021/bi7019035.

Rao, M., Smith, B. C., & Marletta, M. A. (2015). Nitric oxide mediates biofilm formation and symbiosis in silicibacter sp. Strain TrichCH4B. *MBio*, *6*(3), e00206–e00215. http://dx.doi.org/10.1128/mBio.00206-15.

Roberts, M. A., Papachristodoulou, A., & Armitage, J. P. (2010). Adaptation and control circuits in bacterial chemotaxis. *Biochemical Society Transactions*, *38*(5), 1265–1269. http://dx.doi.org/10.1042/BST0381265.

Ross, P., Weinhouse, H., Aloni, Y., Michaeli, D., Weinberger-Ohana, P., Mayer, R., et al. (1987). Regulation of cellulose synthesis in Acetobacter xylinum by cyclic diguanylic acid. *Nature*, *325*(6101), 279–281.

Russwurm, M., & Koesling, D. (2004). NO activation of guanylyl cyclase. *The EMBO Journal*, *23*(22), 4443–4450. http://dx.doi.org/10.1038/sj.emboj.7600422.

Schmidt, P. M., Schramm, M., Schroder, H., Wunder, F., & Stasch, J. P. (2004). Identification of residues crucially involved in the binding of the heme moiety of soluble guanylate cyclase. *The Journal of Biological Chemistry*, *279*(4), 3025–3032. http://dx.doi.org/10.1074/jbc.M310141200.

Schmidt, I., Steenbakkers, P. J., op den Camp, H. J., Schmidt, K., & Jetten, M. S. (2004). Physiologic and proteomic evidence for a role of nitric oxide in biofilm formation by Nitrosomonas europaea and other ammonia oxidizers. *Journal of Bacteriology*, *186*(9), 2781–2788.

Senge, M. O., Medforth, C. J., Forsyth, T. P., Lee, D. A., Olmstead, M. M., Jentzen, W., et al. (1997). Comparative analysis of the conformations of symmetrically and asymmetrically deca- and undecasubstituted porphyrins bearing meso-alkyl or -aryl groups. *Inorganic Chemistry*, *36*(6), 1149–1163.

Shaw, A. W., & Vosper, A. J. (1977). Solubility of nitric-oxide in aqueous and non-aqueous solvents. *Journal of the Chemical Society, Faraday Transactions I*, *73*, 1239–1244. http://dx.doi.org/10.1039/f19777301239.

Stevanin, T. M., Moir, J. W., & Read, R. C. (2005). Nitric oxide detoxification systems enhance survival of Neisseria meningitidis in human macrophages and in nasopharyngeal mucosa. *Infection and Immunity, 73*(6), 3322–3329. http://dx.doi.org/10.1128/IAI.73.6.3322-3329.2005.

Stevanin, T. M., Poole, R. K., Demoncheaux, E. A., & Read, R. C. (2002). Flavohemoglobin Hmp protects Salmonella enterica serovar typhimurium from nitric oxide-related killing by human macrophages. *Infection and Immunity, 70*(8), 4399–4405.

Stewart, P. S., & Costerton, J. W. (2001). Antibiotic resistance of bacteria in biofilms. *Lancet, 358*(9276), 135–138.

Stone, J. R., & Marletta, M. A. (1994). Soluble guanylate cyclase from bovine lung: Activation with nitric oxide and carbon monoxide and spectral characterization of the ferrous and ferric states. *Biochemistry, 33*(18), 5636–5640.

Stone, J. R., & Marletta, M. A. (1996). Spectral and kinetic studies on the activation of soluble guanylate cyclase by nitric oxide. *Biochemistry, 35*(4), 1093–1099. http://dx.doi.org/10.1021/bi9519718.

Sun, Y., Benabbas, A., Zeng, W., Muralidharan, S., Boon, E. M., & Champion, P. M. (2016). Kinetic control of $O_2$ reactivity in H-NOX domains. *The Journal of Physical Chemistry. B, 120*(24), 5351–5358. http://dx.doi.org/10.1021/acs.jpcb.6b03348.

Szurmant, H., Bu, L., Brooks, C. L., 3rd, & Hoch, J. A. (2008). An essential sensor histidine kinase controlled by transmembrane helix interactions with its auxiliary proteins. *Proceedings of the National Academy of Sciences of the United States of America, 105*(15), 5891–5896. http://dx.doi.org/10.1073/pnas.0800247105.

Thomas, D. D., Liu, X., Kantrow, S. P., & Lancaster, J. R., Jr. (2001). The biological lifetime of nitric oxide: Implications for the perivascular dynamics of NO and $O_2$. *Proceedings of the National Academy of Sciences of the United States of America, 98*(1), 355–360. http://dx.doi.org/10.1073/pnas.011379598.

Tran, R., Boon, E. M., Marletta, M. A., & Mathies, R. A. (2009). Resonance Raman spectra of an $O_2$-binding H-NOX domain reveal heme relaxation upon mutation. *Biochemistry, 48*(36), 8568–8577. http://dx.doi.org/10.1021/bi900563g.

Tsai, A. L., Berka, V., Martin, F., Ma, X., van den Akker, F., Fabian, M., et al. (2010). Is Nostoc H-NOX a NO sensor or redox switch? *Biochemistry, 49*(31), 6587–6599. http://dx.doi.org/10.1021/bi1002234.

Tsai, A. L., Berka, V., Martin, E., & Olson, J. S. (2012). A "sliding scale rule" for selectivity among NO, CO, and O(2) by heme protein sensors. *Biochemistry, 51*(1), 172–186. http://dx.doi.org/10.1021/bi2015629.

Tsai, A. L., Martin, E., Berka, V., & Olson, J. S. (2012). How do heme-protein sensors exclude oxygen? Lessons learned from cytochrome c′, Nostoc puntiforme heme nitric oxide/oxygen-binding domain, and soluble guanylyl cyclase. *Antioxidants & Redox Signaling, 17*(9), 1246–1263. http://dx.doi.org/10.1089/ars.2012.4564.

Ueno, T., Fischer, J., & Boon, E. (unpublished data).

von Wantoch Rekowski, M., Kumar, V., Zhou, Z., Moschner, J., Marazioti, A., Bantzi, M., et al. (2013). Insights into soluble guanylyl cyclase activation derived from improved heme-mimetics. *Journal of Medicinal Chemistry, 56*(21), 8948–8952. http://dx.doi.org/10.1021/jm400539d.

Wang, Y., Dufour, Y. S., Carlson, H. K., Donohue, T. J., Marletta, M. A., & Ruby, E. G. (2010). H-NOX-mediated nitric oxide sensing modulates symbiotic colonization by Vibrio fischeri. *Proceedings of the National Academy of Sciences of the United States of America, 107*(18), 8375–8380. http://dx.doi.org/10.1073/pnas.1003571107.

Wang, Y., & Ruby, E. G. (2011). The roles of NO in microbial symbioses. *Cellular Microbiology, 13*(4), 518–526. http://dx.doi.org/10.1111/j.1462-5822.2011.01576.x.

Wedel, B., Humbert, P., Harteneck, C., Foerster, J., Malkewitz, J., Bohme, E., et al. (1994). Mutation of His-105 in the beta 1 subunit yields a nitric oxide-insensitive form of soluble guanylyl cyclase. *Proceedings of the National Academy of Sciences of the United States of America, 91*(7), 2592–2596.

Weinert, E. E., Phillips-Piro, C. M., & Marletta, M. A. (2013). Porphyrin pi-stacking in a heme protein scaffold tunes gas ligand affinity. *Journal of Inorganic Biochemistry*, *127*, 7–12. http://dx.doi.org/10.1016/j.jinorgbio.2013.06.004.

Weinert, E. E., Phillips-Piro, C. M., Tran, R., Mathies, R. A., & Marletta, M. A. (2011). Controlling conformational flexibility of an O(2)-binding H-NOX domain. *Biochemistry*, *50*(32), 6832–6840. http://dx.doi.org/10.1021/bi200788x.

Wenzel, R. P. (2007). Health care-associated infections: Major issues in the early years of the 21st century. *Clinical Infectious Diseases*, *45*(Suppl. 1), S85–S88. http://dx.doi.org/10.1086/518136.

Williams, S. B., & Stewart, V. (1999). Functional similarities among two-component sensors and methyl-accepting chemotaxis proteins suggest a role for linker region amphipathic helices in transmembrane signal transduction. *Molecular Microbiology*, *33*(6), 1093–1102.

Winter, M. B., Herzik, M. A., Jr., Kuriyan, J., & Marletta, M. A. (2011). Tunnels modulate ligand flux in a heme nitric oxide/oxygen binding (H-NOX) domain. *Proceedings of the National Academy of Sciences of the United States of America*, *108*(43), E881–E889. http://dx.doi.org/10.1073/pnas.1114038108.

Wu, G., Liu, W., Berka, V., & Tsai, A. L. (2013). The selectivity of Vibrio cholerae H-NOX for gaseous ligands follows the "sliding scale rule" hypothesis. Ligand interactions with both ferrous and ferric Vc H-NOX. *Biochemistry*, *52*(52), 9432–9446. http://dx.doi.org/10.1021/bi401408x.

Wu, G., Liu, W., Berka, V., & Tsai, A. L. (2015). H-NOX from Clostridium botulinum, like H-NOX from Thermoanaerobacter tengcongensis, binds oxygen but with a less stable oxyferrous heme intermediate. *Biochemistry*, *54*(48), 7098–7109. http://dx.doi.org/10.1021/acs.biochem.5b00994.

Zhang, Y., Lu, M., Cheng, Y., & Li, Z. (2010). H-NOX domains display different tunnel systems for ligand migration. *Journal of Molecular Graphics and Modelling*, *28*(8), 814–819. http://dx.doi.org/10.1016/j.jmgm.2010.02.007.

Zhao, Y., Brandish, P. E., Ballou, D. P., & Marletta, M. A. (1999). A molecular basis for nitric oxide sensing by soluble guanylate cyclase. *Proceedings of the National Academy of Sciences of the United States of America*, *96*(26), 14753–14758.

CHAPTER TWO

# Manganese in Marine Microbiology

**Colleen M. Hansel**[1]
Woods Hole Oceanographic Institution, Woods Hole, MA, United States
[1]Corresponding author: e-mail address: chansel@whoi.edu

## Contents



## Abstract

The importance of manganese in the physiology of marine microbes, the biogeochemistry of the ocean and the health of microbial communities of past and present is emerging. Manganese is distributed widely throughout the global ocean, taking the form of an essential antioxidant ($Mn^{2+}$), a potent oxidant ($Mn^{3+}$) and strong adsorbent (Mn oxides) sequestering disproportionately high levels of trace metals and nutrients in comparison to the surrounding seawater. Manganese is, in fact, linked to nearly all other elemental cycles and intricately involved in the health, metabolism and function of the ocean's microbiome. Here, we briefly review the diversity of microbes and pathways responsible for the transformation of Mn within the three Mn pools and their distribution within the marine environment. Despite decades of interrogation, we still have much to learn about the players, mechanisms and consequences of the Mn cycle, and new and exciting discoveries are being made at a rapid rate. What is clear is the dynamic and

*Advances in Microbial Physiology*, Volume 70
ISSN 0065-2911
http://dx.doi.org/10.1016/bs.ampbs.2017.01.005

ever-inspiring complexity of reactions involving Mn, and the acknowledgement that microorganisms are the catalytic engine driving the Mn cycle.

## 1. INTRODUCTION

The cycling of manganese (Mn) among three oxidation states has a direct impact on human and ecosystem health. Manganese is an essential trace nutrient for all forms of life, serving as a cofactor in a broad range of enzymes in both eukaryotes and prokaryotes that are important in antioxidant and physiological functions. Aqueous Mn(III) complexes and Mn(III,IV) oxides are among the strongest oxidants on our planet, rivalling that of oxygen under some conditions, giving them the unique ability to degrade even the most recalcitrant forms of carbon and oxidize toxic organics and inorganics. Mn oxides have also garnered interest in biotechnology because these layered oxide structures contain numerous vacancies making them highly photoconductive (Kwon, Refson, & Sposito, 2008). Mn oxides are receiving increasing attention as a reliable palaeoproxy since the oxidation of Mn(II) to Mn oxides requires oxidants with high reduction potentials, making them a sensitive indicator of oxygen conditions. Mn also has antioxidant properties, whereby it scavenges reactive oxygen species (ROS), a suite of molecules involved in oxidative stress and programmed cell death (apoptosis). On the flipside, however, excessive levels of Mn may be toxic to humans and lead to neurological disorders, including Parkinson's disease (Keen et al., 1999). As such, a greater understanding of the Mn cycle has wide environmental, geobiological, industrial and public health implications. Recognition of Mn as a biogeochemically relevant metal is rapidly growing, in part due to recent discoveries, including for instance that Mn is the litter-decomposing machinery within coniferous forests (Keiluweit et al., 2015), a major antioxidant within some ocean basins (Wuttig, Heller, & Croot, 2013), part of a possible precursor complex to oxygenic photosynthesis (Johnson et al., 2013) and present in concentrated deposits on Mars that are reminiscent of Mn oxides (Lanza et al., 2014). At the root of the Mn redox cycle are microbes. Through their direct and indirect activity, microbes control the cycling of Mn and the formation of Mn oxides within a wide range of environmental and engineered systems.

This chapter provides a brief synopsis of the varied ways in which microbes interact with Mn, with a specific emphasis on marine systems.

As discussed in more detail later, Mn can reside within Mn-specific and non-specific metalloenzymes (see Section 7), undergo redox reactions mediated by microorganisms that are linked to or independent of metabolism (see Sections 5 and 6) and be stored intracellularly for oxidative stress defence (see Section 7). Further, due to the strong oxidation and sorption capacity of Mn(III) and Mn(IV) oxides, the oxidation of Mn(II) may also provide indirect benefits to microbes (see Section 8). This review is necessarily brief, and I refer the readers to the many references throughout, as well as previous seminal and informative reviews on various aspects of manganese in microbiology (Gadd, 2010; Ghiorse, 1988; Hansel & Learman, 2015; Kehres & Maguire, 2003; Lovley, 1991; Morgan, 2000; Nealson & Saffarini, 1994; Tebo et al., 2004; Tebo, Johnson, McCarthy, & Templeton, 2005).

## 2. BRIEF OVERVIEW OF Mn SPECIATION IN MARINE SYSTEMS

Manganese is the third most abundant transition metal in the Earth's crust (~0.019 mol $kg^{-1}$), where it is about 56 times less abundant than iron. Nonetheless, the abundance of Mn is ~60% greater in oceanic compared to continental crust (Morgan, 2000). Mn exists in three dominant oxidation states in marine environments—$Mn^{2+}$, $Mn^{3+}$ and $Mn^{4+}$. Manganese cycles among the three oxidation states via reduction, oxidation and disproportionation reactions (Fig. 1). Most oxidation and reduction redox reactions are believed to be one-step electron transfer reactions (Luther, 2005), whereby the first electron transfer from molecular oxygen ($O_2$) to Mn(II) forming Mn(III) is thermodynamically unfavoured within typical marine waters and pH conditions less than ~8.5 (Luther, 2010).

Despite the fully oxidized Mn(IV) species being thermodynamically favoured in oxic seawater, Mn(II) is the dominant species in surface marine waters (see for instance Bruland, Orans, & Cowen, 1994; Johnson, Coale, Berelson, & Gordon, 1996; Sunda & Huntsman, 1988). This phenomenon is a function, in part, due to the thermodynamic and kinetic constraints of

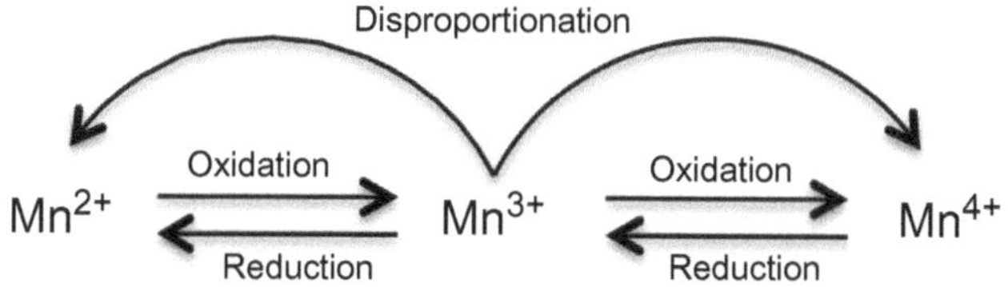


**Fig. 1** Dominant oxidation states of Mn within environmental systems.

the oxidation of Mn(II) to Mn(III) via $O_2$ (Luther, 2010). Further, within the surface ocean, light, in particular, plays an important role in the cycling and speciation of Mn. For instance, sunlight can induce rapid photoreduction of Mn oxides (Sunda & Huntsman, 1994; Sunda, Huntsman, & Harvey, 1983) and light has been observed to inhibit microbial Mn(II) oxidation (Sunda & Huntsman, 1988), leading to preferential accumulation of Mn(II) within surface waters. Nevertheless, Mn(II) oxidation and Mn oxide accumulation have at times been observed in the euphotic zone (Jacobs, Emerson, & Skei, 1985). Mn(II) exists primarily as the hydrolysed ion $Mn(H_2O)_6{}^{2+}$. Divalent manganese in seawater is complexed by ions, such as chloride (Goldberg & Arrhenius, 1958), sulphate and bicarbonate (Hem, 1963), and by organic substances like amino acids (Graham, 1959).

Historically, dissolved Mn was assumed to be solely Mn(II), as the natural occurrence of soluble Mn(III) in aqueous systems was considered rare. However, substantial levels of soluble Mn(III) complexes have more recently been observed in suboxic and anoxic marine waters (Dellwig, Schnetger, Brumsack, Grossart, & Umlauf, 2012; Oldham, Owings, Jones, Tebo, & Luther, 2015; Trouwborst, Clement, Tebo, Glazer, & Luther, 2006; Yakushev, Pakhomova, Sørenson, & Skei, 2009) and porewaters within haemipelagic sediments, where Mn(III) accounted for up to 90% of the total dissolved Mn pool (Madison, Tebo, & Luther, 2011; Madison, Tebo, Mucci, Sundby, & Luther, 2013). While the composition of these ligands is presently unknown, the reactivity of these observed Mn(III) complexes is estimated to be stronger than complexes of Mn(III) with pyrophosphate but weaker than those with desferrioxamine B (Madison et al., 2011). Current models of Mn cycling and speciation within the ocean will no doubt require revising as the broader distribution of Mn(III) is determined.

In the absence of stabilizing, ligands, such as citrate, pyrophosphate and siderophores (Klewicki & Morgan, 1998, 1999; Kostka, Luther, & Nealson, 1995), Mn(III) tends to disproportionate into the +2 and +4 oxidation states (reaction 1). The more oxidized species, Mn(IV), is not stable in solution and instead precipitates as Mn hydroxides, oxyhydroxides or oxides (hereinafter referred to as oxides and written as $MnO_2$).

$$2Mn^{3+} + 2H_2O \rightarrow Mn^{2+} + MnO_{2(s)} + 4H^+ \quad (1)$$

Deeper in the water column, typically below the photic zone, a peak in particulate Mn occurs, representing a combination of Mn(II) oxidation to

Mn oxides and Mn scavenging on particles (Bruland et al., 1994; Johnson et al., 1996; Sunda & Huntsman, 1988). Deeper still, manganese oxides are often found in high concentrations in surface sediments within deep fjords or ocean basins, where geochemical focusing leads to their accumulation (Thamdrup, 2000; Thamdrup, Glud, & Hansen, 1994). For instance, Mn oxide concentrations exceed 150 μmol $cm^{-3}$ to depths of 15–25 cm in the sediments of various coastal fjords and ocean basins globally. Reductive dissolution of these oxides can lead to the formation of rhodochrosite ($MnCO_3$) within marine sediments (Thamdrup, 2000).

At the sediment interface at abyssal depths, a significant portion of manganese is concentrated in ferromanganese concretions (nodules) and crusts (Hein & Koschinsky, 2013). Such concretions have been found in all the oceans of the world (Horn, Horn, & Delach, 1972), where they may cover vast areas or be distributed in patches. Nodules develop around a nucleus, which may be a foraminiferal test, a piece of pumice, a clay particle, an older nodule fragment, among others (Fig. 2). The growth rate of manganese nodules in the deep sea is reportedly very slow, ranging from 1 to 168 mm per $10^6$ years (Kadko & Burckle, 1980; Ku & Broecker, 1969; Reyss, Marchig, & Ku, 1982). The rate of nodule growth, however, is not constant and can include periods of no growth, indicating that conditions must be continually favourable for maintained nodule growth (Heye & Beiersdorf, 1973).

Over 30 different natural and synthetic Mn oxide minerals are known to exist, largely due to the combination of varied valence states of Mn and the

**Fig. 2** A several million-year-old ferromanganese nodule that formed around a *C. megalodon* tooth, an extinct species of shark that lived during the Cenozoic Era (WHOI Seafloor Sediments Data Collection archive).

diverse arrangements of $MnO_6$ octahedra (the building blocks of Mn oxide structures) (Post, 1999). Most Mn oxides are categorized as having either a layer or a tunnel structure (Fig. 3). The common phyllomanganates (layer–structure oxides), such as those within the group of birnessite-like oxides, are composed of stacked sheets of edge-sharing Mn octahedra. Minerals within the birnessite group vary in symmetry (hexagonal, pseudo-orthogonal), Mn(III) content and distribution, and stacking, disorder, strain and number of vacancies within the phyllomanganate layers (Giovanoli, 1980; Manceau, Kersten, Marcus, Geoffroy, & Granina, 2007; Villalobos, Lanson, Manceau, Toner, & Sposito, 2006; Villalobos, Toner, Bargar, & Sposito, 2003).

Hexagonal birnessite phases oftentimes undergo abiotic transformations and ageing to more crystalline (more energetically favourable) Mn phases, such as triclinic birnessite, todorokite and feitknechtite (Bargar et al., 2005; Bodei, Manceau, Geoffroy, Baronnet, & Buatier, 2007; Learman, Voelker, Vazquez-Rodriguez, & Hansel, 2011; Learman, Wankel, et al., 2011; Webb, Dick, Bargar, & Tebo, 2005). Structural evolution of birnessite

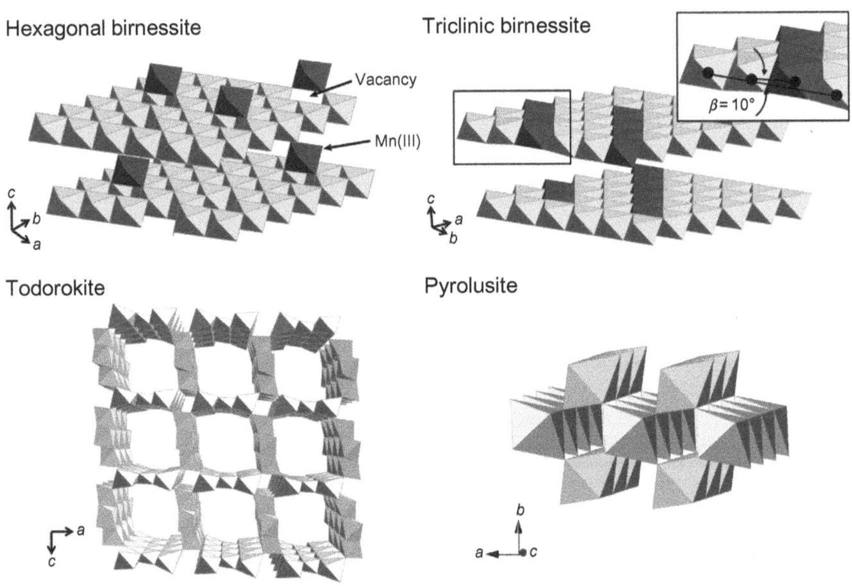


**Fig. 3** Simplified polyhedral representation of layered (birnessite) and tunnel (todorokite, pyrolusite) Mn oxide structures. The *dark shaded octahedra* represent lattice positions of $Mn^{3+}$ octahedra. Interlayer cations are omitted for simplicity. Differences in the $Mn^{3+}$ content, number of vacancies and layer strain (*inset*) are illustrated for the common biogenic hexagonal and triclinic birnessite phases. Simplified illustration of the out-of-plane bending angle ($\beta$) is depicted for triclinic birnessite (*inset*).

is accelerated by redox transfer reactions at the mineral surface, as observed for surface reactions with Mn(II) and organic compounds (Banerjee & Nesbitt, 2001; Bargar et al., 2005; Elzinga & Kustka, 2015; Learman, Voelker, et al., 2011; Learman, Wankel, et al., 2011; Webb et al., 2005) (Fig. 4). A wide variety of cations can be incorporated into the interlayers and tunnels depending largely on physical properties, such as the atomic structure, crystal morphology and mineral size (Hochella et al., 2008). Availability of different ions therefore influences the Mn oxide formed upon nucleation, such as barium-rich waters favouring the formation of romanechite over todorokite and calcium-rich waters favouring triclinic over hexagonal birnessite (Feng, Yanagisawa, & Yamasaki, 1998; Manceau, Lanson, & Geoffroy, 2007; Webb et al., 2005).

The most widespread Mn oxide in ferromanganese deposits is the hexagonal *c*-disordered birnessite phase vernadite ($\delta$-$MnO_2$) (Hein &

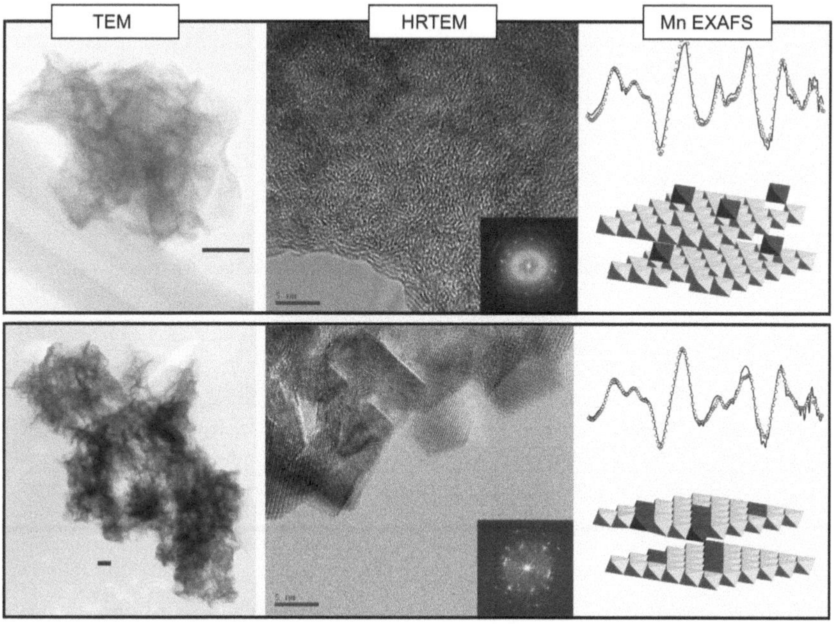


**Fig. 4** Mn oxides produced by biogenic superoxide after 4 (*top*) and 24 h (*bottom*) of reaction. Transmission electron microscopy (TEM) image shows dispersed, individual Mn oxide colloids at 4 h that aggregate at 24 h. Scale bar = 20 nm. HRTEM shows increased crystallinity over time (seen also in the Fourier filtered transform (FFT)—*inset*), and lattice fringes lend support for aggregated growth of 4 h colloids. Scale bar = 5 nm. $k^3$-weighted Mn extended X-ray absorption fine structure (EXAFS) spectroscopy spectra (*solid*) and fit (*dotted*) show change from hexagonal to triclinic birnessite ($k$ = 3–12).

Koschinsky, 2013; Manceau, Kersten, et al., 2007; Peacock & Moon, 2012; Post, 1999; Takahashi, Manceau, Geoffroy, Marcus, & Usui, 2007). Both triclinic birnessite and todorokite are also common in diagenetic Fe–Mn nodules and hydrothermal deposits (Burns & Burns, 1977). Furthermore, enriched barium within hydrothermal waters promotes the formation of either romanechite (lower temperatures) or hollandite (higher temperatures) within ferromanganese deposits (Feng et al., 1998; Manceau, Lanson, & Geoffroy, 2007). The Mn oxide mineralogy within Fe–Mn deposits is oftentimes heterogeneous with layered and/or zoned mineralogy and chemistry, involving both phyllo- and tectomanganates with variable trace metal enrichment (Post, 1999). For instance, recent investigations of ferromanganese nodules revealed a biphasic mineral association with alternating layers of Ni-rich vernadite (a phyllomanganate whose synthetic analogue is $\delta$-$MnO_2$) and barium-rich romanechite (a tectomanganate) (Manceau, Kersten, et al., 2007). This binary Mn oxide banding may represent a two-mode accretionary model, with romanechite formation during episodic hydrothermal activity superimposed on continuous diagenetic formation of vernadite (Manceau, Kersten, et al., 2007). Foreign ions may play a particularly important role in Mn oxide reactivity in ferromanganese precipitates due to their enrichment in metal(loid)s by many orders of magnitude relative to crustal abundances (Goldberg, 1954; Jenne, 1968).

## 3. MANGANESE-TRANSFORMING MICROBES

Mn-oxidizing and -reducing bacteria are widely distributed in the environment and are readily cultured from soils (Yang et al., 2013), desert varnish (Hugate et al., 1987; Taylor-George et al., 1983), freshwater springs (Ehrlich & Zapkin, 1985; Mustoe, 1981), lakes and ponds (Santelli, Chaput, & Hansel, 2014; Sokolova-Dubinina & Deriugina, 1967; Sokolova-Dubinina & Deryugina, 1976; Stabel & Kleiner, 1983), caves (Carmichael, Carmichael, Santelli, Strom, & Bräuer, 2013; Northup et al., 2003; Northup & Lavoie, 2001), hydrothermal environments (Lovley, Holmes, & Nevin, 2004; Tor, Kashefi, & Lovley, 2003), contaminated sediments (Lovley & Anderson, 2000; Lovley & Phillips, 1992; Lovley, Phillips, & Lonergan, 1989) and remediation treatment systems (Lovley & Anderson, 2000; Lovley & Phillips, 1992; Lovley et al., 1989; Santelli et al., 2010). Within marine systems, Mn-oxidizing and -reducing bacteria have been identified and/or isolated from a wide range of environments, including ocean water (Cowen & Silver, 1984; Moffett, 1997), sediments

(Thamdrup, 2000; Vandieken, Finke, & Thamdrup, 2014), ocean floor (Ehrlich, 2000; Ehrlich, Ghiorse, & Johnson, 1972; Templeton, Staudigel, & Tebo, 2005), estuaries/bays (Brauer et al., 2011; Hansel & Francis, 2006; Krumbein, 1971; Tebo & Emerson, 1985; Tebo, Nealson, Emerson, & Jacobs, 1984; Thiel, 1925) and hydrothermal vents (Cowen et al., 1998; Cowen, Massoth, & Feely, 1990; Dick & Tebo, 2010).

To date, only Mn(II)-oxidizing bacteria and fungi have been identified in the literature and it is unknown if organisms within the *Archaea* domain can oxidize Mn(II). Bacteria with the ability to oxidize Mn(II) to Mn oxides have been observed in a diversity of phylogenetic groups, spanning the low G + C Gram-positive bacteria (*Firmicutes*), *Actinobacteria* and *Alpha-*, *Beta-* and *Gammaproteobacteria* (Fig. 5). Some of the common Mn(II)-oxidizing bacterial isolates include *Leptothrix* sp. (Brouwers, Vijgenboom, Corstjens, de Vrind, & de Vrind-de Jong, 2000; Corstjens, de Vrind, Goosen, & de Vrind-de Jong, 1997; Takeda et al., 2012), *Pedomicrobium* (Larsen, Sly, & McEwan, 1999), *Pseudomonas* sp. (Brouwers et al., 1999; de Vrind, Brouwers, Corstjens, den Dulk, & de Vrind-de Jong, 1998; Francis & Tebo, 2001; Geszvain, McCarthy, & Tebo, 2013; Okazaki et al., 1997; Santelli et al., 2014), *Escherichia coli* (Su et al., 2013), *Ralstonia* sp. (Yang et al., 2013), *Variovorax* sp. (Yang et al., 2013), *Albidiferax* sp. (Akob et al., 2014), *Citrobacter* sp. (Tang, Xia, Zeng, Wu, & Ye, 2013), *Flavobacterium* sp. (Carmichael et al., 2013; Santelli et al., 2014) and the marine bacteria *Aurantimonas* sp. (Dick, Podell, et al., 2008), *Erythrobacter* sp. (Francis, Co, & Tebo, 2001), *Roseobacter* clade sp. (Hansel & Francis, 2006; Templeton et al., 2005), *Bacillus* sp. (Francis & Tebo, 2002; Van Waasbergen, Hildebrand, & Tebo, 1996; Van Waasbergen, Hoch, & Tebo, 1993), *Marinobacter* sp. (Templeton et al., 2005), *Pseudoalteromonas* sp. (Ehrlich, 1983; Ehrlich & Salerno, 1990; Templeton et al., 2005) and *Alteromonas* sp. (Ehrlich, 1976; Templeton et al., 2005). While bacteria are the dominant Mn(II)-oxidizing organisms observed in marine systems, known Mn(II)-oxidizing fungi have been observed within deep-sea ferromanganese crusts (Connell, Barret, Templeton, & Staudigel, 2009). Further, many Mn(II)-oxidizing fungi isolated from terrestrial environments (Miyata, Maruo, et al., 2006; Miyata, Tani, et al., 2006; Santelli et al., 2010; Tani et al., 2003) have cosmopolitan distributions, including marine systems, presenting a potential for fungal Mn(II) oxidation within various marine and/or brackish environments.

The diversity of microbes involved in the reduction of manganese has undergone only minimal investigation. Within marine sediments and

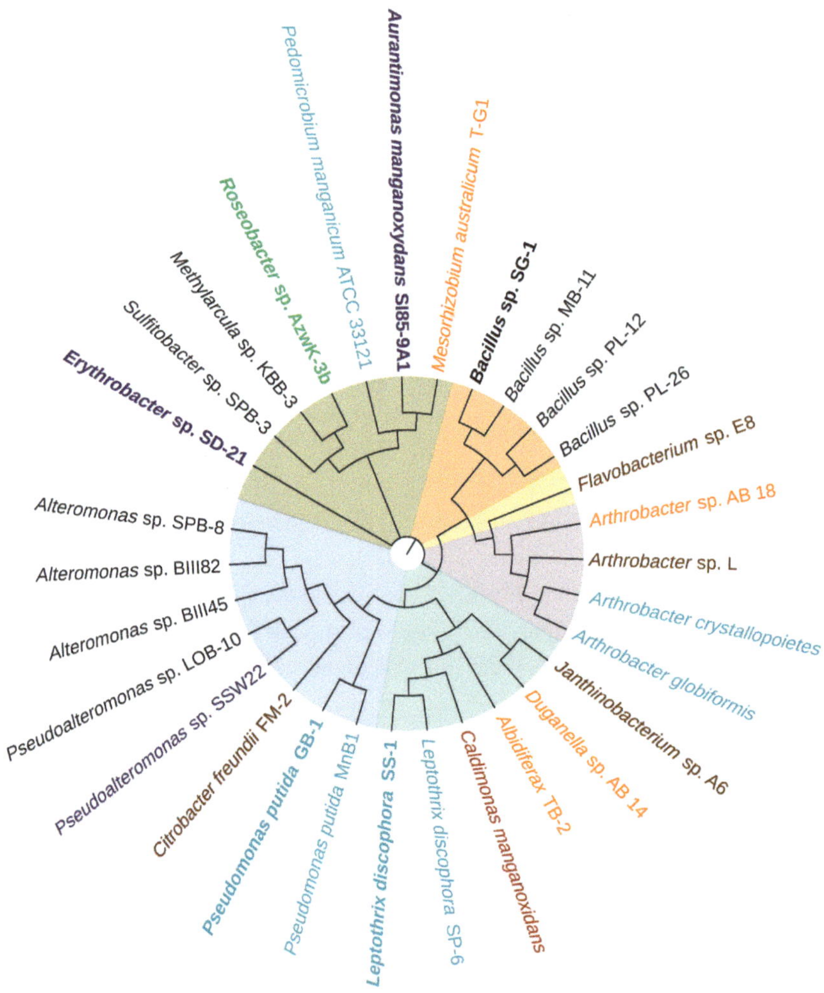


**Fig. 5** Neighbour-joining phylogenetic tree based on 16S ribosomal DNA sequences of representative Mn(II)-oxidizing bacteria. Sequences were aligned over >1300 positions using Clustal Omega (https://www.ebi.ac.uk/tools/msa/clustalo/) and assembled into a tree using the interactive Tree of Life (Lenunic & Bork, 2011; Letunic & Bork, 2007; http://itol.embl.de/). Sequence similarity is only an estimate of evolutionary relationships. Branch colours represent phylogenetic group, including *Alphaproteobacteria* (*olive green*), *Gammaproteobacteria* (*blue*), *Betaproteobacteria* (*aquamarine*), *Actinobacteria* (*purple*), *Bacteroidetes* (*gray*) and *Firmicutes* (*orange*). Organism names are coloured based on the environment from which they were isolated, including marine water (*dark blue*), marine sediments/crust/nodule (*black*), estuarine water (*green blue*), freshwater sediment/water (*light blue*), soil and cave deposits (*brown*), mining-impacted acidic sediments (*orange*) and a hot spring (*red*). *Names in bold* indicate organisms that have served as model organisms for identifying Mn oxidase genes/proteins.

incubations undergoing Mn(IV) reduction, Mn(IV)-reducing bacteria have been identified as *Pelobacter* within the *Deltaproteobacteria*; *Acrobacter* within the *Epsilonproteobacteria*; and *Colwellia*, *Oceanospirillaceae* and *Shewanella* within the *Gammaproteobacteria* (Vandieken et al., 2012; Vandieken & Thamdrup, 2013). These identifications were based on DNA sequencing of enrichment cultures and incubations combined with stable-isotope probing. Whether these organisms have been obtained in pure culture and characterized is unknown.

More broadly, the only bacteria isolated specifically using Mn oxides as the terminal electron acceptor are *Shewanella oneidensis* MR-1 isolated from sediments of Lake Oneida, NY, and the thermophile *Bacillus infernus* isolated from the terrestrial subsurface (Boone et al., 1995; Burdige & Nealson, 1985; Myers & Nealson, 1988a). Instead, most Mn(IV)-reducing organisms were isolated using other electron acceptors, particularly ferric iron. These bacteria include *Geobacter metallireducens* (Lovley & Phillips, 1988), *G. sulfurreducens* (Caccavo et al., 1994), *Geothrix fermentans* (Coates, Ellis, Gaw, & Lovley, 1999), *Sulfurospirillum barnesii* SES-3 (Laverman et al., 1995), *Desulfovibrio desulfuricans*, *Desulfomicrobium baculatum*, *Desulfobacterium autotrophicum* and *Desulfuromonas acetoxidans* (Lovley & Phillips, 1994), *S. oneidensis* MR-4 (Larsen et al., 1998), *Bacillus polymyxa* D1, *Bacillus* MBX1 (Rusin, Quintana, Sinclair, Arnold, & Oden, 1991; Rusin, Sharp, Arnold, & Sinclair, 1991) and *Pantoea agglomerans* SP1 (Francis, Obraztsova, & Tebo, 2000). At present, the only archaeal Mn oxide reducer is *Pyrobaculum islandicum* (Kashefi & Lovley, 2000). Some fungal species also have the capacity to reduce Mn(IV) (Gadd, 2010; Wei, Hillier, & Gadd, 2012).

# 4. OBSERVATIONS OF MICROBIAL Mn CYCLING IN THE MARINE ENVIRONMENT

## 4.1 Mn Oxidation

Microbially mediated Mn(II) oxidation and biogenic Mn oxides have been observed from a number of coastal waters (Moffett, 1994a, 1994b; Sunda & Huntsman, 1987), bays (Krumbein, 1971; Krumbein & Altmann, 1973; Moffett & Ho, 1996), fjords (Jacobs et al., 1985; Tebo et al., 1984), inlets (Emerson et al., 1982; Tebo & Emerson, 1985), estuaries (Hansel & Francis, 2006; Vojak, Edwards, & Jones, 1985a, 1985b) and the deep sea, including sediments and crusts adjacent to hydrothermal sources (Cowen et al., 1990; Fortin, Ferris, & Scott, 1998; Juniper & Tebo, 1995;

Mandernack & Tebo, 1993), hydrothermal vent plumes and surrounding waters (Cowen et al., 1990; Dick et al., 2009; Mandernack & Tebo, 1993; Tambiev & Demina, 1992) and deep-sea basalt surfaces (Hein et al., 1997; Templeton et al., 2009, 2005).

A role for microbes in oxidizing Mn(II) and precipitating Mn oxides in marine water and sediments has been suggested based on shipboard Mn(II) incubation experiments, the presence of and/or isolation of known Mn(II)-oxidizing microbes and/or observations of Mn oxide-encrusted bacteria (Cowen & Silver, 1984). Within the oxic water column of the Sargasso Sea and equatorial Pacific, for instance, Mn removal has been attributed to both biogenic Mn(II) oxidation and nonoxidative Mn update, with the relative proportion of the removal mechanism varying in space and time (Moffett, 1997; Sunda & Huntsman, 1988). Microbial Mn(II) oxidation and Mn oxide formation have also been observed in suboxic marine waters. For instance, in the suboxic zone of the Black Sea, microbially catalysed Mn(II) oxidation (Tebo, 1991) has been attributed to lateral intrusions of $O_2$ into this zone (Schippers, Neretin, Lavik, Leipe, & Pollehne, 2005) and low levels of $O_2$ (Clement, Luther, & Tebo, 2009). The extent and rate of Mn(II) oxidation and Mn oxide formation are influenced by a number of factors, including $O_2$ levels, temperature, light, salinity, Mn(II) concentrations and amount of suspended matter (Emerson et al., 1982; Richardson, Aguilar, & Nealson, 1988; Sunda & Huntsman, 1987, 1988; Tebo & Emerson, 1985; Tebo et al., 1984), thus resulting in seasonal and diel variation in Mn composition and distributions (Sunda & Huntsman, 1990; Thamdrup et al., 1994). In the deep sea, the relative role of biotic vs abiotic mechanisms in Mn(II) oxidation appears to vary among different hydrothermal sites and locations along the ridge axis (Cowen et al., 1990; Dick et al., 2009; Mandernack & Tebo, 1993).

Despite the lack of a universally accepted model, Mn(II)-oxidizing organisms have also been implicated in the formation of marine ferromanganese crusts and nodules. In particular, DNA sequences (Stein, La Duc, Grundl, & Nealson, 2001; Wu et al., 2013; Yli-Hemminki, Jorgensen, & Lehtoranta, 2014) and cellular morphologies (Burnett & Nealson, 1981, 1983; Ehrlich et al., 1972; Ghiorse, 1980; LaRock & Ehrlich, 1975) consistent with known Mn(II)-oxidizing bacteria and fungi have been observed within ferromanganese crusts, nodules and concretions. Mn(II)-oxidizing bacteria (Ehrlich, 1976; Templeton et al., 2005) and fungi (Connell et al., 2009) have also been isolated from ferromanganese crusts and nodules. Further, Mn oxide-encrusted cells are typically observed as part of the endolithic microbial communities within fractures of seafloor basalts (McLoughlin et al., 2011; Thorseth, Pedersen, & Christie, 2003; Thorseth et al., 2001). In addition to directly contributing to Mn oxide formation through enzymatic activity, indirect microbial activity (Kalinenko, Belokopytova, & Nikolaeva, 1962) has also been suggested whereby bacterial destruction of Mn(II) organic complexes in seawater liberates aqueous Mn(II) ions, which then abiotically oxidize and precipitate (Graham, 1959).

## 4.2 Mn Reduction

On a thermodynamic basis, the reduction of Mn(IV) yields more energy than ferric iron and sulphate, and thus microbial Mn reducers are believed to have a competitive advantage over iron- and sulphate-reducing bacteria when metabolizing similar carbon substrates. For instance, Thamdrup, Rossello-Mora, and Amann (2000) found that dissimilatory Mn reduction was the most important means of organic carbon mineralization in the surface layer down to ~1 cm depth in shelf sediments of the Black Sea, whereas dissimilatory sulphate reduction was the main carbon mineralization process below this depth. Similarly, microbial reduction of Mn oxides was found to be the only important anaerobic carbon oxidation process within coastal sediments enriched in manganese (>3%) (Canfield et al., 1993; Canfield, Thamdrup, & Hansen, 1993). Microbial manganese reduction has also been attributed to 25%–99% of carbon oxidation within manganese-rich coastal marine sediments (Aller, 1990; Canfield, Jørgensen, et al., 1993; Canfield, Thamdrup, & Hansen, 1993; Nickel, Vandieken, Bruchert, & Jorgensen, 2008; Thamdrup et al., 2000; Vandieken et al., 2014).

Within three Mn oxide-rich sediments from three geographically and geochemically distinct regions (Gullmar Fjord, Sweden; Skagerrak, Norway; Ulleung Basin, Korea), manganese reduction was the prevailing terminal electron accepting process in anoxic incubations (Vandieken et al., 2012). In fact, addition of carbon substrates within the sediments did not stimulate iron and sulphate reduction, hinting at the role of manganese-reducing organisms in controlling carbon mineralization within these sediments. Further, the dominant anaerobic pathway of carbon oxidation within sediments to 15 cm depth in Gullmar Fjord (Sweden) was identified as manganese reduction (Vandieken et al., 2014). In these sediments, up to 75% of the electron flow for manganese reduction was linked to $H_2$ oxidation, rather than lactate and acetate, which may have resulted in competitive inhibition of iron- and sulphate-reducing communities by low $H_2$.

Within Mn oxide-rich marine sediments, manganese-reducing bacteria are in high abundance with most probable number (MPN) counts as high as $4 \times 10^6$ cells $cm^{-3}$ and up to three orders of magnitude higher than MPNs for iron-reducing organisms (Nealson, Myers, & Wimpee, 1991; Thamdrup et al., 2000; Vandieken et al., 2012). Interesting, MPNs of Mn(IV)-reducing bacteria were similar for these Mn-rich sediments and the surface layer of sediment from Aarhus Bay in Denmark (Vandieken & Thamdrup, 2013), sediments with a lower Mn content more typical of global coastal environments. Further, while the Mn(IV)-reducing communities were phylogenetically distinct from the iron-reducing communities in Mn-rich sediments, the communities had more taxonomic overlap in the lower Mn sediments of Aarhus Bay.

# 5. MANGANESE IN MICROBIAL RESPIRATION

## 5.1 Mn Oxidation

A strong and consistent link between Mn(II) oxidation and energy gain is unclear. In some investigations, Mn(II) oxidation by aerobic bacteria has been attributed to energy gain for the bacterium, including both autotrophic (Beijerinck, 1913) and mixotrophic growth (Lieske, 1919; Sartory & Meyer, 1947). The two most studied of these organisms are two marine Gram-negative bacteria. Specifically, Mn(II) oxidation by the Gram-negative marine bacterium, *Pseudoalteromonas* sp. $SSW_{22}$, has been coupled to ATP synthesis during aerobic respiration (Ehrlich, 1983; Ehrlich & Salerno, 1990). Further, *Alteromonas* sp. BIII45, which was isolated from a marine ferromanganese nodule, is believed to gain energy during mixotrophic growth through the oxidation of Mn(II) coupled to oxygen reduction, but only if the Mn(II) is adsorbed first to Mn oxides (Ehrlich, 1976). Several possible models for the Mn(II) oxidation and energy generation processes have been proposed (see Ehrlich & Newman, 2008 for review); however, the underlying genetic machinery and biochemical reactions remain unknown.

Regardless of whether the Mn(II) is dissolved or adsorbed, however, it has been proposed that energy generation could only occur in the second step, the Mn(III)/Mn(IV) couple (Ehrlich & Newman, 2008). Although oxidation of Mn(II) to $MnO_2$ in a single step (reaction 2) by removal of two electrons is thermodynamically viable (Luther, 2005, 2010), the $E^{o\prime}$ for the Mn(II)/Mn(III) redox couple is too high to allow for ATP synthesis coupled to electron transport (Ehrlich & Newman, 2008). In fact, Mn(III) has been consistently measured as an intermediate in microbial Mn(II) oxidation (Learman, Voelker, et al., 2011; Webb et al., 2005), indicating that it is a one-electron transfer reaction (simplified reactions 3 and 4):

$$2Mn^{2+} + O_2 + 2H_2O \rightarrow 2MnO_{2(s)} + 4H^+ \quad (2)$$

$$Mn^{2+} + \frac{1}{4}O_2 + H^+ \rightarrow Mn^{3+} + \frac{1}{2}H_2O \quad (3)$$

$$2Mn^{3+} + \frac{1}{2}O_2 + 3H_2O \rightarrow 2MnO_{2(s)} + 6H^+ \quad (4)$$

While Mn(II) oxidation coupled to denitrification is thermodynamically possible (Luther, 2010; Luther, Sundby, Lewis, Brendel, & Silverberg, 1997) and field observations have hinted at the presence of Mn(II)-based denitrification (Luther et al., 1997; Vandenabeele, De Beer, Germonpré, Van de Sande, & Verstraete, 1995), strong evidence for this reaction mechanism is still lacking.

Instead, the majority of Mn(II)-oxidizing bacteria in culture to date are strict heterotrophic aerobes where Mn(II) oxidation is a side reaction not involved in energy generation (see Section 6).

## 5.2 Mn Reduction

The oxidation of organic carbon or hydrogen can be coupled by marine microorganisms to the anaerobic reduction of soluble Mn(III) complexes (Hui, Szeinbaum, DiChristina, & Taillefert, 2012; Kostka et al., 1995) and solid-phase Mn(III,IV) oxides (Burdige & Nealson, 1985; Lovley et al., 1993, 1989; Lovley & Phillips, 1988; Myers & Nealson, 1988a; Nealson et al., 1991; Thamdrup, 2000; Vandieken et al., 2014) (Fig. 6). Despite observations of unique Mn(IV)-reducing communities within marine sediment incubations (Vandieken et al., 2012), at present there are no cultured isolates of obligate Mn(IV)-reducing microorganisms. Instead, organisms capable of Mn reduction typically have the capacity to utilize a broad range of electron acceptors, including other metals (e.g. iron, uranium, chromate) and at times oxygen, nitrate and/or organic compounds (e.g. fumarate) (Boone et al., 1995; Burdige & Nealson, 1985; DiChristina, 1992; Lovley et al., 2004; Myers & Nealson, 1988a; Nealson & Saffarini, 1994). Depending on the organism, Mn(IV)-reducing bacteria can also respire a variety of sulphur compounds, including elemental sulphur, sulphite, sulphate and thiosulphate. The presence of these

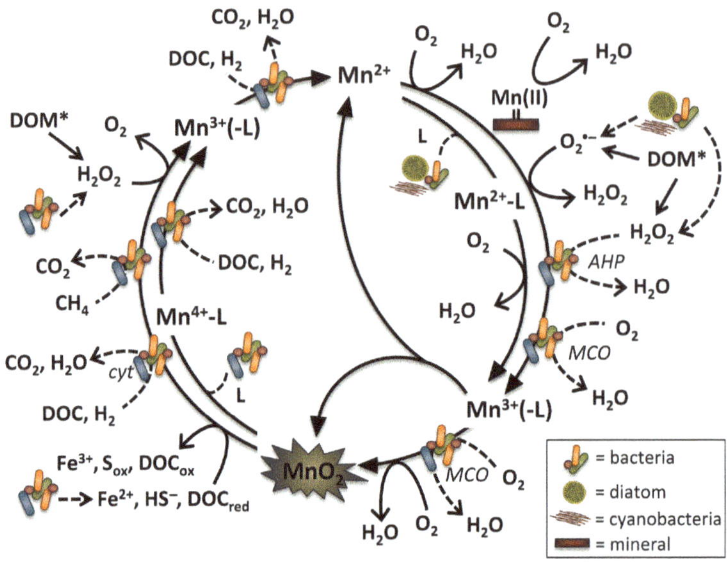


**Fig. 6** Simplified schematic representing some of the major pathways for Mn oxidation and reduction within marine systems. Mn(II) oxidation pathways include direct enzymatic activity by multicopper oxidases (MCO) and animal heme peroxidases (AHP), abiotic reaction with superoxide ($O_2^{\bullet-}$) produced biogenically and via photochemically excited dissolved organic matter (DOM*), and ligand (L)-mediated oxidation. Abiotic oxidation pathways also include direct homogeneous oxidation by molecular oxygen ($O_2$) (at pH > 8) and heterogeneous oxidation by $O_2$ catalysed by Mn(II) adsorption onto mineral surfaces. Mn reduction pathways include microbial reduction coupled to oxidation of $H_2$ or dissolved organic carbon (DOC) mediated by *c*-type cytochromes (*cyt*), microbial-induced coupling with methane oxidation and abiotic reaction with abiotic- and microbial-derived reductants such as hydrogen peroxide ($H_2O_2$), reduced DOC, ferrous iron ($Fe^{2+}$) and sulphide ($HS^-$). Abiotic chemical reactions are *curved solid black lines* and microbially mediated are *curved dashed lines*. Mn(III) formed upon Mn(II) oxidation or Mn(IV) oxide reduction can undergo further redox reactions or disproportionate to both Mn(II) and Mn oxides.

alternative electron acceptors can inhibit Mn reduction. For instance, oxygen, nitrate and fumarate inhibit Mn oxide respiration by *S. oneidensis*, but sulphate, sulphite, molybdate, nitrite and tungstate do not (Myers & Nealson, 1988a). Some bacteria, like *S. oneidensis*, are facultative anaerobes but nevertheless reduce Mn(IV) oxide only anaerobically (Myers & Nealson, 1988a), whereas others, like *G. metallireducens*, are strict anaerobes and reduce Mn oxides only anaerobically (Lovley & Phillips, 1988). Mn(IV)-reducing bacteria also have a broad array of electron donors that can support Mn oxide reduction, including a wide range of organic acids that can be either incompletely oxidized (e.g. many *Shewanella* spp.) or completely oxidized to $CO_2$ (e.g. *Geobacter* spp.).

A great deal of research has been conducted exploring the mechanisms of microbial extracellular electron transfer to solid-phase oxides—yet, these studies have centred on freshwater sediment iron-reducing bacteria. Fe and Mn oxide-reducing microbes are presented with the challenge of transferring electrons outside the cell, as the solid-phase electron acceptor cannot contact the inner membrane-localized electron transport chain (DiChristina, Fredrickson, & Zachara, 2005; Richter, Schicklberger, & Gescher, 2012; Shi, Squier, Zachara, & Fredrickson, 2007). The enzymatic machinery and biochemical pathways employed by bacteria to reduce Mn oxides are believed to be similar to those used for Fe oxide reduction. In both cases, since the organisms are respiring solid-phase electron acceptors, bacteria have developed unique pathways to transfer electrons from the cytoplasm to the oxides outside the cell. To respire solid-phase Fe and Mn oxides, organisms employ various strategies including (i) the use of exogenous (e.g. humics, flavins) or endogenous soluble electron shuttles (e.g. phenazines), (ii) production of molecules to chelate or complex Mn or Fe from the oxide surface (e.g. siderophores) and (iii) direct electron transfer through enzymes localized on the bacterial outer membrane or on extracellular appendages (e.g. nanowires, pili) (DiChristina et al., 2005; Gorby et al., 2006; Gralnick & Newman, 2007; Nevin & Lovley, 2002). Mn(III,IV) oxide reduction is believed to proceed via a two-step, single-electron transfer with Mn(III) as an intermediate (Lin, Szeinbaum, DiChristina, & Taillefert, 2012; Szeinbaum, Burns, & DiChristina, 2014). The initial reduction of solid-phase derived Mn(IV) to Mn(III) may require solubilization of the oxide via an endogenous organic ligand (Fennessey, Jones, Taillefert, & DiChristina, 2010; Taillefert et al., 2007).

The two most studied Fe and Mn oxide-reducing organisms are the freshwater sediment bacteria *S. onedensis* MR-1 and *Geobacter sulfurreducens*. Both bacteria use multiple *c*-type cytochromes to transport electrons generated in the cytoplasm to extracellular electron acceptors. While the molecular mechanisms of extracellular respiration in *Geobacter* spp. have focused on Fe oxides, the pathways for both Mn(III) and Mn(IV) reduction have been specifically explored in *S. onedensis* MR-1. In brief, *S. oneidensis* MR-1 has a branched electron transport chain that includes inner membrane dehydrogenases, menaquinone and a menaquinol-oxidizing *c*-type cytochrome (CymA) that serves as the branching point for electron transfer to either soluble or insoluble electron acceptors (Szeinbaum et al., 2014). Electrons generated from central metabolism are passed through the electron transport chain to CymA and then proceed to soluble, periplasmic *c*-type cytochromes (e.g. MtrA) (Schuetz, Schicklberger, Kuermann, Spormann, & Gescher, 2009) and subsequently to an outer membrane-localized electron conduit composed of a transmembrane β-barrel protein (MtrB) (Wee, Burns, & DiChristina, 2014) and two different outer membrane decahaem cytochromes, MtrC and OmcA (Coursolle, Baron, Bond, & Gralnick, 2010; Shi et al., 2006). A type II secretion protein, GspD, located in the outer membrane has also been shown to be important in the export of exoproteins important for Mn reduction (DiChristina, Moore, & Haller, 2002; Shi et al., 2008). Interestingly, these same enzymes are involved in the reduction of soluble Mn(III) complexes coupled to carbon oxidation (Szeinbaum et al., 2014).

## 6. Mn(II) OXIDATION DECOUPLED FROM ENERGY GENERATION

### 6.1 Mn Oxidation

A wide diversity of Mn(II)-oxidizing organisms are strict heterotrophs that do not gain energy via the oxidation of Mn(II). The mechanisms used by these microorganisms to oxidize Mn(II) include a number of abiotic and biotic processes (Fig. 6). Microbes mediate the oxidation of Mn(II) to Mn(III) and Mn oxides via both direct and indirect processes (see Hansel & Learman, 2015; Tebo et al., 2005 for more extensive reviews). Regardless of the pathway or microbial catalyst, the oxidation of Mn(II) to Mn(IV) oxides occurs via a one-electron transfer process, leading to the formation of a Mn(III) intermediate (Anderson et al., 2009; Dick, Podell, et al., 2008; Johnson & Tebo, 2008; Learman, Voelker, et al., 2011; Learman, Wankel, et al., 2011; Miyata, Maruo, et al., 2006; Miyata, Tani, et al., 2006; Su et al., 2013; Webb et al., 2005). The reaction pathway from Mn(III) to Mn(IV) is less clear and may involve either disproportionation to Mn(II) and Mn(IV) or further oxidation to Mn oxides either through a second electron transfer reaction within the enzyme (Butterfield, Soldatova, Lee, Spiro, & Tebo, 2013) or by the same or secondary oxidant.

A number of enzymes have been linked to Mn oxidation in Mn(II)-oxidizing bacteria. The primary enzymes identified to date are multicopper oxidases (MCOs) and most recently animal haem peroxidases (AHPs); however, genes involved in flagellar synthesis, protein transport, two-component response regulation, cytochrome *c* synthesis and carbon metabolism have also been shown to influence Mn oxidation in some organisms (Andeer, Learman, McIlvin, Dunn, & Hansel, 2015; Anderson et al., 2009; Brouwers, Corstjens, et al., 2000; Brouwers et al., 1999; Brouwers, Vijgenboom, et al., 2000; Corstjens et al., 1997; de Vrind et al., 1998; De Vrind, De Groot, Brouwers, Tommassen, & De Vrind-De Jong, 2003; Tebo et al., 2005; Van Waasbergen et al., 1996). MCOs are a structurally and functionally diverse family of enzymes with the ability to oxidize a number of (in)organic substrates (e.g. Fe(II), diphenolics) (Brouwers, Vijgenboom, et al., 2000; Solomon, Sundaram, & Machonkin, 1996). MCOs have been implicated in Mn(II) oxidation in various freshwater and marine bacteria, including *Leptothrix* species (Bocioaga et al., 2014; Brouwers, Vijgenboom, et al., 2000; Corstjens et al., 1997; El Gheriany et al., 2009; Takeda et al., 2012), *Pseudomonas putida* (Brouwers et al., 1999; de Vrind et al., 1998; Francis et al., 2001; Geszvain et al., 2013; Okazaki et al., 1997) and *Bacillus* species (Butterfield et al., 2013; Dick, Torpey, Beveridge, & Tebo, 2008; Francis & Tebo, 2002; Su et al., 2013; Van Waasbergen et al., 1996). MCO-like enzymes, specifically laccases, have also been linked to Mn(II) oxidation by fungi. Several studies have shown both basidiomycete and ascomycete fungi utilize a laccase to oxidize Mn(II) to Mn(III) and in some cases Mn oxides (Hofer & Schlosser, 1999; Miyata, Maruo, et al., 2006; Miyata, Tani, Iwahori, & Soma, 2004; Miyata, Tani, et al., 2006; Schlosser & Hofer, 2002).

The marine spore former *Bacillus* SG-1 and the freshwater bacterium *P. putida* GB-1 have served as model organisms to define the molecular mechanism of MCO-directed Mn(II) oxidation. *P. putida* generates a Mn (II)-oxidizing protein complex that includes an MCO that appears to reside in the outer membrane (Brouwers, Corstjens, et al., 2000; Okazaki et al., 1997). A recent study of *P. putida* GB-1 found that oxidation in this bacterium is related to two MCO genes, *mnxG* and *mcoA*, which are positively regulated by the response regulator MnxR (Geszvain et al., 2013). Interestingly, each of these MCOs has the ability to oxidize Mn(II) independently. *P. putida* GB-1 was recently found to have a third oxidase (Geszvain, Smesrud, & Tebo, 2016), which is an AHP with sequence similarity to those designated as MopA in other Mn(II)-oxidizing bacteria (see below). The relative expression of these three oxidase genes (*mnxG*, *mcoA*, *mopA*) appears to be influenced by flagella synthesis (Geszvain et al., 2016), possibly pointing to a role for lifestyle (planktonic vs sessile) in operative Mn oxidation pathways.

For the marine *Bacillus* sp. SG-1, the dormant spores, rather than the vegetative cells, oxidize Mn(II) (Dick, Torpey, et al., 2008; Francis & Tebo, 1999; Nealson & Ford, 1980; Rosson & Nealson, 1982). The spores bind and oxidize Mn(II) via a protein component in the exosporium. As in *P. putida*, Mn(II)-oxidizing activity in *Bacillus* is stimulated by Cu(II) and is linked to MCOs and the *mnx* operon (Van Waasbergen et al., 1996, 1993). Butterfield et al. (2013) recently purified the Mn(II)-oxidizing MnxEFG complex. As previously thought, Mn(II) is oxidized to Mn(III) (Webb et al., 2005), but the complex MnxEFG is also able to oxidize Mn(III) to Mn(IV) (Butterfield et al., 2013). Thus, MCOs may mediate oxidation of both Mn(II) and Mn(III). The ability to measure Mn(III) complexes through the addition of exogenous ligands may indicate that the Mn(III)-enzyme complex is somewhat labile and/or exchangeable.

Rather than MCO enzymes, AHPs have been connected to bacterial Mn(II) oxidation and Mn oxide formation in the marine bacteria *Erythrobacter* sp. SD-21, *Aurantimonas manganoxydans* SI85-9A1 (Anderson et al., 2009) and *Roseobacter* AzwK-3b (Andeer et al., 2015). The physiological function of these types of peroxidases has not been experimentally verified, yet certain domains within this group of proteins suggest that these proteins could be used for cell defence (Zamocky, Jakopitsch, Furtmuller, Dunand, & Obinger, 2008). In *A. manganoxydans* SI85-9A1, the AHP was localized to the loosely bound outer membrane fraction. For *Roseobacter* AzwK-3b and *Erythrobacter* sp. SD-21, the AHP and Mn(II)-oxidizing activity are concentrated in the soluble, excreted protein fraction (Andeer et al., 2015; Anderson et al., 2009; Hansel & Francis, 2006; Johnson & Tebo, 2008; Learman, Voelker, et al., 2011; Learman, Wankel, et al., 2011). AHPs accounted for ~1.3% of the global proteome of *Roseobacter* AzwK-3b, where the proportion of spectral counts assigned to AHPs was similar in the presence and absence of Mn(II), suggesting that Mn oxidation is not inducible in this organism (Learman & Hansel, 2014). In contrast, AHP expression in *A. manganoxydans* SI85-9A1 was induced by Mn(II), pointing to fundamental differences in AHP function within these two organisms (Anderson et al., 2009).

Further, the mechanism of Mn(II) oxidation by these haem peroxidases is also different for *A. manganoxydans* SI85-9A1 and *Roseobacter* AzwK-3b. Specifically, direct Mn(II) oxidation by the haem peroxidases was identified for *A. manganoxydans* SI85-9A1 and *Erythrobacter* SD-21 (Anderson et al., 2009), but not for *Roseobacter* AwK-3b (Andeer et al., 2015). Instead, for *Roseobacter* sp. AzwK-3b-soluble extracellular AHPs produce the ROS superoxide ($O_2^{\bullet -}$) which then abiotically oxidizes Mn(II) to Mn(III) (reaction 5) ultimately forming Mn oxides (Andeer et al., 2015; Learman, Voelker, et al., 2011; Learman, Wankel, et al., 2011). In fact, unlike $O_2$, the oxidation of Mn(II) to Mn(III) by superoxide is thermodynamically favoured over a broad pH range (Luther, 2010). The kinetics of the reaction is also fast, including conditions representative of typical marine waters (Hansard, Easter, & Voelker, 2011). Production of extracellular superoxide by both transmembrane and soluble secreted enzymes has been observed in a wide variety of terrestrial and marine organisms, including fungi, diatoms, algae and bacteria (Aguirre, Rios-Momberg, Hewitt, & Hansberg, 2005; Diaz et al., 2013; Hansel et al., 2016; Kustka, Shaked, Milligan, King, & Morel, 2005; Rose, 2012). This extracellular superoxide production has been linked to Mn(II) oxidation in organisms besides *Roseobacter* AzwK-3b, including ascomycete fungi (Hansel, Zeiner, Santelli, & Webb, 2012; Tang, Zeiner, Santelli, & Hansel, 2013) and other bacteria (Bohu et al., 2015).

Yet, despite the widespread ability of organisms to produce extracellular superoxide, this activity alone does not necessarily lead to Mn oxide formation (Learman & Hansel, 2014; Learman, Voelker, Madden, & Hansel, 2013). One likely explanation for this is that Mn oxide formation is inhibited by hydrogen peroxide ($H_2O_2$), a product of the reaction between Mn(II) and superoxide (reaction 5). Specifically, hydrogen peroxide reacts with Mn(III), reducing it back to Mn(II) (Learman et al., 2013) (reaction 6). The precipitation of Mn oxides therefore requires also the scavenging of hydrogen peroxide that is formed upon reaction of superoxide and Mn(II). In fact, *Roseobacter* AzwK-3b rapidly degrades exogenous hydrogen peroxide (Andeer et al., 2015), which points to a role for both superoxide production and hydrogen peroxide degradation by AHPs in the *Roseobacter* AzwK-3b exoproteome. Haem peroxidases are, in fact, known to oscillate between oxidative and peroxidative activities (Scheeline et al., 1997) depending on the environmental conditions (Minibayeva, Gordon, Kolesnikov, & Chasov, 2001).

$$Mn^{2+} + O_2^{\bullet -} + 2H^+ \rightarrow Mn^{3+} + H_2O_2 \tag{5}$$

$$Mn^{3+} + \frac{1}{2}H_2O_2 \rightarrow Mn^{2+} + \frac{1}{2}O_2 + H^+ \tag{6}$$

In addition to or in lieu of hydrogen peroxide degradation, the formation of Mn oxides via biogenic superoxide oxidation of Mn(II) may be aided by

Mn(III)-stabilizing organic ligands (Learman et al., 2013; Tang, Zeiner, et al., 2013) or presence of Mn-binding polymers for Mn oxide nucleation (Estes, Andeer, Nordlund, Wankel, & Hansel, 2016).

Microbes also indirectly induce the oxidation of Mn(II) by producing highly reactive Mn oxides that accelerate the oxidation of Mn(II) by $O_2$ (autocatalysis) (Morgan, 2000, 2005). In particular, Mn oxides produced by microbes are similar in structure to δ-$MnO_2$ (Adams & Ghiorse, 1988; Bargar, Tebo, & Villinski, 2000; Jurgensen et al., 2004; Villalobos et al., 2003), a mineral that catalyses Mn(II) oxidation (Coughlin & Matsui, 1976). In fact, biogenic colloidal oxides induce rapid oxidation of Mn(II) with rates (~2 μM $h^{-1}$) (Learman, Voelker, et al., 2011; Learman, Wankel, et al., 2011) equivalent to other mineral-catalysed Mn(II) oxidation rates (e.g. nanohematite rates ~4 μM $h^{-1}$ (Madden & Hochella, 2005)) and faster than reported biological rates in natural waters (~3–12 nM $h^{-1}$ (Dick et al., 2009; Tebo & Emerson, 1985, 1986)). Thus, rates of microbially mediated Mn(II) oxidation may include also these superimposed biooxide-induced reactions once the mineral has nucleated, thereby complicating the distinction between abiotic and biotic reaction pathways. In fact, the Mn(II) oxidation process by *Roseobacter* AzwK-3b alone includes enzymatically produced superoxide, enzymatic $H_2O_2$ degradation, direct mineral autocatalysis, and mineral-derived superoxide and organic radicals, the formation of which is accelerated in the presence of light (see Learman et al., 2013; Learman, Voelker, et al., 2011; Learman, Wankel, et al., 2011) (Fig. 7). Interestingly, enzyme-independent processes are comparable and in many cases faster than enzymatic Mn(II) oxidation processes, pointing to an important, yet currently unappreciated role for mineral- and radical-mediated processes in the formation of environmental Mn oxides.

Further, indirect Mn(II) oxidation may also be promoted through localized increases in pH and production of metabolic end products, which cause chemical oxidation of Mn(II). These can include reactions with microbial-derived organic carbon and metabolites such as siderophores (Duckworth & Sposito, 2005), organic acids (Klewicki & Morgan, 1998) and polysaccharides (Adams & Ghiorse, 1987; Beveridge, 1989; Boogerd & de Vrind, 1987; Ehrlich, 1983; Ghiorse & Hirsch, 1979; van Veen, Mulder, & Deinema, 1978) (Fig. 6). In particular, under Fe limitation, two *Pseudomonas* strains produce the siderophore pyoverdine (PVD) which complexes and abiotically oxidizes Mn(II) producing PVD–Mn(III) complexes. Further, phytoplankton have been shown to change local pH conditions that allow for Mn(II) oxidation (Richardson et al., 1988). Thus, nonenzymatic microbial exudates and activity may represent other

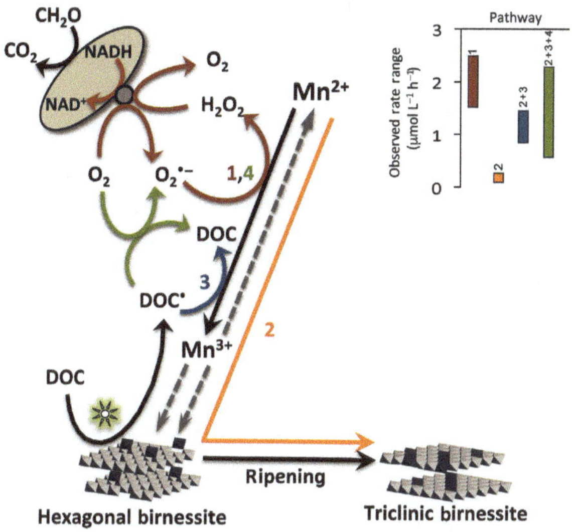


**Fig. 7** Simplified schematic of the network of biotic and abiotic reactions that controls the formation of biogenic Mn oxides mediated by *Roseobacter* sp. AzwK-3b. Pathway 1 (*red arrows*) is initial oxidation of Mn(II) to Mn(III) via biotic production of superoxide. Mn(III) is either oxidized by another unknown factor or disproportionates to colloidal Mn(IV) oxides (hexagonal birnessite) (*dashed gray lines*). Pathway 2 shows the abiotic oxidation of Mn(II) by colloidal hexagonal birnessite (autocatalysis) (*orange arrow*). Pathways 3 (*blue arrow*) and 4 (*green arrows*) are initiated after the initial hexagonal birnessite colloids are formed and react with unidentified organic extracellular metabolites producing organic radicals that either directly oxidize Mn(II) (*blue*; pathway 3) or react with molecular oxygen to produce superoxide (*green*; pathway 4) that oxidizes Mn(II). The combination of these oxidative pathways causes the ripening of hexagonal birnessite to a triclinic birnessite, which greatly diminishes the Mn oxide reactivity. *Inset* displays the rates of some of the individual and combined pathways.

important microbially mediated pathways for Mn(II) oxidation within the environment.

## 6.2 Mn Reduction

A wide diversity of bacteria reduce Mn oxides during fermentative growth (see Lovley, 1991 and references therein). The reduction of Mn oxides, as well as Fe oxides, by fermenting organisms is a side reaction, with less than 5% of the reducing equivalents transferred to the oxides. Further, at present, none of the fermentative organisms are known to gain energy through this process. Rather, the Mn and Fe oxides serve as an electron sink. In fact, thermodynamic calculations indicate that using Fe(III) oxides as an energy sink may result in higher energy yields than fermentation alone. Thus, organisms may benefit from diverting a small portion of electrons to the oxides.

The reduction of Mn oxides has also been proposed to play a role in the anaerobic oxidation of methane (Beal, House, & Orphan, 2009) and ammonium (Hulth, Aller, & Gilbert, 1999; Luther et al., 1997); yet, whether these reactions can be coupled to energy gain within one organism is unknown. Instead, similar to the bacterial–archaeal coupling involved in the anaerobic oxidation of methane coupled to sulphate reduction (Knittel, Losekann, Boetius, Kort, & Amann, 2005; Orphan et al., 2001; Orphan, House, Hinrichs, McKeegan, & DeLong, 2002), Mn reduction and methane oxidation could be performed with a similar consortia or possibly by only bacteria but with a different enzymatic mechanism used for methane oxidation from what is known for archaea (Beal et al., 2009). In fact, the coupling of methane oxidation to Mn reduction has now been implicated in various environments (Crowe et al., 2011; Jones et al., 2011; Segarra, Comerford, Slaughter, & Joye, 2013; Wang et al., 2014). The coupling between Mn reduction and ammonia oxidation has been more difficult to demonstrate (Thamdrup & Dalsgaard, 2000, 2002). Recently, however, a link between the two has been observed, yet the underlying organisms and reaction pathways are not yet resolved (Lin & Taillefert, 2014).

The reduction of Mn(IV) oxides can also occur via indirect microbial activity, such as the production of metabolic products that abiotically reduce Mn(IV) oxides (Fig. 6). A number of organic and inorganic metabolic reductants have been linked to Mn(IV) oxide reduction, including formic acid, pyruvate, $H_2S$, sulphite, $Fe^{2+}$ and hydrogen peroxide (Aller & Rude, 1988; Burdige & Nealson, 1986; Ehrlich, 1966; Ghiorse, 1988; Ghiorse & Ehrlich, 1976; Myers & Nealson, 1988b; Nealson & Saffarini, 1994; Stone, 1987; Trimble & Ehrlich, 1968, 1970). Fungi are not known to gain energy through the reduction of Mn and instead, all fungal Mn oxide reduction appears to occur via reaction with reactive metabolites such as oxalate (Gadd, 2010; Wei et al., 2012).

Lastly, it has been proposed that, in some cases, Mn may be reduced to satisfy a nutritional need for soluble Mn(II) (see de Vrind, Boogerd, & de Vrind-de Jong, 1986; Ehrlich, 1987) or to scavenge excess reducing power, as in some cases of nitrate (Robertson, Van Niel, Torremans, & Kuenen, 1988) and ferric iron reduction (Ghiorse, 1988; Lovley, 1991).

## 7. INTERCELLULAR Mn AND Mn-BASED ENZYMES IN MICROBIAL PHYSIOLOGY

Intracellular Mn has essential physiological roles for microbial life (Jakubovics & Jenkinson, 2001; Kehres & Maguire, 2003). The two most appreciated roles for intracellular Mn in microbial physiology are as metal cofactors in enzymes involved in photosynthesis and antioxidant activity. First, Mn is an essential cofactor within the water-oxidizing complex of photosystem II (PSII) (e.g. Klimov, 1984; Yocum & Pecoraro, 1999). Specifically, a tetra-Mn cluster present in the reaction centre complex of PSII is required for oxygenic photosynthesis by many cyanobacteria and algae in the ocean, thereby exerting a significant control on global primary productivity. Second, the Mn-containing superoxide dismutase (Mn-SOD) is essential for most oxygen-based life by converting superoxide radicals to hydrogen peroxide during ATP synthesis (Leach & Harris, 1997). Mn-SOD is, in fact, widespread in both the Bacteria and the Archaea (Whittaker, 2002).

Aside from its essential role as a cofactor in PSII and SOD, Mn has historically accorded little importance in the physiology of marine microbes. In fact, until recently the mechanisms of Mn acquisition and Mn-responsive gene regulation were unexplored in marine microbes. Recent exploration of the widely distributed and numerically abundant *Roseobacter* clade bacteria, however, indicated that these bacteria contain two different manganese transporters (*sitABCD*, *mntX*) (Green, Todd, & Johnston, 2013). Further, both the SitABCD- and MntX-type transporters were highly represented in nearly all the global ocean sampling expedition sampling sites (Green et al., 2013), illustrating a widespread distribution of Mn transporters within the ocean. In contrast, homologues of MntX were not observed in nonmarine (e.g. soils, humans) metagenomic databases, suggesting that this transporter may be confined and specific to marine bacteria. Indeed, *mntX* was observed in a variety of relevant marine bacterial species, including genera in the Vibrionales, Altermonadales, Oceanospirillales, SAR11 and SAR116. Both *sitABCD* and *mntX* are transcriptionally regulated and repressed under Mn-replete conditions (Green et al., 2013; Learman & Hansel, 2014). This transcriptional regulation is surprising considering the low concentrations of Mn within the ocean and may suggest tightly regulated Mn homeostasis within marine microbes. In fact, Mn limitation can impair the growth, health and photosynthetic activity of some phytoplankton (Brand, Sunda, & Guillard, 1983; Cao, Sun, Wang, Liu, & Liang, 2011). The widespread distribution of Mn transporters within nonphotosynthetic marine microbes under nonstressful conditions likely points to additional essential roles for Mn in marine microbial physiology.

In fact, beyond marine systems, there is now an emerging recognition for a broader relevance of Mn in the physiology of microbes, based on investigations of nonmarine organisms, such as *E. coli*, *Enterobacter* sp. and *Bacillus*

**Table 1** Select Processes and Enzymes Influenced by or Dependent Upon Mn

| **Process or Pathway** | **Example Enzymes or Proteins** |
|---|---|
| Photosynthesis | PSII |
| Gluconeogenesis | PEP synthase; pyruvate carboxylase |
| Glycolysis | Enolase; PEP carboxylase; pyruvate kinase |
| Sugar metabolism | L-Fructose somerase; 6-phospho-β-glucosidase |
| Amino acid metabolism | Arginase; glutamine synthetase; threonine 3-dehydrogenase |
| Protein phosphorylation | Protein kinases |
| Peptide cleavage | Aminopeptidase P (*pep*, *pepQ*) |
| Nucleic acid degradation | Ribonuclease HII; endonuclease IV |
| Polysaccharide synthesis | Polysaccharide polymerase |
| Signal transduction | Protein phosphatases (*prpA*, *prpB*) |
| Conversion of ribonucleotides to deoxyribonucleotides | Ribonucleotide reductase (*nrdAB/EF*) |
| Hydrolyse RNA synthesis regulation | ppGpp hydrolase (*spoT*) |
| Oxidative stress defence | $Mn^{2+}$ superoxide dismutase (*sodA*); nonhaem $Mn^{2+}$ catalase (*katN*) |

Adapted from Kehres, D. G., & Maguire, M. E. (2003). Emerging themes in manganese transport, biochemistry, and pathogenesis in bacteria. *FEMS Microbiology Reviews, 27,* 263–290 and Jakubovics, N. S., & Jenkinson, H. F. (2001). Out of the iron age: New insights into the critical role of manganese homeostasis in bacteria. *Microbiology, 147,* 1709–1718 and references therein.

sp. (Kehres & Maguire, 2003). In addition to novel Mn-dependent enzymes, Mn is also involved in signalling and stimulating a wide range of biochemical processes within the cell that are essential for microbial health and functioning (see Table 1). In particular, discoveries over the past two decades point to a role for Mn in central carbon metabolism, phosphorylation, peptide cleavage, signal transduction, growth, sporulation and cell wall stabilization (Jakubovics & Jenkinson, 2001). Further, within some bacteria, Mn and Mn-dependent enzymes may actually be essential for pathogenesis (Kehres & Maguire, 2003).

Most, if not all, bacteria express multiple $Mn^{2+}$ uptake and efflux systems, and intracellular Mn pools are highly concentrated with respect to the surrounding environment. Intracellular Mn levels oftentimes far exceed the requirements for currently known $Mn^{2+}$-dependent enzymes. In fact, cytoplasmic $Mn^{2+}$ concentrations vary approximately inversely with growth rate and are correlated with growth state (Kehres & Maguire, 2003). Elevated intracellular Mn levels are also linked to bacterial protection from oxidative stress. Specifically, Mn has antioxidant properties, such that Mn(II) reacts with the ROS superoxide and both Mn(III) and Mn oxides with hydrogen peroxide (Archibald & Fridovich, 1982; Barnese, Gralla, Cabelli, & Valentine, 2008; Do, Batchelor, Lee, & Kong, 2009; Hansard et al., 2011; Horsburgh, Wharton, Karavolos, & Foster, 2002; Luther, 2005, 2010). For instance, Mn(II) catalyses the disproportionation (dismutation) of both superoxide (reaction 7) and hydrogen peroxide (reaction 8), as follows:

$$2O_2^{\bullet -} + 2H^+ \rightarrow O_2 + H_2O_2 \quad (7)$$

$$2H_2O_2 \rightarrow O_2 + 2H_2O \quad (8)$$

In fact, accumulation of high intracellular Mn(II) levels has been repeatedly shown to alleviate the $O_2$-senstivity of cells that lack superoxide dismutase enzymes (Barnese, Gralla, Valentine, & Cabelli, 2012; Culotta & Daly, 2013; Latour, 2015). The ability for Mn(II) to effectively scavenge ROS has also been linked to Mn-facilitated ionizing radiation resistance in the bacterium *Deinococcus radiodurans*, where this organism has been shown to accumulate exceedingly high (mM) levels of intracellular Mn (Daly, 2009). Interestingly, formation of Mn oxides by the Mn(II)-oxidizing bacterium *P. putida* GB-1 was recently attributed to increased survival of this bacterium in the presence of exogenous hydrogen peroxide in comparison to survival in the absence of Mn(II) or by mutant strains lacking the ability to oxidize Mn(II) (Banh et al., 2013). Thus, Mn is a critical antioxidant for microbial life, leading to the proposition that Mn could in fact represent the primordial superoxide dismutase allowing life to survive Earth's great oxygenation (Fischer, Hemp, & Valentine, 2016).

Despite a paucity of studies exploring Mn-dependent biochemical processes in marine microbes, the widespread presence of Mn-specific transporters likely reflects similar physiological functions as those described earlier for nonmarine microbes—as well as other processes yet to be discovered.

## 8. OTHER POTENTIAL PHYSIOLOGICAL IMPACTS OF MANGANESE

Aside from being involved in microbial metabolism and as a regulator of and cofactor in enzymes of physiological relevance, the various Mn species possess geochemical properties that may prove otherwise beneficial to microbial communities (Tebo et al., 2004, 2005). Despite a widespread ability for bacteria to oxidize Mn(II) to Mn oxides, most known Mn(II)-oxidizing organisms do not gain energy via the reaction and thus the physiological basis for this activity remains unknown. In all fungal and bacterial systems, the Mn oxide products are formed outside the cell, due to Mn(II)-oxidizing activity being associated with outer membrane proteins, extracellular soluble proteins and/or soluble, extracellular reactive metabolites (e.g. superoxide). In fact, these oxides may encrust portions of the microbial cells (Francis, Casciotti, & Tebo, 2002; Tebo et al., 2004, 2005) or be embedded within an extracellular polymeric substance surrounding the cell (Toner, Fakra, Villalobos, Warwick, & Sposito, 2005). This extracellular and yet proximal distribution of the biogenic Mn oxides allows for interactions between the oxides and the surrounding environment in ways that could potentially benefit the organism.

For example, Mn oxides may influence the composition and/or availability of organic carbon for respiration. Indeed, Mn oxides harbour high concentrations of extracellular proteins or polymers (Fig. 8) (Andeer et al., 2015; Emerson, Garen, & Ghiorse, 1989; Estes et al., 2016; Johnson et al., 2015; Santelli, Webb, Dohnalkova, & Hansel, 2011; Tang, Zeiner, et al., 2013; Toner et al., 2005) and could therefore serve as a carbon reservoir to support continued microbial respiration. Mn(III) and Mn(IV) oxides are also strong oxidants (Bartlett & James, 1979; Johnson et al., 2015; Kostka et al., 1995; Luther, 2005; Murray & Tebo, 2007; Tebo et al., 2004) with the ability to degrade even recalcitrant forms of organic matter (Banerjee & Nesbitt, 2001; Sunda & Kieber, 1994). In fact, Basidiomycota fungi (e.g. wood-rot fungi) employ highly specific Mn(II) oxidation enzymes to degrade recalcitrant forms of carbon, including lignin and cellulose. Specifically, the widely distributed Mn peroxidase enzyme couples the reduction of hydrogen peroxide to the oxidation of Mn(II) to Mn(III) (Wariishi, Valli, & Gold, 1992). The Mn(III) ions form reactive complexes with fungal-derived organic acid metabolites (e.g. oxalate), which then proceed to degrade phenolic units of lignin (Perez & Jeffries, 1992). Thus, in a similar manner, the formation of the strong oxidants Mn(III) and Mn oxides by bacteria could, in theory, serve as a strategy to increase the labile carbon pool for heterotrophic respiration.

Also, Mn oxides are extremely powerful sorbents having demonstrated large capacities for (ad)sorbing both metals and contaminants, including Pb, Ni, Co and Zn (Kay, Conklin, Fuller, & O'Day, 2001; Nelson, Lion,

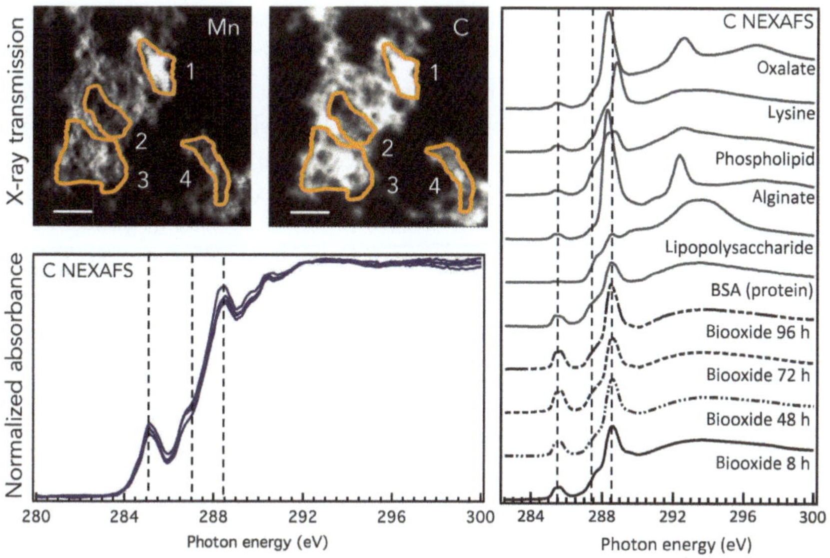


**Fig. 8** Manganese (Mn) and carbon (C) scanning transmission X-ray microscopy (STXM) OD maps of Mn oxides formed after 96 h in cell-free filtrate of the marine M(II)-oxidizing *Alphaproteobacteria Roseobacter* AzwK-3b (*top left*) reveal a strong association of C with the Mn oxides. Carbon 1s NEXAFS spectra (*bottom left*) extracted from regions denoted on the STXM maps illustrate a uniform C composition associated with the Mn oxide particles. Bulk NEXAFS (*right*) further indicate that the carbon composition is stable over time, despite the mineral undergoing a structural transformation from hexagonal (at 4 h) to triclinic birnessite (by 96 h). Guide lines at ~285, 287 and 288.3 eV represent contributions from aromatic, aliphatic and amide/carboxylic functional groups, respectively. The bacterial Mn oxides are dominated by amide/carboxylic C contributions. The concentration of carbon in the biooxides was $8.5 \pm 0.84$ mol OC per kg mineral with a high abundance of proteins ($40.2 \pm 1.1$ mg protein per kg mineral). Parallel nitrogen NEXAFS confirmed that the primary carbon is proteinaceous (see Estes et al., 2016).

Shuler, & Ghiorse, 1999; O'Reilly & Hochella, 2003; Pena, Kwon, Refson, Bargar, & Sposito, 2010; Takahashi et al., 2007; Tebo et al., 2004; Toner, Manceau, Webb, & Sposito, 2006). Metal ions may also be structurally incorporated within the Mn oxide structure, either isomorphically by replacing Mn(III) or Mn(IV) ions or within interstitial vacancies or tunnels (Kay et al., 2001; Manceau, Gorshkov, & Drits, 1992; O'Reilly & Hochella, 2003). The mechanism of attenuation and the strength of the metal binding vary with metal concentration and pH and are a function of both the Mn oxide and metal properties (Manceau, Lanson, & Geoffroy, 2007; Pena et al., 2010; Takahashi et al., 2007; Murray, 1975; O'Reilly & Hochella, 2003; Peacock & Sherman, 2007; Pena et al., 2010). The site and mechanism of metal attenuation within/on Mn oxides will have direct bearing on the stability of the sorbed complex and the propensity for subsequent release back into solution. Owing to this high sorptive capacity, Mn oxide formation has been an attractive strategy for environmental remediation and water treatment. In a similar manner, Mn oxide formation could, in theory, represent a microbial strategy to detoxify metal-laden waters that threaten the organisms' health and function.

On the flipside, formation of Mn(III) may negatively impact Fe acquisition. Mn(III) forms aqueous complexes with various ligands and siderophores (Beyer & Fridovich, 1989; Duckworth & Sposito, 2005). In fact, some siderophores, such as PVD, have a higher affinity for Mn(III) than Fe(III) (Luther, Madison, Mucci, Sundby, & Oldham, 2015; Parker, Sposito, & Tebo, 2004). Further, some ligands, such as pyrophosphate, bind Mn(III) more strongly than Fe (Luther et al., 2015). Thus, ligand competition between Mn(III) and other essential biometals like Fe could under some conditions impact the activity and productivity of microbial communities.

## 9. CONCLUDING REMARKS

Decades of investigation have unveiled the importance of Mn in the physiology of marine microbes, the plethora of microbially mediated Mn transformations and the immensity and complexity of the Mn cycle. Despite this, there is still much to learn about the role of manganese in the microbiology and biogeochemistry of the ocean. Key questions that are central to understanding the marine Mn cycle remain unanswered. For instance, what is the full suite of Mn-dependent enzymes involved in the physiology of marine microbes? What are the dominant organisms and underpinning biochemical processes responsible for Mn(IV) reduction in the ocean? And why do organisms oxidize Mn(II) if energy is not harnessed during the reaction? Relatedly, why haven't organisms evolved to use Mn(II) for autotrophic and/or phototrophic growth? Or do these organisms exist and we just have not discovered them yet? With the recent discovery of the ubiquity and abundance of Mn(III) in the ocean, what is the consequence of this immensely reactive Mn species on the cycling of nutrients and carbon in the ocean? And finally, how essential is Mn for life, both at present and in the past during the oxygenation of our planet?

Although it may appear that there are more questions than answers, what is clear is that the activity of Mn-transforming organisms shapes the ocean landscape. Whether intentional or not, Mn oxidation and reduction link into countless cycles and processes (Fig. 6). Thus, the cycling of Mn cannot be considered in isolation—instead, the cycling of Mn comes to bear on nearly all other elemental cycles (spanning nutrients to radionuclides), touches every possible ecological niche (from the surface to the deep ocean and even below) and involves all domains of life (from bacteria to fungi). As such, Mn is intricately linked to the ecology, geochemistry, biology and health of the ocean. The coming decades will bring answers to the questions above and further shed light on the dynamics and complexity of the biogeochemical processes influenced by Mn.

## ACKNOWLEDGEMENTS

I would like to thank the pioneering Mn biogeochemists and geobiologists who have served as inspirations, mentors and colleagues to me over the years. I also gratefully acknowledge the many students, postdocs and collaborators who conducted and/or contributed to my groups' research in Mn biogeochemistry and for financial support from the NSF (EAR-1322790; OCE-1246174) and NASA (NNX15AM04G).

## REFERENCES

Adams, L. F., & Ghiorse, W. C. (1987). Characterization of an extracellular $Mn^{2+}$-oxidizing activity and isolation of $Mn^{2+}$-oxidizing protein from *Leptothrix discophora* SS-1. *Journal of Bacteriology, 169*, 1279–1285.

Adams, L. F., & Ghiorse, W. C. (1988). Oxidation state of Mn in the Mn oxide produced by *Leptothrix discophora* SS-1. *Geochimica et Cosmochimica Acta, 52*, 2073–2076.

Aguirre, J., Rios-Momberg, M., Hewitt, D., & Hansberg, W. (2005). Reactive oxygen species and development in microbial eukaryotes. *Trends in Microbiology, 13*, 111–118.

Akob, D. M., Bohu, T., Beyer, A., Schäffner, F., Händel, M., Johnson, C. A., et al. (2014). Identification of Mn(II)-oxidizing bacteria from a low-pH contaminated former uranium mine. *Applied and Environmental Microbiology, 80*, 5086–5097.

Aller, R. C. (1990). Bioturbation and manganese cycling in hemipelagic sediments. *Philosophical Transactions of the Royal Society of London A, 331*, 51–68.

Aller, R. C., & Rude, P. D. (1988). Complete oxidation of solid phase sulfides by manganese and bacteria in anoxic marine sediments. *Geochimica et Cosmochimica Acta, 52*, 751–765.

Andeer, P. F., Learman, D. R., McIlvin, M., Dunn, J. A., & Hansel, C. M. (2015). Extracellular heme peroxidases mediate Mn(II) oxidation in a marine *Roseobacter* bacterium via superoxide production. *Environmental Microbiology, 17*, 3925–3936.

Anderson, C. R., Johnson, H. A., Caputo, N., Davis, R. E., Torpey, J. W., & Tebo, B. M. (2009). Mn(II) oxidation is catalyzed by heme peroxidase in "*Aurantimonas manganoxydans*" strain SI85-9A1 and *Erythrobacter* sp. strain SD-21. *Applied and Environmental Microbiology, 75*, 4130–4138.

Archibald, F. S., & Fridovich, I. (1982). The scavenging of superoxide radical by manganous complexes: In vitro. *Archives of Biochemistry and Biophysics, 214*, 452–463.

Banerjee, D., & Nesbitt, H. W. (2001). XPS study of dissolution of birnessite by humate with constraints on reaction mechanism. *Geochimica et Cosmochimica Acta, 65*, 1703–1714.

Banh, A., Chavez, V., Doi, J., Nguyen, A., Hernandez, S., Ha, V., et al. (2013). Manganese (Mn) oxidation increases intracellular Mn in *Pseudomonas putida* GB-1. *PLoS One, 8*, e77835.

Bargar, J. R., Tebo, B. M., Bergmann, U., Webb, S. M., Glatzel, P., Chiu, V. Q., et al. (2005). Biotic and abiotic products of Mn(II) oxidation by spores of the marine *Bacillus* sp. strain SG-1. *American Mineralogist, 90*, 143–154.
Bargar, J. R., Tebo, B. M., & Villinski, J. E. (2000). In situ characterization of Mn(II) oxidation by spores of the marine *Bacillus* sp. strain SG-1. *Geochimica et Cosmochimica Acta, 64*, 2775–2778.
Barnese, K., Gralla, E. B., Cabelli, D. E., & Valentine, J. S. (2008). Manganous phosphate acts as a superoxide dismutase. *Journal of American Chemical Society, 130*, 4604–4606.
Barnese, K., Gralla, E. B., Valentine, J. S., & Cabelli, D. E. (2012). Biologically relevant mechanism for catalytic superoxide removal by simple manganese compounds. *Proceedings of the National Academy of Sciences of the United States of America, 109*, 6892–6897.
Bartlett, R. J., & James, B. (1979). Behavior of chromium in soils: III. Oxidation. *Journal of Environmental Quality, 8*, 31–35.
Beal, E. J., House, C. H., & Orphan, V. J. (2009). Manganese- and iron-dependent marine methane oxidation. *Science, 325*, 184–187.
Beijerinck, M. W. (1913). Oxydation des mangancarbonates durch Bakterien und Schimmelpilze. *Folia Microbiologica, 2*, 213–234.
Beveridge, T. J. (1989). Role of cellular design in bacterial metal accumulation and mineralization. *Annual Review of Microbiology, 43*, 147–171.
Beyer, W. F. J., & Fridovich, I. (1989). Characterization of a superoxide dismutase mimic prepared from desferrioxamine and $MnO_2$. *Archives of Biochemistry and Biophysics, 271*, 149–156.
Bocioaga, D., El Gheriany, I. A., Lion, L., Ghiorse, W. C., Schuler, M. L., & Hay, A. G. (2014). Development of a genetic system for a model Mn-oxidizing proteobacterium, *Leptothrix discophora* SS1. *Microbiology, 160*, 2396–2405.
Bodei, S., Manceau, A., Geoffroy, N., Baronnet, A., & Buatier, M. (2007). Formation of todorokite from vernadite in Ni-rich hemipelagic sediments. *Geochimica et Cosmochimica Acta, 71*, 5698–5716.
Bohu, T., Santelli, C. M., Akob, D. M., Neu, T. R., Ciobota, V., Rösch, P., et al. (2015). Characterization of pH dependent Mn(II) oxidation strategies and formation of a bixbyite-like phase by *Mesorhizobium australicum* T-G1. *Frontiers in Microbiology, 6*, 734.
Boogerd, F. C., & de Vrind, J. P. M. (1987). Manganese oxidation by *Leptothrix discophora*. *Journal of Bacteriology, 169*, 489–494.
Boone, D. R., Liu, Y., Zhao, Z. J., Balkwill, D. L., Drake, G. R., Stevens, T. O., et al. (1995). *Bacillus internus* sp. nov., an Fe(III)- and Mn(IV)-reducing anaerobe from the deep terrestrial subsurface. *International Journal of Systematic Bacteriology, 45*, 441–448.
Brand, L. E., Sunda, W. G., & Guillard, R. R. L. (1983). Limitation of marine phytoplankton reproductive rates by zinc, manganese, and iron. *Limnology and Oceanography, 28*, 1182–1198.
Brauer, S. L., Adams, C., Kranzler, K., Murphy, D., Xu, M., Zuber, P., et al. (2011). Culturable *Rhodobacter* and *Shewanella* species are abundant in estuarine turbidity maxima on the Columbia River. *Environmental Microbiology, 13*, 589–603.
Brouwers, G., Corstjens, P. L. A. M., de Vrind, J. P. M., Verkammen, A., De Kuyper, M., & de Vrind-de Jong, E. W. (2000). Stimulation of $Mn^{2+}$ oxidation in *Leptothrix discophora* SS-1 by $Cu^{2+}$ and sequence analysis of the region flanking the gene encoding putative multicopper oxidase MofA. *Geomicrobiology Journal, 17*, 25–33.
Brouwers, G. J., de Vrind, J. P. M., Corstjens, P. L. A. M., Cornelis, P., Baysse, C., & de Vrind-de Jong, E. W. (1999). *cumA*, a gene encoding a multicopper oxidase, is involved in $Mn^{2+}$ oxidation in *Pseudomonas putida* GB-1. *Applied and Environmental Microbiology, 65*, 1762–1768.
Brouwers, G. J., Vijgenboom, E., Corstjens, P. L. A. M., de Vrind, J. P. M., & de Vrind-de Jong, E. W. (2000). Bacterial $Mn^{2+}$ oxidizing systems and multicopper oxidases: An overview of mechanisms and functions. *Geomicrobiology Journal, 17*, 1–24.

Bruland, K. W., Orans, K. J., & Cowen, J. P. (1994). Reactive trace metals in the stratified central North Pacific. *Geochimica et Cosmochimica Acta, 58*, 3171–3182.

Burdige, D. J., & Nealson, K. H. (1985). Microbiological manganese reduction by enrichment cultures from coastal marine sediments. *Applied and Environmental Microbiology, 50*, 491–497.

Burdige, D. J., & Nealson, K. H. (1986). Microbiological manganese reduction by enrichment cultures from coastal marine sediments. *Applied and Environmental Microbiology, 50*, 491–497.

Burnett, B. R., & Nealson, K. H. (1981). Organic films and microorganisms associated with manganese nodules. *Deep-Sea Research, 28A*, 637–645.

Burnett, B. R., & Nealson, K. H. (1983). Energy dispersive X-ray analysis of the surface of a deep-sea ferromanganese nodule. *Marine Geology, 53*, 313–329.

Burns, R. G., & Burns, V. M. (1977). Mineralogy of ferromanganese nodules. In G. P. Glasby (Ed.), *Marine manganese deposits*. Amsterdam: Elsevier.

Butterfield, C. N., Soldatova, A. V., Lee, S.-W., Spiro, T. G., & Tebo, B. M. (2013). Mn(II, III) oxidation and MnO2 mineralization by an expressed bacterial multicopper oxidase. *Proceedings of the National Academy of Sciences of the United States of America, 110*, 11731–11735.

Caccavo, F., Jr., Lonergan, D. J., Lovley, D. R., Davis, M., Stolz, S. F., & McInerney, M. J. (1994). *Geobacter sulfurreducens* sp. nov., a new hydrogen- and acetate-oxidizing dissimilatory metal-reducing microorganism. *Applied and Environmental Microbiology, 60*, 3752–3759.

Canfield, D. E., Jørgensen, B. B., Fossing, H., Glud, R. N., Gundersen, J., Ramsing, N. B., et al. (1993). Pathways of organic carbon oxidation in three continental margin sediments. *Marine Geology, 113*, 27–40.

Canfield, D. E., Thamdrup, B., & Hansen, J. W. (1993). The anaerobic degradation of organic matter in Danish coastal sediments: Iron reduction, manganese reduction, and sulfate reduction. *Geochimica et Cosmochimica Acta, 57*, 3867–3885.

Cao, C., Sun, S., Wang, X., Liu, W., & Liang, Y. (2011). Effects of manganese on the growth, photosystem II and SOD activity of the dinoflagellate *Amphidinium* sp. *Journal of Applied Phycology, 23*, 1039–1043.

Carmichael, M. J., Carmichael, S. K., Santelli, C., Strom, A., & Bräuer, S. L. (2013). Environmentally relevant members of ferromanganese deposits in caves of the upper Tennessee River Basin. *Geomicrobiology Journal, 30*, 779–800.

Clement, B. G., Luther, G. W., III, & Tebo, B. M. (2009). Rapid, oxygen-dependent microbial Mn(II) oxidation kinetics at sub-micromolar oxygen concentrations in the Black Sea suboxic zone. *Geochimica et Cosmochimica Acta, 73*, 1878–1889.

Coates, J. D., Ellis, D. J., Gaw, C. V., & Lovley, D. R. (1999). *Geothrix fermentans* gen nov., sp. nov., a novel Fe(III)-reducing bacterium from a hydrocarbon contaminated aquifer. *International Journal of Systematic Bacteriology, 49*, 1615–1622.

Connell, L., Barret, A., Templeton, A. S., & Staudigel, H. (2009). Fungal diversity associated with active deep sea volcano: Vailulu'u Seamount, Samoa. *Geomicrobiology Journal, 26*, 597–605.

Corstjens, P. L. A. M., de Vrind, J. P. M., Goosen, T., & de Vrind-de Jong, E. W. (1997). Identification and molecular analysis of the *Leptothrix discophora* SS-1 *mofA* gene, a gene putatively encoding a manganese oxidizing protein with copper domains. *Geomicrobiology Journal, 14*, 91–108.

Coughlin, R. W., & Matsui, I. (1976). Catalytic-oxidation of aqueous Mn(II). *Journal of Catalysis, 41*, 108–123.

Coursolle, D., Baron, D. B., Bond, D. R., & Gralnick, J. A. (2010). The Mtr respiratory pathway is essential for reducing flavins and electrodes in *Shewanella oneidensis*. *Journal of Bacteriology, 192*, 467–474.

Cowen, J. P., Bertram, M. A., Baker, E. T., Feely, R. A., Massoth, G. J., & Summit, M. (1998). Geomicrobial transformation of manganese in Gorda Ridge event plumes. *Deep-Sea Research II*, *45*, 2713–2737.

Cowen, J. P., Massoth, G. J., & Feely, R. A. (1990). Scavenging rates of dissolved manganese in a hydrothermal vent plume. *Deep-Sea Research*, *37*, 1619–1637.

Cowen, J. P., & Silver, M. W. (1984). The association of iron and manganese with bacteria on marine microparticulate material. *Science*, *224*, 1340–1342.

Crowe, S. A., Katsev, S., Leslie, K., Sturm, A., Magen, C., Nomosatryo, S., et al. (2011). The methane cycle in ferruginous Lake Matano. *Geobiology*, *9*, 61–78.

Culotta, V. C., & Daly, M. J. (2013). Manganese complexes: Diverse metabolic routes to oxidative stress resistance in prokaryotes and yeast. *Antioxidants & Redox Signaling*, *19*, 933–944.

Daly, M. J. (2009). A new perspective on radiation resistance based on *Deinococcus radiodurans*. *Nature Reviews in Microbiology*, *7*, 237–245.

Dellwig, O., Schnetger, B., Brumsack, H. J., Grossart, H. P., & Umlauf, L. (2012). Dissolved reactive manganese at pelagic redoxclines (part II): Hydrodynamic conditions for accumulation. *Journal of Marine Systems*, *90*, 31–41.

de Vrind, J. P. M., Boogerd, F. C., & de Vrind-de Jong, E. W. (1986). Manganese reduction by a marine *Bacillus* species. *Journal of Bacteriology*, *167*, 30–34.

de Vrind, J. P. M., Brouwers, G. J., Corstjens, P. L. A. M., den Dulk, J., & de Vrind-de Jong, E. W. (1998). The cytochrome c maturation operon is involved in manganese oxidation in *Pseudomonas putida* GB-1. *Applied and Environmental Microbiology*, *64*, 3556–3562.

De Vrind, J., De Groot, A., Brouwers, G. J., Tommassen, J., & De Vrind-De Jong, E. (2003). Identification of a novel Gsp-related pathway required for secretion of the manganese-oxidizing factor of *Pseudomonas putida* strain GB-1. *Molecular Microbiology*, *47*, 993–1006.

Diaz, J., Hansel, C. M., Voelker, B. M., Mendes, C. M., Andeer, P. F., & Zhang, T. (2013). Widespread production of extracellular superoxide by marine heterotrophic bacteria. *Science*, *340*, 1223–1226.

DiChristina, T. J. (1992). Effects of nitrate and nitrite on dissimilatory iron reduction by *Shewanella putrefaciens* 200. *Journal of Bacteriology*, *174*, 1891–1896.

DiChristina, T. J., Fredrickson, J. K., & Zachara, J. M. (2005). Enzymology of electron transport: Energy generation with geochemical consequences. *Reviews in Mineralogy and Geochemistry*, *59*, 27–52.

DiChristina, T. J., Moore, C. M., & Haller, C. A. (2002). Dissimilatory Fe(III) and Mn(IV) reduction by *Shewanella putrefaciens* requires ferE, a homolog of the pulE (gspE) type II protein secretion gene. *Journal of Bacteriology*, *184*, 142–151.

Dick, G. J., Clement, B. G., Webb, S. M., Fodrie, F. J., Bargar, J. R., & Tebo, B. M. (2009). Enzymatic microbial Mn(II) oxidation and Mn biooxide production in the Guaymas Basin deep-sea hydrothermal plume. *Geochimica et Cosmochimica Acta*, *73*, 6517–6530.

Dick, G. J., Podell, S., Johnson, H. A., Rivera-Espinoza, Y., Bernier-Latmani, R., McCarthy, J. K., et al. (2008). Genomic insights into Mn(II) oxidation by the marine alphaproteobacterium *Aurantimonas* sp. strain SI85-9A1. *Applied and Environmental Microbiology*, *74*, 2646–2658.

Dick, G. J., & Tebo, B. M. (2010). Microbial diversity and biogeochemistry of the Guaymas Basin deep-sea hydrothermal plume. *Environmental Microbiology*, *12*, 1334–1347.

Dick, G. J., Torpey, J. W., Beveridge, T. J., & Tebo, B. M. (2008). Direct identification of a bacterial manganese(II) oxidase, the multicopper oxidase MnxG, from spores of several different marine *Bacillus* species. *Applied and Environmental Microbiology*, *74*, 1527–1534.

Do, S. H., Batchelor, B., Lee, H. K., & Kong, S. H. (2009). Hydrogen peroxide decomposition on manganese oxide (pyrolusite): Kinetics, intermediates, and mechanism. *Chemosphere*, *75*, 8–12.

Duckworth, O. W., & Sposito, G. (2005). Siderophore-manganese(III) interactions. I. Air-oxidation of manganese(II) promoted by desferrioxamine B. *Environmental Science and Technology*, *39*, 6037–6044.
Ehrlich, H. L. (1966). Reactions with manganese by bacteria from ferromanganese nodules. *Developments in Industrial Microbiology*, 7, 43–60.
Ehrlich, H. L. (1976). Manganese as an energy source for bacteria. In J. Nriagu (Ed.), *Environmental biogeochemistry: Vol. 2* (pp. 633–644). Ann Arbor, MI: Ann Arbor Science Publication.
Ehrlich, H. L. (1983). Manganese oxidizing bacteria from a hydrothermally active area on the Galapagos Rift. *The Bulletin of the Ecological Society of America*, *35*, 357–366.
Ehrlich, H. L. (1987). Manganese oxide reduction as a form of anaerobic respiration. *Geomicrobiology Journal*, *5*, 423–431.
Ehrlich, H. L. (2000). Ocean manganese nodules: Biogenesis and bioleaching possibilities. *Mineral Metallurgy Processes*, *17*, 121–128.
Ehrlich, H. L., Ghiorse, W. C., & Johnson, G. L. I. (1972). Distribution of microbes in manganese nodules from the Atlantic and Pacific Oceans. *Developments in Industrial Microbiology*, *13*, 57–65.
Ehrlich, H. L., & Newman, D. K. (2008). *Geomicrobiology* (5th ed.). New York, NY: Marcel Dekker.
Ehrlich, H. L., & Salerno, J. C. (1990). Energy coupling in $Mn^{2+}$ oxidation by a marine bacterium. *Archives of Microbiology*, *154*, 12–17.
Ehrlich, H. L., & Zapkin, M. A. (1985). Manganese-rich layers in calcareous deposits along the western shore of the Dead Sea may have a bacterial origin. *Geomicrobiology Journal*, *4*, 207–221.
El Gheriany, I. A., Bocioaga, D., Hay, A. G., Ghiorse, W. C., Shuler, M. L., & Lion, L. W. (2009). Iron requirement for Mn(II) oxidation by *Leptothrix discophora* SS-1. *Applied and Environmental Microbiology*, *75*, 1229–1235.
Elzinga, E. J., & Kustka, A. B. (2015). A Mn-54 radiotracer study of Mn isotope solid-liquid exchange during reductive transformation of vernadite (δ-$MnO_2$) by aqueous Mn(II). *Environmental Science and Technology*, *49*, 4310–4316.
Emerson, D., Garen, R. E., & Ghiorse, W. (1989). Formation of *Metallogenium*-like structures by a manganese-oxidizing fungus. *Archives of Microbiology*, *151*, 223–231.
Emerson, S., Kalhorn, S., Jacobs, L., Tebo, B. M., Nealson, K. H., & Rosson, R. A. (1982). Environmental oxidation rate of manganese(II): Bacterial catalysis. *Geochimica et Cosmochimica Acta*, *46*, 1073–1079.
Estes, E. R., Andeer, P. F., Nordlund, D., Wankel, S. D., & Hansel, C. M. (2016). Biogenic manganese oxide as reservoirs of organic carbon and proteins in terrestrial and marine environments. *Geobiology*, *15*, 158–172.
Feng, Q., Yanagisawa, K., & Yamasaki, N. (1998). Hydrothermal soft chemical process for synthesis of manganese oxides with tunnel structures. *Journal of Porous Material*, *5*, 153–161.
Fennessey, C. M., Jones, M. E., Taillefert, M., & DiChristina, T. J. (2010). Siderophores are not involved in Fe(III) solubilization during anaerobic Fe(III) respiration by *Shewanella oneidensis* MR-1. *Applied and Environmental Microbiology*, *76*, 2425–2432.
Fischer, W. W., Hemp, J., & Valentine, J. S. (2016). How did life survive Earth's great oxygenation? *Current Opinion in Chemical Biology*, *31*, 166–178.
Fortin, D., Ferris, F. G., & Scott, S. D. (1998). Formation of Fe-silicate and Fe oxides on bacterial surfaces in samples collected near hydrothermal vents on the Southern Explorer Ridge in the northeast Pacific Ocean. *American Mineralogist*, *83*, 1399–1408.
Francis, C. A., Casciotti, K. L., & Tebo, B. M. (2002). Localization of Mn(II)-oxidizing activity and the putative multicopper oxidase, MnxG, to the exosporium of the marine *Bacillus* sp. strain SG-1. *Archives of Microbiology*, *178*, 450–456.
Francis, C. A., Co, E.-M., & Tebo, B. M. (2001). Enzymatic manganese(II) oxidation by a marine α-proteobacterium. *Applied and Environmental Microbiology*, *67*, 4024–4029.

Francis, C. A., Obraztsova, A. Y., & Tebo, B. M. (2000). Dissimilatory metal reduction by the facultative anaerobe *Pantoea agglomerans* SP1. *Applied and Environmental Microbiology, 66*, 543–548.

Francis, C. A., & Tebo, B. M. (1999). Marine *Bacillus* spores as catalysts for oxidative precipitation and sorption of metals. *Journal of Molecular Microbiology and Biotechnology, 1*, 71–78.

Francis, C. A., & Tebo, B. M. (2001). *cumA* multicopper oxidase genes from diverse Mn(II)-oxidizing and non-Mn(II)-oxidizing *Pseudomonas* strains. *Applied and Environmental Microbiology, 67*, 4272–4278.

Francis, C. A., & Tebo, B. M. (2002). Enzymatic manganese(II) oxidation by metabolically dormant spores of diverse *Bacillus* species. *Applied and Environmental Microbiology, 68*, 874–880.

Gadd, G. M. (2010). Metals, minerals and microbes: Geomicrobiology and bioremediation. *Microbiology, 156*, 609–643.

Geszvain, K., McCarthy, J. K., & Tebo, B. M. (2013). Elimination of manganese(II, III) oxidation in Pseudomonas putida GB-1 by a double knockout of two putative multicopper oxidase genes. *Applied and Environmental Microbiology, 79*, 357–366.

Geszvain, K., Smesrud, L., & Tebo, B. M. (2016). Identification of a third Mn(II) oxidase enzyme in *Pseudomonas putida* GB-1. *Applied and Environmental Microbiology, 82*, 3774–3782. http://dx.doi.org/10.1128/AEM.00046-16.

Ghiorse, W. C. (1980). Electron microscopic analysis of metal-depositing microorganisms in surface layers of Baltic Sea ferromanganese concretions. In P. A. Trudinger, M. R. Walter, & B. J. Ralph (Eds.), *Biogeochemistry of ancient and modern environments* (pp. 345–354). Canberra, Australia: Australian Academy of Science.

Ghiorse, W. C. (1988). Biology of iron- and manganese-depositing bacteria. *Annual Review of Microbiology, 38*, 515–550.

Ghiorse, W. C., & Ehrlich, H. L. (1976). Electron transport components of the $MnO_2$ reductase system and the location of the terminal reductase in a marine bacillus. *Applied and Environmental Microbiology, 31*, 977–985.

Ghiorse, W. C., & Hirsch, P. (1979). An ultrastructural study of iron and manganese deposition associated with extracellular polymers of *Pedomicrobium*-like budding bacteria. *Archives of Microbiology, 123*, 213–226.

Giovanoli, R. (1980). Vernadite is random-stacked birnessite. *Mineralium Deposita, 15*, 251–285.

Goldberg, E. D. (1954). Marine geochemistry. I. Chemical scavengers of the sea. *Journal of Geology, 62*, 249–265.

Goldberg, E. D., & Arrhenius, G. O. S. (1958). Chemistry of Pacific pelagic sediments. *Geochimica et Cosmochimica Acta, 13*, 153–212.

Gorby, Y. A., Yanina, S., McLean, J. S., Rosso, K. M., Moyles, D., Dohnalkova, A., et al. (2006). Electrically conductive bacterial nanowires produced by *Shewanella oneidensis* strain MR-1 and other microorganisms. *Proceedings of the National Academy of Sciences of the United States of America, 103*, 11358–11363.

Graham, J. W. (1959). Metabolically induced precipitation of trace elements from sea water. *Science, 129*, 1428–1429.

Gralnick, J. A., & Newman, D. K. (2007). Extracellular respiration. *Molecular Microbiology, 65*, 1–11.

Green, R. T., Todd, J. D., & Johnston, A. W. B. (2013). Manganese uptake in marine bacteria: The novel MntX transporter is widespread in *Roseobacters*, *Vibrios*, *Alteromonadales* and the SAR11 and SAR16 clades. *ISME Journal*, 7, 581–591.

Hansard, S. P., Easter, H. D., & Voelker, B. M. (2011). Rapid reaction of nanomolar Mn(II) with superoxide radical in seawater and simulated freshwater. *Environmental Science and Technology, 45*, 2811–2817.

Hansel, C. M., Buchwald, C., Diaz, J. M., Ossolinski, J. E., Dyhrman, S. T., Van Mooy, B. A. S., et al. (2016). Dynamics of superoxide production by natural *Trichodesmium* colonies from the Sargasso Sea. *Limnology and Oceanography, 61*, 1188–1200.

Hansel, C. M., & Francis, C. A. (2006). Coupled photochemical and enzymatic Mn(II) oxidation pathways of a planktonic *Roseobacter*-like bacterium. *Applied and Environmental Microbiology, 72*, 3543–3549.

Hansel, C. M., & Learman, D. R. (2015). Geomicrobiology of manganese. In H. L. Ehrlich, D. K. Newman, & A. Kappler (Eds.), *Vol. 6. Ehrlich's geomicrobiology* (pp. 401–452). New York, NY: CRC Press.

Hansel, C. M., Zeiner, C. A., Santelli, C. M., & Webb, S. M. (2012). Mn(II) oxidation linked to superoxide production during asexual reproduction in an ascomycete fungi. *Proceedings of the National Academy of Sciences of the United States of America, 109*, 12621–12625.

Hein, J. R., & Koschinsky, A. (2013). Deep-ocean ferromanganese crusts and nodules. *The Treatise on Geochemistry*, 273–290.

Hein, J. R., Koschinsky, A., Halbach, P., Manheim, F. T., Bau, M., Kang, J.-K., et al. (1997). Iron and manganese oxide mineralization in the Pacific. *Geological Society of London, Special Publication, 119*, 123–138.

Hem, J. D. (1963). *Chemical equilibria and rate of manganese oxidation*. Washington, DC: US Geological Survey Water Supply Paper. 1667-A.

Heye, D., & Beiersdorf, H. (1973). Radioaktive und magnetische Untersuchungen an Manganknollen zur Ermittlung der Wachstumsgeschwindigkeit bzw. Zur Altersbestimmung. *Z Geophysik, 39*, 703–726.

Hochella, M. F., Jr., Lower, S., Maurice, P., Penn, R., Sahai, N., Sparks, D. L., et al. (2008). Nanominerals, mineral nanoparticles, and earth systems. *Science, 319*, 1631.

Hofer, C., & Schlosser, D. (1999). Novel enzymatic oxidation of $Mn^{2+}$ to $Mn^{3+}$ catalyzed by a fungal laccase. *FEBS Letters, 451*, 186–190.

Horn, D. R., Horn, B. M., & Delach, M. N. (1972). Distribution of ferromanganese deposits in the world ocean. In D. R. Horn (Ed.), *Ferromanganese deposits on the ocean floor* (pp. 9–17). Washington, DC: The Office of the International Decade of Ocean Exploration, National Science Foundation.

Horsburgh, M. J., Wharton, S. J., Karavolos, M., & Foster, S. J. (2002). Manganese: Elemental defence for a life with oxygen? *Trends in Microbiology, 10*, 496–501.

Hugate, B., Danin, A., Pellerin, N. B., Stemmler, J., Kjellander, P., Adams, B. J., et al. (1987). Characterization of manganese oxidizing ($Mn^{II} \rightarrow Mn^{IV}$) bacteria from Negev Desert rock varnish: Implications in desert varnish formation. *Canadian Journal of Microbiology, 33*, 939–943.

Hui, L., Szeinbaum, N., DiChristina, T. J., & Taillefert, M. (2012). Microbial Mn(IV) reduction requires an initial one-electron reductive stabilization step. *Geochimica et Cosmochimica Acta, 99*, 179–192.

Hulth, S., Aller, R. C., & Gilbert, F. (1999). Coupled anoxic nitrification manganese reduction in marine sediments. *Geochimica et Cosmochimica Acta, 63*, 49–66.

Jacobs, L., Emerson, S., & Skei, J. (1985). Partitioning and transport of metals across an $O_2$/$H_2S$ interface in a permanently anoxic basin: Framvaren Fjord, Norway. *Geochimica et Cosmochimica Acta, 49*, 1433–1444.

Jakubovics, N. S., & Jenkinson, H. F. (2001). Out of the iron age: New insights into the critical role of manganese homeostasis in bacteria. *Microbiology, 147*, 1709–1718.

Jenne, E. A. (1968). Controls on Mn, Co, Ni, Cu, and Zn concentrations in soil and water: The significant role of hydrous Mn and Fe oxides. In R. F. Gould (Ed.), *Trace inorganics in water* (pp. 337–387). Washington, DC: American Chemical Society.

Johnson, K. S., Coale, K. H., Berelson, W. M., & Gordon, R. M. (1996). On the formation of the manganese maximum in the oxygen minimum. *Geochimica et Cosmochimica Acta, 60*, 1291–1299.

Johnson, K., Purvis, G., Lopez-Capel, E., Peacock, C. L., Gray, N., Wagner, T., et al. (2015). Towards a mechanistic understanding of carbon stabilization in manganese oxides. *Nature Communications*, *6*, 7628.

Johnson, H. A., & Tebo, B. M. (2008). In vitro studies indicate a quinone is involved in bacterial Mn(II) oxidation. *Archives of Microbiology*, *189*, 59–69.

Johnson, J. E., Webb, S. M., Thomas, K., Ono, S., Kirschvink, J. L., & Fischer, W. W. (2013). Manganese-oxidizing photosynthesis before the rise of cyanobacteria. *Proceedings of the National Academy of Sciences of the United States of America*, *110*, 11238–11243.

Jones, C., Crowe, S. A., Sturm, A., Leslie, K. L., MacLean, L. C. W., Katsev, S., et al. (2011). Biogeochemistry of manganese in ferruginous Lake Matano, Indonesia. *Biogeoscience*, *8*, 2977–2991.

Juniper, S. K., & Tebo, B. M. (1995). Microbe-metal interactions and mineral deposition at hydrothermal vents. In D. M. Karl (Ed.), *The microbiology of deep-sea hydrothermal vents* (pp. 219–254). Boca Raton, FL: CRC Press.

Jurgensen, A., Widmeyer, J. R., Gordon, R. A., Bendell-Young, L. I., Moore, M. M., & Crozier, E. D. (2004). The structure of the manganese oxide on the sheath of the bacterium *Leptothrix discophora*: An XAFS study. *American Mineralogist*, *89*, 1110–1118.

Kadko, D., & Burckle, L. H. (1980). Manganese nodule growth rates determined by fossil diatom dating. *Nature*, *287*, 725–726.

Kalinenko, V. O., Belokopytova, O. V., & Nikolaeva, G. G. (1962). Bacteriogenic formation of iron-manganese concretions in the Indian Ocean. *Okeanologiya*, *11*, 1050–1059.

Kashefi, K., & Lovley, D. R. (2000). Reduction of Fe(III), Mn(IV), and toxic metals at 100° C by *Pyrobaculum islandicum*. *Applied and Environmental Microbiology*, *66*, 1050–1056.

Kay, J. T., Conklin, M. H., Fuller, C. C., & O'Day, P. A. (2001). Processes of nickel and cobalt uptake by a manganese oxide forming sediment in Pinal Creek, globe mining district, Arizona. *Environmental Science and Technology*, *35*, 4719–4725.

Keen, C. L., Ensunsa, J. L., Watson, M. H., Baly, D. L., Donovan, S. M., Monaco, M. H., et al. (1999). Nutritional aspects of manganese from experimental studies. *Neurotoxicology*, *20*, 213–223.

Kehres, D. G., & Maguire, M. E. (2003). Emerging themes in manganese transport, biochemistry, and pathogenesis in bacteria. *FEMS Microbiology Reviews*, *27*, 263–290.

Keiluweit, M., Nico, P., Harmon, M. E., Mao, J., Pett-Ridge, J., & Kleber, M. (2015). Long-term litter decomposition controlled by manganese redox cycling. *Proceedings of the National Academy of Sciences of the United States of America*, *112*, E5253–E5260.

Klewicki, J. K., & Morgan, J. J. (1998). Kinetic behavior of Mn(III) complexes of pyrophosphate, EDTA and citrate. *Environmental Science and Technology*, *32*, 2916–2922.

Klewicki, J. K., & Morgan, J. J. (1999). Dissolution of α-MnOOH particles by ligand: Pyrophosphate, ethylenediaminetetraacetate, and citrate. *Geochimica et Cosmochimica Acta*, *63*, 2017–3024.

Klimov, V. V. (1984). Charge separation in photosystem II reaction centers. The role of phaeophytin and manganese. In C. Sybesma (Ed.), *Proceedings of the VIth international congress on photosynthesis: Vol. 1. Advances in photosynthetic research* (pp. 131–138). The Hague: Nijhoff.

Knittel, K., Losekann, T., Boetius, A., Kort, R., & Amann, R. (2005). Diversity and distribution of methanotrophic archaea at cold seeps. *Applied and Environmental Microbiology*, *71*, 467–479.

Kostka, J. E., Luther, G. W. I., & Nealson, K. H. (1995). Chemical and biological reduction of Mn(III) pyrophosphate complexes: Potential importance of dissolved Mn(III) as an environmental oxidant. *Geochimica et Cosmochimica Acta*, *59*, 885–894.

Krumbein, W. E. (1971). Manganese oxidizing fungi & bacteria in recent shelf sediments of the Bay of Biscay and the North Sea. *Naturwissenschaften*, *58*, 56–57.

Krumbein, W. E., & Altmann, H. J. (1973). A new method for the detection and enumeration of manganese-oxidizing and reducing microorganisms. *Helgoländer Wissenschaftliche Meeresuntersuchungen, 25*, 347–356.

Ku, T. L., & Broecker, W. S. (1969). Radiochemical studies on manganese nodules of deep-sea origin. *Deep-Sea Research Oceanography Abstracts, 16*, 625–635.

Kustka, A. B., Shaked, Y., Milligan, A. J., King, D. W., & Morel, F. M. M. (2005). Extracellular production of superoxide by marine diatoms: Contrasting effects on iron redox chemistry and bioavailability. *Limnology and Oceanography, 50*, 1172–1180.

Kwon, K. D., Refson, K., & Sposito, G. (2008). Defect-induced photoconductivity in layered manganese oxides: A density functional theory study. *Physical Review Letters, 100*, 146601.

Lanza, N. L., Fischer, W. W., Wiens, R. C., Grotzinger, J., Ollila, A. M., Cousin, A., et al. (2014). High manganese concentrations in rocks at Gale crater, Mars. *Geophysical Research Letters, 41*, 5755–5763.

LaRock, P. A., & Ehrlich, H. L. (1975). Observations of bacterial microcolonies on the surface of ferromanganese nodules from Blake Plateau by scanning electron microscopy. *Microbial Ecology, 2*, 84–96.

Larsen, E. I., Little, B., Nealson, K. H., Ray, R., Stone, A., & Tian, J. (1998). Manganite reduction by *Shewanella putrefaciens* MR-4. *American Mineralogist, 83*, 1564–1572.

Larsen, E. I., Sly, L. I., & McEwan, A. G. (1999). Manganese(II) adsorption and oxidation by whole cells and a membrane fraction of *Pedomicrobium* sp. ACM 3067. *Archives of Microbiology, 171*, 257–264.

Latour, J.-M. (2015). Manganese, the stress reliever. *Metallomics: Integrated Biometal Science, 7*, 25–28.

Laverman, A. M., Switzer Blum, J., Schaefer, J. K., Phillips, E. J. P., Lovley, D. R., & Oremland, R. S. (1995). Growth of strain SES-3 with arsenate and other diverse electron acceptors. *Applied and Environmental Microbiology, 61*, 3556–3561.

Leach, R. M., & Harris, E. D. (1997). Manganese. In B. L. O'Dell & R. A. Sunde (Eds.), *Handbook of nutritionally essential minerals* (pp. 335–355). New York: Marcel Dekker, Inc.

Learman, D. R., & Hansel, C. M. (2014). Comparative proteomics of Mn(II)-oxidizing and non-oxidizing *Roseobacter* clade bacteria reveal an operative manganese transport system but minimal Mn(II)-induced expression of manganese oxidation and antioxidant enzymes. *Environmental Microbiology Reports, 6*, 501–509.

Learman, D. R., Voelker, B. M., Madden, A. S., & Hansel, C. M. (2013). Constraints on superoxide mediated formation of manganese oxides. *Frontiers in Microbiology, 4*, 262.

Learman, D. R., Voelker, B. M., Vazquez-Rodriguez, A. I., & Hansel, C. M. (2011). Formation of manganese oxides by bacterially generated superoxide. *Nature Geoscience, 4*, 95–98.

Learman, D. R., Wankel, S. D., Webb, S. M., Martinez, N., Madden, A. S., & Hansel, C. M. (2011). Coupled biotic-abiotic Mn(II) oxidation pathway mediates the formation and structural evolution of biogenic Mn oxides. *Geochimica et Cosmochimica Acta, 75*, 6048–6063.

Lenunic, I., & Bork, P. (2011). Interactive tree of life v2: Online annotation and display of phylogenetic trees made easy. *Nucleic Acids Research, 39*, W475.

Letunic, I., & Bork, P. (2007). Interactive tree of life (iTOL): An online tool for phylogenetic tree display and annotation. *Bioinformatics, 23*, 127.

Lieske, R. (1919). Zur Ernährungsphysiologie der Eisenbakterien. *Zentralbl Bakteriol Parasitenkd Infektionskr Hyg Abt II, 49*, 413–425.

Lin, H., Szeinbaum, N., DiChristina, T. J., & Taillefert, M. (2012). Microbial Mn(IV) reduction requires an initial one-electron reductive solubilization step. *Geochimica et Cosmochimica Acta, 99*, 179–192.

Lin, H., & Taillefert, M. (2014). Key geochemical factors regulating Mn(IV)-catalyzed anaerobic nitrification in coastal marine sediments. *Geochimica et Cosmochimica Acta, 133*, 17–33.
Lovley, D. R. (1991). Dissimilatory Fe(III) and Mn(IV) reduction. *Microbiological Reviews, 55*, 259–287.
Lovley, D. R., & Anderson, R. T. (2000). The influence of dissimilatory metal reduction on the fate of organic and metal contaminants in the subsurface. *Journal of Hydrology, 8*, 77–88.
Lovley, D. R., Giovannoni, S. J., White, D. C., Champine, J. E., Phillips, E. J. P., Gorby, Y. A., et al. (1993). *Geobacter metallireducens* gen nov. sp. nov., a microorganism capable of coupling the complete oxidation of organic compounds to the reduction of iron and other metals. *Archives of Microbiology, 159*, 336–344.
Lovley, D. R., Holmes, D. E., & Nevin, K. P. (2004). Dissimilatory Fe(III) and Mn(IV) reduction. *Advances in Microbial Physiology, 49*, 219–286.
Lovley, D. R., & Phillips, E. J. P. (1988). Novel mode of microbial energy metabolism: Organic carbon oxidation coupled to dissimilatory reduction of iron and manganese. *Applied and Environmental Microbiology, 54*, 1472–1480.
Lovley, D. R., & Phillips, E. J. P. (1992). Reduction of uranium by *Desulfovibrio desulfuricans*. *Applied and Environmental Microbiology, 58*, 850–856.
Lovley, D. R., & Phillips, E. J. P. (1994). Novel process for anaerobic sulfate production from elemental sulfur by sulfate-reducing bacteria. *Applied and Environmental Microbiology, 60*, 2394–2399.
Lovley, D. R., Phillips, E. J. P., & Lonergan, D. J. (1989). Hydrogen and formate oxidation coupled to dissimilatory reduction of iron and manganese by *Alteromonas putrefaciens*. *Applied and Environmental Microbiology, 55*, 700–706.
Luther, G. W., III (2005). Manganese(II) oxidation and Mn(IV) reduction in the environment—Two one-electron transfer steps versus a single two-electron step. *Geomicrobiology Journal, 22*, 195–203.
Luther, G. W., III (2010). The role of one- and two-electron transfer reactions in forming thermodynamically unstable intermediates as barriers in multi-electron redox reactions. *Aquatic Geochemistry, 16*, 395–420.
Luther, G. W., III, Madison, A. S., Mucci, A., Sundby, B., & Oldham, V. E. (2015). A kinetic approach to assess the strengths of ligands bound to soluble Mn(III). *Marine Chemistry, 173*, 93–99.
Luther, G. W., III, Sundby, B., Lewis, B. L., Brendel, P. J., & Silverberg, N. (1997). Interactions of manganese with the nitrogen cycle: Alternative pathways to dinitrogen. *Geochimica et Cosmochimica Acta, 61*, 4043–4052.
Madden, A. S., & Hochella, M. F. (2005). A test of geochemical reactivity as a function of mineral size: Manganese oxidation promoted by hematite nanoparticles. *Geochimica et Cosmochimica Acta, 69*, 389–398.
Madison, A. S., Tebo, B. M., & Luther, G. W., III. (2011). Simultaneous determination of soluble manganese(III), manganese(II) and total manganese in natural (pore)waters. *Talanta, 84*, 374–381.
Madison, A. S., Tebo, B. M., Mucci, A., Sundby, B., & Luther, G. W., III. (2013). Abundant porewater Mn(III) is a major component of the sedimentary redox system. *Science, 341*, 875–878.
Manceau, A., Gorshkov, A. I., & Drits, V. A. (1992). Structural chemistry of Mn, Fe, Co, and Ni in manganese hydrous oxides: Part II. Information from EXAFS spectroscopy and electron and X-ray diffraction. *American Mineralogist, 77*, 1144–1157.
Manceau, A., Kersten, M., Marcus, M. A., Geoffroy, N., & Granina, L. (2007). Ba and Ni speciation in a nodule of binary Mn oxide phase composition from Lake Baikal. *Geochimica et Cosmochimica Acta, 71*, 1967–1981.

Manceau, A., Lanson, M., & Geoffroy, N. (2007). Natural speciation of Ni, Zn, Ba, and As in ferromanganese coatings on quartz using X-ray fluorescence, absorption, and diffraction. *Geochimica et Cosmochimica Acta, 71,* 95–128.

Mandernack, K. W., & Tebo, B. M. (1993). Manganese scavenging and oxidation at hydrothermal vents and in vent plumes. *Geochimica et Cosmochimica Acta, 57,* 3907–3923.

McLoughlin, N., Wacey, D., Kruber, C., Kilburn, M. R., Thorseth, H. I., & Pedersen, R. B. (2011). A combined TEM and NanoSIMS study of endolithic microfossils in altered seafloor basalt. *Chemical Geology, 289,* 154–162.

Minibayeva, F. V., Gordon, L. K., Kolesnikov, O. P., & Chasov, A. V. (2001). Role of extracellular peroxidase in the superoxide production by wheat root cells. *Protoplasma, 217,* 125–128.

Miyata, N., Maruo, K., Tani, Y., Tsuno, H., Seyama, H., Soma, M., et al. (2006). Production of biogenic manganese oxides by anamorphic ascomycete fungi isolated from streambed pebbles. *Geomicrobiology Journal, 23,* 63–73.

Miyata, N., Tani, Y., Iwahori, K., & Soma, M. (2004). Enzymatic formation of manganese oxides by an *Acremonium*-like hyphomycete fungus, strain KR21-2. *FEMS Microbial Ecology, 47,* 101–109.

Miyata, N., Tani, Y., Maruo, K., Tsuno, H., Sakata, M., & Iwahori, K. (2006). Manganese (IV) oxide production by *Acremonium* sp. strain KR21-2 and extracellular Mn(II) oxidase activity. *Applied and Environmental Microbiology, 72,* 6467–6473.

Moffett, J. W. (1994a). A radiotracer study of Ce and Mn uptake onto suspended particles in Chesapeake Bay. *Geochimica et Cosmochimica Acta, 58,* 695–703.

Moffett, J. W. (1994b). The relationship between Ce and Mn oxidation in seawater. *Limnology and Oceanography, 39,* 1309–1318.

Moffett, J. W. (1997). The importance of microbial Mn oxidation in the upper ocean: A comparison of the Sargasso Sea and equatorial Pacific. *Deep-Sea Research I, 44,* 1277–1291.

Moffett, J. W., & Ho, J. (1996). Oxidation of cobalt and manganese in seawater is a common microbially catalyzed pathway. *Geochimica et Cosmochimica Acta, 60,* 3415–3424.

Morgan, J. J. (2000). Manganese in natural waters and earth's crust: Its availability to organisms. *Metal Ions in Biological Systems, 37,* 1–34.

Morgan, J. J. (2005). Kinetics of reaction between $O_2$ and Mn(II) species in aqueous solutions. *Geochimica et Cosmochimica Acta, 69,* 35–48.

Murray, J. W. (1975). The interaction of metal ions at the manganese dioxide-solution interface. *Geochimica et Cosmochimica Acta, 39,* 505–519.

Murray, K. J., & Tebo, B. M. (2007). Cr(IIII) is indirectly oxidized by the Mn(II)-oxidizing bacterium *Bacillus* sp. strain SG-1. *Environmental Science and Technology, 41,* 528–533.

Mustoe, G. E. (1981). Bacterial oxidation of manganese and iron in a modern cold spring. *Geological Society of America Bulletin Part 1, 92,* 147–153.

Myers, C. R., & Nealson, K. H. (1988a). Bacterial manganese reduction and growth with manganese oxide as the sole electron acceptor. *Science, 240,* 1319–1321.

Myers, C. R., & Nealson, K. H. (1988b). Microbial reduction of manganese oxides: Interaction with iron and sulfur. *Geochimica et Cosmochimica Acta, 52,* 2727–2732.

Nealson, K. H., & Ford, J. (1980). Surface enhancement of bacterial manganese oxidation: Implications for aquatic environments. *Geomicrobiology Journal, 2,* 21–37.

Nealson, K. H., Myers, C. R., & Wimpee, R. R. (1991). Isolation and identification of manganese-reducing bacteria and estimates of microbial Mn(IV)-reducing potential in the Black Sea. *Deep-Sea Research, 38,* S907–S920.

Nealson, K. H., & Saffarini, D. (1994). Iron and manganese in anaerobic respiration: Environmental significance, physiology, regulation. *Annual Review of Microbiology, 48,* 311–343.

Nelson, Y. M., Lion, L. W., Shuler, M. L., & Ghiorse, W. C. (1999). Lead binding to metal oxide and organic phases of natural aquatic biofilms. *Limnology*, *44*, 1715–1729.
Nevin, K. P., & Lovley, D. R. (2002). Mechanisms of Fe(III) oxide reduction in sedimentary environments. *Geomicrobiology Journal*, *19*, 141–159.
Nickel, M., Vandieken, V., Bruchert, V., & Jorgensen, B. B. (2008). Microbial Mn(IV) and Fe(III) reduction in northern Barents Sea sediments under different conditions of ice cover and organic carbon deposition. *Deep-Sea Research II*, *55*, 2390–2398.
Northup, D. E., Barns, S. M., Yu, L. E., Spilde, M. N., Schelble, R. T., Dano, K. E., et al. (2003). Diverse microbial communities inhabiting ferromanganese deposits in Lechuguilla and Spider Caves. *Environmental Microbiology*, *5*, 1071–1086.
Northup, D. E., & Lavoie, K. H. (2001). Geomicrobiology of caves: A review. *Geomicrobiology Journal*, *18*, 199–222.
Okazaki, M., Sugita, T., Shimizu, M., Ohode, Y., Iwamoto, K., de Vrind-de Jong, E. W., et al. (1997). Partial purification and characterization of manganese-oxidizing factors of *Pseudomonas fluorescens* GB-1. *Applied and Environmental Microbiology*, *63*, 4793–4799.
Oldham, V. E., Owings, S. M., Jones, M. R., Tebo, B. M., & Luther, G. W., III. (2015). Evidence for the presence of strong Mn(III)-binding ligands in the water column. *Marine Chemistry*, *171*, 58–66.
O'Reilly, S. E., & Hochella, M. F. (2003). Lead sorption efficiencies of natural and synthetic Mn and Fe-oxides. *Geochimica et Cosmochimica Acta*, *67*, 4471–4487.
Orphan, V. J., Hinrichs, K. U., Ussler, W., Paull, C. K., Taylor, L. T., Sylva, S. P., et al. (2001). Comparative analysis of methane-oxidizing archaea and sulfate-reducing bacteria in anoxic marine sediments. *Applied and Environmental Microbiology*, *67*, 1922–1934.
Orphan, V. J., House, C. H., Hinrichs, K. U., McKeegan, K. D., & DeLong, E. F. (2002). Multiple archaeal groups mediate methane oxidation in anoxic cold seep sediments. *Proceedings of the National Academy of Sciences of the United Sttaes of America*, *99*, 7663–7668.
Parker, D. L., Sposito, G., & Tebo, B. M. (2004). Manganese(III) binding to a pyoverdine siderophore produced by a manganese(II)-oxidizing bacterium. *Geochimica et Cosmochimica Acta*, *68*, 4809–4820.
Peacock, C. L., & Moon, E. M. (2012). Oxidative scavenging of thallium by birnessite: Controls on thallium sorption and stable isotope fractionation in marine ferromanganese precipitates. *Geochimica et Cosmochimica Acta*, *84*, 297–313.
Peacock, C. L., & Sherman, D. M. (2007). Crystal-chemistry of Ni in marine ferromanganese crusts and nodules. *American Mineralogist*, *92*, 1087–1092.
Pena, J., Kwon, K. D., Refson, K., Bargar, J. R., & Sposito, G. (2010). Mechanisms of nickel sorption by a bacteriogenic birnessite. *Geochimica et Cosmochimica Acta*, *74*, 3076–3089.
Perez, J., & Jeffries, T. W. (1992). Roles of manganese and organic acid chelators in regulating lignin degradation and biosynthesis of peroxidases by *Phanerochaete chrysosporium*. *Applied and Environmental Microbiology*, *58*, 2402–2409.
Post, J. (1999). Manganese oxide minerals: Crystal structures and economic and environmental significance. *Proceedings of the National Academy of Sciences of the United States of America*, *96*, 3447–3454.
Reyss, J. L., Marchig, V., & Ku, T. L. (1982). Rapid grown of a deep-sea manganese nodule. *Nature*, *295*, 401–403.
Richardson, L. L., Aguilar, C., & Nealson, K. H. (1988). Manganese oxidation in pH and $O_2$ microenvironments produced by phytoplankton. *Limnology and Oceanography*, *33*, 352–363.
Richter, K., Schicklberger, M., & Gescher, J. (2012). Dissimilatory reduction of extracellular electron acceptors in anaerobic respiration. *Applied and Environmental Microbiology*, *78*, 913–921.

Robertson, L. A., Van Niel, E. W. J., Torremans, R. A. M., & Kuenen, G. J. (1988). Simultaneous nitrification and denitrification in aerobic chemostat cultures of *Thiosphaera pantotropha*. *Applied and Environmental Microbiology, 54*, 2812–2818.

Rose, A. L. (2012). The influence of extracellular superoxide on iron redox chemistry and bioavailability to aquatic microorganisms. *Frontiers in Microbiological Chemistry, 3*, 1–21.

Rosson, R. A., & Nealson, K. H. (1982). Manganese binding and oxidation by spores of a marine bacillus. *Journal of Bacteriology, 151*, 1027–1034.

Rusin, P. A., Quintana, L., Sinclair, N. A., Arnold, R. G., & Oden, K. L. (1991). Physiology and kinetics of manganese-reducing *Bacillus polymyxa* strain D1 isolated from manganiferous silver ore. *Geomicrobiology Journal, 9*, 13–25.

Rusin, P. A., Sharp, J. E., Arnold, R. G., & Sinclair, N. A. (1991). Enhanced recovery of manganese and silver from refractory ores. In R. W. Smith & M. Mishra (Eds.), *Minerals bioprocessing* (pp. 207–218). Warrendale, PA: Minerals, Metals, and Materials Society.

Santelli, C. M., Chaput, D. L., & Hansel, C. M. (2014). Microbial communities promoting Mn(II) oxidation in Ashumet Pond, a historically polluted freshwater pond undergoing remediation. *Geomicrobiology Journal, 31*, 605–616.

Santelli, C. M., Pfister, D. H., Lazarus, D., Sun, L., Burgos, W. D., & Hansel, C. M. (2010). Promotion of Mn(II) oxidation and remediation of coal mine drainage in passive treatment systems by diverse fungal and bacterial communities. *Applied and Environmental Microbiology, 76*, 4871–4875.

Santelli, C. M., Webb, S. M., Dohnalkova, A. C., & Hansel, C. M. (2011). Diversity of Mn oxides produced by Mn(II)-oxidizing fungi. *Geochimica et Cosmochimica Acta, 75*, 2762–2776.

Sartory, A., & Meyer, J. (1947). Contribution à l'étude du métabolisme hydrocarboné des bactéries ferrugineuses. *Comptes Rendus Académie des Sciences, 225*, 541–542.

Scheeline, A., Olson, D. L., Williksen, E. P., Horras, G. A., Klein, M. L., & Larter, R. (1997). The peroxidase-oxidase oscillator and its constituent chemistries. *Chemical Reviews, 97*, 739–756.

Schippers, A., Neretin, L. N., Lavik, G., Leipe, T., & Pollehne, F. (2005). Manganese(II) oxidation driven by lateral oxygen intrusions in the western Black Sea. *Geochimica et Cosmochimica Acta, 69*, 2241–2252.

Schlosser, D., & Hofer, C. (2002). Laccase-catalyzed oxidation of $Mn^{2+}$ in the presence of natural $Mn^{3+}$ chelators as a novel source of extracellular $H_2O_2$ production and its impact on manganese peroxidase. *Applied and Environmental Microbiology, 68*, 3514–3521.

Schuetz, B., Schicklberger, M., Kuermann, J., Spormann, A. M., & Gescher, J. (2009). Periplasmic electron transfer via the c-type cytochromes MtrA and FccA of *Shewanella oneidensis* MR-1. *Applied and Environmental Microbiology, 75*, 7789–7796.

Segarra, K. E. A., Comerford, C., Slaughter, J., & Joye, S. B. (2013). Impact of electron acceptor availability on the anaerobic oxidation of methane in coastal freshwater and brackish wetland sediments. *Geochimica et Cosmochimica Acta, 115*, 15–30.

Shi, L., Chen, B. W., Wang, Z. M., Elias, D. A., Mayer, M. U., Gorby, Y. A., et al. (2006). Isolation of a high affinity functional protein complex between OmcA and MtrC: Two outer membrane decaheme c-type cytochromes of *Shewanella oneidensis* MR-1. *Journal of Bacteriology, 188*, 4705–4714.

Shi, L., Deng, S., Marshall, M. J., Wang, Z., Kennedy, D. W., Dohnalkova, A. C., et al. (2008). Direct involvement of type II secretion system in extracellular translocation of *Shewanella oneidensis* outer membrane cytochromes MtrC and OmcA. *Journal of Bacteriology, 190*, 5512–5516.

Shi, L., Squier, T. C., Zachara, J. M., & Fredrickson, J. K. (2007). Respiration of metal (hydr) oxides by *Shewanella* and *Geobacter*: A key role for multihaem c-type cytochromes. *Molecular Microbiology, 65*, 12–20.

Sokolova-Dubinina, G. A., & Deriugina, Z. P. (1967). On the role of microorganisms in the formation of rodokhrozite in Punnus-Jahrvi Lake. *Mikrobiologiia, 36*, 435–442.

Sokolova-Dubinina, G. A., & Deryugina, Z. P. (1976). Process of iron-manganese concretion formation in Lake Punnus-Yarvi. *Mikrobiologie, 36*, 1066–1076.

Solomon, E. I., Sundaram, U. M., & Machonkin, T. E. (1996). Multicopper oxidases and oxygenases. *Chemical Reviews, 96*, 2563–2605.

Stabel, H. H., & Kleiner, J. (1983). Endogenic flux of manganese to the bottom of Lake Constance. *Archiv für Hydrobiologie, 98*, 307–316.

Stein, L. Y., La Duc, M. T., Grundl, T. J., & Nealson, K. H. (2001). Bacterial and archaeal populations associated with freshwater ferromanganous micronodules and sediments. *Environmental Microbiology, 3*, 10–18.

Stone, A. T. (1987). Microbial metabolites and the reductive dissolution of manganese oxides: Oxalate and pyruvate. *Geochimica et Cosmochimica Acta, 51*, 919–925.

Su, J., Bao, T., Deng, L., Wu, H., Liu, F., & He, J. (2013). CotA, a multicopper oxidase from *Bacillus pumilus* WH4, exhibits manganese-oxidase activity. *PLoS One, 8*, e60573.

Sunda, W. G., & Huntsman, S. A. (1987). Microbial oxidation of manganese in a North Carolina estuary. *Limnology and Oceanography, 32*, 552–564.

Sunda, W. G., & Huntsman, S. A. (1988). Effect of sunlight on redox cycles of manganese in the southwestern Sargasso Sea. *Deep-Sea Research, 35*, 1297–1317.

Sunda, W. G., & Huntsman, S. A. (1990). Diel cycles in microbial manganese oxidation and manganese redox speciation in coastal waters of the Bahama Islands. *Limnology and Oceanography, 35*, 325–338.

Sunda, W. G., & Huntsman, S. A. (1994). Photoreduction of manganese oxides in seawater. *Marine Chemistry, 46*, 133–152.

Sunda, W. G., Huntsman, S. A., & Harvey, G. R. (1983). Photoreduction of manganese oxides in seawater and its geochemical and biological implications. *Nature, 301*, 234–236.

Sunda, W. G., & Kieber, D. J. (1994). Oxidation of humic substances by manganese oxides yields low-molecular-weight organic substrates. *Nature, 367*, 62–64.

Szeinbaum, N., Burns, J. L., & DiChristina, T. J. (2014). Electron transport and protein secretion pathways involved in Mn(III) reduction by *Shewanella oneidensis*. *Environmental Microbiology Reports, 6*, 490–500.

Taillefert, M., Beckler, J. S., Carey, E., Burns, J. L., Fennessey, C. M., & DiChristina, T. J. (2007). *Shewanella putrefaciens* produces an Fe(III)-solubilizing organic ligand during anaerobic respiration on insoluble Fe(III) oxides. *Journal of Inorganic Biochemistry, 101*, 1760–1767.

Takahashi, Y., Manceau, A., Geoffroy, N., Marcus, M. A., & Usui, A. (2007). Chemical and structural control of the partitioning of Co, Ce, and Pb in marine ferromanganese oxides. *Geochimica et Cosmochimica Acta, 71*, 984–1008.

Takeda, M., Kawasaki, Y., Umezu, T., Shimura, S., Hasegawa, M., & Koizumi, J. I. (2012). Patterns of sheath elongation, cell proliferation, and manganese(II) oxidation in *Leptothrix cholodnii*. *Archives of Microbiology, 94*, 667–673.

Tambiev, S. B., & Demina, L. S. (1992). Biogeochemistry and fluxes of manganese and some other metals in regions of hydrothermal activities (Axial Mountain, Juan de Fuc Ridge and Guaymas Basin, Gulf of California). *Deep-Sea Research, 39*, 687–703.

Tang, W., Xia, J., Zeng, X., Wu, L., & Ye, G. (2013). Biological characteristics and oxidation mechanism of a new manganese-oxidizing bacteria FM-2. *Bio-Medical Materials Engineering, 24*, 703–709.

Tang, Y., Zeiner, C. A., Santelli, C. M., & Hansel, C. M. (2013). Fungal oxidative dissolution of the Mn(II)-bearing mineral rhodochrosite and the role of metabolites in manganese oxide formation. *Environmental Microbiology, 15*, 1063–1077.

Tani, Y., Miyata, N., Iwahori, K., Soma, M., Tokuda, H., Seyama, H., et al. (2003). Biogeochemistry of manganese oxide coatings on pebble surfaces in the Kikukawa River System, Shizuoka, Japan. *Applied Geochemistry*, *18*, 1541–1554.

Taylor-George, S., Palmer, F., Staley, J. T., Borns, D. J., Curtiss, B., & Adams, J. B. (1983). Fungi and bacteria involved in desert varnish formation. *Microbial Ecology*, *9*, 227–245.

Tebo, B. M. (1991). Manganese(II) oxidation in the suboxic zone of the Black Sea. *Deep-Sea Research*, *38*, S883–S905.

Tebo, B. M., Bargar, J. R., Clement, B. G., Dick, G. J., Murray, K. J., Parker, D. L., et al. (2004). Biogenic manganese oxides: Properties and mechanisms of formation. *Annual Review of Earth and Planetary Science*, *32*, 287–328.

Tebo, B. M., & Emerson, S. (1985). Effect of oxygen-tension, Mn(II) concentration, and temperature on the microbially catalyzed Mn(II) oxidation rate in a marine fiord. *Applied and Environmental Microbiology*, *50*, 1268–1273.

Tebo, B. M., & Emerson, S. (1986). Microbial manganese(II) oxidation in the marine-environment—A quantitative study. *Biogeochemistry*, *2*, 149–161.

Tebo, B. M., Johnson, H. A., McCarthy, J. K., & Templeton, A. S. (2005). Geomicrobiology of manganese(II) oxidation. *Trends in Microbiology*, *13*, 421–428.

Tebo, B. M., Nealson, K. H., Emerson, S., & Jacobs, L. (1984). Microbial mediation of Mn(II) and Co(II) precipitation at the $O_2/H_2S$ interfaces in two anoxic fjords. *Limnology and Oceanography*, *29*, 1247–1258.

Templeton, A. S., Knowles, E. J., Eldridge, D. L., Arey, B. W., Dohnalkova, A. C., Webb, S. M., et al. (2009). A seafloor microbial biome hosted within incipient ferromanganese crust. *Nature Geoscience*, *2*, 872–876.

Templeton, A. S., Staudigel, H., & Tebo, B. M. (2005). Diverse Mn(II)-oxidizing bacteria isolated from submarine basalts at Loihi Seamount. *Geomicrobiology Journal*, *22*, 127–139.

Thamdrup, B. (2000). Microbial manganese and iron reduction in aquatic sediments. *Advances in Microbial Ecology*, *16*, 41–84.

Thamdrup, B., & Dalsgaard, T. (2000). The fate of ammonium in anoxic manganese oxide-rich marine sediment. *Geochimica et Cosmochimica Acta*, *64*, 4157–4164.

Thamdrup, B., & Dalsgaard, T. (2002). Production of $N_2$ through anaerobic ammonium oxidation coupled to nitrate reduction in marine sediments. *Applied and Environmental Microbiology*, *68*, 1312–1318.

Thamdrup, B., Glud, R. N., & Hansen, J. W. (1994). Manganese oxidation and in situ fluxes from coastal sediment. *Geochimica et Cosmochimica Acta*, *58*, 2563–2570.

Thamdrup, B., Rossello-Mora, R., & Amann, R. (2000). Microbial manganese and sulfate reduction in Black Sea shelf sediments. *Applied and Environmental Microbiology*, *66*, 2888–2897.

Thiel, G. A. (1925). Manganese precipitated by microorganisms. *Economic Geology*, *20*, 301–310.

Thorseth, H. I., Pedersen, R. B., & Christie, D. M. (2003). Microbial alteration of 0–30 Ma seafloor and subseafloor basaltic glasses from the Australian-Antarctic Discordance. *Earth and Planetary Science Letters*, *215*, 237–247.

Thorseth, H. I., Torsvik, T., Torsvik, V., Daae, F. L., Pedersen, R. B., & Party, K.-S. (2001). Diversity of life in ocean floor basalt. *Earth and Planetary Science Letters*, *194*, 31–37.

Toner, B., Fakra, S., Villalobos, M., Warwick, T., & Sposito, G. (2005). Spatially resolved characterization of biogenic manganese oxide production within a bacterial biofilm. *Applied and Environmental Microbiology*, *71*, 1300–1310.

Toner, B., Manceau, A., Webb, S. M., & Sposito, G. (2006). Zinc sorption to biogenic hexagonal-birnessite particles within a hydrated bacterial biofilm. *Geochimica et Cosmochimica Acta*, *70*, 27–43.

Tor, J. M., Kashefi, K., & Lovley, D. R. (2003). Potential importance of dissimilatory Fe(III)-reducing microorganisms in hot sedimentary environments. In W. S. Wilcock & E. F. D. Long (Eds.), *The subseafloor biosphere at mid-ocean ridges.* Washington, DC: American Geophysical Union.

Trimble, R. B., & Ehrlich, H. L. (1968). Bacteriology of manganese nodules. III. Reduction of $MnO_2$ by two strains of nodule bacterial. *Applied Microbiology, 16,* 695–702.

Trimble, R. B., & Ehrlich, H. L. (1970). Bacteriology of manganese nodules. IV. Induction of an MnO2-reductase system in *Bacillus. Applied Microbiology, 19,* 966–972.

Trouwborst, R. E., Clement, B. G., Tebo, B. M., Glazer, B. T., & Luther, G. W., III (2006). Soluble Mn(III) in suboxic zones. *Science, 313,* 1955–1957.

van Veen, W. L., Mulder, E. G., & Deinema, M. H. (1978). The *Sphaerotilus-Leptothrix* group of bacteria. *Microbiology Reviews, 42,* 329–356.

Van Waasbergen, L. G., Hildebrand, M., & Tebo, B. M. (1996). Identification and characterization of a gene cluster involved in manganese oxidation by spores of the marine *Bacillus* sp. SG-1. *Journal of Bacteriology, 178,* 3517–3530.

Van Waasbergen, L. G., Hoch, J. A., & Tebo, B. M. (1993). Genetic analysis of the marine manganese-oxidizing *Bacillus* sp. SG-1: Protoplast transformation, Tn917 mutagenesis, and identification of chromosomal loci involved in manganese oxidation. *Journal of Bacteriology, 175,* 7594–7603.

Vandenabeele, J., De Beer, D., Germonpré, R., Van de Sande, R., & Verstraete, W. (1995). Influence of nitrate on manganese removing microbial consortia from sand filters. *Water Research, 29,* 579–587.

Vandieken, V., Finke, N., & Thamdrup, B. (2014). Hydrogen, acetate, and lactate as electron donors for microbial manganese reduction in a manganese-rich coastal marine sediment. *FEMS Microbial Ecology, 87,* 733–745.

Vandieken, V., Pester, M., Finke, N., Hyun, J.-H., Friedrich, M. W., Loy, A., et al. (2012). Three manganese oxide-rich marine sediments harbor similar communities of acetate-oxidizing manganese-reducing bacteria. *ISME Journal, 6,* 2078–2090.

Vandieken, V., & Thamdrup, B. (2013). Identification of acetate-oxidizing bacteria in a coastal marine surface sediment by RNA-stable isotope probing in anoxic slurries and intact cores. *FEMS Microbial Ecology, 84,* 373–386.

Villalobos, M., Lanson, B., Manceau, A., Toner, B., & Sposito, G. (2006). Structural model for the biogenic Mn oxide produced by *Pseudomonas putida. American Mineralogist, 91,* 489–502.

Villalobos, M., Toner, B., Bargar, J., & Sposito, G. (2003). Characterization of the manganese oxide produced by *Pseudomonas putida* strain MnB1. *Geochimica et Cosmochimica Acta, 67,* 2649–2662.

Vojak, P. W. L., Edwards, C., & Jones, M. R. (1985a). Evidence for microbiological manganese oxidation in the River Tamar estuary, South West England. *Estuarine, Coastal and Shelf Science, 20,* 661–671.

Vojak, P. W. L., Edwards, C., & Jones, M. V. (1985b). A note on the enumeration of manganese-oxidizing bacteria in estuarine water and sediment samples. *Journal of Applied Microbiology, 59,* 375–379.

Wang, P. L., Chiu, Y. P., Cheng, T. W., Chang, Y. H., Tu, W. X., & Lin, L. H. (2014). Spatial variations of community structures and methane cycling across a transect of Lei-Gong-Hou mud volcanoes in eastern Taiwan. *Frontiers in Microbiology, 5,* 121.

Wariishi, H., Valli, K., & Gold, M. H. (1992). Manganese(II) oxidation by manganese peroxidase from the Basidiomycete *Phanerochaete chrysosporium. Journal of Biological Chemistry, 267,* 23688–23695.

Webb, S. M., Dick, G. J., Bargar, J. R., & Tebo, B. M. (2005). Evidence for the presence of Mn(III) intermediates in the bacterial oxidation of Mn(II). *Proceedings of the National Academy of Sciences of the United States of America, 102,* 5558–5563.

Wee, S., Burns, J. L., & DiChristina, T. J. (2014). Identification of a molecular signature unique to metal-reducing Gammaproteobacteria. *FEMS Microbiology Letters*, *350*, 90–99.
Wei, Z., Hillier, S., & Gadd, G. M. (2012). Biotransformation of manganese oxides by fungi: Solubilization and production of manganese oxalate minerals. *Environmental Microbiology*, *14*, 1744–1752.
Whittaker, J. W. (2002). Prokaryotic manganese superoxide dismutases. *Methods in Enzymology*, *349*, 80–90.
Wu, Y.-H., Liao, L., Wang, C.-S., Ma, W.-L., Meng, F.-X., Wu, M., et al. (2013). A comparison of microbial communities in deep-sea polymetallic nodules and the surrounding sediments in the Pacific Ocean. *Deep-Sea Research I*, *79*, 40–49.
Wuttig, K., Heller, M. I., & Croot, P. L. (2013). Pathways of superoxide ($O_2^-$) decay in the Eastern Tropical North Atlantic. *Environmental Science and Technology*, *47*, 10249–10256.
Yakushev, E., Pakhomova, S., Sørenson, K., & Skei, J. (2009). Importance of the different manganese species in the formation of water column redox zones: Observations and modeling. *Marine Chemistry*, *117*, 59–70.
Yang, W., Zhang, Z., Zhang, A., Chen, H., Liu, J., Ali, M., et al. (2013). Population structure of manganese-oxidizing bacteria in stratified soils and properties of manganese oxide aggregates under manganese-complex medium enrichment. *PLoS One*, *8*, e73778.
Yli-Hemminki, P., Jorgensen, K. S., & Lehtoranta, J. (2014). Iron-manganese concretions sustaining microbial life in the Baltic Sea: The structure of the bacterial community and enrichments in metal-oxidizing conditions. *Geomicrobiology Journal*, *31*, 263–275.
Yocum, C. F., & Pecoraro, V. L. (1999). Recent advances in the understanding of the biological chemistry of manganese. *Current Opinion in Chemical Biology*, *3*, 182–187.
Zamocky, M., Jakopitsch, C., Furtmuller, P. G., Dunand, C., & Obinger, C. (2008). The peroxidase-cyclooxygenase superfamily: Reconstructed evolution of critical enzymes of the innate immune system. *Proteins*, *72*, 589–605.

CHAPTER THREE

# Nutritional Immunity and Fungal Pathogenesis: The Struggle for Micronutrients at the Host–Pathogen Interface

**Dhara Malavia, Aaron Crawford, Duncan Wilson[1]**
Aberdeen Fungal Group, MRC Centre for Medical Mycology, School of Medicine, Medical Sciences and Nutrition, University of Aberdeen, Institute of Medical Sciences, Foresterhill, United Kingdom
[1]Corresponding author: e-mail address: duncan.wilson@abdn.ac.uk

## Contents



## Abstract

All living organisms require certain micronutrients such as iron, zinc, manganese and copper for cellular function and growth. For human pathogens however, the maintenance of metal ion homeostasis is particularly challenging. This is because the mammalian host actively enforces extremes of micronutrient availability on potential microbial invaders—processes collectively termed nutritional immunity. The role of iron sequestration in controlling microbial infections is well established and, more recently, the importance of other metals including zinc, manganese and copper has been recognised. In this chapter, we explore the nutritional immune mechanisms that defend the human body against fungal infections and the strategies that these important pathogens exploit to counteract nutritional immunity and thrive in the infected host.

*Advances in Microbial Physiology*, Volume 70
ISSN 0065-2911
http://dx.doi.org/10.1016/bs.ampbs.2017.01.006

## 1. AN INTRODUCTION TO HUMAN FUNGAL PATHOGENS

Pathogenic fungi have an enormous impact on human health. Most people are aware of some of the superficial infections caused by fungi. These include skin and nail infections such as athlete's foot and ringworm, predominantly caused by dermatophytes (*Trichophyton*, *Microsporum* and *Epidermophyton* species). In fact, these superficial mycoses represent one of the most prevalent forms of human infection, as they affect 20%–25% of the world's population (Havlickova, Czaika, & Friedrich, 2008). Mucosal infections are also extremely common; for example, vulvovaginal candidiasis (thrush) affects 75% of women of childbearing age and 5%–8% suffer from recurrent infections (Sobel, 2007). More seriously, certain fungal species can cause invasive diseases in humans. Collectively, these infections have been estimated to kill 1.5 million people per year (Brown et al., 2012). These life-threatening infections are caused primarily by species of the *Candida*, *Aspergillus*, *Cryptococcus* and *Pneumocystis* genera, which cause life-threatening infections in individuals with impaired immunity or other underlying conditions.

These infections can be broadly divided into environmentally (e.g. *Aspergillus* and *Cryptococcus*) or endogenously (e.g. *Candida*) acquired. For example, *Aspergillus fumigatus* grows as a saprophyte in environmental niches such as compost and soil and patients become infected via inhalation of spores. In contrast, *Candida albicans* is a member of the human microbiota (or, in this case, mycobiota) and invasive infections typically originate from colonising cells of the patient's own gastrointestinal tract. Therefore, these different human fungal pathogens have evolved pathogenic potential independently, either in environmental niches or as colonisers of mucosal surfaces. As discussed later, this has led to a fascinating array of micronutrient acquisition strategies employed by human fungal pathogens.

## 2. THE INFECTED HOST AS A NUTRITIONALLY RESTRICTIVE ENVIRONMENT

All living organisms must acquire certain metals, such as iron, zinc, manganese and copper and these essential trace nutrients are also acutely toxic at higher concentrations. Therefore, in order to proliferate, microbes must assimilate these micronutrients from their environment. Mammalian immunity has evolved highly sophisticated mechanisms to manipulate microbial access to key micronutrients: processes collectively termed "nutritional immunity". The study of nutritional immunity has historically focussed on restriction of microbial access to iron; however, more recently, an important role for other metals has emerged, not only in terms of sequestration/limitation but also in host-directed overload and toxicity against the microbe. Therefore, pathogens of humans must have evolved mechanisms to deal with extremes in micronutrient availability. In the following sections, we review the current state of knowledge regarding iron and zinc nutritional immunity and the mechanisms human pathogenic fungi employ to counteract these limitations. It should be noted that whilst manganese sequestration is now recognised as playing a critical role in controlling bacterial infections (Kelliher & Kehl-Fie, 2016), at this time the impact of manganese homeostasis on fungal infections remains poorly understood.

## 3. IRON NUTRITIONAL IMMUNITY AND FUNGAL ASSIMILATION

Although iron is the most abundant transition metal in the human body, its availability to potential pathogens is highly limited due to intracellular storage and the expression of high-affinity iron-binding proteins. The majority of body iron is sequestered intracellularly within haemoglobin (~67%) and ferritin (~25%), and the remaining extracellular iron is tightly coupled to high-affinity binding proteins such as transferrin, with "free" iron present at $10^{-24}$ M, equating to virtually no free iron in circulation (Raymond, Dertz, & Kim, 2003; Winter, Bazydlo, & Harris, 2014).

Upon inflammation, the levels of extracellular iron are further limited, predominantly via the action of hepcidin—a circulating peptide hormone and master regulator of iron homeostasis. Hepcidin functions via binding to the cellular iron efflux transporter ferroportin, and inducing its degradation (Drakesmith & Prentice, 2012).

Armitage et al. (2011) found that in vitro stimulation of human peripheral blood mononuclear cells with heat-inactivated *C. albicans* hyphae stimulated robust hepcidin production. Moreover, mice systemically infected with *C. albicans* exhibited increased hepatic transcription of hepcidin mRNA (Armitage et al., 2011).

Leal, Roy, Vareechon, et al. (2013) used corneal infections with *A. fumigatus* to study the nutritional immune response against fungal infections. They found that infection with this filamentous fungal pathogen resulted in Dectin-1-dependent induction of IL-6. This proinflammatory cytokine is known to positively regulate hepcidin (Hentze, Muckenthaler, Galy, & Camaschella, 2010), and the authors went on to find robust upregulation of hepatic hepcidin in wild type, but not IL-6-deficient mice, providing further evidence of an operational hepcidin axis during fungal infections. In terms of direct antimicrobial activity, hepcidin 20 has been shown to be active against the yeast *Candida glabrata* (Del Gaudio et al., 2013; Tavanti et al., 2011). However, a small clinical study by Chen and Zhong suggested a negative correlation between patient hepcidin levels and susceptibility to invasive fungal infection following transplant (Chen & Zhong, 2013).

A detailed portrait of iron nutritional immunity during systemic fungal infection has recently been reported by Potrykus et al. (2013). These authors used an elegant combination of laser ablation-inductively coupled plasma-mass spectrometry, MALDI imaging, immunohistochemistry and microtranscriptomics to study the effect of systemic *C. albicans* infection on both local and global iron homeostasis in a mouse model of haematogeneously disseminated candidiasis. In this type of experimental infection, the kidney is the primary target of infecting cells (MacCallum & Odds, 2005). The authors found that whilst iron accumulated in the renal medulla, zones of iron exclusion were observed around fungal lesions in the renal cortex. This mobilisation of iron away from the infecting fungal cells appeared to be a coordinated, host-driven response, because increased levels of ferritin and haemoglobin were observed in the medulla, whilst haem oxygenase accumulated around the fungal lesions in the cortex.

Iron availability has been found to be directly linked to fungal disease severity in several studies. For example, Leal et al. (2013) found that administration of exogenous iron or the siderophore deferoxamine (which can be utilised as an iron source by the fungus) enhanced *A. fumigatus* fungal burden and severity of infection in their corneal model of aspergillosis. In contrast, treatment with iron chelators inhibited fungal growth in vivo. Administration of exogenous iron or deferoxamine also worsens the severity of infections caused by *Rhizopus* (Ibrahim, Edwards, Fu, & Spellberg, 2006; Reed, Ibrahim, Edwards, Walot, & Spellberg, 2006).

Therefore, mounting evidence shows that iron nutritional immunity does indeed occur during fungal infections. As many fungal pathogens frequently cause invasive infections in patients which are not iron-overloaded, how do these species acquire iron for growth under the restrictive conditions of nutritional immunity? Interestingly, although certain iron acquisition systems in fungi are conserved, different fungal pathogens appear to rely on different iron assimilation strategies during infection.

Possibly one of the most fascinating cases is that of *C. albicans*—unlike most human fungal pathogens, which normally exist in environmental niches like the soil, this fungus has evolved as a coloniser of human mucosal surfaces and appears to have developed quite specific iron assimilation strategies. It is important to highlight that *C. albicans* is a polymorphic fungus, which can grow in various cellular morphologies. From a pathogenicity perspective, the most important morphological transition is that from unicellular yeast to multicellular hyphae. Hyphae are filamentous structures which play important roles in the pathogenic lifestyle of *C. albicans*. It is hyphal cells which invade host tissues and the regulation of many *C. albicans* virulence factors is coupled to hypha formation. And as we will see in the following sections, this is true for many *C. albicans* host-specific micronutrient assimilation activities.

## 3.1 A Multistage Haemoglobin Iron Assimilation Pathway in *C. albicans*

As the majority of body iron is sequestered within haemoglobin, this represents an attractive nutrient source for pathogens and *C. albicans* possesses an elegant assimilation pathway for the utilisation of haemoglobin-derived iron, summarised in Fig. 1. *C. albicans* hyphae are able to bind to human erythrocytes in an opsonisation-dependent manner (Moors, Stull, Blank, Buckley, & Mosser, 1992), although the fungal cell surface receptor is as yet unknown. Following haemolysis (Manns, Mosser, & Buckley, 1994), released haemoglobin/haem can be bound at the fungal cell surface. This association is mediated by a multimember family of haem-binding proteins. The archetypal member of this family, Rbt5 was the first member to be shown to possess haemoglobin-binding activity and the gene itself was originally named because it is repressed by Tup1. Tup1 is a major transcriptional repressor, which negatively regulates hyphal development in *C. albicans*, and is responsible for the repression of many hypha-coexpressed genes (Braun, Head, Wang, & Johnson, 2000). This association between the hyphal morphology and expression of host-specific iron acquisition systems appears to be a common theme in *C. albicans* pathobiology.

The molecular basis of the *C. albicans* haemoglobin assimilation pathway was originally identified by Weissman and Kornitzer (2004). These authors transformed a *C. albicans* genomic DNA library into a *Saccharomyces cerevisiae*

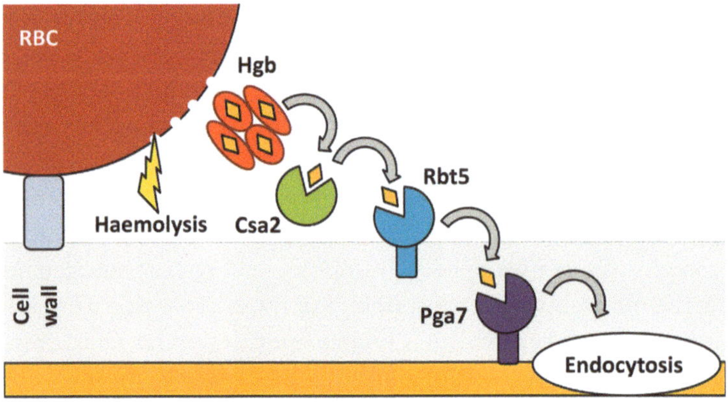


**Fig. 1** Multistage haem assimilation pathway in *Candida albicans*. *C. albicans* hyphae bind erythrocytes by an opsonisation-dependent mechanism and lyse the red blood cell via a hypha-specific secreted factor. Haem (*orange diamonds*) can be extracted from released haemoglobin by fungal haem-binding proteins. The haemophore Csa2 (*green*) is secreted to the culture supernatant; the cell surface-associated proteins, Rbt5 (*blue*) and Pga7 (*purple*) form a haem relay network, delivering haem to the plasma membrane where it is internalised via endocytosis.

strain defective in reductive iron assimilation (*ccc2*Δ) and looked for genes which permit growth of this yeast with haemoglobin as the iron source. They found that the *RBT5* orthologue, *RBT51/PGA10*, permitted *S. cerevisiae ccc2*Δ utilisation of haemoglobin as an iron source and that deletion of *RBT5* in *C. albicans* inhibited haemoglobin utilisation by this pathogenic species. Both Rbt5 and Rbt51/Pga10, together with Csa1, Csa2 and Pga7 belong to a family of cell surface and secreted proteins characterised by the presence of the eight-cysteine CFEM domain (common in several fungal extracellular membrane proteins).

Whilst Rbt5 appeared to be the dominant cell surface haem-binding protein in *C. albicans*, other members of the family have also been implicated in haem-iron assimilation. Csa2, which does not contain a predicted glycosylphosphatidylinositol anchor, has been shown to be secreted to the culture supernatant by *C. albicans* hyphae (Sorgo et al., 2010) and was proposed to act as a soluble haem-binding molecule and pass haem on to other proteins at the cell surface (Okamoto-Shibayama, Kikuchi, Kokubu, Sato, & Ishihara, 2014). A recent study by Nasser et al. has confirmed that *C. albicans* Csa2 does indeed function as a secreted "haemophore" and can extract haem from haemoglobin (Nasser et al., 2016).

At the fungal cell surface, Kuznets et al. have proposed an elegant model of haem passage from the outer reaches of the fungal cell wall to the plasma membrane. The authors reported that whilst both Rbt5 and Pga7 can efficiently extract haem from haemoglobin, the two proteins exhibit different affinities for haem, and reside in discreet regions in the fungal cell wall. They found that Rbt5 primarily existed as a GPI-anchored protein of the outer cell wall, whereas Pga7 resided in the inner cell wall. Whilst both proteins effectively bound haem, Pga17 exhibited higher affinity. Finally, they demonstrated transfer of haem between Rbt5 and Pga7. Together these data suggest a model whereby Rbt5 extracts haem from haemoglobin (or receives haem from Csa2; Nasser et al., 2016) at the outer envelope of the fungal cell wall and this is then passed through the cell wall via an Rbt5-Pga7 relay system towards the plasma membrane. At the plasma membrane, internalisation appears to be via an endocytic mechanism, involving the ESCRT complex, vacuolar acidification (Weissman, Shemer, Conibear, & Kornitzer, 2008) and the haem oxygenase Hmx1 (Santos et al., 2003).

Haem iron utilisation by *C. albicans* appears to play an important role in pathogenicity. Potrykus et al. performed transcriptome analysis of *C. albicans* kidney lesions isolated by laser capture microscopy and found significant upregulation of *CSA1*, *CSA2*, *PGA10* and *HMX1* during the late phases of kidney infection (Potrykus et al., 2013). Kuznets et al. demonstrated that deletion *PGA7* attenuated virulence in a disseminated model of candidiasis, indicating that *C. albicans* does indeed take advantage of haem as an iron source during experimental infection (Kuznets et al., 2014). Importantly, this assimilation pathway likely has direct clinical consequence: Mochon et al. performed serological profiling of candidaemia patients and identified both Rbt5 and Csa1, as well as Flc1 (which is also implicated in haem utilisation; Protchenko, Rodriguez-Suarez, Androphy, Bussey, & Philpott, 2006) and Cfl91 (a putative ferric reductase) as antigen biomarkers of the convalescent stage of candidaemia (Mochon et al., 2010).

Therefore, *C. albicans* hyphae are able to bind to erythrocytes, elicit haemolysis, express a series of haemoglobin/haem-binding proteins for haem extraction from haemoglobin and passage through the fungal cell wall for assimilation. This represents an elaborate host-iron assimilation pathway which is expressed not only during experimental animal infections in the laboratory but also in candidaemia patients in the clinic.

*C. albicans* can also utilise other host-iron sources during infection. A study by Knight et al. suggested that *C. albicans* can utilise iron-loaded transferrin, and that this was dependent on physical cellular contact and the reductive pathway (Knight, Vilaire, Lesuisse, & Dancis, 2005). However, apotransferrin has antifungal activity via iron sequestration, and a role for transferrin iron utilisation by *C. albicans* during infection is currently unclear.

## 3.2 Ferritin Iron Utilisation by *C. albicans*

The second most abundant iron pool in the human body is ferritin. This large multisubunit protein nanocage can accommodate up to 4500 ferric iron ions and is a major intracellular iron store (Torti & Torti, 2002). Almeida and coworkers found that the cellular ferritin content of human epithelial cells correlated with the capacity of *C. albicans* hyphae to damage these cells, suggesting that the fungus may be able to utilise host ferritin as an iron source (Almeida et al., 2008). They went on to show that *C. albicans* could grow with ferritin as a sole iron source in vitro. However, when the neutral alkaline culture medium used for the ferritin utilisation assays was strongly buffered to pH 7.4, *C. albicans* failed to grow, suggesting that fungal-driven acidification is required for assimilation.

Interestingly, hyphal, but not yeast cells of *C. albicans* bound to ferritin, and regulatory mutants with defects in hypha formation also could not bind ferritin. This suggested that a hypha-coexpressed cell surface factor may facilitate the observed interaction with ferritin. They therefore carried out transcriptional profiling under ferritin-binding culture conditions and looked for genes upregulated in ferritin-binding competent strains, but not in the hypha-defective regulatory mutants. From this analysis, they identified *ALS3*, which is predominantly expressed by hyphae and encodes a cell wall protein. Deletion of *ALS3* prevented ferritin binding by *C. albicans* hyphae in vitro and from human epithelial cells in a cell culture infection model. Notably, heterologous expression of *C. albicans* Als3 in *S. cerevisiae* permitted ferritin binding by this normally nonpathogenic model yeast, suggesting that Als3 can directly function as a ferritin receptor without the action of auxiliary *C. albicans* proteins.

The Als3 protein had previously been shown to facilitate *C. albicans* adhesion and invasion of human epithelial cells (Hoyer, 2001; Phan et al., 2007). It is of interest to note that *ALS3* expression is hardwired to the hyphal morphology (Martin et al., 2013), and expression can be independent of environmental iron status. As hyphae are the invasive form of *C. albicans*, the hyphal coexpression of a ferritin-binding factor may represent an example of "predictive adaptation" by this pathogen. This concept describes the capacity of a microorganism to learn certain sequences of events that occur in their natural habitat (Tagkopoulos, Liu, & Tavazoie, 2008). For example, as *Escherichia coli* passes through the mammalian digestive tract it is first exposed to lactose, and then to maltose. The bacterium responds to lactose by switching on genes involved in both lactose and maltose metabolism, improving its subsequent fitness when it encounters the second carbon source: it has "learned" that exposure to lactose precedes maltose availability and prepares appropriately (Mitchell et al., 2009). In the hostile environment of the infected host, preexpression of the ferritin receptor before host cell invasion occurs likely endows *C. albicans* with a similar fitness advantage (Brunke & Hube, 2014).

## 3.3 Conserved and Contrasting Behaviour in Environmentally Acquired Pathogens

Beyond *Candida*, most other invasive human fungal pathogens are environmentally acquired and are not generally thought to have evolved in close association with a mammalian host. Yet these environmental species also cause devastating infections, so must have evolved to resist immunity and thrive in the host. Some species, such as *Coccidioides* are thought to associate with rodents, and may therefore experience the selective pressure of mammalian immunity. For others, the concept of the environmental "virulence school" has emerged. This posits that certain environmental niches may select for attributes which translate to pathogenic potential within a susceptible human host. For example, Casadevall has highlighted the similarities between predatory amoebae in the environment and mammalian phagocytic immune cells, suggesting that the capacity of an environmentally acquired pathogen to survive phagocytosis by a macrophage may have been originally selected within an environmental context. In support of this, Steenbergen and coworkers demonstrated that passage of an avirulent strain of *Histoplasma capsulatum* with the soil amoeba *Acanthameobae castellanii* resulted in increased fungal persistence in a subsequent mouse lung infection model (Steenbergen, Nosanchuk, Malliaris, & Casadevall, 2004). The concept of the environmental virulence school likely extends to nutritional immunity and the struggle for micronutrients by environmentally acquired pathogens. On a fundamental level, simple adaptation to niches of poor trace metal availability would select for higher affinity assimilation mechanisms. Amoebae may also enforce a primordial form of nutritional immunity. The model amoeba *Dictyostelium discoideum* encodes a homologue of the human natural resistance-associated membrane protein 1 (Nramp1), which, in macrophages pumps iron out of the phagosome. Moreover, deletion or overexpression of Nramp1 in *D. discoideum*, respectively, promoted and inhibited intracellular growth of phagocytosed bacteria (Peracino et al., 2006). This may well represent a form of nutritional immunity operating at the single cell level, as the amoebic Nramp1 does indeed facilitate phagosomal iron efflux (Buracco et al., 2015).

Although the selective pressures which shaped them are currently largely speculative, environmentally acquired fungal pathogens express an impressive array of micronutrient assimilation strategies. Like *C. albicans*, *Paracoccidioides* can also assimilate haemoglobin iron and was found to express a cell surface haemoglobin receptor with similarities to *C. albicans* Rbt5 (Bailao, Parente, Pigosso, et al., 2014). *Cryptococcus neoformans* is much more distantly related to *C. albicans*, as this yeast belongs to the Basidiomycete phylum, which diverged from the Ascomycota at least 452 million years ago (Taylor & Berbee, 2006). Nevertheless, this pathogenic yeast has also evolved to utilise haem iron. *A. fumigatus*, on the other hand, does not appear to exploit host-specific iron sources like haemoglobin and relies heavily on siderophore-mediated iron scavenging during infection.

## 3.4 Siderophore-Mediated Iron Assimilation by *A. fumigatus*

Siderophores are high-affinity small molecule chelators which are produced by numerous microorganisms to sequester iron. Although produced by many fungal species, the major pathogenic yeasts *C. albicans* and *C. neoformans* do not synthesise siderophores, but can utilise xenosiderophores from other species. In *C. albicans*, assimilation of siderophores is mediated by the transporter Sit1 (also known as Arn1). Deletion of *SIT1* did not affect virulence in systemic models of disseminated candidiasis, but did reduce *C. albicans* capacity to invade keratinocytes in tissue culture (Heymann et al., 2002; Hu, Bai, Zheng, Wang, & Wang, 2002). In another medically important *Candida* species, *C. glabrata*, preloading Sit1-expressing cells with ferrichrome enhanced fungal survival following macrophage phagocytosis. Xenosiderophores may be relevant to *C. albicans* and *C. glabrata* pathogenicity as these yeasts are members of the human mucosal microbiota. It is possible that *Candida* species may assimilate xenosiderophore iron from other members of the microbiota and utilise these intracellular reserves following subsequent translocation to normally sterile organs. Alternatively, polymicrobial infections are of clinical significance (Allison et al., 2016) and *Candida* may directly utilise siderophores produced by other species during these coinfections. However, in experimental models of candidiasis, deletion of *SIT1* does not attenuate *C. albicans* virulence (Heymann et al., 2002; Hu et al., 2002). Similarly, in the distantly related basidiomycete yeast *C. neoformans* deletion of the *SIT1* siderophore transporter gene does not attenuate virulence in an experimental murine infection model of cryptococcosis (Tangen, Jung, Sham, Lian, & Kronstad, 2007).

In striking contrast, in the pathogenic mould, *A. fumigatus* siderophore iron assimilation is critical for virulence. This fungus synthesises both extracellular and intracellular siderophores for iron scavenging and storage, respectively. Schrettl and coworkers found that disruption of siderophore biosynthesis via deletion of the *sidA* gene, resulted in *A. fumigatus* avirulence in a mouse model of aspergillosis (Schrettl et al., 2004). *sidA* encodes an L-ornithine-$N^5$-monooxygenase, which is responsible for the first committed step in hydroxamate siderophore biosynthesis (Fig. 2), and thus its deletion blocks production of all hydroxamate siderophores in *A. fumigatus*. Interestingly, the same study found no role for Ftr1 (reductive iron assimilation) in *A. fumigatus* virulence, in contrast to *C. albicans* where an *ftr1*Δ deletion mutant is avirulent (Ramanan & Wang, 2000). A subsequent study by the same group further delineated the biosynthetic pathways of siderophore production. The current model for *Aspergillus* siderophore synthesis is shown in Fig. 2: fusarinine C is produced by the acyl transferase SidF and nonribosomal peptide synthetase SidD; and fusarinine C is subsequently converted to triacetylfusarinine C via the $N^2$-transacetylase SidG. Both fusarinine and triacetylfusarinine C are secreted to the culture supernatant

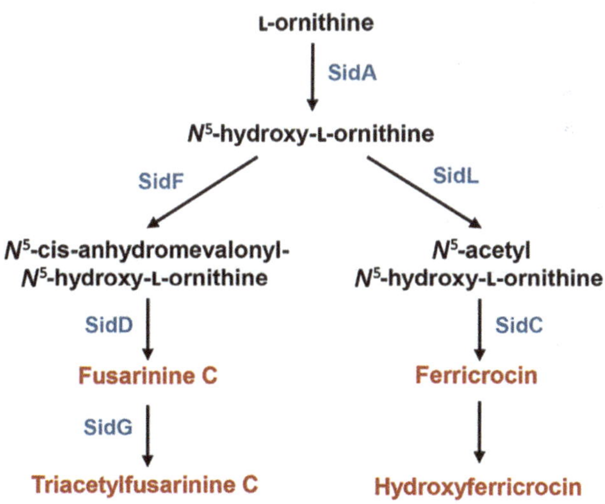


**Fig. 2** Siderophore biosynthesis in *Aspergillus fumigatus*. The proposed model for siderophore biosynthesis from Schrettl et al. (2007) and Blatzer et al. (2011).

by *A. fumigatus*. *A. fumigatus* also produces two intracellular siderophores, hydroxyferricrocin and its precursor, ferricrocin, the synthesis of which relies on the nonribosomal peptide synthase, SidC and transacetylase, SidL (Blatzer et al., 2011; Schrettl et al., 2007).

Interestingly, deletion of *sidF*, *sid*D or *sidC*, but not *sidG*, attenuated *A. fumigatus* virulence, suggesting that both extracellular and intracellular siderophores are important for *A. fumigatus* iron homeostasis and growth in vivo and that fusarinine C is sufficient for extracellular iron scavenging during infection (Schrettl et al., 2007).

Therefore, *A. fumigatus* appears to rely profoundly on siderophore-dependent, but not reductive iron homeostasis for infection and lacks the host-specific iron assimilation pathways (e.g. haem), present in other pathogens such as *C. albicans*.

## 4. AN EMERGING ROLE FOR ZINC ASSIMILATION IN FUNGAL PATHOGENESIS

Whilst the host–pathogen battle for iron remains the best-studied area of nutritional immunity and pathogenicity, the importance of other metals has begun to be recognised. Zinc is the second most abundant transition metal in the human body and, like iron, its availability to potential invaders is strictly limited. Zincaemia, along with ferraemia and fever, is a hallmark of the acute phase inflammatory response. One of the major mechanisms driving removal of zinc from the plasma is via sequestration in the liver, which occurs via induction of the cellular zinc importer Zip14 by the proinflammatory cytokine IL-6 (Liuzzi et al., 2005).

At a more local level, extracellular zinc-binding molecules are produced to sequester this metal, making it unavailable to extracellular pathogens. The dominant immune zinc chelator is calprotectin. This heterodimer is highly associated with inflammation as it represents around half the cytoplasmic protein content of neutrophils (Edgeworth, Gorman, Bennett, Freemont, & Hogg, 1991) and has potent antifungal activity (Urban et al., 2009). Indeed, calprotectin represents one of the key antifungal components of neutrophil extracellular traps (NETs). These web-like structures are released from neutrophils during a novel form of programmed cell death called NETosis (Fuchs et al., 2007). NETs consist of chromatin decorated with a number of antimicrobial factors and Urban and coworkers found that calprotectin-mediated zinc sequestration represented the major antifungal activity of NETs. Calprotectin is a heterodimer made up of S100A8 and S100A9 subunits. Both of S100A8 and S100A9 belong to the S100 family of EF-hand calcium-binding proteins (Zackular, Chazin, & Skaar, 2015). Many S100 proteins have regulatory function, but, in addition to calprotectin, S100A7 (psoriasin) and S100A12 (calgranulin C) have also been found to chelate zinc (Brodersen, Nyborg, & Kjeldgaard, 1999; Moroz, Blagova, Wilkinson, Wilson, & Bronstein, 2009).

Psoriasin also has a demonstrated role in nutritional immunity. Originally shown to rapidly kill *E. coli* (Glaser et al., 2005), psoriasin has recently been shown to possess potent antifungal activity via zinc chelation. Hein and coworkers demonstrated that the reduced form of psoriasin strongly inhibited the growth of a range of pathogenic fungi including *A. fumigatus*, *Malassezia furfur*, *Microsporum canis*, *Rhizopus oryzae*, *Trichophyton mentagrophytes*, as well as the model yeast *S. cerevisiae*, but, interestingly, not *C. albicans*. The mode of action appears to be via intracellular zinc chelation and induction of apoptosis (Hein, Takahashi, Tsumori, et al., 2015).

Some fungal pathogens including *C. neoformans*, *C. glabrata* and *H. capsulatum* can persist, and even replicate intracellularly within macrophages (Seider, Heyken, Luttich, Miramon, & Hube, 2010) and, in this context, zinc nutritional immunity operates at the single cell level. Following engulfment of *H. capsulatum*, activated macrophages attempt to starve this yeast by shuttling zinc away from the phagosome. This is a highly regulated process, involving the STAT3 and STAT5 transcription factors and intramacrophage zinc is made unavailable via multiple mechanisms including the induction of metal-binding metallothioneins and sequestration in the Golgi apparatus.

Yet despite systemic, local and intracellular zinc nutritional immunity, human fungal pathogens grow within their infected hosts and must therefore possess effective zinc homeostatic mechanisms. Fungal zinc homeostasis has been studied extensively in the model yeast *S. cerevisiae* and this has proved a useful framework for the identification of zinc transporters in human fungal pathogens; however, as discussed later, care should be taken in the interpretation of phylogenetic relationships and function, especially in the context of pathogenicity.

*S. cerevisiae* encodes two plasma membrane zinc transporters of the ZIP protein family: the high-affinity Zrt1 and low-affinity Zrt2, and Fig. 3 shows the phylogenetic relationship amongst zinc importer orthologues in this yeast and the characterised human fungal pathogens.

Most fungal species encode an orthologue of the *S. cerevisiae* low-affinity transporter Zrt2 (Fig. 3, cluster 1; Wilson, 2015) and the sequence similarity

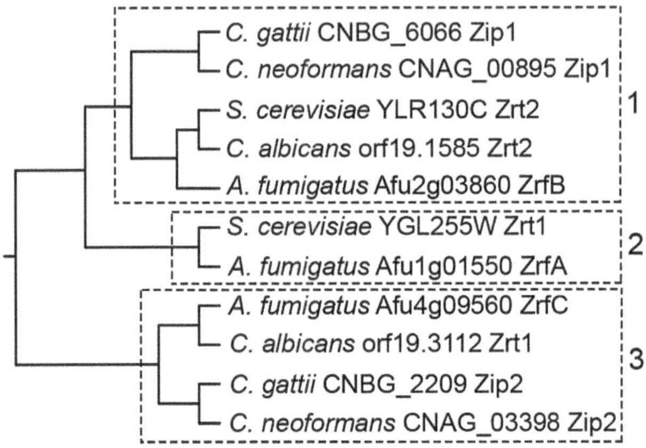


**Fig. 3** Phylogenetic relationship amongst zinc importer in key human fungal pathogens and *Saccharomyces cerevisiae*. Predicted amino acid sequences were down loaded from FungiDB (http://fungidb.org/fungidb/) and alignment and tree drawing performed using Multiple Sequence Alignment by CLUSTALW (http://www.genome.jp/tools/clustalw/).

amongst these orthologues reflects these species' relationship: the yeasts *S. cerevisiae* and *C. albicans* and the mould *A. fumigatus* are all members of ascomycete phylum, whilst both *Cryptococcal* species are basidiomycetes. These are the two major phyla of the fungal kingdom and diverged at least 452 million years ago (Taylor & Berbee, 2006).

In contrast, of the species presented in Fig. 3, only *A. fumigatus* has an orthologue (ZrfA) of *S. cerevisiae* Zrt1 (cluster 2). Finally, all species, with the exception of *S. cerevisiae* encode related zinc transporters with similarity to *A. fumigatus* ZrfC (cluster 3).

In the pathogenic mould *A. fumigatus*, *zrfA* and *zrfB* (clusters 1 and 2, respectively) are transcriptionally upregulated and functionally required under zinc limitation and acidic pH; in contrast, *zrfC* is induced and essential for growth under zinc limitation in neutral alkaline environments (Amich, Vicentefranqueira, Leal, & Calera, 2010; Vicentefranqueira, Moreno, Leal, & Calera, 2005). In *C. albicans* the *zrfC* orthologue *ZRT1* (cluster 3) and *zrfB* orthologue *ZRT2* (cluster 1) are also differentially upregulated in response to alkaline- and acidic pH, respectively (Bensen, Martin, Li, Berman, & Davis, 2004). Moreover, deletion of *ZRT1* inhibits *C. albicans* growth in neutral alkaline zinc limited medium (Citiulo et al., 2012). It is important to note that despite a closer evolutionary relationship with *C. albicans*, pH regulation of *ZRT1* or *ZRT2* has not been observed in *S. cerevisiae*. In both *Cryptococcal* species, deletion of *ZIP1* (cluster 1) but not *ZIP2* (cluster 3) inhibited fungal growth under zinc limitation.

In relevant fungal infection models, the importance of zinc transporters also appears to be very different. In *A. fumigatus* deletion of *zrfC* significantly attenuated virulence, whilst a *zrfA*Δ/*zrfB*Δ/*zrfC*Δ triple mutant was avirulent, and genetic complementation of the triple mutant with *zrfC* restored wild-type virulence (Amich et al., 2014), indicating that the neutral alkaline-specific ZrfC (cluster 3) is the dominant zinc importer operating in *A. fumigatus* pathogenicity. In contrast deletion of *ZRT2* (cluster 1), but not *ZRT1*, inhibited *C. albicans* growth in vivo (unpublished; Noble, French, Kohn, Chen, & Johnson, 2010).

In *C. neoformans*, deletion of *ZIP1* (cluster 1), but not *ZIP2* attenuated virulence, and a *zip1*Δ/*zip2*Δ double deletion mutant was only moderately less virulent than *zip1*Δ. In the related basidiomycete *C. gattii*, deletion of neither *ZIP1* nor *ZIP2* impacted virulence, but a *zip1*Δ/*zip2*Δ double mutant attenuated virulence (Do, Hu, Caza, Kronstad, & Jung, 2016; Schneider Rde et al., 2015). Therefore, analogous to iron, different fungal pathogens rely on different zinc importers for virulence in vivo.

Finally, in addition to transporter-dependent uptake, a "zincophore" has been described in *C. albicans*. This fungus produces a protein called Pra1 which can be secreted, can bind zinc, can reassociate with the fungal cell and is important for zinc scavenging from host tissue. Intriguingly, *PRA1* was originally described as a pH-regulated antigen due to its strong transcriptional induction in response to neutral alkaline pH (Sentandreu, Elorza, Sentandreu, & Fonzi, 1998) and actually shares its promoter with *ZRT1*, which encodes the neutral alkaline zinc transporter (Fig. 3, cluster 3; Citiulo et al., 2012). In support of a conserved function within the fungal kingdom, the *ZRT1* and *PRA1* orthologues in *A. fumigatus* (*zrfC* and *aspF2*, respectively) are also syntenic and are both required for growth under zinc limitation at neutral alkaline pH (Amich et al., 2010). Further phylogenetic analysis suggested that this syntenic relationship occurred before divergence of the two major fungal clades, the Ascomycota and the Basidiomycota, and has been maintained in numerous contemporary species, suggesting positive evolutionary selection. Indeed, in *C. albicans* Pra1 and Zrt1 are not only coexpressed, but also appear to physically interact, as deletion of *ZRT1* prevented the cellular reassociation of Pra1 with *C. albicans*.

In summary, fungal pathogens of humans, like their bacterial counterparts, face extreme iron and zinc limitation due to the action of host nutritional immunity. Yet these species thrive in their hosts causing often devastating diseases. The capacity of certain species to utilise multiple micronutrient sources during infection undoubtedly contributes to their pathogenicity, and this flexibility makes understanding micronutrient assimilation and homeostasis an engaging and highly active field of research. In the future, strategies which target or manipulate pathogen micronutrient access may pave the way for novel forms of therapy.

## REFERENCES


Allison, D. L., Willems, H. M., Jayatilake, J. A., Bruno, V. M., Peters, B. M., & Shirtliff, M. E. (2016). *Candida*-bacteria interactions: Their impact on human disease. *Microbiology Spectrum*, *4*, 1–26.

Almeida, R. S., Brunke, S., Albrecht, A., Thewes, S., Laue, M., Edwards, J. E., et al. (2008). The hyphal-associated adhesin and invasin Als3 of *Candida albicans* mediates iron acquisition from host ferritin. *PLoS Pathogens*, *4*, e1000217.

Amich, J., Vicentefranqueira, R., Leal, F., & Calera, J. A. (2010). *Aspergillus fumigatus* survival in alkaline and extreme zinc-limiting environments relies on the induction of a zinc homeostasis system encoded by the *zrfC* and *aspf2* genes. *Eukaryotic Cell*, *9*, 424–437.

Amich, J., Vicentefranqueira, R., Mellado, E., Ruiz-Carmuega, A., Leal, F., & Calera, J. A. (2014). The ZrfC alkaline zinc transporter is required for *Aspergillus fumigatus* virulence and its growth in the presence of the Zn/Mn-chelating protein calprotectin. *Cellular Microbiology*, *16*, 548–564.

Armitage, A. E., Eddowes, L. A., Gileadi, U., Cole, S., Spottiswoode, N., Selvakumar, T. A., et al. (2011). Hepcidin regulation by innate immune and infectious stimuli. *Blood*, *118*, 4129–4139.

Bailao, E. F., Parente, J. A., Pigosso, L. L., et al. (2014). Hemoglobin uptake by *Paracoccidioides* spp. is receptor-mediated. *PLoS Neglected Tropical Diseases*, *8*, e2856.

Bensen, E. S., Martin, S. J., Li, M., Berman, J., & Davis, D. A. (2004). Transcriptional profiling in *Candida albicans* reveals new adaptive responses to extracellular pH and functions for Rim101p. *Molecular Microbiology*, *54*, 1335–1351.

Blatzer, M., Schrettl, M., Sarg, B., Lindner, H. H., Pfaller, K., & Haas, H. (2011). SidL, an *Aspergillus fumigatus* transacetylase involved in biosynthesis of the siderophores ferricrocin and hydroxyferricrocin. *Applied and Environmental Microbiology*, 77, 4959–4966.

Braun, B. R., Head, W. S., Wang, M. X., & Johnson, A. D. (2000). Identification and characterization of *TUP1*-regulated genes in *Candida albicans*. *Genetics*, *156*, 31–44.

Brodersen, D. E., Nyborg, J., & Kjeldgaard, M. (1999). Zinc-binding site of an S100 protein revealed. Two crystal structures of Ca2+-bound human psoriasin (S100A7) in the Zn2+-loaded and Zn2+-free states. *Biochemistry*, *38*, 1695–1704.

Brown, G. D., Denning, D. W., Gow, N. A., Levitz, S. M., Netea, M. G., & White, T. C. (2012). Hidden killers: Human fungal infections. *Science Translational Medicine*, *4*, 165rv113.

Brunke, S., & Hube, B. (2014). Adaptive prediction as a strategy in microbial infections. *PLoS Pathogens*, *10*, e1004356.

Buracco, S., Peracino, B., Cinquetti, R., Signoretto, E., Vollero, A., Imperiali, F., et al. (2015). Dictyostelium Nramp1, which is structurally and functionally similar to mammalian DMT1 transporter, mediates phagosomal iron efflux. *Journal of Cell Science*, *128*, 3304–3316.

Chen, J., & Zhong, L. (2013). Clinical significance of serum hepcidin-25 levels in predicting invasive fungal disease in patients after transplantation. *European Review for Medical and Pharmacological Sciences*, *17*, 1769–1773.

Citiulo, F., Jacobsen, I. D., Miramon, P., Schild, L., Brunke, S., Zipfel, P., et al. (2012). *Candida albicans* scavenges host zinc via Pra1 during endothelial invasion. *PLoS Pathogens*, *8*, e1002777.

Del Gaudio, G., Lombardi, L., Maisetta, G., Esin, S., Batoni, G., Sanguinetti, M., et al. (2013). Antifungal activity of the non cytotoxic human peptide hepcidin 20 against fluconazole resistant *Candida glabrata* in human vaginal fluid. *Antimicrobial Agents and Chemotherapy, 57*, 4314–4321.

Do, E., Hu, G., Caza, M., Kronstad, J. W., & Jung, W. H. (2016). The ZIP family zinc transporters support the virulence of *Cryptococcus neoformans*. *Medical Mycology: Official Publication of the International Society for Human and Animal Mycology, 54*, 605–615.

Drakesmith, H., & Prentice, A. M. (2012). Hepcidin and the iron-infection axis. *Science, 338*, 768–772.

Edgeworth, J., Gorman, M., Bennett, R., Freemont, P., & Hogg, N. (1991). Identification of p8,14 as a highly abundant heterodimeric calcium binding protein complex of myeloid cells. *The Journal of Biological Chemistry, 266*, 7706–7713.

Fuchs, T. A., Abed, U., Goosmann, C., Hurwitz, R., Schulze, I., Wahn, V., et al. (2007). Novel cell death program leads to neutrophil extracellular traps. *The Journal of Cell Biology, 176*, 231–241.

Glaser, R., Harder, J., Lange, H., Bartels, J., Christophers, E., & Schroder, J. M. (2005). Antimicrobial psoriasin (S100A7) protects human skin from *Escherichia coli* infection. *Nature Immunology, 6*, 57–64.

Havlickova, B., Czaika, V. A., & Friedrich, M. (2008). Epidemiological trends in skin mycoses worldwide. *Mycoses, 51*(Suppl. 4), 2–15.

Hein, K. Z., Takahashi, H., Tsumori, T., et al. (2015). Disulphide-reduced psoriasin is a human apoptosis-inducing broad-spectrum fungicide. *Proceedings of the National Academy of Sciences of the United States of America, 112*, 13039–13044.

Hentze, M. W., Muckenthaler, M. U., Galy, B., & Camaschella, C. (2010). Two to tango: Regulation of mammalian iron metabolism. *Cell, 142*, 24–38.

Heymann, P., Gerads, M., Schaller, M., Dromer, F., Winkelmann, G., & Ernst, J. F. (2002). The siderophore iron transporter of *Candida albicans* (Sit1p/Arn1p) mediates uptake of ferrichrome-type siderophores and is required for epithelial invasion. *Infection and Immunity, 70*, 5246–5255.

Hoyer, L. L. (2001). The ALS gene family of *Candida albicans*. *Trends in Microbiology, 9*, 176–180.

Hu, C. J., Bai, C., Zheng, X. D., Wang, Y. M., & Wang, Y. (2002). Characterization and functional analysis of the siderophore-iron transporter CaArn1p in *Candida albicans*. *The Journal of Biological Chemistry, 277*, 30598–30605.

Ibrahim, A. S., Edwards, J. E., Jr., Fu, Y., & Spellberg, B. (2006). Deferiprone iron chelation as a novel therapy for experimental mucormycosis. *The Journal of Antimicrobial Chemotherapy, 58*, 1070–1073.

Kelliher, J. L., & Kehl-Fie, T. E. (2016). Competition for manganese at the host-pathogen interface. *Progress in Molecular Biology and Translational Science, 142*, 1–25.

Knight, S. A., Vilaire, G., Lesuisse, E., & Dancis, A. (2005). Iron acquisition from transferrin by *Candida albicans* depends on the reductive pathway. *Infection and Immunity, 73*, 5482–5492.

Kuznets, G., Vigonsky, E., Weissman, Z., Lalli, D., Gildor, T., Kauffman, S. J., et al. (2014). A relay network of extracellular heme-binding proteins drives *C. albicans* iron acquisition from hemoglobin. *PLoS Pathogens, 10*, e1004407.

Leal, S. M., Jr., Roy, S., Vareechon, C., et al. (2013). Targeting iron acquisition blocks infection with the fungal pathogens *Aspergillus fumigatus* and *Fusarium oxysporum*. *PLoS Pathogens, 9*, e1003436.

Liuzzi, J. P., Lichten, L. A., Rivera, S., Blanchard, R. K., Aydemir, T. B., Knutson, M. D., et al. (2005). Interleukin-6 regulates the zinc transporter Zip14 in liver and contributes to the hypozincemia of the acute-phase response. *Proceedings of the National Academy of Sciences of the United States of America, 102*, 6843–6848.

MacCallum, D. M., & Odds, F. C. (2005). Temporal events in the intravenous challenge model for experimental *Candida albicans* infections in female mice. *Mycoses*, *48*, 151–161.

Manns, J. M., Mosser, D. M., & Buckley, H. R. (1994). Production of a hemolytic factor by *Candida albicans*. *Infection and Immunity*, *62*, 5154–5156.

Martin, R., Albrecht-Eckardt, D., Brunke, S., Hube, B., Hunniger, K., & Kurzai, O. (2013). A core filamentation response network in *Candida albicans* is restricted to eight genes. *PloS One*, *8*, e58613.

Mitchell, A., Romano, G. H., Groisman, B., Yona, A., Dekel, E., Kupiec, M., et al. (2009). Adaptive prediction of environmental changes by microorganisms. *Nature*, *460*, 220–224.

Mochon, A. B., Jin, Y., Kayala, M. A., Wingard, J. R., Clancy, C. J., Nguyen, M. H., et al. (2010). Serological profiling of a *Candida albicans* protein microarray reveals permanent host-pathogen interplay and stage-specific responses during candidemia. *PLoS Pathogens*, *6*, e1000827.

Moors, M. A., Stull, T. L., Blank, K. J., Buckley, H. R., & Mosser, D. M. (1992). A role for complement receptor-like molecules in iron acquisition by *Candida albicans*. *The Journal of Experimental Medicine*, *175*, 1643–1651.

Moroz, O. V., Blagova, E. V., Wilkinson, A. J., Wilson, K. S., & Bronstein, I. B. (2009). The crystal structures of human S100A12 in Apo form and in complex with zinc: New insights into S100A12 oligomerisation. *Journal of Molecular Biology*, *391*, 536–551.

Nasser, L., Weissman, Z., Pinsky, M., Amartely, H., Dvir, H., & Kornitzer, D. (2016). Structural basis of haem-iron acquisition by fungal pathogens. *Nature Microbiology*, *1*, 16156.

Noble, S. M., French, S., Kohn, L. A., Chen, V., & Johnson, A. D. (2010). Systematic screens of a *Candida albicans* homozygous deletion library decouple morphogenetic switching and pathogenicity. *Nature Genetics*, *42*, 590–598.

Okamoto-Shibayama, K., Kikuchi, Y., Kokubu, E., Sato, Y., & Ishihara, K. (2014). Csa2, a member of the Rbt5 protein family, is involved in the utilization of iron from human hemoglobin during *Candida albicans* hyphal growth. *FEMS Yeast Research*, *14*, 674–677.

Peracino, B., Wagner, C., Balest, A., Balbo, A., Pergolizzi, B., Noegel, A. A., et al. (2006). Function and mechanism of action of *Dictyostelium* Nramp1 (Slc11a1) in bacterial infection. *Traffic*, 7, 22–38.

Phan, Q. T., Myers, C. L., Fu, Y., Sheppard, D. C., Yeaman, M. R., Welch, W. H., et al. (2007). Als3 is a *Candida albicans* invasin that binds to cadherins and induces endocytosis by host cells. *PLoS Biology*, *5*, e64.

Potrykus, J., Stead, D., Maccallum, D. M., Urgast, D. S., Raab, A., van Rooijen, N., et al. (2013). Fungal iron availability during deep seated candidiasis is defined by a complex interplay involving systemic and local events. *PLoS Pathogens*, *9*, e1003676.

Protchenko, O., Rodriguez-Suarez, R., Androphy, R., Bussey, H., & Philpott, C. C. (2006). A screen for genes of heme uptake identifies the FLC family required for import of FAD into the endoplasmic reticulum. *The Journal of Biological Chemistry*, *281*, 21445–21457.

Ramanan, N., & Wang, Y. (2000). A high-affinity iron permease essential for *Candida albicans* virulence. *Science*, *288*, 1062–1064.

Raymond, K. N., Dertz, E. A., & Kim, S. S. (2003). Enterobactin: An archetype for microbial iron transport. *Proceedings of the National Academy of Sciences of the United States of America*, *100*, 3584–3588.

Reed, C., Ibrahim, A., Edwards, J. E., Jr., Walot, I., & Spellberg, B. (2006). Deferasirox, an iron-chelating agent, as salvage therapy for rhinocerebral mucormycosis. *Antimicrobial Agents and Chemotherapy*, *50*, 3968–3969.

Santos, R., Buisson, N., Knight, S., Dancis, A., Camadro, J. M., & Lesuisse, E. (2003). Haemin uptake and use as an iron source by *Candida albicans*: Role of *CaHMX1*-encoded haem oxygenase. *Microbiology*, *149*, 579–588.

Schneider Rde, O., Diehl, C., Dos Santos, F. M., Piffer, A. C., Garcia, A. W., Kulmann, M. I., et al. (2015). Effects of zinc transporters on *Cryptococcus gattii* virulence. *Scientific Reports*, *5*, 10104.

Schrettl, M., Bignell, E., Kragl, C., Joechl, C., Rogers, T., Arst, H. N., Jr., et al. (2004). Siderophore biosynthesis but not reductive iron assimilation is essential for *Aspergillus fumigatus* virulence. *The Journal of Experimental Medicine*, *200*, 1213–1219.

Schrettl, M., Bignell, E., Kragl, C., Sabiha, Y., Loss, O., Eisendle, M., et al. (2007). Distinct roles for intra- and extracellular siderophores during *Aspergillus fumigatus* infection. *PLoS Pathogens*, *3*, 1195–1207.

Seider, K., Heyken, A., Luttich, A., Miramon, P., & Hube, B. (2010). Interaction of pathogenic yeasts with phagocytes: Survival, persistence and escape. *Current Opinion in Microbiology*, *13*, 392–400.

Sentandreu, M., Elorza, M. V., Sentandreu, R., & Fonzi, W. A. (1998). Cloning and characterization of *PRA1*, a gene encoding a novel pH-regulated antigen of *Candida albicans*. *Journal of Bacteriology*, *180*, 282–289.

Sobel, J. D. (2007). Vulvovaginal candidosis. *Lancet*, *369*, 1961–1971.

Sorgo, A. G., Heilmann, C. J., Dekker, H. L., Brul, S., de Koster, C. G., & Klis, F. M. (2010). Mass spectrometric analysis of the secretome of *Candida albicans*. *Yeast*, *27*, 661–672.

Steenbergen, J. N., Nosanchuk, J. D., Malliaris, S. D., & Casadevall, A. (2004). Interaction of *Blastomyces dermatitidis*, *Sporothrix schenckii*, and *Histoplasma capsulatum* with *Acanthamoeba castellanii*. *Infection and Immunity*, *72*, 3478–3488.

Tagkopoulos, I., Liu, Y. C., & Tavazoie, S. (2008). Predictive behavior within microbial genetic networks. *Science*, *320*, 1313–1317.

Tangen, K. L., Jung, W. H., Sham, A. P., Lian, T., & Kronstad, J. W. (2007). The iron- and cAMP-regulated gene *SIT1* influences ferrioxamine B utilization, melanization and cell wall structure in *Cryptococcus neoformans*. *Microbiology*, *153*, 29–41.

Tavanti, A., Maisetta, G., Del Gaudio, G., Petruzzelli, R., Sanguinetti, M., Batoni, G., et al. (2011). Fungicidal activity of the human peptide hepcidin 20 alone or in combination with other antifungals against *Candida glabrata* isolates. *Peptides*, *32*, 2484–2487.

Taylor, J. W., & Berbee, M. L. (2006). Dating divergences in the fungal tree of life: Review and new analyses. *Mycologia*, *98*, 838–849.

Torti, F. M., & Torti, S. V. (2002). Regulation of ferritin genes and protein. *Blood*, *99*, 3505–3516.

Urban, C. F., Ermert, D., Schmid, M., Abu-Abed, U., Goosmann, C., Nacken, W., et al. (2009). Neutrophil extracellular traps contain calprotectin, a cytosolic protein complex involved in host defense against *Candida albicans*. *PLoS Pathogens*, *5*, e1000639.

Vicentefranqueira, R., Moreno, M. A., Leal, F., & Calera, J. A. (2005). The *zrfA* and *zrfB* genes of *Aspergillus fumigatus* encode the zinc transporter proteins of a zinc uptake system induced in an acid, zinc-depleted environment. *Eukaryotic Cell*, *4*, 837–848.

Weissman, Z., & Kornitzer, D. (2004). A family of *Candida cell* surface haem-binding proteins involved in haemin and haemoglobin-iron utilization. *Molecular Microbiology*, *57*, 1209–1220.

Weissman, Z., Shemer, R., Conibear, E., & Kornitzer, D. (2008). An endocytic mechanism for haemoglobin-iron acquisition in *Candida albicans*. *Molecular Microbiology*, *69*, 201–217.

Wilson, D. (2015). An evolutionary perspective on zinc uptake by human fungal pathogens. *Metallomics: Integrated Biometal Science*, 7, 979–985.

Winter, W. E., Bazydlo, L. A., & Harris, N. S. (2014). The molecular biology of human iron metabolism. *Laboratory Medicine*, *45*, 92–102.

Zackular, J. P., Chazin, W. J., & Skaar, E. P. (2015). Nutritional immunity: S100 proteins at the host-pathogen interface. *The Journal of Biological Chemistry*, *290*, 18991–18998.

CHAPTER FOUR

# Metal-Based Combinations That Target Protein Synthesis by Fungi

**Cindy Vallières, Simon V. Avery**[1]

School of Life Sciences, University of Nottingham University Park, Nottingham, United Kingdom

[1]Corresponding author: e-mail address: simon.avery@nottingham.ac.uk

## Contents



## Abstract

A wide range of fungicides (or antifungals) are used in agriculture and medicine, with activities against a spectrum of fungal pathogens. Unfortunately, the evolution of fungicide resistance has become a major issue. Therefore, there is an urgent need for new antifungal treatments. Certain metals have been used for decades as efficient fungicides in agriculture. However, concerns over metal toxicity have escalated over this time. Recent studies have revealed that metals like copper and chromate can impair functions required for the fidelity of protein synthesis in fungi. This occurs through different mechanisms, based on targeting of iron–sulphur cluster integrity or competition for uptake with amino acid precursors. Moreover, chromate at least acts synergistically with other agents known to target translation fidelity, like aminoglycoside antibiotics, causing dramatic and selective growth inhibition of several fungal pathogens of humans and plants. As such synergy allows the application of decreased amounts of metals for effective inhibition, it lessens concerns about nonspecific toxicity and opens new possibilities for metal applications in combinatorial fungicides targeting protein synthesis.

## 1. INTRODUCTION

A wide range of fungi are considered undesirable, including spoilers of foods and commercial products, indoor air inhabitants, human, and plant pathogens. Fungal pathogens of humans such as *Candida*, *Aspergillus*,

*Advances in Microbial Physiology*, Volume 70
ISSN 0065-2911
http://dx.doi.org/10.1016/bs.ampbs.2017.01.001

*Cryptococcus*, and *Pneumocystis* spp. cause life-threatening infections particularly for patients with compromised immune defences. Invasive fungal infections have mortality rates between 20% and 95% causing millions of human deaths every year, comparable to mortality attributable to tuberculosis or malaria (Brown et al., 2012; Calderone et al., 2014; Denning & Bromley, 2015). The number of therapeutic options for treatment of fungal pathogens is limited compared to bacterial infections, with only four classes of drug targeting three distinct metabolic pathways in current use, their efficacy eroded by toxicity or increasing drug resistance (Delarze & Sanglard, 2015; Vandeputte, Ferrari, & Coste, 2012; Walker, Lee, Munro, & Gow, 2015). As fungi are eukaryotes like their human or plant hosts, finding fungal-specific treatments is challenging. Fungal infections also devastate food crops and other plants. Reflecting this, the global fungicide market is worth more than $7 billion (Cools & Hammond-Kosack, 2013; Oliver & Hewitt, 2014). A wide range of fungicides have been developed and approved to counter phytopathogens compared to human pathogens. However, the agrichemical industry faces similar concerns surrounding evolution of resistance to fungicides combined with tightening of fungicide regulations (Coste & Vandeputte, 2015; Oliver & Hewitt, 2014), underscoring the urgent need for development of new effective treatments against fungi.

The relatively broad range of agents used for phytopathogen control includes copper-containing fungicides. Several pesticides which were or still are used in agriculture or other applications contain metals; copper as mentioned (fungicide, algicide) but also chromium (wood preservative), zinc (wood preservative), and cadmium (turf fungicide). Fungicides containing copper have historical significance. Over 250 years ago, it was discovered that seed grains soaked in a weak solution of copper sulphate inhibited seed-borne fungi. A later breakthrough was the development of the 'Bordeaux mixture', a blend of copper sulphate and slaked lime used to inhibit plant disease caused by downy mildew, powdery mildew, and other fungi in vineyards, fruit-farms, and gardens. The mixture was discovered by accident and its original purpose was to dissuade passers-by from picking grapes (copper gives the vines a blue colour). Thousands of tonnes of copper salts are used annually worldwide to protect food crops. However, as a heavy metal, copper causes concern regarding its toxicity to nonpathogenic microorganisms, plants, and humans when used in excess (Husak, 2015). In humans, alterations in copper levels or copper-dependent functions have been associated with the pathogenesis of neurodegenerative disorders such as Wilson's disease (Bandmann, Weiss, & Kaler, 2015; Manto, 2014). There is a delicate balance between the beneficial and toxic effect of copper. At low concentrations, copper serves as an essential micronutrient with important roles in different physiological processes and metabolic pathways. Indeed, copper serves as a cofactor for multiple enzymes, including superoxide dismutases (for superoxide-radical scavenging), cytochrome *c* oxidase (a complex of the mitochondrial electron transport chain), and tyrosinase (an enzyme involved in the pigmentation process) (Pena, Lee, & Thiele, 1999; Uauy, Olivares, & Gonzalez, 1998). Copper becomes toxic at elevated concentrations. Metal ions taken up by cells may coordinate with different chemical groups on macromolecules (e.g. imidazoles, phosphates, sulphydryls, hydroxyls) and disrupt function. The toxicity of copper can also arise from the generation of reactive oxygen species (ROS) mainly through the Fenton and Haber–Weiss reactions (Bremner, 1998; Gaetke & Chow, 2003). These ROS may damage proteins, nucleic acids, and lipids potentially leading to growth inhibition and death. Copper may also block protein functional groups or substitute for essential ions, notably other essential metals, inactivating protein function (Macomber & Imlay, 2009; Pena et al., 1999). Although, a large range of such actions of copper has been described in the literature, the primary target(s) of copper toxicity (i.e. the mechanism primarily responsible for copper-dependent growth inhibition) remains uncertain. It was recently shown that copper impairs the function of Rli1, an essential protein required for mRNA translation (Alhebshi, Sideri, Holland, & Avery, 2012). Interestingly, another metal, chromium, is also now known to impair mRNA translation (Holland et al., 2007).

Since its discovery in 1797 chromium has been used for diverse applications, including leather tanning and in wood preservatives. Chromium is also toxic at elevated concentrations but, like copper, the primary molecular mechanism(s) responsible for cell killing is not fully resolved. One complication is that the two abundant oxidation states of chromium, hexavalent Cr(VI), and trivalent Cr(III), have quite different properties. Cr(VI) is considered highly toxic whereas Cr(III) is a micronutrient and can be found in health supplements (Liu et al., 2014; Urbano, Ferreira, & Alpoim, 2012; Wu et al., 2016). The hexavalent state in aqueous solution exists as oxyanions, such as chromate ($CrO_4^{2-}$) and dichromate anion ($Cr_2O_7^{2-}$). Chromate is chemically similar to sulphate ($SO_4^{2-}$) which allows the chromate to enter cells via sulphate transporters in the plasma membrane, where both ions may compete for uptake (Holland & Avery, 2011; Pereira et al., 2008). Under physiological conditions, Cr(VI) is reduced to Cr(III) through the action of antioxidant enzymes or molecules such as ascorbic acid, glutathione (GSH), or cysteine which are abundant in cells. During this reduction process, unstable intermediates, Cr(V) and Cr(IV), and harmful free radicals are generated which may provoke oxidative stress and damage to cellular constituents, including direct attack on GSH, free cysteine, and cysteine residues in proteins (Guttmann, Poage, Johnston, & Zhitkovich, 2008; Ning & Grant, 2000; Quievryn, Goulart, Messer, & Zhitkovich, 2001). Like copper, several mechanisms of chromium toxicity have been proposed, including DNA damage by the trivalent ion, protein oxidation, and mRNA mistranslation (Holland et al., 2007; Jin et al., 2008; Johnson et al., 2016; Poljsak, Pocsi, Raspor, & Pesti, 2010). Furthermore, recent work has established a link between the competition for uptake between chromate and sulphate ions and a novel, major mechanism of Cr toxicity—loss of fidelity in the process of mRNA translation during protein synthesis—and this action may have application in fungal control.

This review focuses on the effect of metals copper and chromium on protein synthesis function in fungi, and the new possibilities for metal applications in effective and selective fungal inhibition.

## 2. COPPER ACTION ON FUNCTIONS ESSENTIAL FOR mRNA TRANSLATION

Copper is widely reported to cause damage to different cellular macromolecules and cellular processes. Several such actions discussed earlier have been suggested to be primary causes of growth inhibition exerted by copper, but it has been difficult to dissect cause or effect in most cases. A recent study identified a novel, primary target of the metal, an iron–sulphur (FeS) protein termed Rli1 in the yeast model (ABCE1 in humans and other organisms) (Alhebshi et al., 2012). FeS clusters are protein cofactors with diverse functions that are known to be among the most ROS-sensitive structures in the cell (Imlay, 2006). Despite this, FeS clusters have been conserved through evolution and are essential for function of a number of proteins (Imlay, 2006; Lill, 2009; Py, Moreau, & Barras, 2011). Most FeS cluster-containing proteins are not required for cell viability, with Rli1 being the first to be identified as essential (Kispal et al., 2005; Yarunin et al., 2005). Rli1 is a highly conserved (Barthelme et al., 2007; Becker et al., 2012), multifunctional ABC-family protein localized primarily to the cytosol and with diverse roles in protein synthesis: ribosome biogenesis and maturation, translation initiation, translation termination, and ribosome recycling (Nurenberg & Tampe, 2013). The N-terminal [4Fe–4S]-cluster domain of Rli1 plays a crucial role in these functions. This and other considerations about Rli1 outlined earlier lead to the hypothesis that Rli1 could be a primary cellular target of ROS and of redox-active agents like copper (Alhebshi et al., 2012). Key criteria diagnostic of a primary cellular target were considered, namely that the candidate protein should exhibit decreased function (which cannot be accounted for by decreased expression) during treatment with the agent; that knockdown of the target protein should produce a hypersensitive phenotype and, moreover, that overexpression should confer resistance (Avery, 2011). The last two criteria have been investigated using a regulatable *tet-RLI1* construct, with results establishing that copper resistance scaled with the level of *RLI1* expression (Alhebshi et al., 2012). Given the essentiality of Rli1, a *rli1Δ* deletion strain is not viable, but copper sensitivity was discernible in a diploid mutant heterozygous for *RLI1* (where retention of one genomic copy of the gene is sufficient to support viability). Besides Rli1, other essential FeS proteins have now been identified such as the nuclear-localized Rad3, Pri2, Pol1, Pol2, and Pol3 (Klinge, Hirst, Maman, Krude, & Pellegrini, 2007; Netz et al., 2012; Rudolf, Makrantoni, Ingledew, Stark, & White, 2006). The possibility that these proteins may also be candidate targets of copper action has been countered by the demonstration that *TET*-driven manipulation of their expression does not alter copper resistance (Alhebshi et al., 2012). The implication that Rli1 could be a particular target of copper was further supported with evidence based on Rli1 function. Rli1 activity is measurable according to nuclear-export activity of Rps2, a small ribosomal subunit that can be tagged with a fluorescent marker (e.g. GFP) to monitor cytosolic vs nuclear localization (Fig. 1A) (Kispal et al., 2005; Yarunin et al., 2005). This export function requires FeS cluster integrity of Rli1 (Kispal et al., 2005). Exposure to a mild concentration of copper produces a >10-fold increase in the proportion of cells with defective nuclear Rps2-GFP export (Fig. 1B) (Alhebshi et al., 2012). Tying this phenotype more specifically to Rli1 function, it was shown that Rli1 overexpression largely rescued the Rps2-GFP export defect seen during copper exposure (Fig. 1B). The hypothesis that growth inhibition by copper was centred on FeS cluster integrity of Rli1 was supported with a $Rli1^{C58A}$ construct, lacking one of the cysteine residues needed to bind one of the two clusters within the protein. $RLI1^{C58A}$-expressing yeasts are copper hypersensitive according to both growth and assays of essential Rli1 function (i.e. Rps2-GFP export).

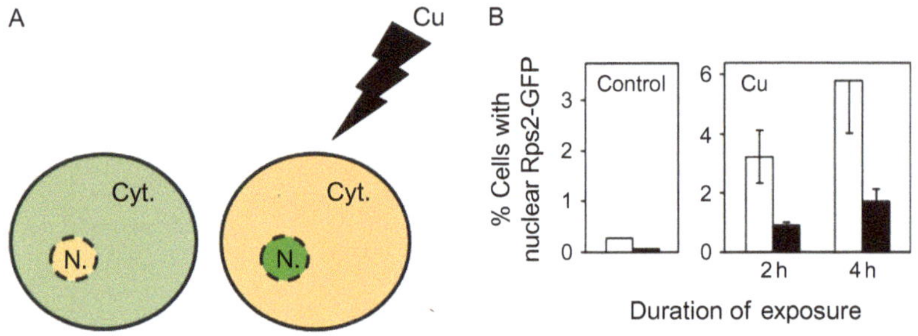


**Fig. 1** Inhibition of Rli1 function by copper. (A) In the absence of copper Rps2-GFP (localization in *green*) is exported to the cytoplasm (Cyt.), but copper decreases Rli1-dependent export and the ribosomal subunit becomes localized to the nucleus (shown bounded by *dashed line*, N.). (B) Wild-type (*white bars*) or *RLI1*-overproducing (*black bars*) yeast expressing a *RPS2-eGFP* fusion construct were incubated in the absence (control) or presence of a mild copper concentration (0.35 mM, which slowed cell doubling time <15%) before analysis of Rps2-GFP localization. *Figure adapted from Alhebshi, A., Sideri, T. C., Holland, S. L., & Avery, S. V. (2012) The essential iron–sulfur protein Rli1 is an important target accounting for inhibition of cell growth by reactive oxygen species.* Molecular Biology of the Cell, 23*(18), 3582–3590.*

Copper has been shown to damage nonoxidatively the [4Fe–4S] clusters of bacterial iron–sulphur dehydratases (Macomber & Imlay, 2009). In that case, damage to the proteins involved displacement of iron atoms from the solvent exposed cofactor, suggesting that copper perturbed the [4Fe–4S] cluster by liganding to the coordinating sulphur atoms (Macomber & Imlay, 2009). In contrast, the relationship between copper toxicity and Rli1 function in yeast requires the presence of oxygen, as Rli1-dependent copper resistance is abolished under anaerobic conditions, suggesting an oxidative mode of copper action in this case (Alhebshi et al., 2012). The two [4Fe–4S] clusters of Rli1p (ABCE1) are predicted to be well shielded from solvent (Karcher, Schele, & Hopfner, 2008) which would suggest that inaccessibility of Cu may explain an absence of direct Fe displacement by Cu, but a similar argument could be made about ROS access to the clusters. This was resolved with the following evidence that it is clusters enroute to Rli1 incorporation that are targeted. The hypothesis that the Rli1-centred action of copper is targeted to FeS clusters was corroborated by analysis of incorporation and turnover of radiolabelled $^{55}$Fe to Rli1. This approach established that FeS clusters of Rli1 are copper labile in vitro, but FeS cluster supply to Rli1 is the critical target in vivo. The collective evidence has established a model in which copper targeting of FeS clusters prior to their assembly into

Rli1 leads to decreased Rli1 function and so defective growth in elevated copper, as Rli1 function is essential for viability (Alhebshi et al., 2012).

The earlier evidence from yeast revealed Rli1 as a primary (not necessarily only) target that could account for the growth inhibitory effects of copper. This opens new possibilities for understanding and potentially exploiting copper action against fungal pathogens. As mentioned earlier, Rli1 is highly conserved across the eukaryotes (and the archaea). Sequence identity with the yeast Rli1 protein is 76%, 74%, and 71% in the phytopathogens *Botrytis cinerea*, *Septoria tritici*, and *Rhizoctonia solani*, respectively, and 83% and 76% in the human pathogens *Candida albicans* and *Aspergillus fumigatus*. Therefore, Rli1 could be a primary target of copper in diverse fungal pathogens. A further key point of interest centres on the essential function of Rli1 in the process of mRNA translation (Khoshnevis et al., 2010; Nurenberg & Tampe, 2013). Section 3 explains how recent studies with another toxic metal have highlighted mRNA mistranslation as a candidate novel target for antifungal action.

## 3. CHROMIUM ACTION ON TRANSPORT PROCESSES LEADING TO mRNA MISTRANSLATION

Genetic and genomic tools available for yeast have been widely exploited to help understand the biological actions of metals and other toxicants (Bleackley, Young, Loewen, & MacGillivray, 2011; Giaever & Nislow, 2014; Johnson et al., 2016). Following a genome-wide screen of the heterozygous yeast deletion strain collection to identify genes required for chromate resistance, it was found that chromium exposure provokes the potentially harmful aggregation of cellular proteins (Holland et al., 2007). Protein oxidation, known to arise during chromium stress (Sumner et al., 2005), is one mechanism that can trigger protein aggregation (Bucciantini et al., 2002; Nystrom, 2005), which in turn may exacerbate ROS production (Weids & Grant, 2014). Moreover, the aggregates generated during chromium stress were comprised of proteins translated de novo during chromium exposure rather than proteins that preexisted the stress; this indicated an underlying action at the point of protein synthesis. It was observed that chromium toxicity was synergistically increased in the presence of the aminoglycoside paromomycin, an agent known to cause mRNA mistranslation by inducing a structural alteration in the small ribosomal subunit decoding centre, allowing entry of near-cognate tRNAs (Carter et al., 2000; Fan-Minogue & Bedwell, 2008). Furthermore, genetic manipulation of translational accuracy in yeast modulated chromium toxicity (Holland et al., 2007). This work established mRNA mistranslation as a major mechanism of chromium toxicity.

The mechanism by which chromium causes mRNA mistranslation was subsequently found to relate to its action in sulphur starvation (see Section 1). Specifically, sulphate uptake (inhibitable by chromate) is required in yeast for synthesis of the *S*-containing amino acids cysteine and methionine, essential for protein synthesis (Holland & Avery, 2011). Depletion of particular amino acids within cells can alter the competition between cognate and noncognate aminoacyl tRNAs leading to translational infidelity (Farabaugh & Bjork, 1999; Gallant & Lindsley, 1998; Sorensen, 2001). Deletion of the sulphate transporters of yeast, Sul1 and Sul2, suppressed mRNA mistranslation caused by chromium (Fig. 2). This was despite continued Cr entry to cells by alternative pathways, indicating that the mistranslation was due to chromium entering cells specifically via the sulphate transporters. Chromate-induced mistranslation could be mimicked by starving yeast for cysteine and methionine, and could be suppressed in a mutant defective for synthesis of cysteine and methionine (Holland,

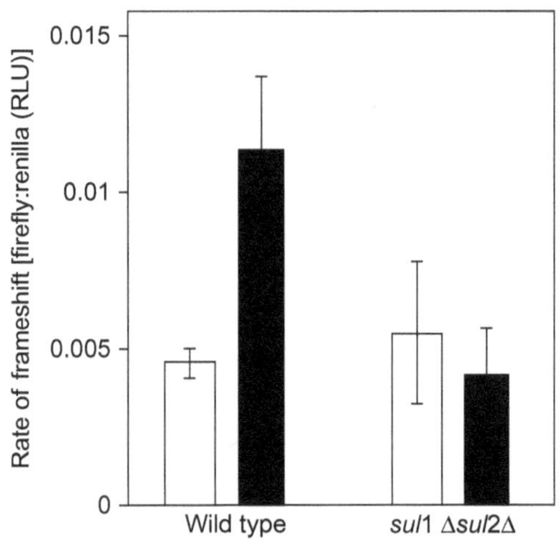


**Fig. 2** Dependency of chromate-induced mistranslation on sulphate transporters. Wild-type and *sul1Δ/sul2Δ* deletant yeast strains transformed with a dual-luciferase frameshift reporter construct were incubated in the absence (*open bars*) or presence (*filled bars*) of 100 μM $CrO_3$, before determination of firefly and *Renilla* luciferase activities in protein extracts. The ratio of the activities indicates the rate of frameshift. RLU, relative light units. *Figure adapted from Holland, S. L., Ghosh, E., & Avery, S. V. (2010). Chromate-induced sulfur starvation and mRNA mistranslation in yeast are linked in a common mechanism of Cr toxicity.* Toxicology in Vitro, 24(6), 1764–1767.

Ghosh, & Avery, 2010). The latter mutant relies on direct uptake of cysteine and methionine from the medium, so the availability of these amino acids for protein synthesis is uncoupled from the process of sulphate uptake. Together, the evidence outlined the following novel pathway of chromium toxicity: chromate ion competition with sulphate at the Sul1 and Sul2 transporters leads to sulphate starvation and, consequently, cysteine and methionine depletion in cells. This amino acid depletion impairs translational fidelity, a major cause of growth inhibition by chromium. Therefore, competition for uptake between chromate and sulphate is at the root of this mode of Cr toxicity.

The Sul1 and/or Sul2 proteins, predicted sites of the chromate/sulphate competition in yeast, are conserved in the fungi. Among pathogenic filamentous fungi, for example, orthologue identity with the yeast protein sequences is 43% and 44% for *R. solani*, and between 38% and 41% for *B. cinerea* and *A. fumigatus*. Therefore, this action of chromium characterized in yeast could be relevant to many undesirable fungi and this is explored further in Section 4. It should also be noted that the Rli1 protein, target of copper as described in Section 2 and required for maintenance of translational fidelity (Khoshnevis et al., 2010), also appears to be a target of chromium. Rli1 activity (in nuclear ribosomal subunit export) is impaired by mild chromium stress in yeast and *RLI1* overexpression improved resistance to Cr (Alhebshi et al., 2012). Therefore, chromate could exert a dual effect on the fidelity of protein synthesis, by competing with sulphate for uptake leading to cysteine and methionine starvation (Holland et al., 2010) and by impairing Rli1 function in mRNA translation (Khoshnevis et al., 2010; Nurenberg & Tampe, 2013). Because copper, like chromium, targets Rli1 it may not be unexpected to find synergistic effects on fungal growth also between copper and agents known to cause mistranslation. Indeed, preliminary observations with yeast indicate synergy with certain aminoglycosides (C. Vallieres & S. V. Avery, unpublished data).

## 4. EXPLOITATION OF NEW INSIGHTS TO METAL ACTION, FOR FUNGAL CONTROL

As outlined earlier, copper and chromium are two metals which can target protein synthesis either by inhibiting Rli1, essential for multiple steps of this process, and/or sulphate uptake necessary for cysteine and methionine biosynthesis. Given also that Rli1 and sulphate-transport mechanisms are well conserved in fungi, translation fidelity is an attractive antifungal target as effects of mistranslation are not simply loss of essential protein functions, but also gain of toxic function through the formation of inhibitory protein aggregates (Holland et al., 2007). However, concern around nonspecific toxicity for any intended antifungal applications can be particularly acute for metals given their potential toxicities to nonpathogenic microorganisms, plants, and humans. One approach to ameliorating such concerns is the use of combination treatments, as these often allow use of lower doses of the individual agents comprising the combination, particularly where these may have a synergistic action due to related cellular targets. In addition, combination treatments reduce the likelihood of resistance, i.e. emergence of resistance to more than one agent is typically slower than to a single agent. Strong synergies between antimicrobial agents are not very common, but their detection can be facilitated by chemical or biological screening approaches. In a recent study, sublethal concentrations of existing antifungals (fluconazole, caspofungin, amphotericin B, terbinafine, benomyl, and cyprodinil) were combined pairwise with compounds from a ~3600-strong chemical library and tested against diverse fungal species (Robbins et al., 2015). The authors generated a deep reservoir of chemical–chemical interactions termed the antifungal combinations matrix. Some compounds like the antimycobacterial drug clofazimine showed synergy with several antifungals. Synergistic combinations have also been found using the yeast model. The minimal inhibitory concentration (MIC) of the lipopeptide antifungal caspofungin was decreased by up to fivefold in the presence of the antimalarial chloroquine (Islahudin et al., 2013). Moreover, in the context of this review's focus, the MIC of chromium was dramatically decreased (by 64-fold) in the presence of the aminoglycoside paromomycin (Moreno-Martinez, Vallieres, Holland, & Avery, 2015).

The latter example of strong synergy has been further considered for its antifungal potential (Moreno-Martinez et al., 2015). Whereas aminoglycosides are commonly employed as antibacterials due to the sensitivity of prokaryotic ribosomes, at higher doses they also target eukaryotic ribosomes (Chang et al., 2010; Fan-Minogue & Bedwell, 2008; Shrestha, Grilley, Fosso, Chang, & Takemoto, 2013). Despite concerns about environmental selection of antibiotic resistance, certain antibiotics related to paromomycin such as the aminoglycoside streptomycin are currently approved as fungicides for crop use in a number of countries (FRAC, 2016). As discussed earlier, certain metal-based compounds have also been approved for crop protection. Studies have established that the synergistic combination involving chromium and paromomycin is also effective against human pathogens such as *C. albicans* and *Candida glabrata* (Moreno-Martinez et al., 2015). Interestingly, paromomycin or chromium could be substituted, respectively, with alternative aminoglycosides (e.g. hygromycin, streptomycin) or alternative metallic sulphate-transport inhibitors potentially less toxic than chromate (e.g. molybdate, vanadate). These alternative combinations also gave strong synergistic growth inhibition. For example, yeast growth yield was decreased >90% when hygromycin + molybdate or streptomycin + vanadate were supplied in combinations, at doses where the individual agents had negligible effects on growth. The mode of synergistic action could also be linked to sulphur starvation combined with the aminoglycoside action: synergy between molybdate and paromomycin was suppressed in mutants resistant to sulphate-transport inhibition. Synergistic effects between different agents are commonly seen where they target a common process (translation fidelity in this case) but by different mechanisms or pathways (Cokol et al., 2011). Accordingly, the translation error rate was synergistically increased by combining aminoglycoside with sulphate-transport inhibitor (e.g. chromate) (Moreno-Martinez et al., 2015). Certain major pathogenic fungi were susceptible to the alternative combinations (Fig. 3). These included the pathogenic yeast *Cryptococcus neoformans*, and the phytopathogens *R. solani* and *S. tritici*. Importantly, neither a human cell line nor the bacterium *Escherichia coli* showed any sensitivity to the synergistic action of these combinations of agents. As sulphate uptake is not

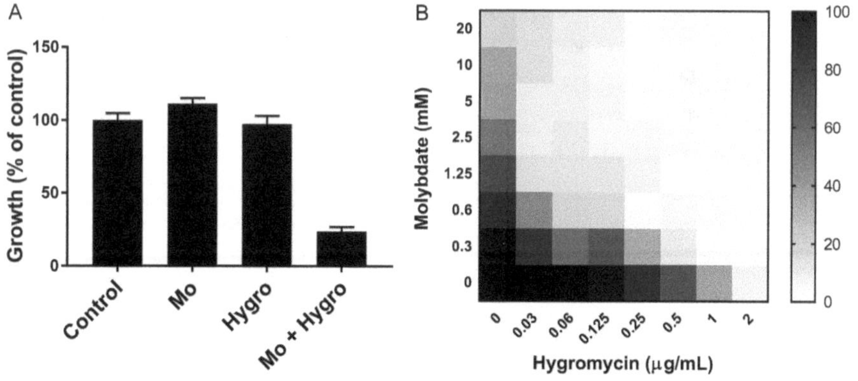


**Fig. 3** Sulphate-mimetic molybdate combined with aminoglycoside hygromycin synergistically inhibits phytopathogen *R. solani*. (A) Growth (derived from $OD_{600}$) of *R. solani* after 5 days in PDB broth supplemented with 1 mM sodium molybdate and/or 0.25 μg $mL^{-1}$ hygromycin B. (B) Checkerboard assay with *R. solani* in PDB medium at the indicated concentrations. The greyscale (shown on right) depicts growth as a percentage of control growth (OD) determined in the absence of added agents.

necessary for supply of the sulphur containing amino acids cysteine and methionine in humans, indeed there should be no expectation that chromate or other sulphate mimetics should cause mistranslation in this case. As such, amino acid biosynthesis could be a considered an exploitable frailty for fungal control (Jastrzebowska & Gabriel, 2015), akin to cell wall or ergosterol biosynthesis. Application of metals in novel combination treatments targeting translation fidelity opens promising new possibilities for control of undesirable fungi, the synergistic combinations requiring substantially decreased metal doses for effective growth inhibition compared to traditional metal treatment (e.g. Bordeaux mixture) used against fungal pathogens.

## 5. CONCLUDING REMARKS

New insights to mechanisms of metal action against fungi may offer new options for effective fungal control. Copper targets the function of Rli1, an essential FeS protein required for protein synthesis including translation fidelity (Alhebshi et al., 2012; Khoshnevis et al., 2010). Chromium (and other metallic sulphate mimetics) compete with sulphate for uptake, leading to depletion of cysteine and methionine, and resultant impairment of translation fidelity (Holland et al., 2010; Pereira et al., 2008). Aminoglycosides are known to provoke mistranslation by targeting the ribosome directly (Carter et al., 2000; Fan-Minogue & Bedwell, 2008). Therefore, metals and aminoglycosides may target a common aspect of protein synthesis (fidelity) through different mechanisms, providing a classic scenario for observing synergy that may be exploitable for fungal control (Fig. 4) (Moreno-Martinez et al., 2015). Translation fidelity is a particularly attractive target as mistranslation provokes loss of essential protein functions (commonly associated with cyto-static growth effects), but also gain of toxic function through the formation of protein aggregates (commonly cidal) (Holland et al., 2007). Sulphate mimetics such as chromate, molybdate, and vanadate proved much more effective inhibitors when combined with aminoglycosides, against major fungal pathogens like *S. tritici*, *R. solani*, and *C. albicans* (Moreno-Martinez et al., 2015). The diversity of fungi susceptible to this synergy is reflected by the conservation of sulphate-transport proteins in fungi. The facts that such synergistic combinations decrease the risk of resistance emergence and, moreover, allow use of lower dosages of agents, could help to reverse previous concerns about metal use in agriculture and toxicity to nontarget organisms. Regarding humans, alterations in copper

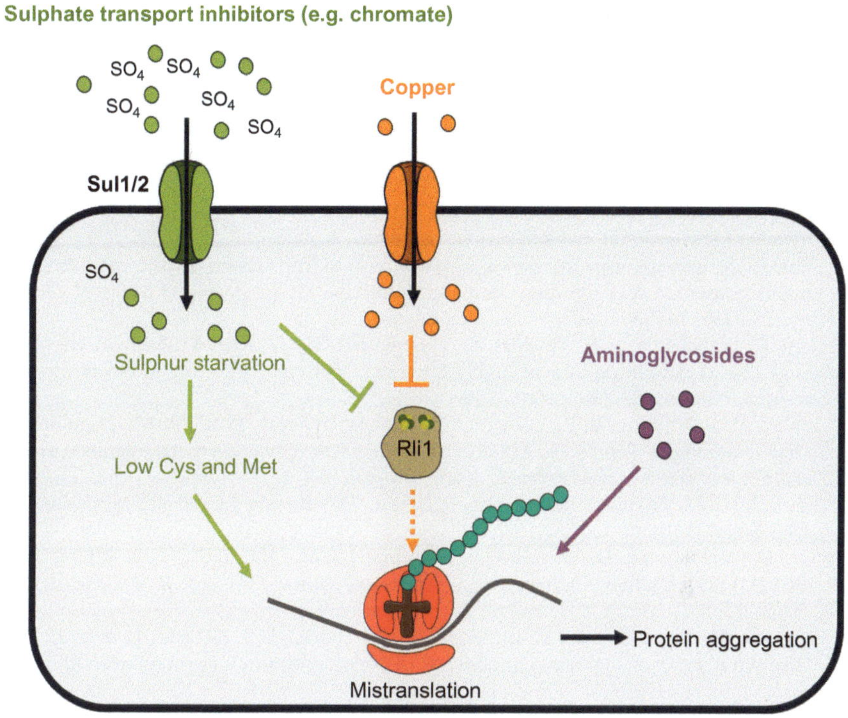


**Fig. 4** Scheme depicting the effects of chromium, copper, and aminoglycosides on fidelity of protein synthesis. Steps that are less well characterized are indicated by a *dashed line*. See the main text for full accounts.

levels are linked with neurodegenerative disorders such as Wilson's disease (Bandmann et al., 2015; Manto, 2014), and the application of copper (or other metals) to combat human fungal pathogens is less likely than for combating other undesirable fungi like phytopathogens. However, and perhaps more importantly, the new findings have allowed identification of translation fidelity as a novel target in human pathogens such as *C. albicans* and *C. neoformans* (and in fungi more generally). This paves the way for finding other classes of synergistic combinations targeting fidelity, besides metallic sulphate mimetics and aminoglycosides. This will offer further alternatives to those described here, with potentially better properties for application in fungal control.

## ACKNOWLEDGEMENT

This work was supported by the Biotechnology and Biological Sciences Research Council (grant numbers BB/M022161/1 and BB/I000852/1).

## REFERENCES


Alhebshi, A., Sideri, T. C., Holland, S. L., & Avery, S. V. (2012). The essential iron–sulfur protein Rli1 is an important target accounting for inhibition of cell growth by reactive oxygen species. *Molecular Biology of the Cell*, *23*(18), 3582–3590.

Avery, S. V. (2011). Molecular targets of oxidative stress. *Biochemical Journal*, *434*(2), 201–210.

Bandmann, O., Weiss, K. H., & Kaler, S. G. (2015). Wilson's disease and other neurological copper disorders. *Lancet Neurology*, *14*(1), 103–113.

Barthelme, D., Scheele, U., Dinkelaker, S., Janoschka, A., Macmillan, F., Albers, S. V., et al. (2007). Structural organization of essential iron–sulfur clusters in the evolutionarily highly conserved ATP-binding cassette proteinABCE1. *The Journal of Biological Chemistry*, *282*(19), 14598–14607.

Becker, T., Franckenberg, S., Wickles, S., Shoemaker, C. J., Anger, A. M., Armache, J. P., et al. (2012). Structural basis of highly conserved ribosome recycling in eukaryotes and archaea. *Nature*, *482*(7386), 501–506.

Bleackley, M. R., Young, B. P., Loewen, C. J., & MacGillivray, R. T. (2011). High density array screening to identify the genetic requirements for transition metal tolerance in Saccharomyces cerevisiae. *Metallomics*, *3*(2), 195–205.

Bremner, I. (1998). Manifestations of copper excess. *The American Journal of Clinical Nutrition*, *67*(5 Suppl.), 1069S–1073S.

Brown, G. D., Denning, D. W., Gow, N. A., Levitz, S. M., Netea, M. G., & White, T. C. (2012). Hidden killers: Human fungal infections. *Science Translational Medicine*, *4*(165), 165rv113.

Bucciantini, M., Giannoni, E., Chiti, F., Baroni, F., Formigli, L., Zurdo, J., et al. (2002). Inherent toxicity of aggregates implies a common mechanism for protein misfolding diseases. *Nature*, *416*(6880), 507–511.

Calderone, R., Sun, N., Gay-Andrieu, F., Groutas, W., Weerawarna, P., Prasad, S., et al. (2014). Antifungal drug discovery: The process and outcomes. *Future Microbiology*, *9*(6), 791–805.

Carter, A. P., Clemons, W. M., Brodersen, D. E., Morgan-Warren, R. J., Wimberly, B. T., & Ramakrishnan, V. (2000). Functional insights from the structure of the 30S ribosomal subunit and its interactions with antibiotics. *Nature*, *407*(6802), 340–348.

Chang, C. W., Fosso, M., Kawasaki, Y., Shrestha, S., Bensaci, M. F., Wang, J., et al. (2010). Antibacterial to antifungal conversion of neamine aminoglycosides through alkyl modification. Strategy for reviving old drugs into agrofungicides. *The Journal of Antibiotics (Tokyo)*, *63*(11), 667–672.

Cokol, M., Chua, H. N., Tasan, M., Mutlu, B., Weinstein, Z. B., Suzuki, Y., et al. (2011). Systematic exploration of synergistic drug pairs. *Molecular Systems Biology*, 7, 544.

Cools, H. J., & Hammond-Kosack, K. E. (2013). Exploitation of genomics in fungicide research: Current status and future perspectives. *Molecular Plant Pathology*, *14*(2), 197–210.

Coste, A. T., & Vandeputte, P. (2015). *Antifungals: From genomics to resistance and the development of novel agents*. Poole, UK: Caister Academic Press.

Delarze, E., & Sanglard, D. (2015). Defining the frontiers between antifungal resistance, tolerance and the concept of persistence. *Drug Resistance Updates*, *23*, 12–19.

Denning, D. W., & Bromley, M. J. (2015). Infectious disease. How to bolster the antifungal pipeline. *Science*, *347*(6229), 1414–1416.

Fan-Minogue, H., & Bedwell, D. M. (2008). Eukaryotic ribosomal RNA determinants of aminoglycoside resistance and their role in translational fidelity. *RNA*, *14*(1), 148–157.

Farabaugh, P. J., & Bjork, G. R. (1999). How translational accuracy influences reading frame maintenance. *EMBO Journal*, *18*(6), 1427–1434.

FRAC. (2016). *FRAC classification on mode of action*. Available at: http://www.frac.info/docs/default-source/publications/frac-mode-of-action-poster/frac-moa-poster-2016.pdf?sfvrsn=2.

Gaetke, L. M., & Chow, C. K. (2003). Copper toxicity, oxidative stress, and antioxidant nutrients. *Toxicology*, *189*(1–2), 147–163.

Gallant, J. A., & Lindsley, D. (1998). Ribosomes can slide over and beyond "hungry" codons, resuming protein chain elongation many nucleotides downstream. *Proceedings of the National Academy of Sciences*, *95*(23), 13771–13776.

Giaever, G., & Nislow, C. (2014). The yeast deletion collection: A decade of functional genomics. *Genetics*, *197*(2), 451–465.

Guttmann, D., Poage, G., Johnston, T., & Zhitkovich, A. (2008). Reduction with glutathione is a weekly mutagenic pathway in chromium(VI) metabolism. *Chemical Research in Toxicology*, *21*(11), 2188–2194.

Holland, S. L., & Avery, S. V. (2011). Chromate toxicity and the role of sulfur. *Metallomics*, *3*(11), 1119–1123.

Holland, S. L., Ghosh, E., & Avery, S. V. (2010). Chromate-induced sulfur starvation and mRNA mistranslation in yeast are linked in a common mechanism of Cr toxicity. *Toxicology In Vitro*, *24*(6), 1764–1767.

Holland, S., Lodwig, E., Sideri, T., Reader, T., Clarke, I., Gkargkas, K., et al. (2007). Application of the comprehensive set of heterozygous yeast deletion mutants to elucidate the molecular basis of cellular chromium toxicity. *Genome Biology*, *8*(12), R268.

Husak, V. (2015). Copper and copper-containing pesticides: Metabolism, toxicity and oxidative stress. *Journal of Vasyl Stefanyk Precarpathian National University*, *2*(1), 38–50.

Imlay, J. A. (2006). Iron–sulphur clusters and the problem with oxygen. *Molecular Microbiology*, *59*(4), 1073–1082.

Islahudin, F., Khozoie, C., Bates, S., Ting, K. N., Pleass, R. J., & Avery, S. V. (2013). Cell wall perturbation sensitizes fungi to the antimalarial drug chloroquine. *Antimicrobial Agents and Chemotherapy*, *57*(8), 3889–3896.

Jastrzebowska, K., & Gabriel, I. (2015). Inhibitors of amino acids biosynthesis as antifungal agents. *Amino Acids*, *47*(2), 227–249.

Jin, Y. H., Dunlap, P. E., McBride, S. J., Al-Refai, H., Bushel, P. R., & Freedman, J. H. (2008). Global transcriptome and deletome profiles of yeast exposed to transition metals. *PLoS Genetics*, *4*(4), e1000053.

Johnson, A. J., Veljanoski, F., O'Doherty, P. J., Zaman, M. S., Petersingham, G., Bailey, T. D., et al. (2016). Revelation of molecular basis for chromium toxicity by phenotypes of Saccharomyces cerevisiae gene deletion mutants. *Metallomics*, *8*(5), 542–550.

Karcher, A., Schele, A., & Hopfner, K. P. (2008). X-ray structure of the complete ABC enzyme ABCE1 from Pyrococcus abyssi. *The Journal of Biological Chemistry*, *283*(12), 7962–7971.

Khoshnevis, S., Gross, T., Rotte, C., Baierlein, C., Ficner, R., & Krebber, H. (2010). The iron–sulphur protein RNase L inhibitor functions in translation termination. *EMBO Reports*, *11*(3), 214–219.

Kispal, G., Sipos, K., Lange, H., Fekete, Z., Bedekovics, T., Janaky, T., et al. (2005). Biogenesis of cytosolic ribosomes requires the essential iron–sulphur protein Rli1p and mitochondria. *EMBO Journal*, *24*(3), 589–598.

Klinge, S., Hirst, J., Maman, J. D., Krude, T., & Pellegrini, L. (2007). An iron–sulfur domain of the eukaryotic primase is essential for RNA primer synthesis. *Nature Structural & Molecular Biology*, *14*(9), 875–877.

Lill, R. (2009). Function and biogenesis of iron–sulphur proteins. *Nature*, *460*(7257), 831–838.

Liu, L., Cui, W. M., Zhang, S. W., Kong, F. H., Pedersen, M. A., Wene, Y., et al. (2014). Effect of glucose tolerance factor (GTF) from high chromium yeast on glucose metabolism in insulin-resistant 3T3-L1 adipocytes. *RSC Advances*, *5*, 3482–3490.

Macomber, L., & Imlay, J. A. (2009). The iron–sulfur clusters of dehydratases are primary intracellular targets of copper toxicity. *Proceedings of the National Academy of Sciences*, *106*(20), 8344–8349.

Manto, M. (2014). Abnormal copper homeostasis: Mechanisms and roles in neurodegeneration. *Toxics*, *2*, 327–345.

Moreno-Martinez, E., Vallieres, C., Holland, S. L., & Avery, S. V. (2015). Novel, synergistic antifungal combinations that target translation fidelity. *Scientific Reports*, *5*, 16700.

Netz, D. J., Pierik, A. J., Stumpfig, M., Bill, E., Sharma, A. K., Pallesen, L. J., et al. (2012). A bridging [4Fe–4S] cluster and nucleotide binding are essential for function of the Cfd1–Nbp35 complex as a scaffold in iron–sulfur protein maturation. *The Journal of Biological Chemistry*, *287*(15), 12365–12378.

Ning, J., & Grant, M. H. (2000). The role of reduced glutathione and glutathione reductase in the cytotoxicity of chromium (VI) in osteoblasts. *Toxicology In Vitro*, *14*(4), 329–335.

Nurenberg, E., & Tampe, R. (2013). Tying up loose ends: Ribosome recycling in eukaryotes and archaea. *Trends in Biochemical Sciences*, *38*(2), 64–74.

Nystrom, T. (2005). Role of oxidative carbonylation in protein quality control and senescence. *EMBO Journal*, *24*(7), 1311–1317.

Oliver, R. P., & Hewitt, H. G. (2014). Fungicides in crop protection (2nd ed.). Wallingford, UK: CABI.

Pena, M. M., Lee, J., & Thiele, D. J. (1999). A delicate balance: Homeostatic control of copper uptake and distribution. *Journal of Nutrition*, *129*(7), 1251–1260.

Pereira, Y., Lagniel, G., Godat, E., Baudouin-Cornu, P., Junot, C., & Labarre, J. (2008). Chromate causes sulfur starvation in yeast. *Toxicological Sciences*, *106*(2), 400–412.

Poljsak, B., Pocsi, I., Raspor, P., & Pesti, M. (2010). Interference of chromium with biological systems in yeasts and fungi: A review. *Journal of Basic Microbiology*, *50*(1), 21–36.

Py, B., Moreau, P. L., & Barras, F. (2011). Fe–S clusters, fragile sentinels of the cell. *Current Opinion in Microbiology*, *14*(2), 218–223.

Quievryn, G., Goulart, M., Messer, J., & Zhitkovich, A. (2001). Reduction of Cr (VI) by cysteine: Significance in human lymphocytes and formation of DNA damage in reactions with variable reduction rates. *Molecular and Cellular Biochemistry*, *222*(1–2), 107–118.

Robbins, N., Spitzer, M., Yu, T., Cerone, R. P., Averette, A. K., Bahn, Y. S., et al. (2015). An antifungal combination matrix identifies a rich pool of adjuvant molecules that enhance drug activity against diverse fungal pathogens. *Cell Reports*, *13*(7), 1481–1492.

Rudolf, J., Makrantoni, V., Ingledew, W. J., Stark, M. J., & White, M. F. (2006). The DNA repair helicases XPD and FancJ have essential iron–sulfur domains. *Molecular Cell*, *23*(6), 801–808.

Shrestha, S., Grilley, M., Fosso, M. Y., Chang, C. W., & Takemoto, J. Y. (2013). Membrane lipid-modulated mechanism of action and non-cytotoxicity of novel fungicide aminoglycoside FG08. *PLoS One*, *8*(9), e73843.

Sorensen, M. A. (2001). Charging levels of four tRNA species in Escherichia coli Rel(+) and Rel(−) strains during amino acid starvation: A simple model for the effect of ppGpp on translational accuracy. *Journal of Molecular Biology*, *307*(3), 785–798.

Sumner, E. R., Shanmuganathan, A., Sideri, T. C., Willetts, S. A., Houghton, J. E., & Avery, S. V. (2005). Oxidative protein damage causes chromium toxicity in yeast. *Microbiology*, *151*(Pt. 6), 1939–1948.

Uauy, R., Olivares, M., & Gonzalez, M. (1998). Essentiality of copper in humans. *The American Journal of Clinical Nutrition*, *67*(5 Suppl.), 952S–959S.

Urbano, A. M., Ferreira, L. M., & Alpoim, M. C. (2012). Molecular and cellular mechanisms of hexavalent chromium-induced lung cancer: An updated perspective. *Current Drug Metabolism*, *13*(3), 284–305.

Vandeputte, P., Ferrari, S., & Coste, A. T. (2012). Antifungal resistance and new strategies to control fungal infections. *International Journal of Microbiology*, *2012*, 713687.

Walker, L. A., Lee, K. K., Munro, C. A., & Gow, N. A. (2015). Caspofungin treatment of Aspergillus fumigatus results in ChsG-dependent upregulation of chitin synthesis and the formation of chitin-rich microcolonies. *Antimicrobial Agents and Chemotherapy*, *59*(10), 5932–5941.

Weids, A. J., & Grant, C. M. (2014). The yeast peroxiredoxin Tsa1 protects against protein-aggregate-induced oxidative stress. *Journal of Cell Science*, *127*(Pt. 6), 1327–1335.

Wu, L. E., Levina, A., Harris, H. H., Cai, Z., Lai, B., Vogt, S., et al. (2016). Carcinogenic chromium(VI) compounds formed by intracellular oxidation of chromium(III) dietary supplements by adipocytes. *Angewandte Chemie, International Edition*, *55*(5), 1742–1745.

Yarunin, A., Panse, V. G., Petfalski, E., Dez, C., Tollervey, D., & Hurt, E. C. (2005). Functional link between ribosome formation and biogenesis of iron–sulfur proteins. *EMBO Journal*, *24*(3), 580–588.

CHAPTER FIVE

# Transition Metal Homeostasis in *Streptococcus pyogenes* and *Streptococcus pneumoniae*

**Andrew G. Turner, Cheryl-lynn Y. Ong, Mark J. Walker, Karrera Y. Djoko, Alastair G. McEwan[1]**

School of Chemistry and Molecular Biosciences and Australian Infectious Diseases Research Centre, The University of Queensland, Brisbane, QLD, Australia

[1]Corresponding author: e-mail address: mcewan@uq.edu.au

## Contents



## Abstract

Trace metals such as Fe, Mn, Zn and Cu are essential for various biological functions including proper innate immune function. The host immune system has complicated and coordinated mechanisms in place to either starve and/or overload invading pathogens with various metals to combat the infection. Here, we discuss the roles of Fe, Mn and Zn in terms of nutritional immunity, and also the roles of Cu and Zn in metal overload in relation to the physiology and pathogenesis of two human streptococcal species, *Streptococcus pneumoniae* and *Streptococcus pyogenes*. *S. pneumoniae* is a major

*Advances in Microbial Physiology*, Volume 70
ISSN 0065-2911
http://dx.doi.org/10.1016/bs.ampbs.2017.01.002

human pathogen that is carried asymptomatically in the nasopharynx by up to 70% of the population; however, transition to internal sites can cause a range of diseases such as pneumonia, otitis media, meningitis and bacteraemia. *S. pyogenes* is a human pathogen responsible for diseases ranging from pharyngitis and impetigo, to severe invasive infections. Both species have overlapping capacity with respect to metal acquisition, export and regulation and how metal homeostasis relates to their virulence and ability to invade and survive within the host. It is becoming more apparent that metals have an important role to play in the control of infection, and with further investigations, it could lead to the potential use of metals in novel antimicrobial therapies.

## 1. ROLE OF TRANSITION METALS IN BIOLOGY

Trace metal ions are essential nutrients for life and it has been estimated that nearly 50% of all enzymes in cells require a metal cofactor (Andreini, Bertini, Cavallaro, Holliday, & Thornton, 2008). It is now understood that cells use dedicated pathways for metal acquisition and trafficking to ensure adequate supply and correct insertion of metal ions into these metalloproteins (Atkinson & Winge, 2009; Finney & O'Halloran, 2003; Waldron & Robinson, 2009; Waldron, Rutherford, Ford, & Robinson, 2009). These systems are faced with a considerable challenge since the binding affinity of proteins to transition metal ions follows the Irving–Williams series (Mn(II) < Fe(II) < Co(II) < Ni(II) < Cu(II) or Cu(I) > Zn(II)) (Irving & Williams, 1953). It is becoming clear that tuning the properties of metal-sensing regulators of gene expression and control of metal homeostasis allows the cell to operate within the thermodynamic constraints imposed by the Irving–Williams series (Waldron & Robinson, 2009). Nevertheless, under conditions of metal starvation or metal overload more competitive metals (higher in the series) may outcompete weaker binding metals ions (lower in the Irving–Williams series) leading to mismetallation, toxicity and, ultimately, cell death (Foster, Osman, & Robinson, 2014).

In mammalian hosts, there is significant movement of some transition metal ions in response to infection (Hood & Skaar, 2012; Johnstone & Nolan, 2015; Ma, Terwilliger, & Maresso, 2015; McDevitt et al., 2011; Nies, 2007; Shankar & Prasad, 1998; Sobocinski, Canterbury, Mapes, & Dinterman, 1978). This movement is usually linked to the restriction of nutrient metal availability (Fe, Zn and Mn) (Hood & Skaar, 2012; Kehl-Fie & Skaar, 2010; Weinberg, 2009) but emerging evidence suggests that in some cases, highly competitive (and thus potentially toxic) metals

(Cu and Zn) are delivered to sites of infection, where they may exert a direct antimicrobial role (Djoko, Ong, Walker, & McEwan, 2015; Stafford et al., 2013).

All free-living and host-associated bacteria possess systems for trace metal ion homeostasis. While they share several key features (Waldron & Robinson, 2009), organism-specific variations do exist. This review focuses on two pathogenic streptococci, *Streptococcus pyogenes* and *Streptococcus pneumoniae.* Based on the knowledge from fundamental discoveries, as well as observations made in other closely related species, we describe the key metal ion homeostasis systems in these two pathogenic streptococci and relate this to their physiology, pathogenesis and interaction with their host.

## 2. GENERAL ASPECTS OF THE BIOLOGY OF STREPTOCOCCI

The genus *Streptococcus* is a diverse group of oxidase-negative, coagulase-negative, Gram-positive and lactic acid bacteria. This genus includes a variety of commensal and pathogenic types (Hardie & Whiley, 1997). Human pathogenic streptococci include *S. pyogenes* (Group A), *S. agalactiae* (Group B), *S. bovis* (Group D), *S. anginosus* (Group F), *S. canis* (Group G), *S. equisimilis* (Group C), *S. mitis*, *S. mutans*, *S. pneumoniae*, *S. sanguinis* (Group H) and *S. suis* (Group RS). This grouping relies on classification by Lancefield antigen type on cell surface carbohydrate (Lancefield, 1933).

Some streptococci such as group B *Streptococcus* occupy a broad host range, while others, such as *S. pneumoniae* and *S. pyogenes* are exclusively human adapted. Both *S. pneumoniae* and *S. pyogenes* might be considered as commensal pathogens in that they are usually associated with an ecological niche within the human host and their interaction with the host does not always result in disease. This phenomenon, known as asymptomatic carriage, is a potential route for development and transmission of infection (Bogaert, De Groot, & Hermans, 2004). Infection by both pathogens can cause severe and life-threatening diseases (Henriques-Normark & Normark, 2010; Mitchell, 2003; Walker et al., 2014).

### 2.1 Pathogenesis

*S. pneumoniae* colonizes the nasopharynx of healthy individuals and rates of asymptomatic carriage are estimated to be 8% in healthy children (Marchisio et al., 2002). *S. pneumoniae* infection is responsible for a higher rate of morbidity and mortality than any other pathogen (Forrest, McIntyre, & Burgess, 2000).

Pneumococcal diseases, such as otitis media, pneumonia, septicaemia and meningitis (AlonsoDeVelasco, Verheul, Verhoef, & Snippe, 1995), cause close to 1 million deaths annually in children under 5 years old worldwide (O'Brien et al., 2009). Cases of nonfatal *S. pneumoniae* meningitis may also result in significant postinfection sequelae such as hearing loss and neurological impairment (Mook-Kanamori, Geldhoff, van der Poll, & van de Beek, 2011). Vaccines are available but coverage is limited (Bogaert et al., 2004; Kyaw et al., 2006). Adding to the substantial disease burden is the problem of rising resistance to antibiotics (Charpentier & Tuomanen, 2000). The first case of penicillin resistance was documented in 1967 (Hansman, Glasgow, Sturt, Devitt, & Douglas, 1971) but treatment failures have now reported with a variety of other antibiotics, such as macrolides (Lonks et al., 2002) and fluoroquinolones (Chen, McGeer, de Azavedo, Low, & Network, 1999).

*S. pyogenes* is carried asymptomatically in the pharynx (at rates between 2.2% and 37%) (Durmaz et al., 2003; Hoffmann, 1985; Levy, Leyden, & Margolis, 2005; Roberts et al., 2012) and on the skin in the general population (Cunningham, 2000; Shaikh, Leonard, & Martin, 2010). Carriers also exist where treatment failure results in continued presence of the bacterium without disease symptoms (Gidengil, Kruskal, & Lee, 2013). The spectrum of disease states caused by *S. pyogenes* ranges from mild suppurative infections of the pharynx and skin, to invasive diseases such as streptococcal toxic shock syndrome, and necrotizing fasciitis. Prior *S. pyogenes* infections may also cause significant postinfectious immune sequelae, such as poststreptococcal glomerulonephritis and rheumatic heart disease (Walker et al., 2014). *S. pyogenes* is estimated to cause 2 million new cases of disease and 500,000 deaths annually worldwide (Carapetis, Steer, Mulholland, & Weber, 2005). There is no licensed vaccine against *S. pyogenes* (Henningham, Gillen, & Walker, 2013) and, despite the sensitivity of this bacterium to penicillin (Albrich, Monnet, & Harbarth, 2004), there have been reports of increased resistance to macrolides (Davies et al., 2015; Martin, Green, Barbadora, & Wald, 2002; Michos et al., 2009).

Both *S. pyogenes* and *S. pneumoniae* produce a variety of virulence factors essential for colonization and survival in the host (Bisno, Brito, & Collins, 2003; Mitchell & Mitchell, 2010; Walker et al., 2014). Among these are streptococcal toxins that abrogate immune responses and interfere with host cell signalling pathways (Barnett et al., 2015), as well as factors responsible for evasion of the innate immune system and promotion of invasive infection (Cunningham, 2000; Walker et al., 2014). Recently, evidence has indicated that disruption of metal homeostasis also affect virulence and survival of these bacteria and their ability to evade the host immune system. Therefore, metals have a greater than previously expected role in the clearance of pathogens in the host immune system.

## 2.2 Cellular Biochemistry of Metals

### 2.2.1 Iron

Although neither *S. pyogenes* nor *S. pneumoniae* possesses a respiratory chain, Fe is an important component of a number of key streptococcal enzymes. *S. pneumoniae* contains 16 predicted iron–sulphur ([Fe–S]) cluster containing proteins, of which half are uncharacterized (Estellon, de Choudens, Smadja, Fontecave, & Vandenbrouck, 2014). In addition to a ferredoxin and an L-serine dehydratase that connects the metabolism of this amino acid to pyruvate production (Grabowski, Hofmeister, & Buckel, 1993), there are several putative [Fe–S]-containing radical *S*-adenosyl methionine enzymes (Broderick, Duffus, Duschene, & Shepard, 2014). The role of such proteins is to maintain stable thiyl or glycyl radicals in key enzymes such as anaerobic ribonucleotide reductase (RNR) and pyruvate formate lyase, respectively. *S. pyogenes* also possesses the complement of enzymes described earlier (Estellon et al., 2014) but it also contains endonuclease III that is involved in DNA repair (Demple & Harrison, 1994). Biogenesis of [Fe–S] proteins in the two streptococci depends on the sulphur formation (Suf) system (Ayala-Castro, Saini, & Outten, 2008; Bandyopadhyay, Chandramouli, & Johnson, 2008).

The possibility that Fe might be a prosthetic group in some mononuclear metalloenzymes has emerged recently (Anjem & Imlay, 2012). Such enzymes include ribulose-5-phosphate epimerase (RPE), peptide deformylase, and threonine dehydrogenase. In *Escherichia coli*, it is estimated that there are more than 100 of these mononuclear metalloenzymes. Imlay and coworkers have shown that Fe-containing RPE is inactivated by hydrogen peroxide ($H_2O_2$) and that this enzyme was protected when it was loaded with Mn rather than Fe (Sobota & Imlay, 2011). Although the physiological significance of this Fe/Mn interplay is only beginning to be recognized, it may well be important in the physiology of streptococci, especially in relation to $H_2O_2$ stress. What is established is that *S. pneumoniae* and *S. pyogenes* use the ferritin family protein Dps/Dpr to protect against Fe and peroxide stress (see Section 4.1.1).

Although Fe is a key nutrient for bacteria, an excess of this metal ion is highly toxic, particularly when oxygen is available. Fe-catalysed production of reactive oxygen species (ROS) is a key source of this toxicity in bacteria. The Haber–Weiss reaction refers to the generation of the highly damaging hydroxyl radical by superoxide $O_2^{-\bullet}$ and $H_2O_2$, a reaction catalysed by redox of Fe (Haber & Weiss, 1932; Walling, 1971). Hydroxyl radical is capable of causing DNA damage and membrane stress within biological systems. Furthermore, superoxide attack at Fe sites within proteins has been shown to result in the dissociation of metal cofactor and incorporation of inappropriate metals (Gu & Imlay, 2013; Imlay, 2014).

### 2.2.2 Manganese

Mn is a key antioxidant in cells (Coassin, Ursini, & Bindoli, 1992; Horsburgh, Wharton, Karavolos, & Foster, 2002; McEwan, 2009). It is thus not surprising that the import of this metal ion during conditions of oxidative stress is a conserved survival strategy in prokaryotes (Faulkner & Helmann, 2011), and that blockage of Mn acquisition leads to greater sensitivity to oxidative stress (Counago et al., 2014; Eijkelkamp et al., 2014). The primary antioxidant role for Mn in streptococci is to act as a prosthetic group in SodA (Aguirre & Culotta, 2012; Archibald & Fridovich, 1981; Cheton & Archibald, 1988). This enzyme is present in both *S. pyogenes* and *S. pneumoniae* (Gerlach, Reichardt, & Vettermann, 1998; Martin et al., 1984; Yesilkaya et al., 2000) but the *S. pneumoniae* homologue may also function in the Fe-form (Eijkelkamp et al., 2014).

The beneficial nature of Mn to bacteria is highlighted by observations in the radiation-tolerant organism, *Deinococcus radiodurans*. This bacterium is able to withstand high levels of ionizing radiation due to accumulation of Mn (Cox & Battista, 2005; Daly et al., 2004; Sharma et al., 2013). Furthermore, the causative organism of Lyme disease, *Borrelia burgdorferi*, possesses no Fe-containing enzymes, but instead uses Mn as a cofactor (Posey & Gherardini, 2000).

During oxidative stress, Mn can also provide a protective effect by temporarily substituting for Fe in mononuclear Fe enzymes to reduce Fenton chemistry (Anjem & Imlay, 2012; Anjem, Varghese, & Imlay, 2009; Sobota & Imlay, 2011). In fact, bacterial physiology of Mn is closely associated with that of Fe, and maintaining the correct balance between Mn and Fe levels is an important function of metal homeostasis. A low Fe, high Mn state may confer a protective advantage during oxidative stress because the potential for Haber–Weiss chemistry is low and at the same time, Mn-SOD is functional. Conversely, a high Fe, low Mn state may be toxic.

Mn also elicits direct antioxidant effects independent of SOD or other Mn-binding enzymes. For instance, simple salts and complexes of Mn(II) ions are efficient superoxide scavengers (Archibald & Fridovich, 1982; Barnese, Gralla, Cabelli, & Valentine, 2008). Work with *Lactobacillus plantarum*, a SOD-deficient bacterium, showed that addition of Mn salts to the growth medium conferred SOD-like activity that could be quenched by the chelator EDTA or by dialysis (Archibald & Fridovich, 1981). Similar findings were obtained in *sodA sodB* knockout mutants of *E. coli*, in which chelatable Mn restored tolerance to oxidative stress and provided direct scavenging of superoxide (Al-Maghrebi, Fridovich, & Benov, 2002). The physiological significance of such nonenzymatic reactions remains unclear.

### 2.2.3 Zinc

Up to 9% of all proteins in *S. pneumoniae* may contain Zn as a cofactor (Sun et al., 2011). Many of these Zn-containing proteins play key roles in normal metabolism (McCall, Huang, & Fierke, 2000; Wątły, Potocki, & Rowińska-Żyrek, 2016). Some regulators containing structural Zn, such as PerR, are also required for virulence (Lee & Helmann, 2006a; Makthal et al., 2013). Although it is incapable of direct redox activity, Zn can exert strong antioxidant effects, likely through binding to proteins or providing protection of sulphydryl groups (Eide, 2011; Powell, 2000). In *S. pyogenes*, research has indicated Zn acquisition proteins are crucial for survival and virulence (Weston, Brenot, & Caparon, 2009) but a link to oxidative stress protection has not been established.

Research examining how Zn may become toxic to *S. pyogenes* has indicated that high Zn concentrations may exert toxic effects through interference with glycolytic enzymes and hyaluronic capsule biosynthesis (Ong, Walker, & McEwan, 2015), while other studies have suggested that high Zn concentrations may also inhibit the *S. pyogenes* cysteine protease, SpeB (Krishnan, Mukundan, Figueroa, Caruso, & Kotb, 2014), which could potentially diminish *S. pyogenes* virulence within the host. In *S. pneumoniae*, it has been shown that Zn-dependent inhibition of Mn uptake via MtsABC leads to increased susceptibility to oxidative stress (Eijkelkamp et al., 2014).

Dpr has a ferroxidase centre which may incorporate Zn ions as well as at a distinct surface location. This Zn binding has been suggested to provide protection against Zn stress (Haikarainen, Tsou, Wu, & Papageorgiou, 2010b; Tsou et al., 2008) (see Section 4.1.1).

### *2.2.4 Copper*

Cu is crucial for many enzymatic functions usually associated with oxygenases and oxidases (MacPherson & Murphy, 2007; Solomon, Sundaram, & Machonkin, 1996). Since streptococci are aerotolerant anaerobes with essentially a fermentative metabolism, it is not surprising that Cu-dependent enzymes have not been found in *S. pyogenes* or *S. pneumoniae*. However, this does not mean that streptococci do not interact with Cu ions, and there is evidence that suggests killing of streptococci in some innate immune cells through use of Cu overload (Johnson, Kehl-Fie, Klein, et al., 2015; Shafeeq, Yesilkaya, et al., 2011).

Cu is a redox-active metal and, for decades, it was thought that Cu exerts its toxic effect by acting as a prooxidant and promoting severe oxidative stress (Aguirre & Culotta, 2012; Valko, Morris, & Cronin, 2005; Dupont, Grass, & Rensing, 2011; Touati, 2000). However, new understanding of the cellular biochemistry of metal ions has suggested that such redox reactions are unlikely to occur inside the cytoplasm. The current paradigm is that Cu exerts an antimicrobial effect by adventitious binding and mismetallation in enzymes. This is because Cu is the most competitive metal on the Irving–Williams series and thus can exert significant toxicity by outcompeting weaker binding metals from metalloenzymes (Foster et al., 2014). Various in vitro studies have demonstrated that Cu can mismetallate in the binding sites for Fe, Mn and Zn (Anjem & Imlay, 2012; Cotruvo & Stubbe, 2012; Djoko & McEwan, 2013; Foster et al., 2014; Johnson, Kehl-Fie, & Rosch, 2015; Macomber & Hausinger, 2011; Macomber & Imlay, 2009; Turner et al., 2015).

It was recently shown that Cu toxicity in *S. pneumoniae* is not primarily associated with oxidative stress, and instead arose through inactivation of Mn-dependent enzymes. In *S. pneumoniae*, the primary target of Cu toxicity was the enzyme responsible for aerobic nucleotide synthesis, NrdD, which is a Mn-cofactored enzyme (Johnson, Kehl-Fie, & Rosch, 2015). This notion of mismetallation of enzymes fits with the paradigm of the Irving–Williams series where Cu, the metal highest in the series is capable of displacing cofactors of metals lower on the series.

## 2.3 Interaction With Innate Immune System

### *2.3.1 Neutrophils*

Upon bacterial entry to the host, innate immune cells, such as neutrophils, migrate to the site of infection to eliminate invading pathogens (Janeway & Medzhitov, 2002; Wagner & Roth, 2000). Cells of the innate immune system use oxidative mechanisms to kill phagocytosed microorganisms through the 'respiratory burst', where NADPH oxidase generates superoxide (Roos, van Bruggen, & Meischl, 2003). Within phagosomes, internalized bacteria encounter hydrogen peroxide, superoxide and the hydroxyl radical (Babior, 2000; Fang, 2004; Winterbourn, Kettle, & Hampton, 2016). Neutrophils are short-lived innate immune cells that undergo rapid cell death in a process that is distinct from apoptosis and is dependent upon generation of ROS (Brinkmann & Zychlinsky, 2007; Fox, Leitch, Duffin, Haslett, & Rossi, 2010) This process results in the release of DNA containing chromatin and granule proteins, elastase, cathepsin G and myeloperoxidase (Brinkmann et al., 2004). Collectively these neutrophil-derived components are known as neutrophil extracellular traps (NETs) on account of their ability to bind and kill bacterial pathogens (Wartha, Beiter, Normark, & Henriques-Normark, 2007).

*S. pyogenes* has been observed to evade killing by NETs (Buchanan et al., 2006; Wartha et al., 2007). This involves the phage-acquired *S. pyogenes* DNase, Sda1, which degrades exogenous DNA and NETs, allowing *S. pyogenes* to escape (Buchanan et al., 2006). This NET entrapment and escape have been shown to provide a key selective pressure for dissemination of hypervirulent *S. pyogenes* within the host (Walker et al., 2007). Phagocytosed *S. pyogenes* can be killed by neutrophils through low pH and bactericidal enzymes, in addition to superoxide generated by NADPH oxidase (Mayer-Scholl, Averhoff, & Zychlinsky, 2004; Nauseef, 2007; Segal, 2005). Recent evidence has indicated metal overload killing may also occur within neutrophils infected with *S. pyogenes* (Ong, Gillen, Barnett, Walker, & McEwan, 2014), and this will be discussed later in this review.

In the process of colonizing the host, *S. pneumoniae* also has to contend with migratory host neutrophils at the site of infection (Craig, Mai, Cai, & Jeyaseelan, 2009; Kadioglu & Andrew, 2004). Neutrophils are a crucial component of the host immune response to otitis media, and exudate from infected children has been shown to contain large quantities of NETs (Thornton et al., 2013), while other observations have indicated that children with recurrent otitis media may have a defect in neutrophil chemotaxis (Hakansson, Foucard, Hallgren, & Venge, 1980; Hill, Book, Hemming, & Herbst, 1977; Ichimura, 1982). The nature of killing of *S. pneumoniae* by neutrophils is less clear. Initial research led to the conclusion that this killing occurs in a NADPH oxidase-independent manner, suggesting that oxidative stress is likely not the route of killing (Standish & Weiser, 2009). This finding was further validated by the observation that patients with chronic granulomatous disease, which causes decreased production of ROS due to defects in NADPH oxidase, have normal killing responses against streptococci, including *S. pneumoniae*, despite defects in killing of other bacteria (Kaplan, Laxdal, & Quie, 1968; Mandell & Hook, 1969; Quie, White, Holmes, & Good, 1967). Instead, it has been proposed that *S. pneumoniae* is killed through the use of serine proteases produced within neutrophils granules (Standish & Weiser, 2009). These proteases are suspected to degrade surface exposed virulence factors and other proteins of *S. pneumoniae* and further facilitate exposure to microbicidal components within neutrophil granules (Standish & Weiser, 2009). Other work has indicated that *S. pneumoniae* is capable of reducing human neutrophil ROS production through undefined mechanisms, as evidenced by killing of *S. pneumoniae* D39 or R6 strains in comparison with deletion mutants in capsule polysaccharide locus (Barbuti, Moschioni, Fumarulo, Censini, & Montemurro, 2010). However, pneumolysin (PLY) induces ROS production within neutrophils via NADPH oxidase (Martner, Dahlgren, Paton, & Wold, 2008). The role of NETs in pathogenesis of *S. pneumoniae* has also been investigated and revealed that a cell surface endonuclease, EndA, is crucial for the degradation of NETs by *S. pneumoniae*. Furthermore, this study indicated that *S. pneumoniae* are trapped in but not killed by NETs (Beiter et al., 2006).

### 2.3.2 Macrophages

Macrophages are a subset of mononuclear phagocyte and are selectively differentiated from monocytes due to cellular responses to inflammation and tissue damage (Hume, 2006). Macrophages play a key role in the phagocytosis and killing of pathogens as well as removal of cellular debris during injury and infection (Dale, Boxer, & Liles, 2008). Different subpopulations of macrophages or monocytes are known to exist in the human host, these are bone marrow-derived circulating monocytes, infiltratory macrophages and resident tissue macrophages, a group which comprises interstitial macrophages, Langerhans cells, alveolar macrophages and Kupffer cells (Wermuth & Jimenez, 2015). A number of bacteria have been shown to subvert and/or coopt the host macrophage as a route to establish disease. This has been well documented for *Mycobacterium tuberculosis* (Pieters, 2008), *Salmonella* (de Jong, Parry, van der Poll, & Wiersinga, 2012; Helaine et al., 2014), *Legionella pneumophila* (Escoll, Rolando, Gomez-Valero, & Buchrieser, 2013) and *Listeria monocytogenes* (Birmingham, Higgins, & Brumell, 2008; Portnoy, Auerbuch, & Glomski, 2002).

Comparatively less is known about *S. pyogenes* phagocytosis, persistence and killing by macrophages. Early work examining *S. pyogenes* and macrophages interactions elucidated the importance of M protein in uptake of *S. pyogenes* within alveolar macrophages, a process which was dependent on serum opsonisation (Gemmell et al., 1981). One study has illustrated that in biopsy samples of severe invasive disease, viable *S. pyogenes* was found within macrophages as determined by immunostaining (Thulin et al., 2006). Additionally, this study indicated that *S. pyogenes* may be capable of escape from macrophages, although it was not established whether this occurs through induction of macrophage apoptosis or alternate escape pathways. Hertzén and colleagues also corroborated previous findings that macrophages can act as a reservoir of viable *S. pyogenes* and demonstrated that *S. pyogenes* was capable of multiplying within macrophages and ultimately escaping in an M1 protein-dependent manner (Hertzen et al., 2010). Further studies have examined gene expression of both *S. pyogenes* (Hertzen et al., 2012) and host cells (Goldmann et al., 2007), revealing the complex nature of *S. pyogenes* infection of innate immune cells.

*S. pneumoniae* is known to interact with resident alveolar macrophages during the establishment of pneumonia (Gordon & Read, 2002). Alveolar macrophage function has been shown to have a protective effect in a murine model of *S. pneumoniae* pneumonia, examining wild-type mice and macrophage-depleted mice (Knapp et al., 2003). While it is recognized that macrophages are capable of killing phagocytosed *S. pneumoniae*, there has also been indications that macrophages undergo apoptosis in order to kill phagocytosed *S. pneumoniae* (Aberdein, Cole, Bewley, Marriott, & Dockrell, 2013; Dockrell, Lee, Lynch, & Read, 2001).

## 2.4 Physiology and Metabolism

Streptococcal species are a member of the 'lactic acid bacteria' superfamily, which generate lactate as the terminal product of sugar catabolism (Farrell, 1935; Hewitt, 1932). Although they grow under aerobic conditions, *S. pyogenes* and *S. pneumoniae* do not carry out aerobic respiration (Farrell, 1935), even in the presence of an external supply of menaquinone and/or haem (Brooijmans, de Vos, & Hugenholtz, 2009; Yamamoto et al., 2005). Both pathogens can reduce molecular oxygen ($O_2$) to water using an NADH oxidase (Nox) but this reaction is not coupled to energy production (Gibson, Mallett, Claiborne, & Caparon, 2000; Zitzelsberger, Gotz, & Schleifer, 1984). A *nox* mutant of *S. pyogenes* was unable to grow under aerobiosis and carbon limitation (Gibson et al., 2000). In *S. pneumoniae*, a *nox* mutant displayed a defect in competence and reduced virulence in a mouse model of infection (Auzat et al., 1999). These observations illustrate the importance of aerobic metabolism in the biology and pathogenesis of these organisms.

Glucose is considered as the default carbon source for *S. pneumoniae*. It is converted to lactate during homolactic fermentation or to ethanol, formate, and acetate during mixed acid fermentation (Yesilkaya et al., 2009). The latter occurs during aerobiosis and is associated with availability of galactose as an alternative energy source, for example in the upper respiratory tract (Philips, Meguer, Redman, & Baker, 2003). It is likely that similar pathways are also used by *S. pyogenes* (Ong et al., 2015).

A hallmark of carbohydrate metabolism in *S. pneumoniae* is the generation of copious quantities of $H_2O_2$ as a byproduct. This is a consequence of pyruvate oxidase (SpxB) activity, which catalyses the oxidative decarboxylation of pyruvate to acetyl phosphate. $H_2O_2$ is also generated during oxidation of lactate to pyruvate by lactate oxidase (Lox) (Seki, Iida, Saito, Nakayama, & Yoshida, 2004). This production of $H_2O_2$ in the cytoplasm poses an unusual challenge for *S. pneumoniae*. This ROS is a strong oxidant of reactive thiols and may cause protein oxidation, DNA damage, as well as peroxidation of membrane lipids (Cabiscol, Tamarit, & Ros, 2000; Imlay, 2003, 2013). Production of $H_2O_2$ also leads to autolysis of *S. pneumoniae* (Regev-Yochay, Trzcinski, Thompson, Lipsitch, & Malley, 2007). The defensive and adaptive responses to $H_2O_2$ are thus crucial to the physiology of *S. pneumoniae*.

Despite its potential for toxicity, production of $H_2O_2$ by *S. pneumoniae* can be beneficial to this organism, particularly during colonization of the host. This ROS is thought to exert bactericidal effects on competing bacteria that also colonize the upper respiratory tract (Pericone, Overweg, Hermans, & Weiser, 2000). It is also toxic to host epithelial cells and hence its production by *S. pneumoniae* may be a factor in pathogenesis (Duane, Rubins, Weisel, & Janoff, 1993; Rai et al., 2015). For instance, a mutant lacking *spxB* showed reduced virulence in a murine model of infection (Ramos-Montanez et al., 2008; Spellerberg et al., 1996, #1800); whether this virulence defect stems from diminished $H_2O_2$ production or altered carbon metabolism has not been determined.

SpxB is absent from *S. pyogenes* but Nox is present. Some *S. pyogenes* isolates also produce $H_2O_2$ (Seki et al., 2004), though not in the large quantities seen with *S. pneumoniae* (Pericone, Bae, Shchepetov, McCool, & Weiser,

2002). Caperon and coworkers also found that higher levels of $H_2O_2$ accumulated in the *S. pyogenes nox* mutant, which was hypothesized to occur due to an increase in cellular NADH concentration, may keep Fe reduced as Fe(II), and thus produce superoxide anion which can be dismutated to $H_2O_2$ by superoxide dismutase (Gibson et al., 2000).

## 2.5 Oxidative Stress Responses

Bacterial response to $H_2O_2$ and other ROS can occur in two ways: (i) directly via enzymatic detoxification of ROS or (ii) indirectly via enzymatic protection or repair of damaged cellular components. These processes often overlap with homeostasis of metal ions, particularly Fe and Mn, but these metal-linked aspects of oxidative stress response will be described later in this review.

The oxidative stress response in *S. pyogenes* is governed by the Fur-paralogue PerR, which is a $H_2O_2$ sensor (Brenot, King, & Caparon, 2005; Grifantini, Toukoki, Colaprico, & Gryllos, 2011; King, Horenstein, & Caparon, 2000; Ricci, Janulczyk, & Bjorck, 2002). PerR controls expression of *nrdFIE* (encoding for a RNR), *ahpCF* (encoding for an alkyl hydroperoxidase), *dpr* (encoding for a Dps-like protein), *pmtA* (encoding for a metal efflux pump) (Bates, Toukoki, Neely, & Eichenbaum, 2005; Beres et al., 2006; Gryllos et al., 2008; Toukoki, Gold, McIver, & Eichenbaum, 2010) and a variety of additional genes that are linked with metal homeostasis. PerR is widely distributed in Gram-positive bacterial species (Herbig & Helmann, 2001; Ji et al., 2015; Wang, Tong, & Dong, 2014; Fuangthong, Herbig, Bsat, & Helmann, 2002, #2240; Zhang et al., 2012). Many of these PerR-containing species also possess additional Fur paralogues including Fur (Fe uptake) and Zur (Zn uptake) (Escolar, Perez-Martin, & De Lorenzo, 1999; Gaballa, Wang, Ye, & Helmann, 2002), but these additional proteins are absent from *S. pyogenes*.

PerR exists as a dimer and features one structural Zn(II) site (Traore et al., 2006) and one sensor Fe(II) site. During normal growth conditions, PerR binds to per boxes and represses gene expression (Lee & Helmann, 2006a; Makthal et al., 2013). During conditions of oxidative stress, oxidation of the Fe(II) site by $H_2O_2$ results in a conformational change, dissociation of PerR from the per boxes, which induces gene transcription (Herbig & Helmann, 2001; Lee & Helmann, 2006b; Traore et al., 2009). Thus the Fe(II) site is essential to the appropriate sensing of $H_2O_2$. When this site is occupied by less reactive Mn, the PerR protein is no longer responsive to $H_2O_2$ (Makthal et al., 2013) and gene transcription remains repressed.

Unlike *S. pyogenes*, *S. pneumoniae* does not possess an inducible oxidative stress response (Pericone, Park, Imlay, & Weiser, 2003) because it lacks a PerR. In addition, homologues of other known oxidative stress regulators, such as OxyR, OhrR, and SoxRS (Green & Paget, 2004) are also absent (Faulkner & Helmann, 2011; Yesilkaya, Andisi, Andrew, & Bijlsma, 2013).

### 2.5.1 Direct Detoxification

Like most bacteria, streptococci can employ low molecular weight thiols such as glutathione (GSH) as an antioxidant in the cytoplasm (Masip, Veeravalli, & Georgioui, 2006). GSH exert its antioxidant effects directly by quenching a variety of ROS species (Franco & Cidlowski, 2009; Lu & Holmgren, 2014; Masip et al., 2006) and indirectly by acting as a cofactor for GSH-dependent detoxification enzymes. These processes generate oxidized glutathione (GSSG). Both *S. pneumoniae* and *S. pyogenes* possesses an NADPH-dependent glutathione reductase (Gor) that reduces GSSG and regenerates GSH (Masip et al., 2006; Potter, Trappetti, & Paton, 2012). *S. pyogenes* also possesses a glutathione peroxidase (GpoA), which catalyses the detoxification of peroxides (Masip et al., 2006) and confers virulence in a murine model (Brenot, King, Janowiak, Griffith, & Caparon, 2004).

Intriguingly, both *S. pyogenes* and *S. pneumoniae* lack the glutamate–cysteine ligase GshA, which catalyse the first of the two steps in GSH biosynthesis, and thus these pathogens were thought to obtain GSH exclusively from external sources such as the host. However, a recent study suggested that bacterial species lacking *gshA* may synthesise γ-glutamyl-cysteine (the product of GshA) via an alternative pathway (Veeravalli, Boyd, Iverson, Beckwith, & Georgiou, 2011). It is unclear if this intermediate is further converted to GSH in *S. pyogenes* and *S. pneumoniae* as the enzyme that usually catalyses this step, GshB, is also absent. In *S. pneumoniae*, GSH is imported into the cytoplasm via the transporter GshT and a *gshT* mutant was attenuated in a mouse model of infection (Potter et al., 2012). A GshT homologue is also present in *S. pyogenes*.

All streptococci lack a catalase and instead use peroxidases to detoxify $H_2O_2$. *S. pneumoniae* possess a thioredoxin-dependent peroxidase (PsaD or TxpD) (Hajaj et al., 2012). PsaD may operate under the control of the Mn sensor PsaR and the link with oxidative stress is discussed in Section 4.2.1.1. In addition to GpoA, *S. pyogenes* possesses an alkyl hydroperoxidase (AhpCF), which is upregulated by PerR during oxidative stress and an additional regulator, Rgg (Brenot et al., 2005; King et al., 2000; Pulliainen, Hytonen, Haataja, & Finne, 2008). Both of these organisms also contain a Mn-dependent superoxide dismutase (SOD or SodA), which catalyses the disproportionation of $O_2^{-\bullet}$ into $H_2O_2$ and water (Gerlach et al., 1998; Miller, 2012; Yesilkaya et al., 2000).

### 2.5.2 Indirect Detoxification

Fe-containing enzymes catalyse a variety of key cellular processes and represent major targets of ROS toxicity as a consequence of localized Fenton chemistry (Imlay, 2013; Walling, 1971). Thus, the protection and repair of these enzymes is a key feature of oxidative stress response in many bacteria. To avoid loss of function during conditions of oxidative stress, Fe-containing enzymes may be substituted with a Fe-free homologue. For instance, RNRs catalyse the formation of deoxyribonucleotides for DNA synthesis (Gon & Beckwith, 2006). *S. pyogenes* possesses two RNRs, NrdHEF and NrdFIE. NrdHEF acts as the primary RNR during normal metabolism. It binds Fe as a cofactor and thus this enzyme is likely ROS sensitive. By contrast, NrdFIE does not require Fe for function and thus it is presumably ROS insensitive. Under conditions of oxidative stress, where NrdHEF may be inactivated, NrdFIE is upregulated by PerR (Grifantini et al., 2011; Gryllos et al., 2008; Roca, Torrents, Sahlin, Gibert, & Sjoberg, 2008; Toukoki et al., 2010). In *E. coli*, substitution of Fe by Mn as a protective mechanism is also known to occur (Sobota & Imlay, 2011), but this has not been demonstrated in streptococci. *S. pyogenes* also feature the DNA polymerase PolA1, which is cotranscribed with and regulated by PerR. PolA1 is suggested to repair oxidative stress-induced DNA damage (Toukoki & Gryllos, 2013).

Proteases such as the HtrA family of serine proteases may also participate in oxidative stress response by digesting and recycling ROS-damaged proteins (Jones, Bolken, Jones, Zeller, & Hruby, 2001). An *htrA* mutant of *S. pneumoniae* lacking a functional HtrA protease activity (Cassone, Gagne, Spruce, Seeholzer, & Sebert, 2012) was unable to tolerate $H_2O_2$ and was attenuated in a murine model of infection (Cassone et al., 2012; Gasc, Giammarinaro, Richter, & Sicard, 1998; Ibrahim, Kerr, McCluskey, & Mitchell, 2004). HrtA in *S. pyogenes* is important in the processing of virulence factors SpeB and streptolysin S (Cole et al., 2007; Lyon & Caparon, 2004). *S. pneumoniae* encodes an additional ATP-dependent serine protease denoted ClpP, which is also thought to play a role in the degradation of damaged or misfolded proteins during conditions of oxidative stress. Like the *htrA* mutant, a *clpP* mutant was sensitive to $H_2O_2$ stress and it was attenuated in a murine model of pneumonia (Ibrahim, Kerr, Silva, & Mitchell, 2005; Robertson, Ng, Foley, Gilmour, & Winkler, 2002). More subtle metabolic activities can also contribute to the overall ability of pathogenic streptococci to tolerate excess ROS. For instance, in vitro data has shown that SpxB is a key determinant in the ability of *S. pneumoniae* to tolerate exogenously generated $H_2O_2$. This study found that *spxB* null mutants generate reduced quantities of acetyl phosphate, and accumulate lower levels of ATP during conditions of oxidative stress. These findings suggested that SpxB functions to provide an oxidative stress-resistant energy source (Echlin et al., 2016; Pericone et al., 2003).

Secondary mechanisms for oxidative stress defence in *S. pyogenes* and *S. pneumoniae* include several electron transfer enzymes, such as flavodoxin, thioredoxin, periplasmic methionine reductases (*msrAB*) (Delaye, Becerra, Orgel, & Lazcano, 2007; Kim, Shin, Lee, Kim, & Hwang, 2009), thioredoxin-like lipoprotein (*tlpA*) (Andisi et al., 2012), thioredoxin reductase (*trxB*), ferredoxin, and glutaredoxin (Collet & Messens, 2010), but their precise roles in conferring resistance to a variety of ROS have yet to be determined. For recent in-depth reviews of the oxidative stress response in *S. pyogenes* and *S. pneumoniae*, see the following excellent reviews (Henningham, Dohrmann, Nizet, & Cole, 2015; Yesilkaya et al., 2013).

## 3. METAL IONS AND THEIR ROLE IN INFECTION CONTROL WITHIN THE HOST

### 3.1 Metal Starvation Within the Host

#### *3.1.1 Iron*

Starving invading pathogens of essential nutrient metal ions is a key strategy of host innate immune defence (Kehl-Fie & Skaar, 2010). This concept, also termed 'nutritional immunity', is best understood in the case of Fe. Modulation of circulation and homeostasis of host Fe in response to bacterial infection is a well-documented phenomenon. In this process, infection induces production of hepcidin in the liver via IL-6 (Nemeth et al., 2004) and subsequently promotes hepcidin-dependent degradation of the Fe efflux transporter ferroportin. This action reduces release of Fe from liver stores (Nemeth et al., 2004) and results in a rapid reduction in serum Fe levels. Unsurprisingly, individuals with haemochromatosis or thalassemia major, two disorders in Fe metabolism that are characterized by systemic Fe overload, are more susceptible to infection by some opportunistic bacterial pathogens (Bullen, Spalding, Ward, & Gutteridge, 1991; Khan, Fisher, & Khakoo, 2007; Teawtrakul, Jetsrisuparb, Sirijerachai, Chansung, & Wanitpongpun, 2015; Vento, Cainelli, & Cesario, 2006; Zurlo et al., 1989).

Host Fe is stored and transported by transferrin, ferritin, and lactoferrin (Conrad & Umbreit, 2000). Transferrin is the major Fe carrier in the blood plasma and its levels range between 2 and 3.6 mg/mL (Skikne, Flowers, & Cook, 1990; Wish, 2006). Ferritin is primarily intracellular but serum levels can fluctuate between 12 and 200 ng/mL (Jacobs, Miller, Worwood, Beamish, & Wardrop, 1972; Koperdanova & Cullis, 2015; Walters, Miller, & Worwood, 1973), while lactoferrin is present mainly in secretory fluids (Jonas, Kruger, & Tessenow, 1993; Levay & Viljoen, 1995; Weinberg, 2007). These Fe carriers display extremely high binding affinities for Fe(III) (Aisen, Leibman, & Zweier, 1978; Harris, 1986) and therefore limit 'free' Fe concentrations in their relevant niches to subzeptomolar ($10^{-20}$ M) concentrations (Aisen et al., 1978; Harris, 1986). In vitro, the action of these proteins as strong Fe chelating agents has been shown to exert antimicrobial activity (Arnold, Cole, & McGhee, 1977; Bezkorovainy, 1982), implying that systemic Fe restriction in the host may be an important defensive strategy during bacteremia and sepsis (Weinberg, 2009).

In the context of nutritional immunity, it would seem logical that Fe is also withdrawn from the phagolysosomal compartments of innate immune cells. In macrophages, transport of divalent cations such as Fe(II), Mn(II), and Zn(II) between phagosomal compartments and the cytosol occurs via Nramp1 (Blackwell, Searle, Goswami, & Miller, 2000; Cellier, Courville, & Campion, 2007; Forbes & Gros, 2001; Nevo & Nelson, 2006) but the direction of metal transport by Nramp1 is still contentious. There is evidence that, contrary to the prevalent model of Fe limitation, Fe accumulates within the phagosomes of Nramp1$^{+}$ murine macrophages in response to infection (Kuhn, Baker, Lafuse, & Zwilling, 1999; Kuhn, Lafuse, & Zwilling, 2001; Zwilling, Kuhn, Wikoff, Brown, & Lafuse, 1999). It is plausible that excess Fe here may potentiate the action of host-derived ROS. However, it was later shown that accumulation of Fe in these phagosomes was conditional on acidification of the phagosome (Goswami et al., 2001). It was subsequently proposed that Nramp1 is capable of bidirectional transport, depending on the conditions of the host niche and the context of infection, although these details remain to be elucidated (Fritsche et al., 2007).

### *3.1.2 Zinc*

Zn plays a complex role in innate and adaptive immune cell development, signalling and differentiation (Prasad, 2008; Wirth, Fraker, & Kierszenbaum, 1984) but it is known that individuals with low dietary Zn intake display reduced immune cell function and are often more susceptible to infection

(Ibs & Rink, 2003; Prasad, 2000). Around 12–16 μM of Zn has been detected in human serum, where it is bound predominantly to albumin (Rink & Gabriel, 2000, #1917; Scott & Bradwell, 1983), and between 2.8 and 12 μM Zn has been observed in sputum (Harlyk, Mccourt, Bordin, Rodriguez, & Van der Eeckhout, 1997; Rizzato et al., 1986; Sabbioni et al., 1987; Sulotto et al., 1986). Zn transport and trafficking in mammalian host cells involve 24 different transporters, 10 from the ZnT family and 14 from the ZIP family of proteins, that often overlap in function and location (Eide, 2006; Guerinot, 2000; Jeong & Eide, 2013; Kambe, Hashimoto, & Fujimoto, 2014). During infection, Zn is redistributed from serum to liver stores (Hoeger et al., 2015; Liuzzi et al., 2005; Sobocinski et al., 1978) as a consequence of elevated transcription of hepatic Zn transporters and Zn-binding metallothioneins (Lichten & Cousins, 2009; Liuzzi et al., 2005; Suhy, Simon, Linzer, & O'Halloran, 1999). This decrease in systemic Zn concentrations is important for the control of the inflammatory response (Haase & Rink, 2009) and is likely to help limit availability of 'free' Zn to invading bacteria.

At the cellular level, there is evidence that sequestration of Zn away from invading pathogens may occur via ZIP8, which mobilizes Zn out of the lysosome and into the cytosol (Haase et al., 2008; Liu et al., 2013), or ZnT4 and ZnT7, which export cytosolic Zn to the Golgi (Vignesh, Figueroa, Porollo, Caruso, & Deepe, 2013). However, it is thought that calprotectin plays a more dominant role (Clohessy & Golden, 1995; Kehl-Fie et al., 2011, 2013). Calprotectin binds Zn(II) and Mn(II) strongly in a Ca(II)-dependent manner (Brophy, Hayden, & Nolan, 2012; Hayden, Brophy, Cunden, & Nolan, 2013) but it may also bind Fe(II) (Nakashige, Zhang, Krebs, & Nolan, 2015). The action of calprotectin as a metal chelator exerts a potent in vitro antimicrobial activity against a variety of pathogens, including *Candida albicans*, *Staphylococcus aureus* and *E. coli* (Damo et al., 2013; Kehl-Fie et al., 2013; Loomans, Hahn, Li, Phadnis, & Sohnle, 1998; Sohnle, Collinslech, & Wiessner, 1991; Steinbakk et al., 1990; Urban et al., 2009).

Between 0.65 and 1.1 μg/mL of calprotectin is found in human plasma (Hetland, Berntzen, & Fagerhol, 1992) but these normal circulating levels are elevated in infectious tissue abscesses (Corbin et al., 2008). Calprotectin is also found in the cytoplasm of activated macrophages (Striz & Trebichavsky, 2004) and neutrophils, where it constitutes nearly half of the total cytosolic protein (Hessian, Edgeworth, & Hogg, 1993; Striz & Trebichavsky, 2004). It is thought that calprotectin is secreted by activated neutrophils in NETs (Hessian et al., 1993; Striz & Trebichavsky, 2004) and by monocytes following protein kinase C stimulation (Rammes et al., 1997). A key feature of the metal chelation property of calprotectin is the requirement for the binding of Ca(II) ions in order to coordinate Zn(II) ions (Brophy et al., 2012). The concentration of Ca ions intracellularly is known to be substantially lower than in extracellular spaces (Brini, Ottolini, Cali, & Carafoli, 2013). Hence, calprotectin would only exert its effect once it is released from the cell, for example when it is released from neutrophils in NETs.

### *3.1.3 Manganese*

As described earlier, calprotectin also restricts availability of Mn from invading pathogens (Davidsson, Lonnerdal, Sandstrom, Kunz, & Keen, 1989; Herrera, Pettiglio, & Bartnikas, 2014; Kehl-Fie et al., 2011). Serum levels of Mn in healthy adults range between 110 and 250 nM (Sullivan, Blotcky, Jetton, Hahn, & Burch, 1979; Versieck, Barbier, Speecke, & Hoste, 1974). However, relatively little is known regarding the systemic movement of host Mn in response to inflammation. It is known that in infectious tissue abscesses there are elevated levels of calprotectin that contributes to the starvation of pathogens for Zn and Mn at these sites (Corbin et al., 2008).

## 3.2 Metal Overload

### *3.2.1 Copper*

The concept of metal overload, as opposed to metal starvation, as an antimicrobial strategy in the host is relatively new and is best exemplified in the case of Cu. This metal ion is an essential micronutrient for the innate immune system and dietary deficiency (hypocupremia) results in defects in immune function, including decreases in circulating neutrophil levels and increased susceptibility to infection (Dunlap, James, & Hume, 1974; Heresi, Castilloduran, Munoz, Arevalo, & Schlesinger, 1985; Jones & Suttle, 1983; Mulhern & Koller, 1988; Percival, 1995; Prohaska & Lukasewycz, 1981; Samanovic, Ding, Thiele, & Darwin, 2012). Cu in this context was always assumed to play a regulatory role but an additional concept has recently emerged in which Cu is employed as a direct antimicrobial agent, particularly within phagolysosomal compartments of innate immune cells. However, the molecular mechanisms for this process are yet to be determined and the few available details are not clearly distinguishable from indirect effects on Fe metabolism (Djoko & McEwan, 2013).

So far, it has been demonstrated that Cu trafficking pathways in murine macrophages are modulated in response to inflammation (White, Kambe, et al., 2009). Infection by *Salmonella* upregulated the expression of CTR1 and CTR2, two high-affinity Cu uptake transporters in the plasma membrane (Lee, Pena, Nose, & Thiele, 2002), as well as ceruloplasmin and ATP7A (Achard et al., 2012). The latter usually resides in the Golgi, where it plays a role in the transport of Cu for biogenesis of ceruloplasmin, but there is evidence that ATP7A is trafficked to the phagosome in response to IFN-γ and lipopolysaccharide (Petris et al., 1996; White, Lee, Kambe, Fritsche, & Petris, 2009). Consequently, RNAi silencing of ATP7A in murine macrophages attenuated bacterial killing (White, Lee, et al., 2009).

#### 3.2.2 Zinc

The first evidence for a role of host Zn in the direct overload killing of invading pathogens was obtained by examination of human macrophages infected with either *E. coli* or *M. tuberculosis* (Botella et al., 2011). Using a Zn-responsive fluorescent probe, it was observed that infection led to a global increase in the levels of probe-detectable Zn within the macrophage (Botella et al., 2011). This study also showed that bacterial mutants defective in Zn export (*zntA* in *E. coli* and *ctpC* in *M. tuberculosis*) were impaired in their ability to survive within these human macrophages. An increase in global Zn pools was also observed in murine macrophages upon infection with the fungus *Histoplasma capsulatum* in the presence of granulocyte macrophage-colony stimulating factor (Vignesh et al., 2013) and in human neutrophils upon infection by *S. pyogenes* (Ong et al., 2014). Likewise, mutation of the *czcD* gene encoding Zn efflux in *S. pyogenes* led to a decrease in survival within these neutrophils. The precise intracellular locations of these elevated Zn pools are unknown and whether they represent the release of Zn ions from cytosolic stores and/or an increased import of extracellular Zn is unclear.

## 4. MECHANISMS FOR METAL ION HOMEOSTASIS

### 4.1 Iron

#### 4.1.1 Dpr

Dps proteins are members of the ferritin family and have been shown to provide protection against oxidative stress through binding to DNA nonspecifically (Arnold & Barton, 2013; Calhoun & Kwon, 2011). *S. pyogenes* and *S. pneumoniae* possess a Dps-like protein, Dpr, which does not bind DNA like other Dps proteins (Martinez & Kolter, 1997), but rather acts to protect against oxidative stress through Fe sequestration (Tsou et al., 2008).

Dpr in streptococci has been most extensively studied in *S. suis* where it has been shown to reduce oxidative stress. This likely occurs through the binding of free intracellular Fe(II) which is then sequestered, thus preventing Fe-catalysed oxidative stress and ROS formation (Kauko, Haataja, Pulliainen, Finne, & Papageorgiou, 2004; Kauko et al., 2006; Pulliainen, Kauko, Haataja, Papageorgiou, & Finne, 2005). In *S. pyogenes*, a similar observation regarding $H_2O_2$ sensitivity of a *dpr* mutant has been made and it was also shown that resistance to oxidative stress could be partially restored through treatment with a Fe(II) chelator (Tsou et al., 2008). Additionally, the *dpr* mutant was sensitive to excess Fe and Zn. While recombinant Dpr protein binds Fe(II), it did not bind DNA (Tsou et al., 2008), unlike archetypal Dps proteins which bind DNA nonspecifically (Martinez & Kolter, 1997). Further biochemical characterization of *S. pyogenes* Dpr revealed that the protein exhibits ferroxidase activity which is likely important in preventing the Fenton reaction and exists as a dodecamer with Fe binding at sites between dimers at a ferroxidase centre (Haikarainen, Tsou, Wu, & Papageorgiou, 2010a). Surprisingly, a *dpr* (King et al., 2000) mutant exhibited full virulence in a murine model of infection despite greater sensitivity to challenge with $H_2O_2$ (Brenot et al., 2005; King et al., 2000).

In *S. pneumoniae*, Dpr has also been shown to provide protection from $H_2O_2$ stress (Hua, Howard, Malley, & Lu, 2014). A *dpr* mutant was more sensitive to heat, high Fe, and changes in pH (Hua et al., 2014). Furthermore, the *dpr* mutant was more susceptible than wild-type strain to killing by macrophages in an ex vivo assay and was also attenuated for survival in a murine model of nasopharyngeal colonization. Additionally, a competition experiment with wild type showed that the *dpr* mutant was completely attenuated for survival in this niche (Hua et al., 2014) which may be related to production of hydrogen peroxide by *S. pneumoniae*. In *S. pyogenes*, expression of Dpr is regulated by PerR (Brenot et al., 2005), and by RitR in *S. pneumoniae* (Ulijasz, Andes, Glasner, & Weisblum, 2004).

### *4.1.2 Uptake*

*S. pyogenes* can acquire Fe in either ferric (Fe(III)) or ferrous (Fe(II)) form from a variety of sources, including 'free' haem, ferritin, haemoglobin, myoglobin, lactoferrin and catalase (Eichenbaum, Muller, Morse, & Scott, 1996), but not transferrin (Francis, Booth, & Becker, 1985). Similarly, *S. pneumoniae* can use haem and haemoglobin as Fe sources, but not lactoferrin or transferrin (Romero-Espejel, Gonzalez-Lopez, & Olivares-Trejo, 2013; Tai, Lee, & Winter, 1993). Although the release of haemoglobin from erythrocytes via β- or α-haemolysis is a key pathogenic mechanism for both pathogenic streptococci (Betschel, Borgia, Barg, Low, & De Azavedo, 1998; Molloy, Cotter, Hill, Mitchell, & Ross, 2011; Nizet et al., 2000; Paton, 1996; Shewell et al., 2014; Tilley, Orlova, Gilbert, Andrew, & Saibil, 2005), whether this process is coupled to haem acquisition is unclear. Both *S. pyogenes* and *S. pneumoniae* do not synthesize known siderophores but are able to use ferric ferrichrome, a hydroxamate siderophore produced by other bacteria in the niche, as an Fe source (Hanks, Liu, McClure, & Lei, 2005; Krewulak & Vogel, 2008; Pramanik & Braun, 2006).

Pneumococcal surface protein A (PspA) is a highly immunogenic protein and it has been examined as a potential vaccine target (Briles et al., 1998; Crain et al., 1990). Additionally, PspA has been shown to inhibit activation of the complement cascade (Tu, Fulgham, McCrory, Briles, & Szalai, 1999). Other studies on this protein have found that PspA binds human lactoferrin (Hakansson et al., 2001; Hammerschmidt, Bethe, Remane, & Chhatwal, 1999), and that a mutant in *pspA* exhibited diminished survival following incubation with the metal free *apo*-form of lactoferrin and that prior anti-PspA antibody treatment resulted in further enhanced killing by *apo*-lactoferrin (Shaper, Hollingshead, Benjamin, & Briles, 2004). Thus, it can be concluded that PspA may be involved in Fe acquisition from lactoferrin. For a recent review examining aspects of Fe acquisition in streptococci, see Ge and Sun (2014).

#### 4.1.2.1 Sia/Hts Uptake Pathway for Haem

*S. pyogenes* acquires haem directly by the concerted action of Shr, Shp and SiaABC (HtsABC) in the linear Sia/Hts transport pathway (Montanez, Neely, & Eichenbaum, 2005). Haem transport in this system occurs stepwise. Haemoglobin and other haemoproteins are recognized and bound by the surface-associated receptor Shr. Haem is extracted with the concomitant reduction of the Fe(III) centre to Fe(II) by two NEAT domains within this protein and is subsequently transferred to the haem-binding protein Shp (Bates, Montanez, Woods, Vincent, & Eichenbaum, 2003; Lei et al., 2002; Ouattara et al., 2010). Haem is finally shuttled by Shp to SiaA and imported into the cytoplasm by the ABC-type transporter SiaABC (Liu & Lei, 2005; Nygaard et al., 2006; Ran et al., 2007). Consistent with the current mechanistic model for general metal homeostasis, each step in this process occurs via direct and specific protein–protein interaction. Hence, Shp does not obtain haem directly from haemoglobin and Shr does not transfer haem directly to SiaA (Zhu, Liu, & Lei, 2008).

An *shr* mutant of *S. pyogenes* grew poorly in human blood (Dahesh, Nizet, & Cole, 2012). In addition, it was more resistant to killing by hydrogen peroxide or the Fe-dependent antibiotic streptonigrin (Bates et al., 2003). This result would be consistent with a reduced level of cytoplasmic Fe, and thus a lesser potential for Fenton chemistry, as a consequence of impaired uptake. Shr in *S. pyogenes* M49 was shown to be cell surface exposed and also interacts with human laminin and fibronectin, aiding in adherence to immortalized human epithelial cells. An *shr* mutant displayed reduced virulence in a zebrafish model of infection (Fisher et al., 2008). Further characterization of Shr in *S. pyogenes* M1T1 showed that a null mutant was attenuated for survival in a murine model of infection, growth in human blood, but not survival in neutrophils. This mutant did not display a defect in adherence to HEp-2 cells, contrary to findings using the M49 strain from Fisher et al. (2008), which may be due to differences in cell-surface exposed adhesions between these two M-types (Dahesh et al., 2012). Subsequent experimentation revealed that Shp is also expressed at the cell surface, where it is accessible by antibodies and furthermore is expressed during in vivo infection as assessed by ELISA. Biochemical characterization of purified Shp confirmed a 1:1 haem binding stoichiometry (Lei et al., 2002) and the process of haem transfer to SiaA, results in the formation of a transient complex between the two proteins (Liu & Lei, 2005; Nygaard et al., 2006; Ran et al., 2007).

*S. pyogenes* encodes an putative haem oxygenase, denoted HupZ, which is capable of the oxidative degradation of haem and is suggested to play a role in Fe release from haem in *S. pyogenes* (Sachla et al., 2016). *S. pneumoniae* lacks this enzyme and may utilize another enzyme for the catabolism of haem.

The ABC transporter component SiaABC has been shown to bind haemoglobin, myoglobin, haem–albumin and haemoglobin–haptoglobin. Furthermore, it has been shown that SiaA is surface exposed in *S. pyogenes* (Bates et al., 2003). Biochemical characterization of recombinant SiaA suggested a relatively high reduction potential of the haem protein complex and that haem transfer may be linked to changes to haem redox state (Sook et al., 2008). Further study has reported that SiaA may undergo a partial unfolding and refolding in the process of trafficking haem (Akbas et al., 2016). These acquisition systems are depicted in Fig. 1.

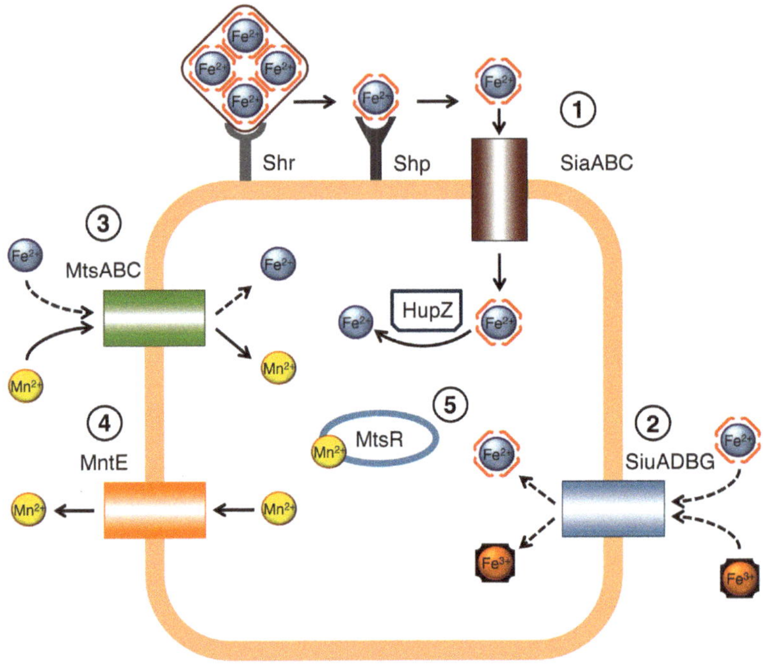


**Fig. 1** Model of Fe and Mn homeostasis in *Streptococcus pyogenes*. *Solid arrows* indicate demonstrated metal transport and *dashed arrows* indicate putative metal transport. Haem = Fe(II) in *red brackets*; haemoglobin = four coordinated haem; ferric ferrichrome = *black boxed* Fe(III). (1) Haemoglobin is bound to Shr and a single haem molecule is extracted. The haem is then transferred it Shp, for direct uptake into the cytoplasm via SiaABC. Fe is then extracted from haem by the haem oxygenase, HupZ. (2) SiuADBG potentially imports either haem or ferric ferrichrome into the cytoplasm. (3) MtsABC imports Mn(II) and potentially Fe(II) into the cytoplasm. (4) MntE exports Mn(II) out of the cytoplasm. (5) The regulator MtsR exists in a Mn-bound form under normal conditions and regulates the *shr*, *shp*, *sia* and *mts* operon.

#### 4.1.2.2 Siu/Fhu/Fts

The Siu (Fts) operon (*siuADBG*) was first identified as a potential Fe transport operon that was upregulated in Fe-limited growth conditions (Smoot et al., 2001). Further study of the Siu operon has indicated two different functions for this set of genes. Work by Eichenbaum and colleagues first showed that the Siu system is an ABC transporter consisting of ATP-binding protein SiuA, a substrate-binding protein SiuD, and two intrinsic membrane components SiuBG (Montanez et al., 2005). Study of a *siuG* mutant revealed a reduction in ferric $^{55}Fe$ accumulation when compared to wild-type NZ131. However, a defect in intracellular Fe accumulation was also evident when the *siuG* and *siuB* mutants were grown in Fe-depleted media supplemented with haem. With increasing concentrations of haem, these single mutants reached wild-type levels of $^{55}Fe$ accumulation; however, a double *siuB siuG* mutant accumulated lower $^{55}Fe$, suggesting that Siu is responsible for haem acquisition and that haem may block the uptake of Fe by other transporters. Furthermore, it was revealed that the *siuG* mutant was defective for virulence in a zebrafish model of infection (Montanez et al., 2005).

Lei and colleagues then identified that SiuADBG may be able to acquire Fe through use of exogenous siderophores, with ferric ferrichrome as the primary substrate (Hanks et al., 2005). This work highlighted that *S. pyogenes* is capable of utilizing ferric ferrichrome as an Fe source in Fe-limited growth media and that inactivation mutants of both *siuB* and *siuC* are defective for $^{55}Fe$ ferric ferrichrome uptake (Hanks et al., 2005). Despite the finding of hydroxymate siderophore-type acquisition proteins in *S. pneumoniae* and *S. pyogenes*, it is understood that both organisms lack the necessary biosynthetic genes required to synthesize hydroxamate siderophores (Hanks et al., 2005; Pramanik & Braun, 2006). Thus, it has been suggested that *S. pneumoniae* and *S. pyogenes* may use exogenously synthesized ferrichrome produced by other resident bacteria within the nasopharynx as has been suggested to occur with *L. monocytogenes* (Simon, Coulanges, Andre, & Vidon, 1995), *Enterococcus faecalis* and *Enterococcus faecium* (Szarapinska-Kwaszewska & Mikucki, 2001), a genera which is known to lack ferrichrome (Burnham & Neilands, 1961). Taken together, these data seem to suggest that the Siu operon plays a role in the uptake of Fe, but the relationship to haem is unclear at this stage (Fig. 1).

#### 4.1.2.3 Pia(Fhu) and Piu

The first examination of Fe uptake in *S. pneumoniae* identified two distinct ABC-type transporters encoded by the *piaABCD* and *piuBCDA* operons (Brown, Gilliland, & Holden, 2001). Growth of single *piuB* or *piaA* (Fhu) mutants was not significantly altered in metal ion-depleted medium but a double *piuA/piuB* mutant was attenuated for growth rate relative to wild type, and growth of this mutant could be partially restored through supplementation with Fe(II) or Fe(III), but not by haemoglobin, lactoferrin or ferritin (Brown, Gilliland, & Holden, 2001). This corroborates previous work suggesting that *S. pneumoniae* is unable to use lactoferrin or transferrin as sources of Fe, but it is able to use haemin and haemoglobin (Tai et al., 1993), suggesting that the Piu or Pia transport systems may be involved in Fe uptake from haemoglobin. Furthermore, mutants in either *piuB* or *piaA* showed enhanced survival in the presence of the Fe-dependent antibiotic, streptonigrin, suggesting that the mutants have lower intracellular Fe accumulation. In examination of ferric $^{55}Fe$ accumulation after challenge with $^{55}FeCl_3$, neither of the single mutants in *piuB* or *piaA* were attenuated for $^{55}Fe$ uptake, however the double mutant had approximately 50% lower intracellular $^{55}Fe$ and was also found to be avirulent in a murine model of pulmonary infection (Brown, Gilliland, & Holden, 2001). Following on from this work, Brown and colleagues examined the efficacy of the lipoprotein portions of the Piu and Pia operons, PiuA and PiaA, respectively, as potential vaccine targets. Immunization of mice with either protein was found to be approximately equally protective in a murine model of invasive infection, and a summative effect was achieved from immunization with both proteins together (Brown, Ogunniyi, Woodrow, Holden, & Paton, 2001). It has also been documented that the PiuA and PiaA lipoproteins are widespread among *S. pneumoniae* strains as well as *S. oralis* and *S. mitis* strains (Whalan et al., 2006). Furthermore, RT-PCR revealed that there is the highest abundance of the *piaA* transcript within wild-type *S. pneumonia*, which, together with findings examining the relative inhibition of the three mutants, suggests that Pia may be the dominant Fe transporter for *S. pneumoniae* (Brown, Gilliland, Ruiz-Albert, & Holden, 2002). PiuA has since been revealed to bind haemin and haemoglobin and found to be cytoplasmic membrane-associated and sequestered from the cell surface (Tai, Yu, & Lee, 2003). These iron acquisition systems are detailed in Fig. 2.

Recombinant PiaA bound the antibiotics albomycin and salmycin and was shown to mediate the binding and uptake of Fe from ferric hydroxymate (ferrichrome) and ferrioxamine B (Pramanik & Braun, 2006). This type of system has also been described in *S. agalactiae* (Clancy et al., 2006). The crystal structure of PiaA has been elucidated, revealing a structural site for ferrichrome that undergoes structural changes during the binding and release of ferrichrome (Cheng, Li, Jiang, Zhou, & Chen, 2013).

#### 4.1.2.4 Pit

A third Fe uptake system encoded by an ABC transporter, *pitADBC* has also been identified in *S. pneumoniae*. Deletion of the lipoprotein component *pitA*, led to a defect in Fe uptake and a lower sensitivity to streptonigrin (Brown et al., 2002), while deletion of *pitA* in a *piuB/piaA* double mutant further increased streptonigrin sensitivity and growth in Fe-limited conditions (Brown et al., 2002). Both phenotypes are consistent with a lower level of intracellular Fe import.

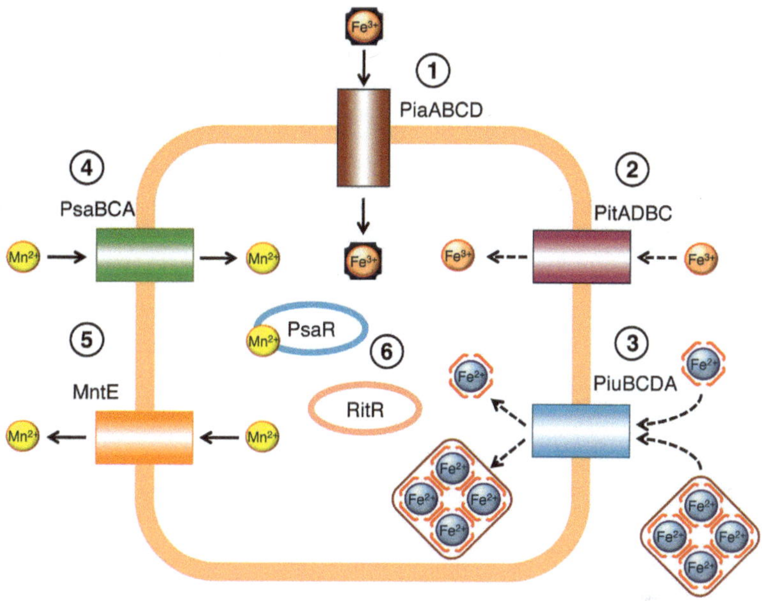


**Fig. 2** Model of Fe and Mn homeostasis in *Streptococcus pneumoniae. Solid arrows* indicate demonstrated metal transport and *dashed arrows* indicate putative metal transport. Haem = Fe(II) in *red brackets*; haemoglobin = four coordinated haem; ferric ferrichrome = *black boxed* Fe(III). (1) PiaABCD imports ferric ferrichrome into the cytoplasm. (2) PitADBC potentially imports Fe(III) into the cytoplasm. (3) PiuBCDA potentially imports either haem or haemoglobin into the cytoplasm. (4) Mn is imported into the cytoplasm via PsaBCA complex and (5) exported from the cytoplasm via MntE. (6) PsaR likely exists in the Mn-bound form and regulates the *psa* operon, while RitR regulates the *piu* operon.

### *4.1.3 Efflux*

Although it has hitherto always been assumed that bacteria would store excess Fe, there is emerging evidence for Fe(II) efflux in Gram-positive bacteria. This was first revealed to occur in *Bacillus subtilis* via the PerR-regulated $P_{1\text{-}B4}$ P-type ATPase, peroxide-inducible ferrous efflux transporter, PfeT (formerly YkvW/ZosA), which was previously thought to be a Zn acquisition protein (Gaballa & Helmann, 2002; Guan et al., 2015). Following this discovery, a similar $P_{1\text{-}B4}$ P-type ATPase was found in *L. monocytogenes*, denoted Fur-regulated virulence locus FrvA, which also effluxes Fe(II) (Pi, Patel, Arguello, & Helmann, 2016). A homologue, PmtA has been identified in *S. pyogenes* but it was thought to function in the transport of Zn (Brenot, Weston, & Caparon, 2007). We have recently shown that PmtA functions as a PerR-regulated Fe(II) transporter in

*S. pyogenes* which plays a role in resistance to oxidative stress (Turner et al., in preparation).

#### 4.1.3.1 Pef

β-Haemolysis of erythrocytes by *S. pyogenes* may liberate large quantities of haemoglobin and as such there is evidence that bacteria may require haem homeostasis. The first insight into this phenomenon was described in *S. aureus* and the proteins involved denoted HrtAB, which encodes a haem-regulated exporter belonging to the ABC family (Stauff et al., 2008). Since this study, a similar system has been discovered in *S. agalactiae* which comprises a MarR-family regulator PefR, and the adjacent operons of PefAB and PefRCD which seem to constitute a member of the major facilitator superfamily (Fernandez et al., 2010). In *S. agalactiae*, a *pefAB* or *pefCD* mutant is sensitive to haem and displays increased haem accumulation (Fernandez et al., 2010). It has also been shown that *S. pyogenes* has a strong requirement for the maintenance of haem homeostasis, as high haem concentrations are toxic and can result in membrane lipid damage and protein oxidation (Anzaldi & Skaar, 2010). Furthermore, high haem exposure results in the upregulation of a homologue of the *pefRCD* system (Sachla, Le Breton, Akhter, McIver, & Eichenbaum, 2014). Further characterization in *S. pyogenes* revealed that the PefCD is from the class-1 ABC transporter family and PefAB is a putative multidrug resistance system belonging to the drug/proton antiporter family (Sachla & Eichenbaum, 2016). In plate-based assays, a *pefCD* mutant is more sensitive to extracellular haem as well as a number of toxic compounds such as ampicillin, erythromycin, doxorubicin and DNA-binding agents, ethidium bromide and Hoechst 33342 (Sachla & Eichenbaum, 2016). Thus, it is thought that Pef functions to defend against excess haem and protoporphyrin IX accumulation through the efflux of these compounds (Sachla & Eichenbaum, 2016). This system does not appear to be present in *S. pneumoniae* (Sachla et al., 2014).

### *4.1.4 Regulation*

#### 4.1.4.1 PerR

As discussed earlier, *S. pyogenes* uses the Fur family regulator, PerR to coordinate the oxidative stress response through sensing of $H_2O_2$. PerR has been shown to regulate *dpr* (Brenot et al., 2005) and it additionally regulates the *siu* operon as well as the *sia* operon together with *shr* and *shp* (Grifantini et al., 2011). These observations are corroborated by findings that a *perR* mutant accumulates reduced $^{55}$Fe, which may also be due to reduced *mtsA*

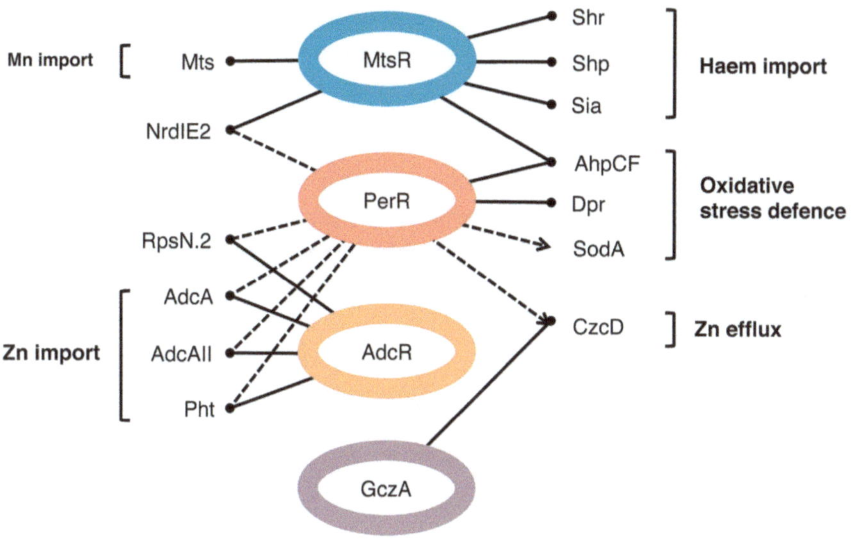


**Fig. 3** Summary of gene regulation of metal homeostasis proteins in *Streptococcus pyogenes*. *Arrows* denote upregulation during normal conditions, while *circular ends* denote repression during normal conditions. *Solid lines* indicate confirmed, direct regulation while *dashed lines* denote indirect regulation.

transcription (Ricci et al., 2002). In *S. pyogenes*, PerR is known to contain a structural Fe site whose occupancy is sensitive to hydrogen peroxide (Herbig & Helmann, 2001; Makthal et al., 2013). The metal homeostasis and oxidative stress defence genes regulated by PerR are depicted in Fig. 3. It is interesting to note that *S. pneumoniae* does not encode PerR. It seems likely that *S. pneumoniae* instead uses the orphan response regulator RitR to control Fe homeostasis (Ong et al., 2013). This will be discussed in Section 4.1.4.4.

#### 4.1.4.2 MtsR

The DtxR family regulator MtsR of *S. pyogenes* may act as a sensor of intracellular Mn and also function in the regulation of intracellular Fe/Mn ratio in *S. pyogenes*. Examination of transcriptional changes in *S. pyogenes* M1 illustrated that *siaA* is constitutively upregulated in a *mtsR* null mutant and is downregulated through addition of Fe(III), an effect that seems to be dependent on MtsR (Hanks et al., 2006). MtsR also negatively regulates the expression of the putative metal efflux protein PmtA in normal growth conditions (Toukoki et al., 2010). The metal homeostasis proteins regulated by MtsR are depicted in Fig. 3. The haem oxygenase-like protein (HupZ) is also regulated by MtsR in *S. pyogenes* (Sachla et al., 2016).

#### 4.1.4.3 PefR

The Pef operon for haem efflux in *S. agalactiae* is controlled by the MarR-family repressor PefR (Fernandez et al., 2010). This regulator has recently been described in *S. pyogenes* (Sachla & Eichenbaum, 2016; Sachla et al., 2014). Recombinant PefR binds haem in a 1:2 stoichiometry and protoporphyrin IX in equimolar ratio (Sachla et al., 2014). It was proposed that haem binding to PefR leads to derepression of *pefCD* (Sachla et al., 2014). PefR binds a putative PefR recognition sequence of AGTTATCTAGAGAACTA, which is similar to the sequence found in *S. agalactiae* (Fernandez et al., 2010; Sachla et al., 2014). Research examining transcriptional responses of haem toxicity in *S. pyogenes* revealed down-regulation of the *sag* operon encoding streptolysin S (Sachla et al., 2014), suggesting a possible feedback loop with β-haemolysis during conditions of haem abundance.

#### 4.1.4.4 RitR

*S. pneumoniae* does not possess a PerR and its MtsR-like regulator (PsaR) is not thought to participate in Fe homeostasis. Instead, Fe transport in *S. pneumoniae* is regulated by RitR, an orphan response regulator without a cognate sensor kinase (Ulijasz et al., 2004). In Fe-replete conditions, RitR downregulates expression of *piu* genes for Fe uptake and activates expression of *dpr* for Fe storage (Ulijasz et al., 2004), presumably to avoid Fe-linked toxicity in the cytoplasm. The regulation of these genes by RitR is shown in Fig. 4. RitR also regulates a range of genes not involved in metal homeostasis, such as putative ABC transporters involved in transport of sugars and phosphate, as well as a variety of metabolic genes (Ulijasz et al., 2004). Loss of regulation in a *ritR* mutant led to the accumulation of intracellular Fe and impaired growth in the presence of supplemental Fe or the Fe-dependent antibiotic, streptonigrin. However, normal Fe homeostasis and growth were restored by addition of Mn (Ong et al., 2013; Ulijasz et al., 2004). The various protective effects of Mn have been discussed in Section 2.2.2. The *ritR* mutant was attenuated for survival in the mouse lung but not in a deep tissue model of invasive disease or a mouse model of systemic infection (Ong et al., 2013; Ulijasz et al., 2004). The results from these studies may point to niche-specific metal concentrations and requirements for metal homeostasis.

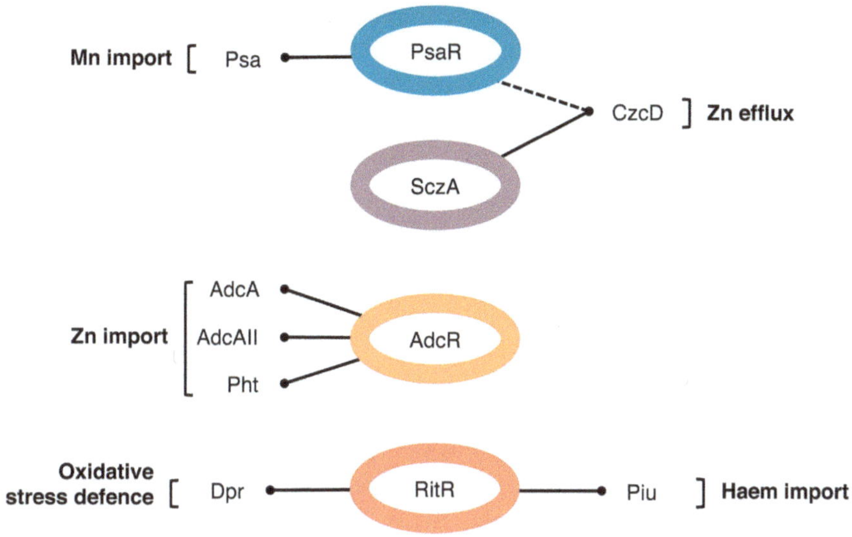


**Fig. 4** Summary of gene regulation of metal homeostasis proteins in *Streptococcus pyogenes*. *Arrows* denote upregulation during normal conditions, while *circular ends* denote repression during normal conditions. *Solid lines* indicate confirmed, direct regulation while *dashed lines* denote indirect regulation.

## 4.2 Manganese

### *4.2.1 Uptake*

#### 4.2.1.1 PsaBCA/PsaD

The Psa system for Mn acquisition in *S. pneumoniae* consists of the ABC family transporter PsaBCA and the adjacent thiol peroxidase PsaD (Novak, Braun, Charpentier, & Tuomanen, 1998). Mutants of *psaB*, *psaC*, or *psaA* were all Mn-deficient (McAllister et al., 2004) and required supplemental Mn for normal growth (Johnston et al., 2004; Tseng, McEwan, Paton, & Jennings, 2002). These mutants were also unable to survive challenge by $H_2O_2$ or the superoxide generator paraquat unless exogenous Mn was added (Johnston et al., 2004; McAllister et al., 2004; Tseng et al., 2002). The alternative pathway for Mn import under these conditions has yet to be identified.

Several lines of evidence suggest that the Psa system is important for *S. pneumoniae* virulence. Firstly, *psa* promoters were activated during a murine lung model of pneumonia (Marra, Lawson, Asundi, Brigham, & Hromockyj, 2002) and *psa* knockout mutants were attenuated in a variety of intranasal and invasive murine models of infection, as well as a gerbil model of otitis media (Berry & Paton, 1996; Marra et al., 2002;

McAllister et al., 2004). In a murine model of systemic infection, the $LD_{50}$ dose for these *psa* mutant strains was 10,000-fold greater compared to the wild type, but virulence was restored if the bacterial inoculum was supplemented with Mn (Marra et al., 2002).

The PsaBCA permease also appears play a role in *S. pneumoniae* physiology and pathogenesis that may or may not be independent of Mn. For instance, *psa* mutants were diminished in competence (Dintilhac, Alloing, Granadel, & Claverys, 1997; Ogunniyi et al., 2010). In addition, PsaA, the solute-binding component of the PsaBCA transporter, was originally characterized as a surface adhesin that promotes adherence to lung epithelial cells (Gosink, Mann, Guglielmo, Tuomanen, & Masure, 2000; Johnston et al., 2004; Novak et al., 1998). Structural examination has indicated that PsaA may not protrude significantly beyond the *S. pneumoniae* capsule (Jedrzejas, 2001; Lawrence et al., 1998; Rajam, Anderton, Carlone, Sampson, & Ades, 2008; Tomasz, 1981). As a consequence, surface availability of PsaA varies with capsule thickness, which is linked to phase variation between opaque and transparent phenotypes (Weiser, Austrian, Sreenivasan, & Masure, 1994) and also virulence (Berry & Paton, 1996; Kim & Weiser, 1998; Romero-Steiner et al., 2003). Mn import by the PsaBCA complex is shown in Fig. 2.

**4.2.1.1.1 PsaR** Uptake of Mn via the PsaBCA permease is regulated by PsaR, a member of the DtxR/MntR family of transcriptional repressors. PsaR is a homodimer with one structural Zn(II) site and one Mn(II) site that directly senses the intracellular concentrations of this metal ion (Johnston, Briles, Myers, & Hollingshead, 2006; Lisher, Higgins, Maroney, & Giedroc, 2013). Derepression of *psa* genes by PsaR occurs during Mn(II) limitation which results in the dissociation of the Mn(II) ion from PsaR and generating *apo*-PsaR. This has also been shown to occur in response to excess Zn (Kloosterman, Witwicki, van der Kooi-Pol, Bijlsma, & Kuipers, 2008) but this observation is linked to Mn starvation as a consequence of Zn(II)-dependent blockage of PsaBCA (Counago et al., 2014; Eijkelkamp et al., 2014). PsaBCA and PsaD are thought to be regulated by PsaR in a separate manner. PsaBCA features a PsaR consensus sequence denoted P1, while PsaD features a site denoted P2 (Novak et al., 1998). This differential regulation is observed in a *psaR* null mutant which has constitutive upregulation of *psaBCA*, while *psaD* is not differentially regulated (Johnston et al., 2006; McAllister et al., 2004).

PsaR appears to control multiple physiological processes in *S. pneumoniae* that are not immediately linked with Mn uptake or trace metal homeostasis. Deletion of *psaR* led to a reduction of virulence in a murine model of lung infection (Johnston et al., 2006), but strain-specific differences in phenotypes have been reported. Deletion of *psaR* in both strains resulted in constitutive upregulation of three genes; the *psaBCA* operon, *pcpA* and *prtA*. However D39 or TIGR4 also had 31 and 14 differentially regulated genes, respectively. It was observed that in a murine model of pneumonia, a D39 *psaR* mutant had higher bacterial load at 12 h postinfection, whereas a *psaR* mutant in TIGR4 had lower bacterial load than wild type at 24 h post-infection (Hendriksen et al., 2009).

Expression of *psaBCA* is also controlled by a two-component system (TCS), TCS04 (McCluskey, Hinds, Husain, Witney, & Mitchell, 2004), which shows some similarity to the PhoP/PhoQ system described in *Salmonella typhimurium* (Miller, Kukral, & Mekalanos, 1989), *B. subtilis* (Hulett, 1996) and *E. coli* (Kato, Tanabe, & Utsumi, 1999). This TCS senses low availability of divalent metals Ca(II), Mg(II) and Mn(II) (Groisman, 2001). Furthermore, a knockout of the response regulator in *S. pneumoniae* TIGR4 was attenuated for virulence in a murine model of infection, whereas mutants in D39 or serotype 3 *S. pneumoniae* were equally virulent as the wild-type strain (McCluskey et al., 2004). This serotype-dependent phenotype was also evident in microarray analysis of the three strains illustrating that the response regulator knockout in TIGR4 had diminished regulation of the *psaBCA* operon and only moderate down-regulation in D39 (McCluskey et al., 2004). Furthermore, it was observed that the mutant in TIGR4 was more sensitive than wild-type strain to challenge by hydrogen peroxide (McCluskey et al., 2004), which is consistent with other findings examining Mn restriction in *S. pneumoniae* (Eijkelkamp et al., 2014).

#### 4.2.1.2 Mts

The PsaBCA-like system for Mn acquisition in *S. pyogenes* is named MtsBCA. An *mtsA* mutant strain grew poorly in aerobic conditions but growth was restored partially if cultured in a microaerobic environment and fully if the culture medium also contained supplemental Mn(II) (Janulczyk, Ricci, & Bjorck, 2003). These findings are consistent with the notion that Mn is involved in defence against excess $H_2O_2$, which is generated as a by product of aerobic metabolism (see Section 2.4). Like the *psaA* mutant of *S. pneumoniae*, an *mtsA* mutant of *S. pyogenes* was susceptible to killing by $H_2O_2$ and paraquat. This ROS-sensitive phenotype correlated with a reduced SodA activity, which was restored readily by Mn supplementation (Janulczyk et al., 2003), suggesting that SodA was not adequately metallated as a consequence of a defect in Mn uptake in the *mtsA* mutant.

There is a suggestion that MtsBCA also facilitates import of Fe, based primarily on a higher binding affinity of MtsA for Fe(II) (Janulczyk, Pallon, & Bjorck, 1999; Sun, Ge, Chiu, Sun, & He, 2008) and the observation that exogenous Fe ions can compete with Mn import (Janulczyk et al., 2003). In addition, an *mtsA* mutant was shown to be Fe-deficient (Janulczyk et al., 1999). Given what we now know about the importance of Fe/Mn balance in the cytoplasm, this phenotype may be a consequence of decreased Fe import or increased Fe export, either of which may occur to protect the cytoplasm from Fenton chemistry in the absence of Mn. It is important to note that no Fe-transporting role has been identified for PsaBCA in *S. pneumoniae*. Nevertheless, uptake of Fe(II) via these Mn(II) permeases is plausible given the size, as well as comparable ionic radius and other physicochemical properties of these two ions. This Fe(II)/Mn uptake by MtsABC is shown in Fig. 1. Whichever the metal substrate, its import into the cytoplasm via MtsBCA is important for virulence, as determined in an invasive murine model of infection (Janulczyk et al., 2003).

**4.2.1.2.1 MtsR** Under Mn-replete conditions, uptake of Mn via MtsBCA is repressed by the PsaR-like Mn sensor MtsR. This transcriptional regulator also represses 64 additional genes, many of which are co-regulated by PerR (e.g. *nrdFIE*, *ahpC* and *pmtA*) (Bates et al., 2005; Beres et al., 2006; Gryllos et al., 2008; Toukoki et al., 2010). This multilayered regulation is highlighted in Fig. 3. Intriguingly, the MtsR consensus sequence (ATTAAGTTNAGTTAAT) is also found upstream of *sia* and *siu* genes for Fe uptake, and thus MtsR is also thought to act as a cellular Fe sensor (Hanks et al., 2006). Mutation of *mtsR* resulted in sensitivity to streptonigrin (Bates et al., 2005), consistent with derepression of Fe uptake and the accumulation of Fe in the cytoplasm. There is preliminary evidence that binding of Mn(II) and Fe(II) to the sensor site in MtsR can induce different allosteric mechanisms for DNA binding at different promoters (Hanks et al., 2006; Toukoki et al., 2010). Hence, MtsR may coordinate Fe and Mn homeostasis systems in response to the cellular Fe/Mn ratio but this proposal is yet to be fully tested.

MtsR also controls expression of several genes that are not implicated in metal homeostasis, including *ska* and *psrA*. The *ska* gene encodes for streptokinase, which converts host plasminogen to its active form plasmin, a key determinant of invasive disease (Cole, Barnett, Nizet, &

Walker, 2011; Walker, McArthur, McKay, & Ranson, 2005). *psrA* is a chaperonin that catalyses posttranslational modification of SpeB, another global virulence determinant (Olsen et al., 2010). Despite this apparent connection to known virulence factors, loss of *mtsR* does not yield a clear and consistent avirulent phenotype in a variety of animal models (Bates et al., 2005; Beres et al., 2006; Hanks et al., 2006).

### *4.2.2 Efflux*

#### 4.2.2.1 MntE

The efflux of Mn from the cytoplasm of both pathogenic streptococci occurs via a cation diffusion facilitator (CDF) family protein named MntE (Rosch, Gao, Ridout, Wang, & Tuomanen, 2009; Turner et al., 2015). This protein appears to be important for virulence in an in vivo murine model (Rosch et al., 2009) and an ex vivo human neutrophil model (Turner et al., 2015). In view of overwhelming evidence for the beneficial role of Mn, the need to efflux Mn would seem counter-intuitive. Yet, toxicity of Mn was shown to occur and growth of the *mntE* mutant strains of both *S. pneumoniae* and *S. pyogenes* were suppressed by high concentrations of Mn, likely as a result of the trapping of *excess* Mn ions in the cytoplasm (Rosch et al., 2009; Turner et al., 2015). Mn efflux by MntE in *S. pyogenes* and *S. pneumoniae* is shown in Figs. 1 and 2, respectively. In addition, supplemental Mn sensitized, instead of protected, the *mntE* mutant of *S. pyogenes* to oxidative stress (Turner et al., 2015). It was postulated that MntE may protect the peroxide response regulator, PerR, from mismetallation by excess Mn (Turner et al., 2015). The Fe(II) sensor site in PerR also displays a high affinity to Mn(II), leading to the formation of a stable Mn-PerR form that is redox insensitive and is thus unable to sense $H_2O_2$ (Lee & Helmann, 2006b; Makthal et al., 2013). This example illustrates that metal uptake and efflux systems must be tightly controlled and coordinated to maintain adequate nutrition, and at the same time avoid metal poisoning.

## 4.3 Zinc

### *4.3.1 RpsN.2*

Zn-restricted conditions within the host can also result in adaptive physiological changes to deal with low free Zn. One such mechanism involves the two 30S ribosomal subunits, RpsN and RpsN.2 (Natori et al., 2007). It has been best studied in *B. subtilis* and is present in some streptococci. In the

*S. pyogenes*, *rpsN* is encoded adjacent to other genes involved in ribosome synthesis. The protein RpsN features a key Zn-ribbon motif with two conserved CXXC sites. By contrast, its homologue RpsN.2 lacks this motif but features an AdcR-binding site and is chromosomally segregated from other ribosomal protein genes (Makarova, Ponomarev, & Koonin, 2001). During Zn starvation, AdcR relieves the repression of the Zn-free homologue RpsN.2, leading to incorporation of RpsN.2 into the S14 ribosomal subunit. This mechanism is thought to allow the bacterium to reduce the cellular requirement for Zn (Panina, Mironov, & Gelfand, 2003; Sanson et al., 2015). Interestingly, *S. pneumoniae* lacks this Zn-free subunit, suggesting a different strategy in coordinating responses to Zn starvation (Shafeeq, Kloosterman, & Kuipers, 2011). The *S. pyogenes rpsN.2* gene appears to be indirectly positively regulated by the peroxide response regulator, PerR (Gryllos et al., 2008), but the physiological significance of this regulation is yet to be elucidated.

### *4.3.2 Uptake*

#### 4.3.2.1 Adc

Zn uptake in *S. pneumoniae* and *S. pyogenes* occurs via an ABC transporter, which consists of the substrate (Zn)-binding domain AdcA, transmembrane domain AdcB, ATP-binding domain AdcC and the regulator AdcR. The genetic organization of the *adc* genes differs between both species. In *S. pneumoniae*, the *adc* genes are encoded together on the chromosome and are arranged as *adcABC* followed by *adcR* (Plumptre et al., 2014). In *S. pyogenes*, *adcRBC* is spatially segregated from *adcA* (Sanson et al., 2015). Furthermore, both species contain a second homologue of AdcA, termed AdcAII, and the gene encoding this protein is situated on a separate site to the *adc* operon within the chromosome with adjacent Pht family protein (*phtDE* in *S. pneumoniae* and *phtD* in *S. pyogenes*).

In *S. pneumoniae*, it was determined that both AdcA and AdcAII were functionally redundant (Bayle et al., 2011). Additionally, only the *adcA*/*adcAII* double deletion mutant was unable to grow without supplemental Zn (Bayle et al., 2011). Furthermore, a defect in Zn uptake in vitro was only observed when both *adcA* and *adcAII* were deleted. The resultant *adcA*/*adcAII* double mutant accumulated approximately 25% less Zn than the wild type (Plumptre, Eijkelkamp, et al., 2014). The Zn-binding ability of AdcAII has also been confirmed through metal-binding studies with the purified protein (Loisel et al., 2008). Additionally, an *adcAII* mutant in

*S. pneumoniae* is defective in invasion of airway epithelial cells when compared to wild type and is more efficiently cleared from the nasopharynx in a murine model of infection (Brown et al., 2016).

In *S. pyogenes*, AdcAII was originally denoted Lmb or Lsp, based on its characterization as an adhesin involved in the attachment and invasion of immortalized mammalian cell lines through binding to laminin (Elsner et al., 2002; Terao, Kawabata, Kunitomo, Nakagawa, & Hamada, 2002). This protein showed high sequence identity to AdcAII in *S. pneumoniae* and structural characterization of the recombinant protein indicated the presence of a Zn(II) site and concluded that this protein functioned as a Zn receptor (Linke, Caradoc-Davies, Young, Proft, & Baker, 2009). In *S. pyogenes*, an *adcAII* null mutant was unable to grow in Zn(II)-depleted media and was attenuated in virulence in a murine model of infection based on assessment of lesion size (Weston et al., 2009). Similar findings were observed for an *adcA* mutant when grown under Zn-limiting conditions (Tedde, Rosini, & Galeotti, 2016). This study also showed that AdcAII is expressed in higher levels in an *adcA* mutant when grown in Zn(II)-depleted conditions as determined by Western blot analysis, suggesting that AdcA and AdcAII are also functionally redundant proteins in *S. pyogenes* (Tedde et al., 2016).

#### 4.3.2.2 Pht

The Pht (Pneumococcal histidine triad) proteins are immunogenic, surface-associated proteins that are characterized by the presence of 4–6 histidine triad motif's (HxxHxH) (Adamou et al., 2001). This family of proteins is best studied in *S. pneumoniae* where it plays a key role in Zn homeostasis as part of the AdcR regulon by upstream AdcR consensus sequences, act as cellular adhesins and assist in evasion of the complement system (Ogunniyi et al., 2009; Plumptre, Ogunniyi, & Paton, 2012). *S. pneumoniae* encodes PhtA, PhtB, PhtD, and PhtE while *S. pyogenes* encodes PhtD and PhtY. Initially, structural characterization of recombinant PhtA containing a Zn ion suggested that these proteins may be involved in Zn binding (Riboldi-Tunnicliffe, Bent, Isaacs, & Mitchell, 2004; Riboldi-Tunnicliffe, Isaacs, & Mitchell, 2005). However, characterization of PhtD in *S. pneumoniae* indicated that it was able to physically interact with AdcAII in vitro (Loisel et al., 2011). This lead to the hypothesis that the Pht proteins function to scavenge available Zn outside the cell and shuttle it to the AdcA and AdcAII transporters for direct acquisition and this bound Zn within Pht proteins may also provide a reservoir of Zn to cells during infection. A quadruple *phtABDE* deletion mutant in *S. pneumoniae* was unable to grow in Zn-restricted media (through use of the chelator TPEN) and growth was restored to *S. pneumoniae* wild type through supplementation with Zn (Plumptre, Eijkelkamp, et al., 2014).

There is evidence that PhtD may provide Zn(II) ions to AdcAII for subsequent deliver into AdcBC and the cytoplasm. PhtD and AdcAII immunoprecipitate together and the transfer of Zn(II) ions from PhtD to AdcAII was shown to occur in vitro by NMR (Bersch et al., 2013). Other Pht proteins may perform a similar function under Zn starvation conditions (Plumptre et al., 2014). Furthermore, this work highlighted the impact of multiple AdcR-binding sites upstream of Zn acquisition genes, showing that multiple consensus sequences upstream of *phtE* resulted in greater fold induction of gene expression by low Zn conditions (Plumptre, Eijkelkamp, et al., 2014). In *S. pyogenes*, bioinformatic analysis has shown that the genome has genes that encode for four homologues of the Pht proteins, although only PhtY and PhtD possess upstream AdcR-binding sites (Sanson et al., 2015). However, there are no studies to date that have confirmed if these function similarly to the *S. pneumoniae* proteins.

#### 4.3.2.3 AdcR

The global Zn starvation response in both *S. pyogenes* and *S. pneumoniae* is controlled by the MarR-family regulator AdcR (adhesin competence repressor) (Reyes-Caballero et al., 2010; Shafeeq, Kloosterman, & Kuipers, 2011). AdcR likely functions as a homodimer in vivo with two Zn(II) sensing sites per monomer (Guerra, Dann, & Giedroc, 2011; Reyes-Caballero et al., 2010). Dissociation of the Zn ion from the sensing site under Zn-limiting conditions leads to conformational changes in the AdcR structure (Guerra et al., 2011) and the derepression of approximately 70 genes, including those involved in Zn acquisition such as *adcA*, *adcBC*, *adcAII*, *phtD*, and *phtY* (Brenot et al., 2007; Sanson et al., 2015; Weston et al., 2009). AdcR also regulates the expression of *phtA* and *phtB*, which is part of the same *phtABDY* operon in *S. pneumoniae* (Ogunniyi et al., 2009) but not *S. pyogenes* (Plumptre, Eijkelkamp, et al., 2014; Sanson et al., 2015). Regulation by AdcR in *S. pyogenes* and *S. pneumoniae* is shown in Figs. 3 and 4, respectively.

The ability of AdcR to sense cellular Zn status is crucial to virulence in a murine model of infection as assessed through examination of *S. pyogenes adcR* mutants with abolished Zn-binding site (Sanson et al., 2015). However, it has not been identified which Zn uptake proteins under its control are essential to survival.

### 4.3.3 Efflux

#### 4.3.3.1 CzcD

Excess Zn ions from the cytoplasm of *S. pyogenes* and *S. pneumoniae* are exported via CzcD, a CDF (Jacobsen, Kazmierczak, Lisher, Winkler, & Giedroc, 2011; Kloosterman, van der Kooi-Pol, Bijlsma, & Kuipers, 2007). The *czcD* gene was first identified in *Cupriavidus metallidurans* as part of a large *czc* operon for metal resistance that also encodes for putative metal-sensing regulators, ABC-type metal export transporters and accessory proteins (Anton, Grosse, Reissmann, Pribyl, & Nies, 1999). In *S. pyogenes* and *S. pneumoniae*, *czcD* is found upstream, adjacent and divergent from the gene that encodes for its regulator (*sczA* in *S. pneumoniae* and *gczA* in *S. pyogenes*) (Kloosterman et al., 2007; Ong et al., 2014). The SczA (GczA) regulator is a member of the TetR superfamily of transcriptional repressors that also regulates the expression of a Zn-dependent alcohol dehydrogenase (AdhB) in response to Zn (Kloosterman et al., 2007). Microarray analysis of a *sczA* null mutant of *S. pneumoniae* indicated that SczA may control seven additional genes, including those encoding for enzymes involved in anaerobic nucleotide reduction (NrdD and NrdG), a putative MerR-family transcriptional regulator and a Zn-containing alcohol dehydrogenase (Kloosterman et al., 2007).

In both *S. pneumoniae* and *S. pyogenes*, deletion of *czcD* resulted in attenuated growth in the presence of excess extracellular Zn (Jacobsen et al., 2011; Kloosterman et al., 2007; Ong et al., 2014). As mentioned in Section 3.2, intracellular pathogens may encounter increased levels of Zn within neutrophils and macrophages (Botella et al., 2011; Ong et al., 2014). Consistently, the *czcD* mutant of *S. pyogenes* was attenuated in an ex vivo model of infections of human neutrophils but survival was rescued if the Zn chelator TPEN was present (Ong et al., 2014). Furthermore, both the *czcD* and *sczA* deletion mutants of *S. pyogenes* displayed reduced virulence in the murine model of invasive infection (Ong et al., 2014).

## 4.4 Copper

### 4.4.1 Uptake

While distinct pathways for the uptake of Cu in most prokaryotes have not been identified, it is largely thought that high extracellular Cu may leak into cells via porins (Porcheron, Garénaux, Proulx, Sabri, & Dozois, 2013; Rensing & Grass, 2003; Speer, Rowland, Haeili, Niederweis, & Wolschendorf, 2013). Nevertheless, it has been shown that a knockout of the *S. pneumoniae* stringent response regulator, $rel_{spn}$ in *S. pneumoniae* results in a requirement for supplemental Cu as well as Zn and Mn in chemically defined media (Kazmierczak, Wayne, Rechtsteiner, & Winkler, 2009). The biochemical significance of this observation is not well understood.

### 4.4.2 Efflux

#### 4.4.2.1 Cop

The primary mechanism for Cu detoxification in streptococci is the Cop system that is responsible for Cu efflux (Arguello, Raimunda, & Padilla-Benavides, 2013). In *S. pneumoniae*, the Cop system consists of CopY, a cellular Cu sensor and regulator; CupA, a Cu chaperone; and CopA, the Cu efflux protein (Fu et al., 2013; Portmann, Poulsen, Wimmer, & Solioz, 2006). In *S. pneumoniae*, deletion of the *copA* or *cupA* genes led to sensitivity to high concentrations of Cu, and significant accumulation of intracellular Cu (Fu et al., 2013; Johnson, Kehl-Fie, Klein, et al., 2015; Shafeeq, Yesilkaya, et al., 2011). The *copA* deletion mutant was also attenuated in a murine model and was efficiently cleared from blood of infected mice (Shafeeq, Yesilkaya, et al., 2011). Interestingly, only the loss of *copA*, but not *cupA*, resulted in substantially decreased virulence in a murine model (Johnson, Kehl-Fie, Klein, et al., 2015).

*S. pyogenes* possesses a similar Cop efflux system to *S. pneumoniae*. Transforming the *S. pyogenes copA* gene into an *E. coli* strain deficient in Cu efflux restored its ability to tolerate Cu (Young, Gordon, Fang, Holder, & Reid, 2015). However, the role of the Cop efflux system in terms of the virulence of *S. pyogenes* has not yet been studied.

### 4.4.3 Regulation

Cu homeostasis in bacteria is controlled by high-affinity Cu sensors (Rademacher & Masepohl, 2012). In *S. pneumoniae*, this is CopY, which functions as a constitutive repressor of Cu efflux genes in the *cop* operon (Shafeeq, Yesilkaya, et al., 2011). This form of regulator has been structurally characterized in *Enterococcus hirae* (formerly group D *Streptococcus*) and functional studies have led to the hypothesis that in Cu-restricted conditions, CopY exists in a Cu-free, Zn-bound form and maintains binding to consensus sequences upstream of *cop* genes. However during Cu stress, Cu displaces the Zn ion out of the binding site and triggers release of CopY from the consensus sequence (Portmann et al., 2006; Strausak & Solioz, 1997).

## 5. METALS AS ANTIMICROBIALS

Long before the concepts of homeostasis and nutritional immunity were recognized, trace metals, especially Cu, were used in daily life to control infections (Dollwet & Sorenson, 1985; Grass, Rensing, & Solioz, 2011). This practise continues today and Cu-containing salts and complexes are used in agriculture to inhibit bacterial and fungal infections of crops (Behlau, Belasque, Graham, & Leite, 2010; Large, 1945; Wilson, 1997).

Metal ions potentially offer an opportunity for antimicrobial killing in a health and medical setting. Contact killing on metallic or alloyed (usually with Zn) Cu surfaces has emerged as an important method for the control of nosocomial infections in healthcare settings (Chyderiotis et al., 2015; Grass et al., 2011; Kuhn, 1983; Thurman & Gerba, 1988). The efficacy of Cu can be greatly enhanced using ionophores. Recently, we showed that *S. pneumoniae* was susceptible to killing by Cu(gtsm) (Djoko, Goytia, et al., 2015), although it was not as sensitive as *Neisseria gonorrhoeae* or *Haemophilus influenzae* (Djoko, Goytia, et al., 2015; Djoko, Paterson, Donnelly, & McEwan, 2014).

Zn ionophores are already in use in healthcare products, the best known being Zn pyrithione, which is found in 'Head and Shoulders™' shampoo (Warner, Schwartz, Boissy, & Dawson, 2001) and its potential as a skin disinfectant has been observed (Guthery, Seal, & Anderson, 2005). Although it was originally considered that Zn pyrithione can exert toxicity against bacteria and fungi through metal overload (Dinning, Al-Adham, Eastwood, Austin, & Collier, 1998; Marks, Pearse, & Walker, 1985), it seems more likely that it works following transmetallation with Cu and actually delivers this much more toxic ion to microbial cells (Reeder et al., 2011). Other antimicrobial agents that show promise include 8-hydroxyquinoline (8-HQ), which is active against *M. tuberculosis* (Bidauit & Urbain, 1928; McElroy, 1910; Urbanski, Slopek, & Venulet, 1951) and it is now known to work via a Cu-dependent mechanism (Prachayasittikul, Prachayasittikul, Ruchirawat, & Prachayasittikul, 2013; Shah et al., 2016). Furthermore, new research conducted on derivatives of these compounds and new metal-antibiotic therapies may provide new windows of therapy for multidrug resistant bacteria (Cavet, 2014).

Recently, Zn-containing antibacterial compounds have been put into practice in consumer oral health products such as toothpaste (Newby et al., 2011). These compounds in isolation or in combination with toothpaste components have displayed antimicrobial effects against a variety of bacteria implicated in development of dental caries, such as *S. mutans* and those involved in halitosis, such as *Porphyromonas gingivalis* (Pizzey, Marquis, & Bradshaw, 2011). Treatment of oral biofilms with toothpastes containing Zn ions has also been shown to result in a reduction of volatile sulphur compounds produced by oral bacteria, without subsequent loss in bacterial viability (Burnett, Stephen, Pizzey, & Bradshaw, 2011). This is a development from products which had previously used Zn salts in synergy with the antibacterial agent Triclosan to achieve a reduction in dental calculus levels, although this study did not show statistically significant reduction in bacterial number on the dental surface (Stephen, Saxton, Jones, Ritchie, & Morrison, 1990). However, Zn has since been shown to inhibit acid production by *S. mutans*, which is a mechanism for dental erosion (Phan, Buckner, Sheng, Baldeck, & Marquis, 2004). Other work has examined the potential for Zn oxide nanoparticles as antimicrobial agents, which display potent antimicrobial activity against a variety of bacteria, including *S. pyogenes* (Jones, Ray, Ranjit, & Manna, 2008). These compounds have since been shown to exert bacteriocidal activity against *Campylobacter jejuni* (Xie, He, Irwin, Jin, & Shi, 2011) and *Salmonella enterica* Sv Enteriditis (Vidovic et al., 2015). Silver nanoparticles have also been examined for the efficacy in killing of *S. mutans* for the prospective use in oral formulations (Perez-Diaz et al., 2015).

Given the lack of new antibiotics under development, metal-based antimicrobials may well have value in the control of streptococcal infections. However, their use will require a refined understanding of metal ion homeostasis in the pathogen and the role that transition metal ions play in inflammation and antimicrobial activity in the host.

## 6. CONCLUDING REMARKS

As the late R.J.P. Williams remarked, in relation to the impact of trace metal ions on biology, 'inorganic elements can be likened to governors of society—few in number but powerful in effect' (Williams, 1997). Given the importance of these metal ions, acquisition of transition metal ions by living organisms has been the subject of intense study over several decades with significant progress in understanding the properties of the key systems governing metal ion acquisition and distribution. In this review, we have explored the way in which pathogenic *S. pneumoniae* and *S. pyogenes* acquire and efflux metal during metal within the host, as well as the many ways that metal homeostasis is important for virulence.

Although there exists a wealth of knowledge highlighting the role of metal starvation as a key mechanism to control bacteria within the host, recent evidence into the role of metal overload killing provides an exciting new aspect to the term 'nutritional immunity'. The initial discovery of metal overload killing has reignited interest in the ancient discoveries of metal being used to treat bacterial infection. Furthermore, the susceptibility of bacteria to such killing mechanisms and the subsequent efflux proteins that provide protection against this killing, have provided insight into the possibility of new antimicrobials based on inhibition of efflux pumps and metal-delivering antibiotics, as well as the possibility of metal uptake and efflux systems as new vaccine targets.

Examination of *S. pyogenes* and *S. pneumoniae* offers an unique insight into the multifaceted nature of metal homeostasis during infection. Both organisms can cause a spectrum of disease states within the human host, ranging from mild to severe, and in doing so, are required to modulate metal homeostasis during the infectious process and movement into niches within the host. Thus, the complement of acquisition systems, efflux proteins and regulators are not in play continuously. It seems likely that these organisms both encode multiple metalloregulators for the explicit purpose of differentially regulating metal homeostasis genes.

## ACKNOWLEDGEMENTS

A.G.T. is a recipient of the Australian Postgraduate Award. C.Y.O. is a recipient of a Garnett Passe and Rodney Williams Memorial Fellowship. M.J.W. is a NHMRC Principal Research Fellow. This work was supported by National Health and Medical Research Council (Australia) grant APP1084460 to A.G.M.

## REFERENCES

Aberdein, J. D., Cole, J., Bewley, M. A., Marriott, H. M., & Dockrell, D. H. (2013). Alveolar macrophages in pulmonary host defence—The unrecognized role of apoptosis as a mechanism of intracellular bacterial killing. *Clinical and Experimental Immunology*, *174*(2), 193–202.

Achard, M. E., Stafford, S. L., Bokil, N. J., Chartres, J., Bernhardt, P. V., Schembri, M. A., ... McEwan, A. G. (2012). Copper redistribution in murine macrophages in response to *Salmonella* infection. *Biochemical Journal*, *444*(1), 51–57.

Adamou, J. E., Heinrichs, J. H., Erwin, A. L., Walsh, W., Gayle, T., Dormitzer, M., ... Johnson, S. (2001). Identification and characterization of a novel family of pneumococcal proteins that are protective against sepsis. *Infection and Immunity*, *69*(2), 949–958.

Aguirre, J. D., & Culotta, V. C. (2012). Battles with iron: Manganese in oxidative stress protection. *Journal of Biological Chemistry*, *287*(17), 13541–13548.

Aisen, P., Leibman, A., & Zweier, J. (1978). Stoichiometric and site characteristics of the binding of iron to human transferrin. *Journal of Biological Chemistry*, *253*(6), 1930–1937.

Akbas, N., Draganova, E. B., Block, D. R., Sook, B. R., Chan, Y. F., Zhuo, J., … Dixon, D. W. (2016). Heme-bound SiaA from *Streptococcus pyogenes*: Effects of mutations and oxidation state on protein stability. *Journal of Inorganic Biochemistry*, *158*, 99–109.

Albrich, W. C., Monnet, D. L., & Harbarth, S. (2004). Antibiotic selection pressure and resistance in *Streptococcus pneumoniae* and *Streptococcus pyogenes*. *Emerging Infectious Diseases*, *10*(3), 514–517.

Al-Maghrebi, M., Fridovich, I., & Benov, L. (2002). Manganese supplementation relieves the phenotypic deficits seen in superoxide-dismutase-null *Escherichia coli*. *Archives of Biochemistry and Biophysics*, *402*(1), 104–109.

AlonsoDeVelasco, E., Verheul, A. F., Verhoef, J., & Snippe, H. (1995). *Streptococcus pneumoniae*: Virulence factors, pathogenesis, and vaccines. *Microbiological Reviews*, *59*(4), 591–603.

Andisi, V. F., Hinojosa, C. A., de Jong, A., Kuipers, O. P., Orihuela, C. J., & Bijlsma, J. J. E. (2012). Pneumococcal gene complex involved in resistance to extracellular oxidative stress. *Infection and Immunity*, *80*(3), 1037–1049.

Andreini, C., Bertini, I., Cavallaro, G., Holliday, G. L., & Thornton, J. M. (2008). Metal ions in biological catalysis: From enzyme databases to general principles. *Journal of Biological Inorganic Chemistry*, *13*(8), 1205–1218.

Anjem, A., & Imlay, J. A. (2012). Mononuclear iron enzymes are primary targets of hydrogen peroxide stress. *Journal of Biological Chemistry*, *287*(19), 15544–15556.

Anjem, A., Varghese, S., & Imlay, J. A. (2009). Manganese import is a key element of the OxyR response to hydrogen peroxide in *Escherichia coli*. *Molecular Microbiology*, *72*(4), 844–858.

Anton, A., Grosse, C., Reissmann, J., Pribyl, T., & Nies, D. H. (1999). CzcD is a heavy metal ion transporter involved in regulation of heavy metal resistance in *Ralstonia* sp. strain CH34. *Journal of Bacteriology*, *181*(22), 6876–6881.

Anzaldi, L. L., & Skaar, E. P. (2010). Overcoming the heme paradox: Heme toxicity and tolerance in bacterial pathogens. *Infection and Immunity*, *78*(12), 4977–4989.

Archibald, F. S., & Fridovich, I. (1981). Manganese and defenses against oxygen toxicity in *Lactobacillus plantarum*. *Journal of Bacteriology*, *145*(1), 442–451.

Archibald, F. S., & Fridovich, I. (1982). The scavenging of superoxide radical by manganous complexes—*In vitro*. *Archives of Biochemistry and Biophysics*, *214*(2), 452–463.

Arguello, J. M., Raimunda, D., & Padilla-Benavides, T. (2013). Mechanisms of copper homeostasis in bacteria. *Frontiers in Cellular and Infection Microbiology*, *3*, 73.

Arnold, A. R., & Barton, J. K. (2013). DNA protection by the bacterial ferritin dps via dna charge transport. *Journal of the American Chemical Society*, *135*(42), 15726–15729.

Arnold, R. R., Cole, M. F., & McGhee, J. R. (1977). A bactericidal effect for human lactoferrin. *Science*, *197*(4300), 263–265.

Atkinson, A., & Winge, D. R. (2009). Metal acquisition and availability in the mitochondria. *Chemical Reviews*, *109*(10), 4708–4721.

Auzat, I., Chapuy-Regaud, S., Le Bras, G., Dos Santos, D., Ogunniyi, A. D., Le Thomas, I., … Trombe, M. C. (1999). The NADH oxidase of *Streptococcus pneumoniae*: Its involvement in competence and virulence. *Molecular Microbiology*, *34*(5), 1018–1028.

Ayala-Castro, C., Saini, A., & Outten, F. W. (2008). Fe-S cluster assembly pathways in bacteria. *Microbiology and Molecular Biology Reviews*, *72*(1), 110–125.

Babior, B. M. (2000). Phagocytes and oxidative stress. *American Journal of Medicine*, *109*(1), 33–44.

Bandyopadhyay, S., Chandramouli, K., & Johnson, M. K. (2008). Iron-sulfur cluster biosynthesis. *Biochemical Society Transactions*, *36*, 1112–1119.

Barbuti, G., Moschioni, M., Fumarulo, R., Censini, S., & Montemurro, P. (2010). *Streptococcus pneumoniae* modulates the respiratory burst response in human neutrophils. *FEMS Immunology and Medical Microbiology*, *60*(1), 57–62.

Barnese, K., Gralla, E. B., Cabelli, D. E., & Valentine, J. S. (2008). Manganous phosphate acts as a superoxide dismutase. *Journal of the American Chemical Society*, *130*(14), 4604–4606.

Barnett, T. C., Cole, J. N., Rivera-Hernandez, T., Henningham, A., Paton, J. C., Nizet, V., & Walker, M. J. (2015). Streptococcal toxins: Role in pathogenesis and disease. *Cellular Microbiology*, *17*(12), 1721–1741.

Bates, C. S., Montanez, G. E., Woods, C. R., Vincent, R. M., & Eichenbaum, Z. (2003). Identification and characterization of a *Streptococcus pyogenes* operon involved in binding of hemoproteins and acquisition of iron. *Infection and Immunity*, *71*(3), 1042–1055.

Bates, C. S., Toukoki, C., Neely, M. N., & Eichenbaum, Z. (2005). Characterization of MtsR, a new metal regulator in group A *Streptococcus*, involved in iron acquisition and virulence. *Infection and Immunity*, *73*(9), 5743–5753.

Bayle, L., Chimalapati, S., Schoehn, G., Brown, J., Vernet, T., & Durmort, C. (2011). Zinc uptake by *Streptococcus pneumoniae* depends on both AdcA and AdcAII and is essential for normal bacterial morphology and virulence. *Molecular Microbiology*, *82*(4), 904–916.

Behlau, F., Belasque, J., Jr., Graham, J. H., & Leite, R. P., Jr. (2010). Effect of frequency of copper applications on control of citrus canker and the yield of young bearing sweet orange trees. *Crop Protection*, *29*(3), 300–305.

Beiter, K., Wartha, F., Albiger, B., Normark, S., Zychlinsky, A., & Henriques-Normark, B. (2006). An endonuclease allows *Streptococcus pneumoniae* to escape from neutrophil extracellular traps. *Current Biology*, *16*(4), 401–407.

Beres, S. B., Richter, E. W., Nagiec, M. J., Sumby, P., Porcella, S. F., Deleo, F. R., & Musser, J. M. (2006). Molecular genetic anatomy of inter- and intraserotype variation in the human bacterial pathogen Group A *Streptococcus*. *Proceedings of the National Academy of Sciences of the United States of America*, *103*(18), 7059–7064.

Berry, A. M., & Paton, J. C. (1996). Sequence heterogeneity of PsaA, a 37-kilodalton putative adhesin essential for virulence of *Streptococcus pneumoniae*. *Infection and Immunity*, *64*(12), 5255–5262.

Bersch, B., Bougault, C., Roux, L., Favier, A., Vernet, T., & Durmort, C. (2013). New insights into histidine triad proteins: Solution structure of a *Streptococcus pneumoniae* PhtD domain and zinc transfer to AdcAII. *PLoS One*, *8*(11), e81168.

Betschel, S. D., Borgia, S. M., Barg, N. L., Low, D. E., & De Azavedo, J. C. S. (1998). Reduced virulence of group A streptococcal Tn916 mutants that do not produce streptolysin S. *Infection and Immunity*, *66*(4), 1671–1679.

Bezkorovainy, A. (1982). antimicrobial properties of iron-binding proteins. *Advances in Experimental Medicine and Biology*, *135*, 139–154.

Bidauit, C., & Urbain, A. (1928). The effect of the neutral sulphate of ortho-oxyquinoline on the tuberculous Bacillus. *Comptes Rendus des Seances de la Societe de Biologie et de Ses Filiales*, *99*, 461–464.

Birmingham, C. L., Higgins, D. E., & Brumell, J. H. (2008). Avoiding death by autophagy: Interactions of *Listeria monocytogenes* with the macrophage autophagy system. *Autophagy*, *4*(3), 368–371.

Bisno, A. L., Brito, M. O., & Collins, C. M. (2003). Molecular basis of group A streptococcal virulence. *Lancet Infectious Diseases*, *3*(4), 191–200.

Blackwell, J. M., Searle, S., Goswami, T., & Miller, E. N. (2000). Understanding the multiple functions of Nramp1. *Microbes and Infection*, *2*(3), 317–321.

Bogaert, D., De Groot, R., & Hermans, P. W. (2004). *Streptococcus pneumoniae* colonisation: The key to pneumococcal disease. *Lancet Infectious Diseases*, *4*(3), 144–154.

Botella, H., Peyron, P., Levillain, F., Poincloux, R., Poquet, Y., Brandli, I., ... Neyrolles, O. (2011). Mycobacterial p(1)-type ATPases mediate resistance to zinc poisoning in human macrophages. *Cell Host & Microbe, 10*(3), 248–259.

Brenot, A., King, K. Y., & Caparon, M. G. (2005). The PerR regulon in peroxide resistance and virulence of *Streptococcus pyogenes*. *Molecular Microbiology, 55*(1), 221–234.

Brenot, A., King, K. Y., Janowiak, B., Griffith, O., & Caparon, M. G. (2004). Contribution of glutathione peroxidase to the virulence of *Streptococcus pyogenes*. *Infection and Immunity, 72*(1), 408–413.

Brenot, A., Weston, B. F., & Caparon, M. G. (2007). A PerR-regulated metal transporter (PmtA) is an interface between oxidative stress and metal homeostasis in *Streptococcus pyogenes*. *Molecular Microbiology, 63*(4), 1185–1196.

Briles, D. E., Tart, R. C., Swiatlo, E., Dillard, J. P., Smith, P., Benton, K. A., ... McDaniel, L. S. (1998). Pneumococcal diversity: Considerations for new vaccine strategies with emphasis on pneumococcal surface protein A (PspA). *Clinical Microbiology Reviews, 11*(4), 645–657.

Brini, M., Ottolini, D., Cali, T., & Carafoli, E. (2013). Calcium in health and disease. *Metal Ions in Life Sciences, 13*, 81–137.

Brinkmann, V., Reichard, U., Goosmann, C., Fauler, B., Uhlemann, Y., Weiss, D. S., ... Zychlinsky, A. (2004). Neutrophil extracellular traps kill bacteria. *Science, 303*(5663), 1532–1535.

Brinkmann, V., & Zychlinsky, A. (2007). Beneficial suicide: Why neutrophils die to make NETs. *Nature Reviews Microbiology, 5*(8), 577–582.

Broderick, J. B., Duffus, B. R., Duschene, K. S., & Shepard, E. M. (2014). Radical S-adenosylmethionine enzymes. *Chemical Reviews, 114*(8), 4229–4317.

Brooijmans, R., de Vos, W. M., & Hugenholtz, J. (2009). Electron transport chains of lactic acid bacteria—Walking on crutches is part of their lifestyle. *F1000 Biology Reports, 1*, 34.

Brophy, M. B., Hayden, J. A., & Nolan, E. M. (2012). Calcium ion gradients modulate the zinc affinity and antibacterial activity of human calprotectin. *Journal of the American Chemical Society, 134*(43), 18089–18100.

Brown, J. S., Gilliland, S. M., & Holden, D. W. (2001). A *Streptococcus pneumoniae* pathogenicity island encoding an ABC transporter involved in iron uptake and virulence. *Molecular Microbiology, 40*(3), 572–585.

Brown, J. S., Gilliland, S. M., Ruiz-Albert, J., & Holden, D. W. (2002). Characterization of pit, a *Streptococcus pneumoniae* iron uptake ABC transporter. *Infection and Immunity, 70*(8), 4389–4398.

Brown, L. R., Gunnell, S. M., Cassella, A. N., Keller, L. E., Scherkenbach, L. A., Mann, B., ... Thornton, J. A. (2016). AdcAII of *Streptococcus pneumoniae* affects Pneumococcal invasiveness. *PLoS One, 11*(1), e0146785.

Brown, J. S., Ogunniyi, A. D., Woodrow, M. C., Holden, D. W., & Paton, J. C. (2001). Immunization with components of two iron uptake ABC transporters protects mice against systemic *Streptococcus pneumoniae* infection. *Infection and Immunity, 69*(11), 6702–6706.

Buchanan, J. T., Simpson, A. J., Aziz, R. K., Liu, G. Y., Kristian, S. A., Kotb, M., ... Nizet, V. (2006). DNase expression allows the pathogen Group A *Streptococcus* to escape killing in neutrophil extracellular traps. *Current Biology, 16*(4), 396–400.

Bullen, J. J., Spalding, P. B., Ward, C. G., & Gutteridge, J. M. (1991). Hemochromatosis, iron and septicemia caused by Vibrio vulnificus. *Archives of Internal Medicine, 151*(8), 1606–1609.

Burnett, G. R., Stephen, A. S., Pizzey, R. L., & Bradshaw, D. J. (2011). In vitro effects of novel toothpaste actives on components of oral malodour. *International Dental Journal, 61*, 67–73.

Burnham, B. F., & Neilands, J. B. (1961). Studies on the metabolic function of the ferrichrome compounds. *Journal of Biological Chemistry*, *236*, 554–559.

Cabiscol, E., Tamarit, J., & Ros, J. (2000). Oxidative stress in bacteria and protein damage by reactive oxygen species. *International Microbiology*, *3*(1), 3–8.

Calhoun, L. N., & Kwon, Y. M. (2011). Structure, function and regulation of the DNA-binding protein Dps and its role in acid and oxidative stress resistance in *Escherichia coli*: A review. *Journal of Applied Microbiology*, *110*(2), 375–386.

Carapetis, J. R., Steer, A. C., Mulholland, E. K., & Weber, M. (2005). The global burden of group A streptococcal diseases. *Lancet Infectious Diseases*, *5*(11), 685–694.

Cassone, M., Gagne, A. L., Spruce, L. A., Seeholzer, S. H., & Sebert, M. E. (2012). The HtrA protease from *Streptococcus pneumoniae* digests both denatured proteins and the competence-stimulating peptide. *Journal of Biological Chemistry*, *287*(46), 38449–38459.

Cavet, J. S. (2014). Copper as a magic bullet for targeted microbial killing. *Chemistry and Biology*, *21*(8), 921–922.

Cellier, M. F., Courville, P., & Campion, C. (2007). Nramp1 phagocyte intracellular metal withdrawal defense. *Microbes and Infection*, *9*(14–15), 1662–1670.

Charpentier, E., & Tuomanen, E. (2000). Mechanisms of antibiotic resistance and tolerance in *Streptococcus pneumoniae*. *Microbes and Infection*, *2*(15), 1855–1864.

Chen, D. K., McGeer, A., de Azavedo, J. C., Low, D. E., & Network, C. B. S. (1999). Decreased susceptibility of *Streptococcus pneumoniae* to fluoroquinolones in Canada. *New England Journal of Medicine*, *341*(4), 233–239.

Cheng, W., Li, Q., Jiang, Y. L., Zhou, C. Z., & Chen, Y. X. (2013). Structures of *Streptococcus pneumoniae* PiaA and its complex with ferrichrome reveal insights into the substrate binding and release of high affinity iron transporters. *PLoS One*, *8*(8), e71451.

Cheton, P. L. B., & Archibald, F. S. (1988). Manganese complexes and the generation and scavenging of hydroxyl free-radicals. *Free Radical Biology and Medicine*, *5*(5–6), 325–333.

Chyderiotis, S., Legeay, C., Verjat-Trannoy, D., Le Gallou, F., Astagneau, P., & Lepelletier, D. (2015). Efficacy of copper surfaces in the healthcare environment: A systematic review. *Antimicrobial Resistance and Infection Control*, *4*(Suppl. 1), P45.

Clancy, A., Loar, J. W., Speziali, C. D., Oberg, M., Heinrichs, D. E., & Rubens, C. E. (2006). Evidence for siderophore-dependent iron acquisition in group B *Streptococcus*. *Molecular Microbiology*, *59*(2), 707–721.

Clohessy, P. A., & Golden, B. E. (1995). Calprotectin-mediated zinc chelation as a biostatic mechanism in host-defense. *Scandinavian Journal of Immunology*, *42*(5), 551–556.

Coassin, M., Ursini, F., & Bindoli, A. (1992). Antioxidant effect of manganese. *Archives of Biochemistry and Biophysics*, *299*(2), 330–333.

Cole, J. N., Aquilina, J. A., Hains, P. G., Henningham, A., Sriprakash, K. S., Caparon, M. G., … Walker, M. J. (2007). Role of group A *Streptococcus* HtrA in the maturation of SpeB protease. *Proteomics*, 7(24), 4488–4498.

Cole, J. N., Barnett, T. C., Nizet, V., & Walker, M. J. (2011). Molecular insight into invasive group A streptococcal disease. *Nature Reviews. Microbiology*, *9*(10), 724–736.

Collet, J. F., & Messens, J. (2010). Structure, function, and mechanism of thioredoxin proteins. *Antioxidants & Redox Signaling*, *13*(8), 1205–1216.

Conrad, M. E., & Umbreit, J. N. (2000). Iron absorption and transport—An update. *American Journal of Hematology*, *64*(4), 287–298.

Corbin, B. D., Seeley, E. H., Raab, A., Feldmann, J., Miller, M. R., Torres, V. J., … Skaar, E. P. (2008). Metal chelation and inhibition of bacterial growth in tissue abscesses. *Science*, *319*(5865), 962–965.

Cotruvo, J. A., Jr., & Stubbe, J. (2012). Metallation and mismetallation of iron and manganese proteins in vitro and in vivo: The class I ribonucleotide reductases as a case study. *Metallomics*, *4*(10), 1020–1036.

Counago, R. M., Ween, M. P., Begg, S. L., Bajaj, M., Zuegg, J., O'Mara, M. L., ... McDevitt, C. A. (2014). Imperfect coordination chemistry facilitates metal ion release in the Psa permease. *Nature Chemical Biology*, *10*(1), 35–41.
Cox, M. M., & Battista, J. R. (2005). Deinococcus radiodurans—The consummate survivor. *Nature Reviews. Microbiology*, *3*(11), 882–892.
Craig, A., Mai, J., Cai, S. S., & Jeyaseelan, S. (2009). Neutrophil recruitment to the lungs during bacterial pneumonia. *Infection and Immunity*, 77(2), 568–575.
Crain, M. J., Waltman, W. D., Turner, J. S., Yother, J., Talkington, D. F., Mcdaniel, L. S., ... Briles, D. E. (1990). Pneumococcal surface protein-a (PspA) is serologically highly variable and is expressed by all clinically important capsular serotypes of *Streptococcus pneumoniae*. *Infection and Immunity*, *58*(10), 3293–3299.
Cunningham, M. W. (2000). Pathogenesis of group A streptococcal infections. *Clinical Microbiology Reviews*, *13*(3), 470–511.
Dahesh, S., Nizet, V., & Cole, J. N. (2012). Study of streptococcal hemoprotein receptor (Shr) in iron acquisition and virulence of M1T1 group A *Streptococcus*. *Virulence*, *3*(7), 566–575.
Dale, D. C., Boxer, L., & Liles, W. C. (2008). The phagocytes: Neutrophils and monocytes. *Blood*, *112*(4), 935–945.
Daly, M. J., Gaidamakova, E. K., Matrosova, V. Y., Vasilenko, A., Zhai, M., Venkateswaran, A., ... Ghosal, D. (2004). Accumulation of Mn(II) in *Deinococcus radiodurans* facilitates gamma-radiation resistance. *Science*, *306*(5698), 1025–1028.
Damo, S. M., Kehl-Fie, T. E., Sugitani, N., Holt, M. E., Rathi, S., Murphy, W. J., ... Chazin, W. J. (2013). Molecular basis for manganese sequestration by calprotectin and roles in the innate immune response to invading bacterial pathogens. *Proceedings of the National Academy of Sciences of the United States of America*, *110*(10), 3841–3846.
Davidsson, L., Lonnerdal, B., Sandstrom, B., Kunz, C., & Keen, C. L. (1989). Identification of transferrin as the major plasma carrier protein for manganese introduced orally or intravenously or after *in vitro* addition in the rat. *Journal of Nutrition*, *119*(10), 1461–1464.
Davies, M. R., Holden, M. T., Coupland, P., Chen, J. H. K., Venturini, C., Barnett, T. C., ... Walker, M. J. (2015). Emergence of scarlet fever *Streptococcus pyogenes* emm12 clones in Hong Kong is associated with toxin acquisition and multidrug resistance. *Nature Genetics*, *47*(1), 84–87.
de Jong, H. K., Parry, C. M., van der Poll, T., & Wiersinga, W. J. (2012). Host–pathogen interaction in invasive Salmonellosis. *PLoS Pathogens*, *8*(10), e1002933.
Delaye, L., Becerra, A., Orgel, L., & Lazcano, A. (2007). Molecular evolution of peptide methionine sulfoxide reductases (MsrA and MsrB): On the early development of a mechanism that protects against oxidative damage. *Journal of Molecular Evolution*, *64*(1), 15–32.
Demple, B., & Harrison, L. (1994). Repair of oxidative damage to DNA: Enzymology and biology. *Annual Review of Biochemistry*, *63*(1), 915–948.
Dinning, A. J., Al-Adham, I. S. I., Eastwood, I. M., Austin, P., & Collier, P. J. (1998). Pyrithione biocides as inhibitors of bacterial ATP synthesis. *Journal of Applied Microbiology*, *85*(1), 141–146.
Dintilhac, A., Alloing, G., Granadel, C., & Claverys, J. P. (1997). Competence and virulence of *Streptococcus pneumoniae*: Adc and PsaA mutants exhibit a requirement for Zn and Mn resulting from inactivation of putative ABC metal permeases. *Molecular Microbiology*, *25*(4), 727–739.
Djoko, K. Y., Goytia, M. M., Donnelly, P. S., Schembri, M. A., Shafer, W. M., & McEwan, A. G. (2015). Copper(II)-bis(thiosemicarbazonato) complexes as antibacterial agents: Insights into their mode of action and potential as therapeutics. *Antimicrobial Agents and Chemotherapy*, *59*(10), 6444–6453.
Djoko, K. Y., & McEwan, A. G. (2013). Antimicrobial action of copper is amplified via inhibition of heme biosynthesis. *ACS Chemical Biology*, *8*(10), 2217–2223.

Djoko, K. Y., Ong, C. L., Walker, M. J., & McEwan, A. G. (2015). The role of copper and zinc toxicity and its role in innate immune defense against bacterial pathogens. *Journal of Biological Chemistry*, *290*(31), 18954–18961.

Djoko, K. Y., Paterson, B. M., Donnelly, P. S., & McEwan, A. G. (2014). Antimicrobial effects of copper(II) bis(thiosemicarbazonato) complexes provide new insight into their biochemical mode of action. *Metallomics*, *6*(4), 854–863.

Dockrell, D. H., Lee, M., Lynch, D. H., & Read, R. C. (2001). Immune-mediated phagocytosis and killing of *Streptococcus pneumoniae* are associated with direct and bystander macrophage apoptosis. *Journal of Infectious Diseases*, *184*(6), 713–722.

Dollwet, H. H. A., & Sorenson, J. R. J. (1985). Historic uses of copper-compounds in medicine. *Trace Elements in Medicine*, *2*(2), 80–87.

Duane, P. G., Rubins, J. B., Weisel, H. R., & Janoff, E. N. (1993). Identification of hydrogen peroxide as a *Streptococcus pneumoniae* toxin for rat alveolar epithelial-cells. *Infection and Immunity*, *61*(10), 4392–4397.

Dunlap, W. M., James, G. W., 3rd, & Hume, D. M. (1974). Anemia and neutropenia caused by copper deficiency. *Annals of Internal Medicine*, *80*(4), 470–476.

Dupont, C. L., Grass, G., & Rensing, C. (2011). Copper toxicity and the origin of bacterial resistance-new insights and applications. *Metallomics*, *3*(11), 1109–1118.

Durmaz, R., Durmaz, B., Bayraktar, M., Ozerol, I. H., Kalcioglu, M. T., Aktas, E., & Cizmeci, Z. (2003). Prevalence of group A streptococcal carriers in asymptomatic children and clonal relatedness among isolates in Malatya, Turkey. *Journal of Clinical Microbiology*, *41*(11), 5285–5287.

Echlin, H., Frank, M. W., Iverson, A., Chang, T. C., Johnson, M. D. L., Rock, C. O., & Rosch, J. W. (2016). Pyruvate Oxidase as a critical link between metabolism and capsule biosynthesis in *Streptococcus pneumoniae*. *PLoS Pathogens*, *12*(10), e1005951. http://dx.doi.org/10.1371/journal.ppat.1005951.

Eichenbaum, Z., Muller, E., Morse, S. A., & Scott, J. R. (1996). Acquisition of iron from host proteins by the group A *Streptococcus*. *Infection and Immunity*, *64*(12), 5428–5429.

Eide, D. J. (2006). Zinc transporters and the cellular trafficking of zinc. *Biochimica et Biophysica Acta—Molecular Cell Research*, *1763*(7), 711–722.

Eide, D. J. (2011). The oxidative stress of zinc deficiency. *Metallomics*, *3*(11), 1124–1129.

Eijkelkamp, B. A., Morey, J. R., Ween, M. P., Ong, C. L., McEwan, A. G., Paton, J. C., & McDevitt, C. A. (2014). Extracellular zinc competitively inhibits manganese uptake and compromises oxidative stress management in *Streptococcus pneumoniae*. *PLoS One*, *9*(2), e89427.

Elsner, A., Kreikemeyer, B., Braun-Kiewnick, A., Spellerberg, B., Buttaro, B. A., & Podbielski, A. (2002). Involvement of Lsp, a member of the LraI-lipoprotein family in *Streptococcus pyogenes*, in eukaryotic cell adhesion and internalization. *Infection and Immunity*, *70*(9), 4859–4869.

Escolar, L., Perez-Martin, J., & De Lorenzo, V. (1999). Opening the iron box: Transcriptional metalloregulation by the fur protein. *Journal of Bacteriology*, *181*(20), 6223–6229.

Escoll, P., Rolando, M., Gomez-Valero, L., & Buchrieser, C. (2013). From amoeba to macrophages: Exploring the molecular mechanisms of *Legionella pneumophila* infection in both hosts. *Current Topics in Microbiology and Immunology*, *376*, 1–34.

Estellon, J., de Choudens, S. O., Smadja, M., Fontecave, M., & Vandenbrouck, Y. (2014). An integrative computational model for large-scale identification of metalloproteins in microbial genomes: A focus on iron-sulfur cluster proteins. *Metallomics*, *6*(10), 1913–1930.

Fang, F. C. (2004). Antimicrobial reactive oxygen and nitrogen species: Concepts and controversies. *Nature Reviews Microbiology*, *2*(10), 820–832.

Farrell, M. A. (1935). Studies on the respiratory mechanism of the streptococci. *Journal of Bacteriology*, *29*(4), 411–435.

Faulkner, M. J., & Helmann, J. D. (2011). Peroxide stress elicits adaptive changes in bacterial metal ion homeostasis. *Antioxidants & Redox Signaling*, *15*(1), 175–189.

Fernandez, A., Lechardeur, D., Derre-Bobillot, A., Couve, E., Gaudu, P., & Gruss, A. (2010). Two coregulated efflux transporters modulate intracellular heme and protoporphyrin IX availability in *Streptococcus agalactiae*. *PLoS Pathogens*, *6*(4), e1000860.

Finney, L. A., & O'Halloran, T. V. (2003). Transition metal speciation in the cell: Insights from the chemistry of metal ion receptors. *Science*, *300*(5621), 931–936.

Fisher, M., Huang, Y. S., Li, X. R., McIver, K. S., Toukoki, C., & Eichenbaum, Z. (2008). Shr is a broad-spectrum surface receptor that contributes to adherence and virulence in group A *Streptococcus*. *Infection and Immunity*, *76*(11), 5006–5015.

Forbes, J. R., & Gros, P. (2001). Divalent-metal transport by NRAMP proteins at the interface of host–pathogen interactions. *Trends in Microbiology*, *9*(8), 397–403.

Forrest, J. M., McIntyre, P. B., & Burgess, M. A. (2000). Pneumococcal disease in Australia. *Communicable Diseases Intelligence*, *24*(4), 89–92.

Foster, A. W., Osman, D., & Robinson, N. J. (2014). Metal preferences and metallation. *Journal of Biological Chemistry*, *289*(41), 28095–28103.

Fox, S., Leitch, A. E., Duffin, R., Haslett, C., & Rossi, A. G. (2010). Neutrophil apoptosis: Relevance to the innate immune response and inflammatory disease. *Journal of Innate Immunity*, *2*(3), 216–227.

Francis, R. T., Jr., Booth, J. W., & Becker, R. R. (1985). Uptake of iron from hemoglobin and the haptoglobin–hemoglobin complex by hemolytic bacteria. *International Journal of Biochemistry*, *17*(7), 767–773.

Franco, R., & Cidlowski, J. A. (2009). Apoptosis and glutathione: Beyond an antioxidant. *Cell Death and Differentiation*, *16*(10), 1303–1314.

Fritsche, G., Nairz, M., Theurl, I., Mair, S., Bellmann-Weiler, R., Barton, H. C., & Weiss, G. (2007). Modulation of macrophage iron transport by Nramp1 (Slc11a1). *Immunobiology*, *212*(9–10), 751–757.

Fu, Y., Tsui, H. C. T., Bruce, K. E., Sham, L. T., Higgins, K. A., Lisher, J. P., … Giedroc, D. P. (2013). A new structural paradigm in copper resistance in *Streptococcus pneumoniae*. *Nature Chemical Biology*, *9*(3), 177–183.

Fuangthong, M., Herbig, A. F., Bsat, N., & Helmann, J. D. (2002). Regulation of the Bacillus subtilis fur and perR genes by PerR: Not all members of the PerR regulon are peroxide inducible. *Journal of Bacteriology*, *184*(12), 3276–3286. http://dx.doi.org/10.1128/JB.184.12.3276-3286.2002.

Gaballa, A., & Helmann, J. D. (2002). A peroxide-induced zinc uptake system plays an important role in protection against oxidative stress in *Bacillus subtilis*. *Molecular Microbiology*, *45*(4), 997–1005.

Gaballa, A., Wang, T., Ye, R. W., & Helmann, J. D. (2002). Functional analysis of the *Bacillus subtilis* Zur regulon. *Journal of Bacteriology*, *184*(23), 6508–6514.

Gasc, A. M., Giammarinaro, P., Richter, S., & Sicard, M. (1998). Organization around the dnaA gene of *Streptococcus pneumoniae*. *Microbiology*, *144*, 433–439.

Ge, R., & Sun, X. (2014). Iron acquisition and regulation systems in *Streptococcus* species. *Metallomics*, *6*(5), 996–1003.

Gemmell, C. G., Peterson, P. K., Regelman, W., Schmeling, D., Hoidal, J. R., & Quie, P. G. (1981). Phagocytosis and killing of *Streptococcus pyogenes* by human alveolar macrophages. *Infection and Immunity*, *32*(3), 1298–1300.

Gerlach, D., Reichardt, W., & Vettermann, S. (1998). Extracellular superoxide dismutase from *Streptococcus pyogenes* type 12 strain is manganese-dependent. *FEMS Microbiology Letters*, *160*(2), 217–224.

Gibson, C. M., Mallett, T. C., Claiborne, A., & Caparon, M. G. (2000). Contribution of NADH oxidase to aerobic metabolism of *Streptococcus pyogenes*. *Journal of Bacteriology*, *182*(2), 448–455.

Gidengil, C. A., Kruskal, B. A., & Lee, G. M. (2013). Initial antibiotic choice in the treatment of group a streptococcal pharyngitis and return visit rates. *Journal of the Pediatric Infectious Diseases Society*, *2*(4), 361–367.

Goldmann, O., von Kockritz-Blickwede, M., Holtje, C., Chhatwal, G. S., Geffers, R., & Medina, E. (2007). Transcriptome analysis of murine macrophages in response to infection with *Streptococcus pyogenes* reveals an unusual activation program. *Infection and Immunity*, *75*(8), 4148–4157.
Gon, S., & Beckwith, J. (2006). Ribonucleotide reductases: Influence of environment on synthesis and activity. *Antioxidants & Redox Signaling*, *8*(5–6), 773–780.
Gordon, S. B., & Read, R. C. (2002). Macrophage defences against respiratory tract infections. *British Medical Bulletin*, *61*, 45–61.
Gosink, K. K., Mann, E. R., Guglielmo, C., Tuomanen, E. I., & Masure, H. R. (2000). Role of novel choline binding proteins in virulence of *Streptococcus pneumoniae*. *Infection and Immunity*, *68*(10), 5690–5695.
Goswami, T., Bhattacharjee, A., Babal, P., Searle, S., Moore, E., Li, M., & Blackwell, J. M. (2001). Natural-resistance-associated macrophage protein 1 is an H+/bivalent cation antiporter. *Biochemical Journal*, *354*, 511–519.
Grabowski, R., Hofmeister, A. E. M., & Buckel, W. (1993). Bacterial L-serine dehydratases—A new family of enzymes containing iron–sulfur clusters. *Trends in Biochemical Sciences*, *18*(8), 297–300.
Grass, G., Rensing, C., & Solioz, M. (2011). Metallic copper as an antimicrobial surface. *Applied and Environmental Microbiology*, *77*(5), 1541–1547.
Green, J., & Paget, M. S. (2004). Bacterial redox sensors. *Nature Reviews. Microbiology*, *2*(12), 954–966.
Grifantini, R., Toukoki, C., Colaprico, A., & Gryllos, I. (2011). Peroxide stimulon and role of PerR in Group A *Streptococcus*. *Journal of Bacteriology*, *193*(23), 6539–6551.
Groisman, E. A. (2001). The pleiotropic two-component regulatory system PhoP-PhoQ. *Journal of Bacteriology*, *183*(6), 1835–1842.
Gryllos, I., Grifantini, R., Colaprico, A., Cary, M. E., Hakansson, A., Carey, D. W., … Wessels, M. R. (2008). PerR confers phagocytic killing resistance and allows pharyngeal colonization by Group A *Streptococcus*. *PLoS Pathogens*, *4*(9), e1000145.
Gu, M., & Imlay, J. A. (2013). Superoxide poisons mononuclear iron enzymes by causing mismetallation. *Molecular Microbiology*, *89*(1), 123–134.
Guan, G., Pinochet-Barros, A., Gaballa, A., Patel, S. J., Arguello, J. M., & Helmann, J. D. (2015). PfeT, a P1B4-type ATPase, effluxes ferrous iron and protects *Bacillus subtilis* against iron intoxication. *Molecular Microbiology*, *98*(4), 787–803.
Guerinot, M. L. (2000). The ZIP family of metal transporters. *Biochimica et Biophysica Acta—Biomembranes*, *1465*(1–2), 190–198.
Guerra, A. J., Dann, C. E., 3rd, & Giedroc, D. P. (2011). Crystal structure of the zinc-dependent MarR family transcriptional regulator AdcR in the Zn(II)-bound state. *Journal of the American Chemical Society*, *133*(49), 19614–19617.
Guthery, E., Seal, L. A., & Anderson, E. L. (2005). Zinc pyrithione in alcohol-based products for skin antisepsis: Persistence of antimicrobial effects. *American Journal of Infection Control*, *33*(1), 15–22.
Haase, H., Ober-Bloebaum, J. L., Engelhardt, G., Hebel, S., Heit, A., Heine, H., & Rink, L. (2008). Zinc signals are essential for lipopolysaccharide-induced signal transduction in monocytes. *Journal of Immunology*, *181*(9), 6491–6502.
Haase, H., & Rink, L. (2009). Functional significance of zinc-related signaling pathways in immune cells. *Annual Review of Nutrition*, *29*, 133–152.
Haber, F., & Weiss, J. (1932). Über die katalyse des hydroperoxydes. *Naturwissenschaften*, *20*(51), 948–950.
Haikarainen, T., Tsou, C. C., Wu, J. J., & Papageorgiou, A. C. (2010a). Crystal structures of *Streptococcus pyogenes* Dpr reveal a dodecameric iron-binding protein with a ferroxidase site. *Journal of Biological Inorganic Chemistry*, *15*(2), 183–194.
Haikarainen, T., Tsou, C. C., Wu, J. J., & Papageorgiou, A. C. (2010b). Structural characterization and biological implications of di-zinc binding in the ferroxidase center of

*Streptococcus pyogenes* Dpr. *Biochemical and Biophysical Research Communications*, *398*(3), 361–365.

Hajaj, B., Yesilkaya, H., Benisty, R., David, M., Andrew, P. W., & Porat, N. (2012). Thiol peroxidase is an important component of *Streptococcus pneumoniae* in oxygenated environments. *Infection and Immunity*, *80*(12), 4333–4343.

Hakansson, L., Foucard, T., Hallgren, R., & Venge, P. (1980). Neutrophil function in infection-prone children. *Archives of Disease in Childhood*, *55*(10), 776–781.

Hakansson, A., Roche, H., Mirza, S., McDaniel, L. S., Brooks-Walter, A., & Briles, D. E. (2001). Characterization of binding of human lactoferrin to pneumococcal surface protein A. *Infection and Immunity*, *69*(5), 3372–3381.

Hammerschmidt, S., Bethe, G., Remane, P. H., & Chhatwal, G. S. (1999). Identification of pneumococcal surface protein a as a lactoferrin-binding protein of *Streptococcus pneumoniae*. *Infection and Immunity*, *67*(4), 1683–1687.

Hanks, T. S., Liu, M. Y., McClure, M. J., Fukumura, M., Duffy, A., & Lei, B. F. (2006). Differential regulation of iron- and manganese-specific MtsABC and heme-specific HtsABC transporters by the metalloregulator MtsR of Group A *Streptococcus*. *Infection and Immunity*, *74*(9), 5132–5139.

Hanks, T. S., Liu, M., McClure, M. J., & Lei, B. (2005). ABC transporter FtsABCD of *Streptococcus pyogenes* mediates uptake of ferric ferrichrome. *BMC Microbiology*, *5*, 62.

Hansman, D., Glasgow, H., Sturt, J., Devitt, H. L., & Douglas, R. (1971). Increased resistance to penicillin of pneumococci isolated from man. *New England Journal of Medicine*, *284*(4), 175–177.

Hardie, J. M., & Whiley, R. A. (1997). Classification and overview of the genera *Streptococcus* and *Enterococcus*. *Society for Applied Bacteriology Symposium Series*, *26*, 1S–11S.

Harlyk, C., Mccourt, J., Bordin, G., Rodriguez, A. R., & Van der Eeckhout, A. (1997). Determination of copper, zinc and iron in broncho-alveolar lavages by atomic absorption spectroscopy. *Journal of Trace Elements in Medicine and Biology*, *11*(3), 137–142.

Harris, W. R. (1986). Thermodynamics of gallium complexation by human lactoferrin. *Biochemistry*, *25*(4), 803–808.

Hayden, J. A., Brophy, M. B., Cunden, L. S., & Nolan, E. M. (2013). High-affinity manganese coordination by human calprotectin is calcium-dependent and requires the histidine-rich site formed at the dimer interface. *Journal of the American Chemical Society*, *135*(2), 775–787.

Helaine, S., Cheverton, A. M., Watson, K. G., Faure, L. M., Matthews, S. A., & Holden, D. W. (2014). Internalization of *Salmonella* by macrophages induces formation of nonreplicating persisters. *Science*, *343*(6167), 204–208.

Hendriksen, W. T., Bootsma, H. J., van Diepen, A., Estevao, S., Kuipers, O. P., de Groot, R., & Hermans, P. W. (2009). Strain-specific impact of PsaR of *Streptococcus pneumoniae* on global gene expression and virulence. *Microbiology*, *155*(Pt. 5), 1569–1579.

Henningham, A., Dohrmann, S., Nizet, V., & Cole, J. N. (2015). Mechanisms of group A Streptococcus resistance to reactive oxygen species. *FEMS Microbiology Reviews*, *39*, 488–508.

Henningham, A., Gillen, C. M., & Walker, M. J. (2013). Group A streptococcal vaccine candidates: Potential for the development of a human vaccine. *Current Topics in Microbiology and Immunology*, *368*, 207–242.

Henriques-Normark, B., & Normark, S. (2010). Commensal pathogens, with a focus on *Streptococcus pneumoniae*, and interactions with the human host. *Experimental Cell Research*, *316*(8), 1408–1414.

Herbig, A. F., & Helmann, J. D. (2001). Roles of metal ions and hydrogen peroxide in modulating the interaction of the *Bacillus subtilis* PerR peroxide regulon repressor with operator DNA. *Molecular Microbiology, 41*(4), 849–859.

Heresi, G., Castilloduran, C., Munoz, C., Arevalo, M., & Schlesinger, L. (1985). Phagocytosis and immunoglobulin levels in hypocupremic infants. *Nutrition Research, 5*(12), 1327–1334.

Herrera, C., Pettiglio, M. A., & Bartnikas, T. B. (2014). Investigating the role of transferrin in the distribution of iron, manganese, copper, and zinc. *Journal of Biological Inorganic Chemistry, 19*(6), 869–877.

Hertzen, E., Johansson, L., Kansal, R., Hecht, A., Dahesh, S., Janos, M., … Norrby-Teglund, A. (2012). Intracellular *Streptococcus pyogenes* in human macrophages display an altered gene expression profile. *PLoS One*, 7(4)e35218

Hertzen, E., Johansson, L., Wallin, R., Schmidt, H., Kroll, M., Rehn, A. P., … Norrby-Teglund, A. (2010). M1 protein-dependent intracellular trafficking promotes persistence and replication of *Streptococcus pyogenes* in macrophages. *Journal of Innate Immunity, 2*(6), 534–545.

Hessian, P. A., Edgeworth, J., & Hogg, N. (1993). MRP-8 and MRP-14, two abundant Ca(2+)-binding proteins of neutrophils and monocytes. *Journal of Leukocyte Biology, 53*(2), 197–204.

Hetland, G., Berntzen, H. B., & Fagerhol, M. K. (1992). Levels of calprotectin (leukocyte L1 protein) during apheresis. *Scandinavian Journal of Clinical and Laboratory Investigation, 52*(6), 479–482.

Hewitt, L. F. (1932). Bacterial metabolism. Lactic acid production by haemolytic streptococci. *Biochemical Journal, 26*, 208–217.

Hill, H. R., Book, L. S., Hemming, V. G., & Herbst, J. J. (1977). Defective neutrophil chemotactic responses in patients with recurrent episodes of otitis-media and chronic diarrhea. *American Journal of Diseases of Children, 131*(4), 433–436.

Hoeger, J., Simon, T. P., Doemming, S., Thiele, C., Marx, G., Schuerholz, T., & Haase, H. (2015). Alterations in zinc binding capacity, free zinc levels and total serum zinc in a porcine model of sepsis. *Biometals, 28*(4), 693–700.

Hoffmann, S. (1985). The throat carrier rate of group a and other beta-hemolytic streptococci among patients in general practice. *Acta Pathologica Microbiologica Et Immunologica Scandinavica. Section B—Microbiology, 93*(5), 347–351.

Hood, M. I., & Skaar, E. P. (2012). Nutritional immunity: Transition metals at the pathogen–host interface. *Nature Reviews. Microbiology, 10*(8), 525–537.

Horsburgh, M. J., Wharton, S. J., Karavolos, M., & Foster, S. J. (2002). Manganese: Elemental defence for a life with oxygen? *Trends in Microbiology, 10*(11), 496–501.

Hua, C. Z., Howard, A., Malley, R., & Lu, Y. J. (2014). Effect of nonheme iron-containing ferritin Dpr in the stress response and virulence of pneumococci. *Infection and Immunity, 82*(9), 3939–3947.

Hulett, F. M. (1996). The signal-transduction network for Pho regulation in *Bacillus subtilis*. *Molecular Microbiology, 19*(5), 933–939.

Hume, D. A. (2006). The mononuclear phagocyte system. *Current Opinion in Immunology, 18*(1), 49–53.

Ibrahim, Y. M., Kerr, A. R., McCluskey, J., & Mitchell, T. J. (2004). Role of HtrA in the virulence and competence of *Streptococcus pneumoniae*. *Infection and Immunity, 72*(6), 3584–3591.

Ibrahim, Y. M., Kerr, A. R., Silva, N. A., & Mitchell, T. J. (2005). Contribution of the ATP-dependent protease ClpCP to the autolysis and virulence of *Streptococcus pneumoniae*. *Infection and Immunity, 73*(2), 730–740.

Ibs, K. H., & Rink, L. (2003). Zinc-altered immune function. *Journal of Nutrition, 133*(5 Suppl. 1), 1452S–1456S.

Ichimura, K. (1982). Neutrophil chemotaxis in children with recurrent otitis media. *International Journal of Pediatric Otorhinolaryngology*, *4*(1), 47–55.

Imlay, J. A. (2003). Pathways of oxidative damage. *Annual Review of Microbiology*, *57*, 395–418.

Imlay, J. A. (2013). The molecular mechanisms and physiological consequences of oxidative stress: Lessons from a model bacterium. *Nature Reviews. Microbiology*, *11*(7), 443–454.

Imlay, J. A. (2014). The mismetallation of enzymes during oxidative stress. *Journal of Biological Chemistry*, *289*(41), 28121–28128.

Irving, H., & Williams, R. J. P. (1953). The stability of transition–metal complexes. *Journal of the Chemical Society*, 3192–3210.

Jacobs, A., Miller, F., Worwood, M., Beamish, M. R., & Wardrop, C. A. (1972). Ferritin in the serum of normal subjects and patients with iron deficiency and iron overload. *British Medical Journal*, *4*(5834), 206–208.

Jacobsen, F. E., Kazmierczak, K. M., Lisher, J. P., Winkler, M. E., & Giedroc, D. P. (2011). Interplay between manganese and zinc homeostasis in the human pathogen *Streptococcus pneumoniae*. *Metallomics*, *3*(1), 38–41.

Janeway, C. A., & Medzhitov, R. (2002). Innate immune recognition. *Annual Review of Immunology*, *20*, 197–216.

Janulczyk, R., Pallon, J., & Bjorck, L. (1999). Identification and characterization of a *Streptococcus pyogenes* ABC transporter with multiple specificity for metal cations. *Molecular Microbiology*, *34*(3), 596–606.

Janulczyk, R., Ricci, S., & Bjorck, L. (2003). MtsABC is important for manganese and iron transport, oxidative stress resistance, and virulence of *Streptococcus pyogenes*. *Infection and Immunity*, *71*(5), 2656–2664.

Jedrzejas, M. J. (2001). Pneumococcal virulence factors: Structure and function. *Microbiology and Molecular Biology Reviews*, *65*(2), 187–207.

Jeong, J., & Eide, D. J. (2013). The SLC39 family of zinc transporters. *Molecular Aspects of Medicine*, *34*(2–3), 612–619.

Ji, C.-J., Kim, J.-H., Won, Y.-B., Lee, Y.-E., Choi, T.-W., Ju, S.-Y., … Lee, J.-W. (2015). *Staphylococcus aureus* PerR is a hypersensitive hydrogen peroxide sensor using Fe-mediated histidine oxidation. *Journal of Biological Chemistry*, *290*, 20374–20386.

Johnson, M. D., Kehl-Fie, T. E., Klein, R., Kelly, J., Burnham, C., Mann, B., & Rosch, J. W. (2015a). Role of copper efflux in pneumococcal pathogenesis and resistance to macrophage-mediated immune clearance. *Infection and Immunity*, *83*(4), 1684–1694. http://dx.doi.org/10.1128/IAI.03015-14.

Johnson, M. D., Kehl-Fie, T. E., & Rosch, J. W. (2015b). Copper intoxication inhibits aerobic nucleotide synthesis in *Streptococcus pneumoniae*. *Metallomics*, 7(5), 786–794.

Johnston, J. W., Briles, D. E., Myers, L. E., & Hollingshead, S. K. (2006). Mn2+-dependent regulation of multiple genes in *Streptococcus pneumoniae* through PsaR and the resultant impact on virulence. *Infection and Immunity*, *74*(2), 1171–1180.

Johnston, J. W., Myers, L. E., Ochs, M. A., Benjamin, W. H., Briles, D. E., & Hollingshead, S. K. (2004). Lipoprotein PsaA in virulence of *Streptococcus pneumoniae*: Surface accessibility and role in protection from superoxide. *Infection and Immunity*, *72*(10), 5858–5867.

Johnstone, T. C., & Nolan, E. M. (2015). Beyond iron: Non-classical biological functions of bacterial siderophores. *Dalton Transactions*, *44*(14), 6320–6339.

Jonas, L., Kruger, B., & Tessenow, W. (1993). Immunohistochemical detection of lactoferrin in different human glandular tissues with special reference to the exocrine pancreas. *Acta Histochemica*, *95*(1), 53–59.

Jones, C. H., Bolken, T. C., Jones, K. F., Zeller, G. O., & Hruby, D. E. (2001). Conserved DegP protease in gram-positive bacteria is essential for thermal and oxidative tolerance and full virulence in *Streptococcus pyogenes*. *Infection and Immunity*, *69*(9), 5538–5545.

Jones, N., Ray, B., Ranjit, K. T., & Manna, A. C. (2008). Antibacterial activity of ZnO nanoparticle suspensions on a broad spectrum of microorganisms. *FEMS Microbiology Letters*, *279*(1), 71–76.

Jones, D. G., & Suttle, N. F. (1983). The effect of copper deficiency on the resistance of mice to infection with *Pasteurella haemolytica*. *Journal of Comparative Pathology*, *93*(1), 143–149.

Kadioglu, A., & Andrew, P. W. (2004). The innate immune response to pneumococcal lung infection: The untold story. *Trends in Immunology*, *25*(3), 143–149.

Kambe, T., Hashimoto, A., & Fujimoto, S. (2014). Current understanding of ZIP and ZnT zinc transporters in human health and diseases. *Cellular and Molecular Life Sciences*, *71*(17), 3281–3295.

Kaplan, E. L., Laxdal, T., & Quie, P. G. (1968). Studies of polymorphonuclear leukocytes from patients with chronic granulomatous disease of childhood—Bactericidal capacity for streptococci. *Pediatrics*, *41*(3), 591–599.

Kato, A., Tanabe, H., & Utsumi, R. (1999). Molecular characterization of the PhoQ-PhoQ two-component system in *Escherichia coli* K-12: Identification of extracellular Mg(2+)-responsive promoters. *Journal of Bacteriology*, *181*(17), 5516–5520.

Kauko, A., Haataja, S., Pulliainen, A. T., Finne, J., & Papageorgiou, A. C. (2004). Crystal structure of *Streptococcus suis* Dps-like peroxide resistance protein Dpr: Implications for iron incorporation. *Journal of Molecular Biology*, *338*(3), 547–558.

Kauko, A., Pulliainen, A. T., Haataja, S., Meyer-Klaucke, W., Finne, J., & Papageorgiou, A. C. (2006). Iron incorporation in *Streptococcus suis* Dps-like peroxide resistance protein Dpr requires mobility in the ferroxidase center and leads to the formation of a ferrihydrite-like core. *Journal of Molecular Biology*, *364*(1), 97–109.

Kazmierczak, K. M., Wayne, K. J., Rechtsteiner, A., & Winkler, M. E. (2009). Roles of rel(Spn) in stringent response, global regulation and virulence of serotype 2 *Streptococcus pneumoniae* D39. *Molecular Microbiology*, *72*(3), 590–611.

Kehl-Fie, T. E., Chitayat, S., Hood, M. I., Damo, S., Restrepo, N., Garcia, C., … Skaar, E. P. (2011). Nutrient metal sequestration by calprotectin inhibits bacterial superoxide defense, enhancing neutrophil killing of *Staphylococcus aureus*. *Cell Host & Microbe*, *10*(2), 158–164.

Kehl-Fie, T. E., & Skaar, E. P. (2010). Nutritional immunity beyond iron: A role for manganese and zinc. *Current Opinion in Chemical Biology*, *14*(2), 218–224.

Kehl-Fie, T. E., Zhang, Y. F., Moore, J. L., Farrand, A. J., Hood, M. I., Rathi, S., … Skaar, E. P. (2013). MntABC and MntH contribute to systemic *Staphylococcus aureus* infection by competing with calprotectin for nutrient manganese. *Infection and Immunity*, *81*(9), 3395–3405.

Khan, F. A., Fisher, M. A., & Khakoo, R. A. (2007). Association of hemochromatosis with infectious diseases: Expanding spectrum. *International Journal of Infectious Diseases*, *11*(6), 482–487.

Kim, Y. K., Shin, Y. J., Lee, W. H., Kim, H. Y., & Hwang, K. Y. (2009). Structural and kinetic analysis of an MsrA-MsrB fusion protein from *Streptococcus pneumoniae*. *Molecular Microbiology*, *72*(3), 699–709.

Kim, J. O., & Weiser, J. N. (1998). Association of intrastrain phase variation in quantity of capsular polysaccharide and teichoic acid with the virulence of *Streptococcus pneumoniae*. *Journal of Infectious Diseases*, *177*(2), 368–377.

King, K. Y., Horenstein, J. A., & Caparon, M. G. (2000). Aerotolerance and peroxide resistance in peroxidase and PerR mutants of *Streptococcus pyogenes*. *Journal of Bacteriology*, *182*(19), 5290–5299.

Kloosterman, T. G., van der Kooi-Pol, M. M., Bijlsma, J. J., & Kuipers, O. P. (2007). The novel transcriptional regulator SczA mediates protection against $Zn^{2+}$ stress by activation of the $Zn^{2+}$-resistance gene *czcD* in *Streptococcus pneumoniae*. *Molecular Microbiology*, *65*(4), 1049–1063.

Kloosterman, T. G., Witwicki, R. M., van der Kooi-Pol, M. M., Bijlsma, J. J. E., & Kuipers, O. P. (2008). Opposite effects of Mn2+ and Zn2+ on PsaR-mediated expression of the virulence genes *pcpA*, *prtA*, and *psaBCA* of *Streptococcus pneumoniae*. *Journal of Bacteriology*, *190*(15), 5382–5393.
Knapp, S., Leemans, J. C., Florquin, S., Branger, J., Maris, N. A., Pater, J., ... van der Poll, T. (2003). Alveolar macrophages have a protective antiinflammatory role during murine pneumococcal pneumonia. *American Journal of Respiratory and Critical Care Medicine*, *167*(2), 171–179.
Koperdanova, M., & Cullis, J. O. (2015). Interpreting raised serum ferritin levels. *British Medical Journal*, *351*, h3692.
Krewulak, K. D., & Vogel, H. J. (2008). Structural biology of bacterial iron uptake. *Biochimica et Biophysica Acta*, *1778*(9), 1781–1804.
Krishnan, K. C., Mukundan, S., Figueroa, J. A. L., Caruso, J. A., & Kotb, M. (2014). Metal-mediated modulation of streptococcal cysteine protease activity and its biological implications. *Infection and Immunity*, *82*(7), 2992–3001.
Kuhn, P. J. (1983). Doorknobs: A source of nosocomial infection. *Diagnostic Medicine*, *6*(8), 62–63.
Kuhn, D. E., Baker, B. D., Lafuse, W. P., & Zwilling, B. S. (1999). Differential iron transport into phagosomes isolated from the RAW264.7 macrophage cell lines transfected with Nramp1Gly169 or Nramp1Asp169. *Journal of Leukocyte Biology*, *66*(1), 113–119.
Kuhn, D. E., Lafuse, W. P., & Zwilling, B. S. (2001). Iron transport into *Mycobacterium avium*-containing phagosomes from an Nramp1(Gly169)-transfected RAW264.7 macrophage cell line. *Journal of Leukocyte Biology*, *69*(1), 43–49.
Kyaw, M. H., Lynfield, R., Schaffner, W., Craig, A. S., Hadler, J., Reingold, A., ... Active Bacterial Core Surveillance of the Emerging Infections Program, N. (2006). Effect of introduction of the pneumococcal conjugate vaccine on drug-resistant Streptococcus pneumoniae. *New England Journal of Medicine*, *354*(14), 1455–1463.
Lancefield, R. C. (1933). A serological differentiation of human and other groups of hemolytic streptococci. *Journal of Experimental Medicine*, *57*(4), 571–595.
Large, E. C. (1945). Field trials of copper fungicides for the control of potato blight I. Foliage protection and yield. *Annals of Applied Biology*, *32*(4), 319–329.
Lawrence, M. C., Pilling, P. A., Epa, V. C., Berry, A. M., Ogunniyi, A. D., & Paton, J. C. (1998). The crystal structure of pneumococcal surface antigen PsaA reveals a metal-binding site and a novel structure for a putative ABC-type binding protein. *Structure*, *6*(12), 1553–1561.
Lee, J. W., & Helmann, J. D. (2006a). Biochemical characterization of the structural Zn2+ site in the *Bacillus subtilis* peroxide sensor PerR. *Journal of Biological Chemistry*, *281*(33), 23567–23578.
Lee, J. W., & Helmann, J. D. (2006b). The PerR transcription factor senses $H_2O_2$ by metal-catalysed histidine oxidation. *Nature*, *440*(7082), 363–367.
Lee, J., Pena, M. M. O., Nose, Y., & Thiele, D. J. (2002). Biochemical characterization of the human copper transporter Ctr1. *Journal of Biological Chemistry*, *277*(6), 4380–4387.
Lei, B. F., Smoot, L. M., Menning, H. M., Voyich, J. M., Kala, S. V., Deleo, F. R., ... Musser, J. M. (2002). Identification and characterization of a novel heme-associated cell surface protein made by *Streptococcus pyogenes*. *Infection and Immunity*, *70*(8), 4494–4500.
Levay, P. F., & Viljoen, M. (1995). Lactoferrin: A general review. *Haematologica*, *80*(3), 252–267.
Levy, R. M., Leyden, J. J., & Margolis, D. J. (2005). Colonisation rates of *Streptococcus pyogenes* and *Staphylococcus aureus* in the oropharynx of a young adult population. *Clinical Microbiology and Infection*, *11*(2), 153–155.
Lichten, L. A., & Cousins, R. J. (2009). Mammalian zinc transporters: Nutritional and physiologic regulation. *Annual Review of Nutrition*, *29*, 153–176.

Linke, C., Caradoc-Davies, T. T., Young, P. G., Proft, T., & Baker, E. N. (2009). The laminin-binding protein Lbp from *Streptococcus pyogenes* is a zinc receptor. *Journal of Bacteriology, 191*(18), 5814–5823.

Lisher, J. P., Higgins, K. A., Maroney, M. J., & Giedroc, D. P. (2013). Physical characterization of the manganese-sensing pneumococcal surface antigen repressor (PsaR) from *Streptococcus pneumoniae. Biochemistry, 52*(43), 7689–7701.

Liu, M. J., Bao, S. Y., Galvez-Peralta, M., Pyle, C. J., Rudawsky, A. C., Pavlovicz, R. E., … Knoell, D. L. (2013). ZIP8 regulates host defense through zinc-mediated inhibition of NF-kappa B. *Cell Reports, 3*(2), 386–400.

Liu, M. Y., & Lei, B. F. (2005). Heme transfer from streptococcal cell surface protein Shp to HtsA of transporter HtsABC. *Infection and Immunity, 73*(8), 5086–5092.

Liuzzi, J. P., Lichten, L. A., Rivera, S., Blanchard, R. K., Aydemir, T. B., Knutson, M. D., … Cousins, R. J. (2005). Interleukin-6 regulates the zinc transporter Zip14 in liver and contributes to the hypozincemia of the acute-phase response. *Proceedings of the National Academy of Sciences of the United States of America, 102*(19), 6843–6848.

Loisel, E., Chimalapati, S., Bougault, C., Imberty, A., Gallet, B., Di Guilmi, A. M., … Durmort, C. (2011). Biochemical characterization of the histidine triad protein Phtd as a cell surface zinc-binding protein of Pneumococcus. *Biochemistry, 50*(17), 3551–3558.

Loisel, E., Jacquamet, L., Serre, L., Bauvois, C., Ferrer, J. L., Vernet, T., … Durmort, C. (2008). AdcAII, a new pneumococcal Zn-binding protein homologous with ABC transporters: Biochemical and structural analysis. *Journal of Molecular Biology, 381*(3), 594–606.

Lonks, J. R., Garau, J., Gomez, L., Xercavins, M., de Echaguen, A. O., Gareen, I. F., … Medeiros, A. A. (2002). Failure of macrolide antibiotic treatment in patients with bacteremia due to erythromycin-resistant *Streptococcus pneumoniae. Clinical Infectious Diseases, 35*(5), 556–564.

Loomans, H. J., Hahn, B. L., Li, Q. Q., Phadnis, S. H., & Sohnle, P. G. (1998). Histidine-based zinc-binding sequences and the antimicrobial activity of calprotectin. *Journal of Infectious Diseases, 177*(3), 812–814.

Lu, J., & Holmgren, A. (2014). The thioredoxin antioxidant system. *Free Radical Biology and Medicine, 66*, 75–87.

Lyon, W. R., & Caparon, M. G. (2004). Role for serine protease HtrA (DegP) of *Streptococcus pyogenes* in the biogenesis of virulence factors SpeB and the hemolysin streptolysin S. *Infection and Immunity, 72*(3), 1618–1625.

Ma, L., Terwilliger, A., & Maresso, A. W. (2015). Iron and zinc exploitation during bacterial pathogenesis. *Metallomics*, 7(12), 1541–1554.

Macomber, L., & Hausinger, R. P. (2011). Mechanisms of nickel toxicity in microorganisms. *Metallomics, 3*(11), 1153–1162.

Macomber, L., & Imlay, J. A. (2009). The iron-sulfur clusters of dehydratases are primary intracellular targets of copper toxicity. *Proceedings of the National Academy of Sciences of the United States of America, 106*(20), 8344–8349.

MacPherson, I. S., & Murphy, M. E. P. (2007). Type-2 copper-containing enzymes. *Cellular and Molecular Life Sciences, 64*(22), 2887–2899.

Makarova, K. S., Ponomarev, V. A., & Koonin, E. V. (2001). Two C or not two C: Recurrent disruption of Zn-ribbons, gene duplication, lineage-specific gene loss, and horizontal gene transfer in evolution of bacterial ribosomal proteins. *Genome Biology, 2*(9), research0033.1–research0033.14.

Makthal, N., Rastegari, S., Sanson, M., Ma, Z., Olsen, R. J., Helmann, J. D., … Kumaraswami, M. (2013). Crystal structure of peroxide stress regulator from *Streptococcus pyogenes* provides functional insights into the mechanism of oxidative stress sensing. *Journal of Biological Chemistry, 288*(25), 18311–18324.

Mandell, G. L., & Hook, E. W. (1969). Leukocyte bactericidal activity in chronic granulomatous disease—Correlation of bacterial hydrogen peroxide production and susceptibility to intracellular killing. *Journal of Bacteriology, 100*(1), 531–532.

Marchisio, P., Esposito, S., Schito, G. C., Marchese, A., Cavagna, R., Principi, N., & Hercules Project Collaborative, G. (2002). Nasopharyngeal carriage of Streptococcus pneumoniae in healthy children: Implications for the use of heptavalent pneumococcal conjugate vaccine. *Emerging Infectious Diseases, 8*(5), 479–484.

Marks, R., Pearse, A. D., & Walker, A. P. (1985). The effects of a shampoo containing zinc pyrithione on the control of dandruff. *British Journal of Dermatology, 112*(4), 415–422.

Marra, A., Lawson, S., Asundi, J. S., Brigham, D., & Hromockyj, A. E. (2002). In vivo characterization of the *psa* genes from *Streptococcus pneumoniae* in multiple models of infection. *Microbiology, 148*(Pt. 5), 1483–1491.

Martin, J. M., Green, M., Barbadora, K. A., & Wald, E. R. (2002). Erythromycin-resistant group A streptococci in schoolchildren in Pittsburgh. *New England Journal of Medicine, 346*(16), 1200–1206.

Martin, M. E., Strachan, R. C., Aranha, H., Evans, S. L., Salin, M. L., Welch, B., ... Byers, B. R. (1984). Oxygen toxicity in *Streptococcus mutans*—Manganese, iron, and superoxide-dismutase. *Journal of Bacteriology, 159*(2), 745–749.

Martinez, A., & Kolter, R. (1997). Protection of DNA during oxidative stress by the nonspecific DNA-binding protein Dps. *Journal of Bacteriology, 179*(16), 5188–5194.

Martner, A., Dahlgren, C., Paton, J. C., & Wold, A. E. (2008). Pneumolysin released during *Streptococcus pneumoniae* autolysis is a potent activator of intracellular oxygen radical production in neutrophils. *Infection and Immunity, 76*(9), 4079–4087.

Masip, L., Veeravalli, K., & Georgioui, G. (2006). The many faces of glutathione in bacteria. *Antioxidants & Redox Signaling, 8*(5–6), 753–762.

Mayer-Scholl, A., Averhoff, P., & Zychlinsky, A. (2004). How do neutrophils and pathogens interact? *Current Opinion in Microbiology, 7*(1), 62–66.

McAllister, L. J., Tseng, H. J., Ogunniyi, A. D., Jennings, M. P., McEwan, A. G., & Paton, J. C. (2004). Molecular analysis of the psa permease complex of *Streptococcus pneumoniae. Molecular Microbiology, 53*(3), 889–901.

McCall, K. A., Huang, C. C., & Fierke, C. A. (2000). Function and mechanism of zinc metalloenzymes. *Journal of Nutrition, 130*(5), 1437S–1446S.

McCluskey, J., Hinds, J., Husain, S., Witney, A., & Mitchell, T. J. (2004). A two-component system that controls the expression of pneumococcal surface antigen A (PsaA) and regulates virulence and resistance to oxidative stress in *Streptococcus pneumoniae. Molecular Microbiology, 51*(6), 1661–1675.

McDevitt, C. A., Ogunniyi, A. D., Valkov, E., Lawrence, M. C., Kobe, B., McEwan, A. G., & Paton, J. C. (2011). A molecular mechanism for bacterial susceptibility to zinc. *PLoS Pathogens, 7*(11)e1002357

McElroy, J. (1910). The treatment of pulmonary tuberculosis by intravenous injections of chinosol with formaldehyde. *Lancet, 2*, 1408–1409.

McEwan, A. G. (2009). New insights into the protective effect of manganese against oxidative stress. *Molecular Microbiology, 72*(4), 812–814.

Michos, A. G., Bakoula, C. G., Braoudaki, M., Koutouzi, F. I., Roma, E. S., Pangalis, A., ... Syriopoulou, V. P. (2009). Macrolide resistance in *Streptococcus pyogenes*: Prevalence, resistance determinants, and emm types. *Diagnostic Microbiology and Infectious Disease, 64*(3), 295–299.

Miller, A. F. (2012). Superoxide dismutases: Ancient enzymes and new insights. *FEBS Letters, 586*(5), 585–595.

Miller, S. I., Kukral, A. M., & Mekalanos, J. J. (1989). A two-component regulatory system (phoP phoQ) controls *Salmonella typhimurium* virulence. *Proceedings of the National Academy of Sciences of the United States of America, 86*(13), 5054–5058.

Mitchell, T. J. (2003). The pathogenesis of streptococcal infections: From tooth decay to meningitis. *Nature Reviews Microbiology, 1*(3), 219–230.

Mitchell, A. M., & Mitchell, T. J. (2010). *Streptococcus pneumoniae*: Virulence factors and variation. *Clinical Microbiology and Infection, 16*(5), 411–418.

Molloy, E. M., Cotter, P. D., Hill, C., Mitchell, D. A., & Ross, R. P. (2011). Streptolysin S-like virulence factors: The continuing sagA. *Nature Reviews. Microbiology, 9*(9), 670–681.

Montanez, G. E., Neely, M. N., & Eichenbaum, Z. (2005). The streptococcal iron uptake (Siu) transporter is required for iron uptake and virulence in a zebrafish infection model. *Microbiology, 151*, 3749–3757.

Mook-Kanamori, B. B., Geldhoff, M., van der Poll, T., & van de Beek, D. (2011). Pathogenesis and pathophysiology of pneumococcal meningitis. *Clinical Microbiology Reviews, 24*(3), 557–591.

Mulhern, S. A., & Koller, L. D. (1988). Severe or marginal copper deficiency results in a graded reduction in immune status in mice. *Journal of Nutrition, 118*(8), 1041–1047.

Nakashige, T. G., Zhang, B., Krebs, C., & Nolan, E. M. (2015). Human calprotectin is an iron-sequestering host-defense protein. *Nature Chemical Biology, 11*(10), 765–771.

Natori, Y., Nanamiya, H., Akanuma, G., Kosono, S., Kudo, T., Ochi, K., & Kawamura, F. (2007). A fail-safe system for the ribosome under zinc-limiting conditions in *Bacillus subtilis*. *Molecular Microbiology, 63*(1), 294–307.

Nauseef, W. M. (2007). How human neutrophils kill and degrade microbes: An integrated view. *Immunological Reviews, 219*, 88–102.

Nemeth, E., Rivera, S., Gabayan, V., Keller, C., Taudorf, S., Pedersen, B. K., & Ganz, T. (2004). IL-6 mediates hypoferremia of inflammation by inducing the synthesis of the iron regulatory hormone hepcidin. *Journal of Clinical Investigation, 113*(9), 1271–1276.

Nemeth, E., Tuttle, M. S., Powelson, J., Vaughn, M. B., Donovan, A., Ward, D. M., … Kaplan, J. (2004). Hepcidin regulates cellular iron efflux by binding to ferroportin and inducing its internalization. *Science, 306*(5704), 2090–2093.

Nevo, Y., & Nelson, N. (2006). The NRAMP family of metal-ion transporters. *Biochimica et Biophysica Acta—Molecular Cell Research, 1763*(7), 609–620.

Newby, C. S., Rowland, J. L., Lynch, R. J. M., Bradshaw, D. J., Whitworth, D., & Bosma, M. L. (2011). Benefits of a silica-based fluoride toothpaste containing o-cymen-5-ol, zinc chloride and sodium fluoride. *International Dental Journal, 61*, 74–80.

Nies, D. H. (2007). Bacterial transition metal homeostasis. In D. H. Nies & S. Silver (Eds.), *Molecular microbiology of heavy metals* (pp. 117–142). Berlin, Heidelberg: Springer.

Nizet, V., Beall, B., Bast, D. J., Datta, V., Kilburn, L., Low, D. E., & De Azavedo, J. C. (2000). Genetic locus for streptolysin S production by Group A *Streptococcus*. *Infection and Immunity, 68*(7), 4245–4254.

Novak, R., Braun, J. S., Charpentier, E., & Tuomanen, E. (1998). Penicillin tolerance genes of *Streptococcus pneumoniae*: The ABC-type manganese permease complex Psa. *Molecular Microbiology, 29*(5), 1285–1296.

Nygaard, T. K., Blouin, G. C., Liu, M., Fukumura, M., Olson, J. S., Fabian, M., … Lei, B. (2006). The mechanism of direct heme transfer from the streptococcal cell surface protein Shp to HtsA of the HtsABC transporter. *Journal of Biological Chemistry, 281*(30), 20761–20771.

O'Brien, K. L., Wolfson, L. J., Watt, J. P., Henkle, E., Deloria-Knoll, M., McCall, N., ... Cherian, T. (2009). Burden of disease caused by *Streptococcus pneumoniae* in children younger than 5 years: Global estimates. *The Lancet, 374*(9693), 893–902.

Ogunniyi, A. D., Grabowicz, M., Mahdi, L. K., Cook, J., Gordon, D. L., Sadlon, T. A., & Paton, J. C. (2009). Pneumococcal histidine triad proteins are regulated by the Zn2+-dependent repressor AdcR and inhibit complement deposition through the recruitment of complement factor H. *FASEB Journal, 23*(3), 731–738.

Ogunniyi, A. D., Mahdi, L. K., Jennings, M. P., McEwan, A. G., McDevitt, C. A., Van der Hoek, M. B., ... Paton, J. C. (2010). Central role of manganese in regulation of stress responses, physiology, and metabolism in *Streptococcus pneumoniae*. *Journal of Bacteriology, 192*(17), 4489–4497.

Olsen, R. J., Sitkiewicz, I., Ayeras, A. A., Gonulal, V. E., Cantu, C., Beres, S. B., ... Musser, J. M. (2010). Decreased necrotizing fasciitis capacity caused by a single nucleotide mutation that alters a multiple gene virulence axis. *Proceedings of the National Academy of Sciences of the United States of America, 107*(2), 888–893.

Ong, C. L., Gillen, C. M., Barnett, T. C., Walker, M. J., & McEwan, A. G. (2014). An antimicrobial role for zinc in innate immune defense against Group A *Streptococcus*. *Journal of Infectious Diseases, 209*(10), 1500–1508.

Ong, C. L. Y., Potter, A. J., Trappetti, C., Walker, M. J., Jennings, M. P., Paton, J. C., & McEwan, A. G. (2013). Interplay between manganese and iron in Pneumococcal pathogenesis: Role of the orphan response regulator RitR. *Infection and Immunity, 81*(2), 421–429.

Ong, C. L., Walker, M. J., & McEwan, A. G. (2015). Zinc disrupts central carbon metabolism and capsule biosynthesis in *Streptococcus pyogenes*. *Scientific Reports, 5*, 10799.

Ouattara, M., Cunha, E. B., Li, X. R., Huang, Y. S., Dixon, D., & Eichenbaum, Z. (2010). Shr of group A streptococcus is a new type of composite NEAT protein involved in sequestering haem from methaemoglobin. *Molecular Microbiology, 78*(3), 739–756.

Panina, E. M., Mironov, A. A., & Gelfand, M. S. (2003). Comparative genomics of bacterial zinc regulons: Enhanced ion transport, pathogenesis, and rearrangement of ribosomal proteins. *Proceedings of the National Academy of Sciences of the United States of America, 100*(17), 9912–9917.

Paton, J. C. (1996). The contribution of pneumolysin to the pathogenicity of *Streptococcus pneumoniae*. *Trends in Microbiology, 4*(3), 103–106.

Percival, S. S. (1995). Neutropenia caused by copper deficiency: Possible mechanisms of action. *Nutrition Reviews, 53*(3), 59–66.

Perez-Diaz, M. A., Boegli, L., James, G., Velasquillo, C., Sanchez-Sanchez, R., Martinez-Martinez, R. E., ... Martinez-Gutierrez, F. (2015). Silver nanoparticles with antimicrobial activities against *Streptococcus mutans* and their cytotoxic effect. *Materials Science & Engineering. C-Materials for Biological Applications, 55*, 360–366.

Pericone, C. D., Bae, D., Shchepetov, M., McCool, T., & Weiser, J. N. (2002). Short-sequence tandem and nontandem DNA repeats and endogenous hydrogen peroxide production contribute to genetic instability of *Streptococcus pneumoniae*. *Journal of Bacteriology, 184*(16), 4392–4399.

Pericone, C. D., Overweg, K., Hermans, P. W. M., & Weiser, J. N. (2000). Inhibitory and bactericidal effects of hydrogen peroxide production by *Streptococcus pneumoniae* on other inhabitants of the upper respiratory tract. *Infection and Immunity, 68*(7), 3990–3997.

Pericone, C. D., Park, S., Imlay, J. A., & Weiser, J. N. (2003). Factors contributing to hydrogen peroxide resistance in *Streptococcus pneumoniae* include pyruvate oxidase (SpxB) and avoidance of the toxic effects of the Fenton reaction. *Journal of Bacteriology, 185*(23), 6815–6825.

Petris, M. J., Mercer, J. F. B., Culvenor, J. G., Lockhart, P., Gleeson, P. A., & Camakaris, J. (1996). Ligand-regulated transport of the Menkes copper P-type ATPase efflux pump from the golgi apparatus to the plasma membrane: A novel mechanism of regulated trafficking. *EMBO Journal*, *15*(22), 6084–6095.

Phan, T. N., Buckner, T., Sheng, J., Baldeck, J. D., & Marquis, R. E. (2004). Physiologic actions of zinc related to inhibition of acid and alkali production by oral streptococci in suspensions and biofilms. *Oral Microbiology and Immunology*, *19*(1), 31–38.

Philips, B. J., Meguer, J. X., Redman, J., & Baker, E. H. (2003). Factors determining the appearance of glucose in upper and lower respiratory tract secretions. *Intensive Care Medicine*, *29*(12), 2204–2210.

Pi, H., Patel, S. J., Arguello, J. M., & Helmann, J. D. (2016). The *Listeria monocytogenes* Fur-regulated virulence protein FrvA is an Fe(II) efflux P -type ATPase. *Molecular Microbiology*, *100*, 1066–1079.

Pieters, J. (2008). *Mycobacterium tuberculosis* and the macrophage: Maintaining a balance. *Cell Host & Microbe*, *3*(6), 399–407.

Pizzey, R. L., Marquis, R. E., & Bradshaw, D. J. (2011). Antimicrobial effects of o-cymen-5-ol and zinc, alone & in combination in simple solutions and toothpaste formulations. *International Dental Journal*, *61*, 33–40.

Plumptre, C. D., Eijkelkamp, B. A., Morey, J. R., Behr, F., Counago, R. M., Ogunniyi, A. D., … McDevitt, C. A. (2014). AdcA and AdcAII employ distinct zinc acquisition mechanisms and contribute additively to zinc homeostasis in *Streptococcus pneumoniae*. *Molecular Microbiology*, *91*(4), 834–851.

Plumptre, C. D., Hughes, C. E., Harvey, R. M., Eijkelkamp, B. A., McDevitt, C. A., & Paton, J. C. (2014). Overlapping functionality of the pht proteins in zinc homeostasis of *Streptococcus pneumoniae*. *Infection and Immunity*, *82*(10), 4315–4324.

Plumptre, C. D., Ogunniyi, A. D., & Paton, J. C. (2012). Polyhistidine triad proteins of pathogenic streptococci. *Trends in Microbiology*, *20*(10), 485–493.

Porcheron, G., Garénaux, A., Proulx, J., Sabri, M., & Dozois, C. M. (2013). Iron, copper, zinc, and manganese transport and regulation in pathogenic Enterobacteria: Correlations between strains, site of infection and the relative importance of the different metal transport systems for virulence. *Frontiers in Cellular and Infection Microbiology*, *3*, 90.

Portmann, R., Poulsen, K. R., Wimmer, R., & Solioz, M. (2006). CopY-like copper inducible repressors are putative 'winged helix' proteins. *Biometals*, *19*(1), 61–70.

Portnoy, D. A., Auerbuch, V., & Glomski, I. J. (2002). The cell biology of *Listeria monocytogenes* infection: The intersection of bacterial pathogenesis and cell-mediated immunity. *Journal of Cell Biology*, *158*(3), 409–414.

Posey, J. E., & Gherardini, F. C. (2000). Lack of a role for iron in the Lyme disease pathogen. *Science*, *288*(5471), 1651–1653.

Potter, A. J., Trappetti, C., & Paton, J. C. (2012). *Streptococcus pneumoniae* uses glutathione to defend against oxidative stress and metal ion toxicity. *Journal of Bacteriology*, *194*(22), 6248–6254.

Powell, S. R. (2000). The antioxidant properties of zinc. *Journal of Nutrition*, *130*(5S Suppl.), 1447S–1454S.

Prachayasittikul, V., Prachayasittikul, S., Ruchirawat, S., & Prachayasittikul, V. (2013). 8-Hydroxyquinolines: A review of their metal chelating properties and medicinal applications. *Drug Design, Development and Therapy*, *7*, 1157–1178.

Pramanik, A., & Braun, V. (2006). Albomycin uptake via a ferric hydroxamate transport system of *Streptococcus pneumoniae* R6. *Journal of Bacteriology*, *188*(11), 3878–3886.

Prasad, A. S. (2000). Effects of zinc deficiency on immune functions. *Journal of Trace Elements in Experimental Medicine*, *13*(1), 1–20.

Prasad, A. S. (2008). Zinc in human health: Effect of zinc on immune cells. *Molecular Medicine*, *14*(5–6), 353–357.

Prohaska, J. R., & Lukasewycz, O. A. (1981). Copper deficiency suppresses the immune-response of mice. *Science*, *213*(4507), 559–561.

Pulliainen, A. T., Hytonen, J., Haataja, S., & Finne, J. (2008). Deficiency of the Rgg regulator promotes H2O2 resistance, AhpCF-mediated H2O2 decomposition, and virulence in *Streptococcus pyogenes*. *Journal of Bacteriology*, *190*(9), 3225–3235.

Pulliainen, A. T., Kauko, A., Haataja, S., Papageorgiou, A. C., & Finne, J. (2005). Dps/Dpr ferritin-like protein: Insights into the mechanism of iron incorporation and evidence for a central role in cellular iron homeostasis in Streptococcus suis. *Molecular Microbiology*, *57*(4), 1086–1100.

Quie, P. G., White, J. G., Holmes, B., & Good, R. A. (1967). *In vitro* bactericidal capacity of human polymorphonuclear leukocytes—Diminished activity in chronic granulomatous disease of childhood. *Journal of Clinical Investigation*, *46*(4), 668–679.

Rademacher, C., & Masepohl, B. (2012). Copper-responsive gene regulation in bacteria. *Microbiology*, *158*, 2451–2464.

Rai, P., Parrish, M., Tay, I. J. J., Li, N., Ackerman, S., He, F., … Engelward, B. P. (2015). *Streptococcus pneumoniae* secretes hydrogen peroxide leading to DNA damage and apoptosis in lung cells. *Proceedings of the National Academy of Sciences of the United States of America*, *112*(26), E3421–E3430.

Rajam, G., Anderton, J. M., Carlone, G. M., Sampson, J. S., & Ades, E. W. (2008). Pneumococcal surface adhesin A (PsaA): A review. *Critical Reviews in Microbiology*, *34*(3–4), 131–142.

Rammes, A., Roth, J., Goebeler, M., Klempt, M., Hartmann, M., & Sorg, C. (1997). Myeloid-related protein (MRP) 8 and MRP14, calcium-binding proteins of the S100 family, are secreted by activated monocytes via a novel, tubulin-dependent pathway. *Journal of Biological Chemistry*, *272*(14), 9496–9502.

Ramos-Montanez, S., Tsui, H. C. T., Wayne, K. J., Morris, J. L., Peters, L. E., Zhang, F., … Winkler, M. E. (2008). Polymorphism and regulation of the spxB (pyruvate oxidase) virulence factor gene by a CBS-HotDog domain protein (SpxR) in serotype 2 *Streptococcus pneumoniae*. *Molecular Microbiology*, *67*(4), 729–746.

Ran, Y. C., Zhu, H., Liu, M. Y., Fabian, M., Olson, J. S., Aranda, R. I. V., … Lei, B. (2007). Bis-methionine ligation to heme iron in the streptococcal cell surface protein shp facilitates rapid hemin transfer to HtsA of the HtsABC transporter. *Journal of Biological Chemistry*, *282*(43), 31380–31388.

Reeder, N. L., Kaplan, J., Xu, J., Youngquist, R. S., Wallace, J., Hu, P., … Saunders, C. W. (2011). Zinc pyrithione inhibits yeast growth through copper influx and inactivation of iron-sulfur proteins. *Antimicrobial Agents and Chemotherapy*, *55*(12), 5753–5760.

Regev-Yochay, G., Trzcinski, K., Thompson, C. M., Lipsitch, M., & Malley, R. (2007). SpxB is a suicide gene of *Streptococcus pneumoniae* and confers a selective advantage in an in vivo competitive colonization model. *Journal of Bacteriology*, *189*(18), 6532–6539.

Rensing, C., & Grass, G. (2003). Escherichia coli mechanisms of copper homeostasis in a changing environment. *FEMS Microbiology Reviews*, *27*(2–3), 197–213.

Reyes-Caballero, H., Guerra, A. J., Jacobsen, F. E., Kazmierczak, K. M., Cowart, D., Koppolu, U. M. K., … Giedroc, D. P. (2010). The metalloregulatory zinc site in *Streptococcus pneumoniae* AdcR, a zinc-activated MarR family repressor. *Journal of Molecular Biology*, *403*(2), 197–216.

Riboldi-Tunnicliffe, A., Bent, C. J., Isaacs, N. W., & Mitchell, T. J. (2004). Expression, purification and X-ray characterization of residues 18-230 from the pneumococcal histidine triad protein A (PhtA) from *Streptococcus pneumoniae*. *Acta Crystallographica. Section D—Biological Crystallography*, *60*, 926–928.

Riboldi-Tunnicliffe, A., Isaacs, N. W., & Mitchell, T. J. (2005). 1.2 angstrom crystal structure of the S-pneumoniae PhtA histidine triad domain—A novel zinc binding fold. *FEBS Letters*, *579*(24), 5353–5360.

Ricci, S., Janulczyk, R., & Bjorck, L. (2002). The regulator PerR is involved in oxidative stress response and iron homeostasis and is necessary for full virulence of *Streptococcus pyogenes*. *Infection and Immunity*, *70*(9), 4968–4976.

Rink, L., & Gabriel, P. (2000). Zinc and the immune system. *Proceedings of the Nutrition Society*, *59*(4), 541–552. http://dx.doi.org/10.1017/S0029665100000781.

Rizzato, G., Locicero, S., Barberis, M., Torre, M., Pietra, R., & Sabbioni, E. (1986). Trace of metal exposure in hard metal lung-disease. *Chest*, *90*(1), 101–106.

Roberts, A. L., Connolly, K. L., Kirse, D. J., Evans, A. K., Poehling, K. A., Peters, T. R., & Reid, S. D. (2012). Detection of group A *Streptococcus* in tonsils from pediatric patients reveals high rate of asymptomatic streptococcal carriage. *BMC Pediatrics*, *12*, 3.

Robertson, G. T., Ng, W. L., Foley, J., Gilmour, R., & Winkler, M. E. (2002). Global transcriptional analysis of clpP mutations of type 2 *Streptococcus pneumoniae* and their effects on physiology and virulence. *Journal of Bacteriology*, *184*(13), 3508–3520.

Roca, I., Torrents, E., Sahlin, M., Gibert, I., & Sjoberg, B. M. (2008). NrdI essentiality for class Ib ribonucleotide reduction in *Streptococcus pyogenes*. *Journal of Bacteriology*, *190*(14), 4849–4858.

Romero-Espejel, M. E., Gonzalez-Lopez, M. A., & Olivares-Trejo, J. D. (2013). *Streptococcus pneumoniae* requires iron for its viability and expresses two membrane proteins that bind haemoglobin and haem. *Metallomics*, *5*(4), 384–389.

Romero-Steiner, S., Pilishvili, T., Sampson, J. S., Johnson, S. E., Stinson, A., Carlone, G. M., & Ades, E. W. (2003). Inhibition of pneumococcal adherence to human nasopharyngeal epithelial cells by anti-PsaA antibodies. *Clinical and Diagnostic Laboratory Immunology*, *10*(2), 246–251.

Roos, D., van Bruggen, R., & Meischl, C. (2003). Oxidative killing of microbes by neutrophils. *Microbes and Infection*, *5*(14), 1307–1315.

Rosch, J. W., Gao, G., Ridout, G., Wang, Y. D., & Tuomanen, E. I. (2009). Role of the manganese efflux system MntE for signalling and pathogenesis in *Streptococcus pneumoniae*. *Molecular Microbiology*, *72*(1), 12–25.

Sabbioni, E., Pietra, R., Mousty, F., Colombo, F., Rizzato, G., & Scansetti, G. (1987). Trace metals in human lung as determined by neutron activation analysis of bronchoalveolar lavage. *Journal of Radioanalytical and Nuclear Chemistry*, *110*(2), 595–601.

Sachla, A. J., & Eichenbaum, Z. (2016). The GAS PefCD exporter is a MDR system that confers resistance to heme and structurally diverse compounds. *BMC Microbiology*, *16*, 68.

Sachla, A. J., Le Breton, Y., Akhter, F., McIver, K. S., & Eichenbaum, Z. (2014). The crimson conundrum: Heme toxicity and tolerance in GAS. *Frontiers in Cellular and Infection Microbiology*, *4*, 159.

Sachla, A. J., Ouattara, M., Romero, E., Agniswamy, J., Weber, I. T., Gadda, G., & Eichenbaum, Z. (2016). In vitro heme biotransformation by the HupZ enzyme from group A *Streptococcus*. *Biometals*, *29*(4), 593–609.

Samanovic, M. I., Ding, C., Thiele, D. J., & Darwin, K. H. (2012). Copper in microbial pathogenesis: Meddling with the metal. *Cell Host & Microbe*, *11*(2), 106–115.

Sanson, M., Makthal, N., Flores, A. R., Olsen, R. J., Musser, J. M., & Kumaraswami, M. (2015). Adhesin competence repressor (AdcR) from *Streptococcus pyogenes* controls adaptive responses to zinc limitation and contributes to virulence. *Nucleic Acids Research*, *43*(1), 418–432.

Scott, B. J., & Bradwell, A. R. (1983). Identification of the serum binding proteins for iron, zinc, cadmium, nickel, and calcium. *Clinical Chemistry*, *29*(4), 629–633.

Segal, A. W. (2005). How neutrophils kill microbes. *Annual Review of Immunology, 23*, 197–223.

Seki, M., Iida, K., Saito, M., Nakayama, H., & Yoshida, S. (2004). Hydrogen peroxide production in *Streptococcus pyogenes*: Involvement of lactate oxidase and coupling with aerobic utilization of lactate. *Journal of Bacteriology, 186*(7), 2046–2051.

Shafeeq, S., Kloosterman, T. G., & Kuipers, O. P. (2011). Transcriptional response of *Streptococcus pneumoniae* to Zn2+ limitation and the repressor/activator function of AdcR. *Metallomics, 3*(6), 609–618.

Shafeeq, S., Yesilkaya, H., Kloosterman, T. G., Narayanan, G., Wandel, M., Andrew, P. W., ... Morrissey, J. A. (2011). The cop operon is required for copper homeostasis and contributes to virulence in *Streptococcus pneumoniae*. *Molecular Microbiology, 81*(5), 1255–1270. http://dx.doi.org/10.1111/J.1365-2958.2011.07758.X.

Shah, S., Dalecki, A. G., Malalasekera, A. P., Crawford, C. L., Michalek, S. M., Kutsch, O., ... Wolschendorf, F. (2016). 8-Hydroxyquinolines are boosting-agents of copper related toxicity in Mycobacterium tuberculosis. *Antimicrobial Agents and Chemotherapy, 60*, 5765–5776.

Shaikh, N., Leonard, E., & Martin, J. M. (2010). Prevalence of streptococcal pharyngitis and streptococcal carriage in children: A meta-analysis. *Pediatrics, 126*(3), E557–E564.

Shankar, A. H., & Prasad, A. S. (1998). Zinc and immune function: The biological basis of altered resistance to infection. *American Journal of Clinical Nutrition, 68*(2), 447S–463S.

Shaper, M., Hollingshead, S. K., Benjamin, W. H., & Briles, D. E. (2004). PspA protects *Streptococcus pneumoniae* from killing by apolactoferrin, and antibody to PspA enhances killing of pneumococci by apolactoferrin. *Infection and Immunity, 72*(9), 5031–5040.

Sharma, A., Gaidamakova, E. K., Matrosova, V. Y., Bennett, B., Daly, M. J., & Hoffman, B. M. (2013). Responses of Mn2+ speciation in *Deinococcus radiodurans* and *Escherichia coli* to γ-radiation by advanced paramagnetic resonance methods. *Proceedings of the National Academy of Sciences, 110*(15), 5945–5950.

Shewell, L. K., Harvey, R. M., Higgins, M. A., Day, C. J., Hartley-Tassell, L. E., Chen, A. Y., ... Jennings, M. P. (2014). The cholesterol-dependent cytolysins pneumolysin and streptolysin O require binding to red blood cell glycans for hemolytic activity. *Proceedings of the National Academy of Sciences of the United States of America, 111*(49), E5312–E5320.

Simon, N., Coulanges, V., Andre, P., & Vidon, D. J. M. (1995). Utilization of exogenous siderophores and natural catechols by *Listeria monocytogenes*. *Applied and Environmental Microbiology, 61*(4), 1643–1645.

Skikne, B. S., Flowers, C. H., & Cook, J. D. (1990). Serum transferrin receptor: A quantitative measure of tissue iron deficiency. *Blood, 75*(9), 1870–1876.

Smoot, L. M., Smoot, J. C., Graham, M. R., Somerville, G. A., Sturdevant, D. E., Migliaccio, C. A. L., ... Musser, J. M. (2001). Global differential gene expression in response to growth temperature alteration in group A Streptococcus. *Proceedings of the National Academy of Sciences of the United States of America, 98*(18), 10416–10421.

Sobocinski, P. Z., Canterbury, W. J., Jr., Mapes, C. A., & Dinterman, R. E. (1978). Involvement of hepatic metallothioneins in hypozincemia associated with bacterial infection. *American Journal of Physiology, 234*(4), E399–E406.

Sobota, J. M., & Imlay, J. A. (2011). Iron enzyme ribulose-5-phosphate 3-epimerase in *Escherichia coli* is rapidly damaged by hydrogen peroxide but can be protected by manganese. *Proceedings of the National Academy of Sciences of the United States of America, 108*(13), 5402–5407.

Sohnle, P. G., Collinslech, C., & Wiessner, J. H. (1991). Antimicrobial activity of an abundant calcium-binding protein in the cytoplasm of human neutrophils. *Journal of Infectious Diseases, 163*(1), 187–192.

Solomon, E. I., Sundaram, U. M., & Machonkin, T. E. (1996). Multicopper oxidases and oxygenases. *Chemical Reviews*, *96*(7), 2563–2606.

Sook, B. R., Block, D. R., Sumithran, S., Montanez, G. E., Rodgers, K. R., Dawson, J. H., … Dixon, D. W. (2008). Characterization of SiaA, a streptococcal heme-binding protein associated with a heme ABC transport system. *Biochemistry*, *47*(8), 2678–2688.

Speer, A., Rowland, J. L., Haeili, M., Niederweis, M., & Wolschendorf, F. (2013). Porins increase copper susceptibility of mycobacterium tuberculosis. *Journal of Bacteriology*, *195*(22), 5133–5140.

Spellerberg, B., Cundell, D. R., Sandros, J., Pearce, B. J., Idanpaan-Heikkila, I., Rosenow, C., & Masure, H. R. (1996). Pyruvate oxidase, as a determinant of virulence in *Streptococcus pneumoniae*. *Molecular Microbiology*, *19*(4), 803–813. http://dx.doi.org/10.1046/J.1365-2958.1996.425954.X.

Stafford, S. L., Bokil, N. J., Achard, M. E., Kapetanovic, R., Schembri, M. A., McEwan, A. G., & Sweet, M. J. (2013). Metal ions in macrophage antimicrobial pathways: Emerging roles for zinc and copper. *Bioscience Reports*, *33*, 541–554.

Standish, A. J., & Weiser, J. N. (2009). Human neutrophils kill *Streptococcus pneumoniae* via serine proteases. *Journal of Immunology*, *183*(4), 2602–2609.

Stauff, D. L., Bagaley, D., Torres, V. J., Joyce, R., Anderson, K. L., Kuechenmeister, L., … Skaar, E. P. (2008). Staphylococcus aureus HrtA is an ATPase required for protection against heme toxicity and prevention of a transcriptional heme stress response. *Journal of Bacteriology*, *190*(10), 3588–3596.

Steinbakk, M., Naess-Andresen, C. F., Lingaas, E., Dale, I., Brandtzaeg, P., & Fagerhol, M. K. (1990). Antimicrobial actions of calcium binding leucocyte L1 protein, calprotectin. *Lancet*, *336*(8718), 763–765.

Stephen, K. W., Saxton, C. A., Jones, C. L., Ritchie, J. A., & Morrison, T. (1990). Control of gingivitis and calculus by a dentifrice containing a zinc salt and triclosan. *Journal of Periodontology*, *61*(11), 674–679.

Strausak, D., & Solioz, M. (1997). CopY is a copper-inducible repressor of the *Enterococcus hirae* copper ATPases. *Journal of Biological Chemistry*, *272*(14), 8932–8936.

Striz, I., & Trebichavsky, I. (2004). Calprotectin—A pleiotropic molecule in acute and chronic inflammation. *Physiological Research*, *53*(3), 245–253.

Suhy, D. A., Simon, K. D., Linzer, D. I. H., & O'Halloran, T. V. (1999). Metallothionein is part of a zinc-scavenging mechanism for cell survival under conditions of extreme zinc deprivation. *Journal of Biological Chemistry*, *274*(14), 9183–9192.

Sullivan, J. F., Blotcky, A. J., Jetton, M. M., Hahn, H. K. J., & Burch, R. E. (1979). Serum levels of selenium, calcium, copper magnesium, manganese and zinc in various human diseases. *Journal of Nutrition*, *109*(8), 1432–1437.

Sulotto, F., Romano, C., Berra, A., Botta, G. C., Rubino, G. F., Sabbioni, E., & Pietra, R. (1986). Rare-earth pneumoconiosis—A new case. *American Journal of Industrial Medicine*, *9*(6), 567–575.

Sun, X. S., Ge, R. G., Chiu, J. F., Sun, H. Z., & He, Q. Y. (2008). Lipoprotein MtsA of MtsABC in *Streptococcus pyogenes* primarily binds ferrous ion with bicarbonate as a synergistic anion. *FEBS Letters*, *582*(9), 1351–1354.

Sun, X. S., Xiao, C. L., Ge, R. G., Yin, X. F., Li, H., Li, N., … He, Q. Y. (2011). Putative copper- and zinc-binding motifs in *Streptococcus pneumoniae* identified by immobilized metal affinity chromatography and mass spectrometry. *Proteomics*, *11*(16), 3288–3298.

Szarapinska-Kwaszewska, J., & Mikucki, J. (2001). Utilization of exogenous siderophores by enterococci. *Medycyna Doświadczalna i Mikrobiologia*, *53*(3), 233–243.

Tai, S. S., Lee, C. J., & Winter, R. E. (1993). Hemin utilization is related to virulence of *Streptococcus pneumoniae*. *Infection and Immunity*, *61*(12), 5401–5405.

Tai, S. S., Yu, C. L., & Lee, J. K. (2003). A solute binding protein of *Streptococcus pneumoniae* iron transport. *FEMS Microbiology Letters*, *220*(2), 303–308.

Teawtrakul, N., Jetsrisuparb, A., Sirijerachai, C., Chansung, K., & Wanitpongpun, C. (2015). Severe bacterial infections in patients with non-transfusion-dependent thalassemia: Prevalence and clinical risk factors. *International Journal of Infectious Diseases*, *39*, 53–56.

Tedde, V., Rosini, R., & Galeotti, C. L. (2016). Zn2+ uptake in *Streptococcus pyogenes*: Characterization of *adcA* and *lmb* null mutants. *PLoS One*, *11*(3), e0152835.

Terao, Y., Kawabata, S., Kunitomo, E., Nakagawa, I., & Hamada, S. (2002). Novel laminin-binding protein of *Streptococcus pyogenes*, Lbp, is involved in adhesion to epithelial cells. *Infection and Immunity*, *70*(2), 993–997.

Thornton, R. B., Wiertsema, S. P., Kirkham, L. A. S., Rigby, P. J., Vijayasekaran, S., Coates, H. L., & Richmond, P. C. (2013). Neutrophil extracellular traps and bacterial biofilms in middle ear effusion of children with recurrent acute otitis media—A potential treatment target. *PLoS One*, *8*(2), e53837.

Thulin, P., Johansson, L., Low, D. E., Gan, B. S., Kotb, M., McGeer, A., & Norrby-Teglund, A. (2006). Viable group A streptococci in macrophages during acute soft tissue infection. *PLoS Medicine*, *3*(3), e53.

Thurman, R. B., & Gerba, C. P. (1988). The molecular mechanisms of copper and silver ion disinfection of bacteria and viruses. *Critical Reviews in Environmental Control*, *18*(4), 295–315.

Tilley, S. J., Orlova, E. V., Gilbert, R. J., Andrew, P. W., & Saibil, H. R. (2005). Structural basis of pore formation by the bacterial toxin pneumolysin. *Cell*, *121*(2), 247–256.

Tomasz, A. (1981). Surface components of *Streptococcus pneumoniae*. *Reviews of Infectious Diseases*, *3*(2), 190–211.

Touati, D. (2000). Iron and oxidative stress in bacteria. *Archives of Biochemistry and Biophysics*, *373*(1), 1–6.

Toukoki, C., Gold, K. M., McIver, K. S., & Eichenbaum, Z. (2010). MtsR is a dual regulator that controls virulence genes and metabolic functions in addition to metal homeostasis in the group A *Streptococcus*. *Molecular Microbiology*, *76*(4), 971–989.

Toukoki, C., & Gryllos, I. (2013). PolA1, a putative DNA polymerase I, is coexpressed with PerR and contributes to peroxide stress defenses of group A Streptococcus. *Journal of Bacteriology*, *195*(4), 717–725. http://dx.doi.org/10.1128/JB.01847-12.

Traore, D. A. K., El Ghazouani, A., Ilango, S., Dupuy, J., Jacquamet, L., Ferrer, J. L., … Latour, J. M. (2006). Crystal structure of the apo-PerR-Zn protein from *Bacillus subtilis*. *Molecular Microbiology*, *61*(5), 1211–1219.

Traore, D. A. K., El Ghazouani, A., Jacquamet, L., Borel, F., Ferrer, J. L., Lascoux, D., … Latour, J. M. (2009). Structural and functional characterization of 2-oxo-histidine in oxidized PerR protein. *Nature Chemical Biology*, *5*(1), 53–59.

Tseng, H. J., McEwan, A. G., Paton, J. C., & Jennings, M. P. (2002). Virulence of *Streptococcus pneumoniae*: PsaA mutants are hypersensitive to oxidative stress. *Infection and Immunity*, *70*(3), 1635–1639.

Tsou, C. C., Chiang-Ni, C., Lin, Y. S., Chuang, W. J., Lin, M. T., Liu, C. C., & Wu, J. J. (2008). An iron-binding protein, Dpr, decreases hydrogen peroxide stress and protects *Streptococcus pyogenes* against multiple stresses. *Infection and Immunity*, *76*(9), 4038–4045.

Tu, A. H. T., Fulgham, R. L., McCrory, M. A., Briles, D. E., & Szalai, A. J. (1999). Pneumococcal surface protein A inhibits complement activation by *Streptococcus pneumoniae*. *Infection and Immunity*, *67*(9), 4720–4724.

Turner, A. G., Ong, C. L., Gillen, C. M., Davies, M. R., West, N. P., McEwan, A. G., & Walker, M. J. (2015). Manganese homeostasis in Group A *Streptococcus* is critical for resistance to oxidative stress and virulence. *MBio*, *6*(2), e00278-15.

Ulijasz, A. T., Andes, D. R., Glasner, J. D., & Weisblum, B. (2004). Regulation of iron transport in *Streptococcus pneumoniae* by RitR, an orphan response regulator. *Journal of Bacteriology, 186*(23), 8123–8136.

Urban, C. F., Ermert, D., Schmid, M., Abu-Abed, U., Goosmann, C., Nacken, W., … Zychlinsky, A. (2009). Neutrophil extracellular traps contain calprotectin, a cytosolic protein complex involved in host defense against *Candida albicans*. *PLoS Pathogens, 5*(10), e1000639.

Urbanski, T., Slopek, S., & Venulet, J. (1951). Antitubercular activity of some 8-hydroxyquinoline derivatives. *Nature, 168*(4262), 29.

Valko, M., Morris, H., & Cronin, M. T. D. (2005). Metals, toxicity and oxidative stress. *Current Medicinal Chemistry, 12*(10), 1161–1208. http://dx.doi.org/10.2174/0929867053764635.

Veeravalli, K., Boyd, D., Iverson, B. L., Beckwith, J., & Georgiou, G. (2011). Laboratory evolution of glutathione biosynthesis reveals natural compensatory pathways. *Nature Chemical Biology, 7*(2), 101–105.

Vento, S., Cainelli, F., & Cesario, F. (2006). Infections and thalassaemia. *Lancet Infectious Diseases, 6*(4), 226–233.

Versieck, J., Barbier, F., Speecke, A., & Hoste, J. (1974). Normal manganese concentrations in human serum. *Acta Endocrinologica, 76*(4), 783–788.

Vidovic, S., Elder, J., Medihala, P., Lawrence, J. R., Predicala, B., Zhang, H. X., & Korber, D. R. (2015). ZnO nanoparticles impose a panmetabolic toxic effect along with strong necrosis, inducing activation of the envelope stress response in *Salmonella enterica* serovar *enteritidis*. *Antimicrobial Agents and Chemotherapy, 59*(6), 3317–3328.

Vignesh, K. S., Figueroa, J. A. L., Porollo, A., Caruso, J. A., & Deepe, G. S. (2013). Granulocyte macrophage colony stimulating factor induced Zn sequestration enhances macrophage superoxide and limits intracellular pathogen survival. *Immunity, 39*(4), 697–710.

Wagner, J. G., & Roth, R. A. (2000). Neutrophil migration mechanisms, with an emphasis on the pulmonary vasculature. *Pharmacological Reviews, 52*(3), 349–374.

Waldron, K. J., & Robinson, N. J. (2009). How do bacterial cells ensure that metalloproteins get the correct metal? *Nature Reviews. Microbiology, 7*(1), 25–35.

Waldron, K. J., Rutherford, J. C., Ford, D., & Robinson, N. J. (2009). Metalloproteins and metal sensing. *Nature, 460*(7257), 823–830.

Walker, M. J., Barnett, T. C., McArthur, J. D., Cole, J. N., Gillen, C. M., Henningham, A., … Nizet, V. (2014). Disease manifestations and pathogenic mechanisms of Group A *Streptococcus*. *Clinical Microbiology Reviews, 27*(2), 264–301.

Walker, M. J., Hollands, A., Sanderson-Smith, M. L., Cole, J. N., Kirk, J. K., Henningham, A., … Nizet, V. (2007). DNase Sda1 provides selection pressure for a switch to invasive group A streptococcal infection. *Nature Medicine, 13*(8), 981–985.

Walker, M. J., McArthur, J. D., McKay, F., & Ranson, M. (2005). Is plasminogen deployed as a *Streptococcus pyogenes* virulence factor? *Trends in Microbiology, 13*(7), 308–313.

Walling, C. (1971). Fentons reagent revisited. *Abstracts of Papers of the American Chemical Society, 8*, 125–131.

Walters, G. O., Miller, F. M., & Worwood, M. (1973). Serum ferritin concentration and iron stores in normal subjects. *Journal of Clinical Pathology, 26*(10), 770–772.

Wang, X. H., Tong, H. C., & Dong, X. Z. (2014). PerR-regulated manganese ion uptake contributes to oxidative stress defense in an oral streptococcus. *Applied and Environmental Microbiology, 80*(8), 2351–2359.

Warner, R. R., Schwartz, J. R., Boissy, Y., & Dawson, T. L. (2001). Dandruff has an altered stratum corneum ultrastructure that is improved with zinc pyrithione shampoo. *Journal of the American Academy of Dermatology, 45*(6), 897–903.

Wartha, F., Beiter, K., Normark, S., & Henriques-Normark, B. (2007). Neutrophil extracellular traps: Casting the NET over pathogenesis. *Current Opinion in Microbiology*, *10*(1), 52–56.

Wątły, J., Potocki, S., & Rowińska-Żyrek, M. (2016). Zinc homeostasis at the bacteria/host interface—From coordination chemistry to nutritional immunity. *Chemistry: A European Journal*, *22*, 15992–16010.

Weinberg, E. D. (2007). Antibiotic properties and applications of lactoferrin. *Current Pharmaceutical Design*, *13*(8), 801–811.

Weinberg, E. D. (2009). Iron availability and infection. *Biochimica et Biophysica Acta*, *1790*(7), 600–605.

Weiser, J. N., Austrian, R., Sreenivasan, P. K., & Masure, H. R. (1994). Phase variation in pneumococcal opacity—Relationship between colonial morphology and nasopharyngeal colonization. *Infection and Immunity*, *62*(6), 2582–2589.

Wermuth, P. J., & Jimenez, S. A. (2015). The significance of macrophage polarization subtypes for animal models of tissue fibrosis and human fibrotic diseases. *Clinical and Translational Medicine*, *4*, 2.

Weston, B. F., Brenot, A., & Caparon, M. G. (2009). The metal homeostasis protein, Lsp, of *Streptococcus pyogenes* is necessary for acquisition of zinc and virulence. *Infection and Immunity*, *77*(7), 2840–2848.

Whalan, R. H., Funnell, S. G. P., Bowler, L. D., Hudson, M. J., Robinson, A., & Dowson, C. G. (2006). Distribution and genetic diversity of the ABC transporter lipoproteins PiuA and PiaA within *Streptococcus pneumoniae* and related streptococci. *Journal of Bacteriology*, *188*(3), 1031–1038.

White, C., Kambe, T., Fulcher, Y. G., Sachdev, S. W., Bush, A. I., Fritsche, K., … Petris, M. J. (2009). Copper transport into the secretory pathway is regulated by oxygen in macrophages. *Journal of Cell Science*, *122*(9), 1315–1321.

White, C., Lee, J., Kambe, T., Fritsche, K., & Petris, M. J. (2009). A role for the ATP7A copper-transporting ATPase in macrophage bactericidal activity. *Journal of Biological Chemistry*, *284*(49), 33949–33956.

Williams, R. J. P. (1997). Spice of life. *Nature*, *389*(6647), 147–148.

Wilson, M. (1997). Biocontrol of aerial plant diseases in agriculture and horticulture: Current approaches and future prospects. *Journal of Industrial Microbiology & Biotechnology*, *19*(3), 188–191.

Winterbourn, C. C., Kettle, A. J., & Hampton, M. B. (2016). Reactive oxygen species and neutrophil function. *Annual Review of Biochemistry*, *85*, 765–792.

Wirth, J. J., Fraker, P. J., & Kierszenbaum, F. (1984). Changes in the levels of marker expression by mononuclear phagocytes in zinc-deficient mice. *Journal of Nutrition*, *114*(10), 1826–1833.

Wish, J. B. (2006). Assessing iron status: Beyond serum ferritin and transferrin saturation. *Clinical Journal of the American Society of Nephrology*, *1*(Suppl. 1), S4–S8.

Xie, Y. P., He, Y. P., Irwin, P. L., Jin, T., & Shi, X. M. (2011). Antibacterial activity and mechanism of action of zinc oxide nanoparticles against *Campylobacter jejuni*. *Applied and Environmental Microbiology*, *77*(7), 2325–2331.

Yamamoto, Y., Poyart, C., Trieu-Cuot, P., Lamberet, G., Gruss, A., & Gaudu, P. (2005). Respiration metabolism of Group B Streptococcus is activated by environmental haem and quinone and contributes to virulence. *Molecular Microbiology*, *56*(2), 525–534.

Yesilkaya, H., Andisi, V. F., Andrew, P. W., & Bijlsma, J. J. E. (2013). *Streptococcus pneumoniae* and reactive oxygen species: An unusual approach to living with radicals. *Trends in Microbiology*, *21*(4), 187–195.

Yesilkaya, H., Kadioglu, A., Gingles, N., Alexander, J. E., Mitchell, T. J., & Andrew, P. W. (2000). Role of manganese-containing superoxide dismutase in oxidative stress and virulence of *Streptococcus pneumoniae. Infection and Immunity*, *68*(5), 2819–2826.

Yesilkaya, H., Spissu, F., Carvalho, S. M., Terra, V. S., Homer, K. A., Benisty, R., … Andrew, P. W. (2009). Pyruvate formate lyase is required for pneumococcal fermentative metabolism and virulence. *Infection and Immunity*, 77(12), 5418–5427.

Young, C. A., Gordon, L. D., Fang, Z., Holder, R. C., & Reid, S. D. (2015). Copper tolerance and the characterization of a copper-responsive operon, copYAZ, in an M1T1 clinical strain of Streptococcus pyogenes. *Journal of Bacteriology*, *197*, 2580–2592.

Zhang, T. F., Ding, Y., Li, T. T., Wan, Y., Li, W., Chen, H. C., & Zhou, R. (2012). A Fur-like protein PerR regulates two oxidative stress response related operons dpr and metQIN in *Streptococcus suis*. *BMC Microbiology*, *12*, 85.

Zhu, H., Liu, M. Y., & Lei, B. F. (2008). The surface protein Shr of *Streptococcus pyogenes* binds heme and transfers it to the streptococcal heme-binding protein Shp. *BMC Microbiology*, *8*, 15.

Zitzelsberger, W., Gotz, F., & Schleifer, K. H. (1984). Distribution of superoxide dismutases, oxidases, and NADH peroxidase in various streptococci. *FEMS Microbiology Letters*, *21*(2), 243–246.

Zurlo, M., De Stefano, P., Borgna-Pignatti, C., Di Palma, A., Melevendi, C., Piga, A., … Terzoli, S. (1989). Survival and causes of death in thalassaemia major. *The Lancet*, *334*(8653), 27–30.

Zwilling, B. S., Kuhn, D. E., Wikoff, L., Brown, D., & Lafuse, W. (1999). Role of iron in Nramp1-mediated inhibition of mycobacterial growth. *Infection and Immunity*, *67*(3), 1386–1392.

CHAPTER SIX

# Copper and Antibiotics: Discovery, Modes of Action, and Opportunities for Medicinal Applications

**Alex G. Dalecki, Cameron L. Crawford, Frank Wolschendorf[1]**
The University of Alabama at Birmingham, Birmingham, AL, United States
[1]Corresponding author: e-mail address: fwolsche@uab.edu

## Contents



## Abstract

Copper is a ubiquitous element in the environment as well as living organisms, with its redox capabilities and complexation potential making it indispensable for many cellular functions. However, these same properties can be highly detrimental to prokaryotes

*Advances in Microbial Physiology*, Volume 70
ISSN 0065-2911
http://dx.doi.org/10.1016/bs.ampbs.2017.01.007

and eukaryotes when not properly controlled, damaging many biomolecules including DNA, lipids, and proteins. To restrict free copper concentrations, all bacteria have developed mechanisms of resistance, sequestering and effluxing labile copper to minimize its deleterious effects. This weakness is actively exploited by phagocytes, which utilize a copper burst to destroy pathogens. Though administration of free copper is an unreasonable therapeutic antimicrobial itself, due to insufficient selectivity between host and pathogen, small-molecule ligands may provide an opportunity for therapeutic mimicry of the immune system. By modulating cellular entry, complex stability, resistance evasion, and target selectivity, ligand/metal coordination complexes can synergistically result in high levels of antibacterial activity. Several established therapeutic drugs, such as disulfiram and pyrithione, display remarkable copper-dependent inhibitory activity. These findings have led to development of new drug discovery techniques, using copper ions as the focal point. High-throughput screens for copper-dependent inhibitors against *Mycobacterium tuberculosis* and *Staphylococcus aureus* uncovered several new compounds, including a new class of inhibitors, the NNSNs. In this review, we highlight the microbial biology of copper, its antibacterial activities, and mechanisms to discover new inhibitors that synergize with copper.

# 1. INTRODUCTION

Copper has been linked to the pathology of various diseases such as cancer (Garber, 2015a), neurodegeneration (White, Kanninen, & Crouch, 2015), angiogenic dysfunctions (Urso & Maffia, 2015), and, relatively new to this list, microbial infections (Becker & Skaar, 2014; Djoko, Ong, Walker, & Mcewan, 2015; Ladomersky & Petris, 2015; Samanovic, Ding, Thiele, & Darwin, 2012). Copper ions emerged as an attractive drug target but also as an intriguing tool for the development of novel probes intended for medicinal interventions (Denoyer, Masaldan, La Fontaine, & Cater, 2015; Garber, 2015b; Marzano, Pellei, Tisato, & Santini, 2009; Neyrolles, Wolschendorf, Mitra, & Niederweis, 2015). Therapeutically modulating tissue copper content (i.e. chelation therapy (Fatfat et al., 2014; Fu, Naing, Fu, Kuo, & Kurzrock, 2012)) or utilizing copper ions in target-specific approaches (Hung et al., 2012; Rae et al., 2013; Schimmer, 2011) with respect to cancer or neurodegenerative diseases has shown efficacy in animal models (Paranjpe, Zhang, Ali-Osman, Bobustuc, & Srivenugopal, 2014; Soon et al., 2011).

Given that copper ions are well established as a potential therapeutic element in other disease contexts, they are also attractive for metal-related anti-infective intervention strategies. The central driving component of this approach is the intrinsic antibacterial activity of copper ions, which they can bestow upon their metal complexes (Dalecki et al., 2015; Djoko, Goytia, et al., 2015; Festa, Helsel, Franz, & Thiele, 2014; Shah et al., 2016). Additionally, copper's chelation chemistry is well studied, characterized by the stability of its complexes and its competitiveness over other physiological transition metals (e.g. $Zn^{2+}$, $Fe^{2+}$, $Mn^{2+}$, $Ni^{2+}$, $Co^{2+}$) (Irving & Williams, 1953; Lisher & Giedroc, 2013). Therapeutically, metal complexes have shown selectivity in delivering or removing copper to/from diseased tissue (Szymanski, Fraczek, Markowicz, & Mikiciuk-Olasik, 2012).

More recently, copper has been identified as a crucial component of the innate immune system mounting a first line of defence against invading pathogens. While iron and manganese are withheld by the host to "starve" the pathogen for essential nutrients, antimicrobial metals such as copper and occasionally zinc accumulate at high concentrations (Hood & Skaar, 2012; Potrykus, Ballou, Childers, & Brown, 2014). Accordingly, metal acquisition, homeostasis, and resistance pathways are conserved in most pathogenic bacteria (Solioz, Abicht, Mermod, & Mancini, 2010). Impairment of these pathways often compromises viability and virulence (Djoko, Ong, et al., 2015), suggesting that misplacement of copper ions within the bacterial cell is lethal.

Metal complexes can overcome copper resistance mechanisms by hiding their metal cargo from these bacterial defences, as was proposed recently (Dalecki et al., 2015). Consequently, the ligand may expand the toxicity spectrum of copper ions by assaulting targets typically not affected when bacteria are challenged by free copper alone. As such, the metal ligand is expected to facilitate interaction with a specific target in order to generate an impact (Dalecki et al., 2015). Recently, examples of antimicrobial copper complexes have been detailed against Gram negatives (Djoko, Paterson, Donnelly, & McEwan, 2014), Gram positives (Haeili et al., 2014), and mycobacteria (Dalecki et al., 2015; Shah et al., 2016), as well as pathogenic fungi (Festa et al., 2014). The broad range of potentially targetable pathogens, coupled with a wide array of conceivable modes of action, justifies the interest in exploiting copper's antibiotic properties for medicinal applications. Copper's chemical and biological qualities, its omnipresence in vivo (Darwin, 2015), affiliation with infectious diseases, and the existence of several well-established investigational proof-of-concept compounds support a leading role for copper in the development of metal-related antibiotics.

As copper ions are the activity-defining component of antimicrobial copper complexes, this chapter will first discuss copper's interaction with major biological molecules (Section 2) and then describe microbial strategies to handle copper while avoiding toxicity and how the innate immune system uses copper against pathogens (Section 3). Finally, this chapter will review several proof-of-concept compounds, and strategies employed to harness the antimicrobial properties of copper ions for therapeutic antimicrobial strategies (Section 4).

## 2. COPPER'S REACTIVITY AND INTERACTION WITH BIOLOGICAL MOLECULES

Copper ions are an essential nutrient for all domains of life. The origin of this copper dependency has been traced back to Earth's Great Oxygenation Event approximately 2.7 billion years ago when oxygen levels began to rise. The presence of oxygen increasingly reduced the bioavailability of iron, which in its oxidized form ($Fe^{3+}$) has the tendency to form insoluble minerals. Conversely, oxidation of $Cu^{1+}$ to $Cu^{2+}$ increased copper's bioavailability in oceanic habitats, creating an opportunity for life to adapt to an oxidative world by exploiting the reactivity of copper ions (Crichton & Pierre, 2001; Dupont, Grass, & Rensing, 2011; Sessions, Doughty, Welander, Summons, & Newman, 2009). A variety of copper-utilizing enzymes arose with critical roles in respiration (e.g. cytochrome *c* oxidases, nitric oxide reductase), nitrogen assimilation (e.g. nitrite reductase), or oxidative stress defence (e.g. superoxide dismutases, multicopper oxidases) (Argüello, Raimunda, & Padilla-Benavides, 2013). The respiratory chain haem/copper terminal oxidases were particularly monumental, using molecular oxygen as terminal electron acceptor to build a proton gradient across the plasma membrane that subsequently drives ATP synthase (Lee, Reimann, Huang, & Adelroth, 2012). Another nearly ubiquitous example is copper-containing superoxide dismutase, either in the form of Cu,Zn superoxide dismutases (e.g. bacterial SodC or SodC-F, eukaryotic Sod1) or copper-only superoxide dismutases (e.g. fungal Sod4, Sod5, Sod6, or *Mycobacterium tuberculosis* SodC). These enzymes function as catalysts for the detoxification of superoxide radicals ($^{\bullet}O_2^-$) into oxygen ($O_2$) and less reactive hydrogen peroxide ($H_2O_2$) (Battistoni, 2003; Broxton & Culotta, 2016). Other instances of copper-utilizing protein families include amine oxidases (Finney, Moon, Ronnebaum, Lantz, & Mure, 2014; McPherson, Parsons, & Wilmot, 2006) and galactose oxidases (Whittaker, 2005; Yin et al., 2015). However, despite these beneficial metabolic functions, the dependency of cells on copper was not without risks, as the single electron redox-cycling abilities of copper and its promiscuous coordination chemistry can prove disastrous when occurring spontaneously (Argüello et al., 2013). In the following is an overview of copper's abilities to interact with proteins, lipids, DNA, and some metabolites, as well as a discussion on how copper's coordination and redox chemistry may inflict damage to various cellular locations.

## 2.1 Copper's Coordination Chemistry

The current view on copper homeostasis suggests that cellular proteins are the primary bearers of native copper binding sites inside the cell. The biogenesis of these various copper centres typically involves specific proteins, called copper chaperons, which are specialized for intracellular copper transport and delivery of copper ions to their intended native binding sites (Robinson & Winge, 2010). Depending on their valence, available ligands, and chemical environment, copper ions can engage in different coordination geometries. Oxidized $Cu^{2+}$ favours distorted square planar, square pyramidal, or octahedral geometries, while reduced $Cu^{1+}$ ions have an even greater geometrical flexibility and additionally support coordination to two or three dentate binding sites (Crichton & Pierre, 2001; Haas & Franz, 2009). Besides the inherent catalytic benefits, it has been proposed that the coordination of copper ions (and other metal ions) within proteins could have served as a driving force for the structural evolution of multidomain copper proteins such as nitrous oxide reductase. In such a scenario, the coordination of copper ions would be expected to overpower electrostatic repulsion, stearic hindrance, or other noncovalent forces that would otherwise prevent interactions between monomeric proteins (Lu, 2010).

The benefits gained by coordinating copper to enzymatic catalytic sites are opposed by its propensity to coordinate adventitiously to a variety of nonnative biological sites, which is believed to contribute to copper's toxicity in biological systems when homeostasis is impaired. In comparison to other biologically relevant bivalent transition metal ions (e.g. $Zn^{2+}$, $Fe^{2+}$, $Mn^{2+}$), copper often has a higher affinity for proteinaceous metal binding sites and would therefore outcompete those native metal ions for binding (Foster, Osman, & Robinson, 2014; Foster & Robinson, 2011; Tottey et al., 2008). Ultimately, this may lead to the replacement or displacement of other metal ions from their native binding sites, eliminating enzyme activity. This preference is to a large extent independent of the actual ligand and is well described by the Irving–Williams series (Irving & Williams, 1953), which orders first-row divalent metal ions based upon experimentally determined relative thermodynamic stabilities of their respective complexes ($Mn^{2+} < Fe^{2+} < Co^{2+} < Ni^{2+} < Cu^{2+} > Zn^{2+}$). Much of this phenomenon is explained by decreasing ionic radii of metal ions when moving across the periodic table from manganese to zinc. By itself, though, ionic radii fail to account for copper's exceptional stability, being higher than zinc. This stability is therefore attributed to Jahn–Teller distortion, describing a geometrical asymmetry that, in the octahedral coordination sphere favoured by $Cu^{2+}$, leads to the formation of four shorter bonds with a more covalent character at the square pyramidal base and two longer axial bonds (Fraústo Da Silva & Williams, 2001; Haas & Franz, 2009; Jahn & Teller, 1937). More recently, the relative order of metal complex stabilities of a given ligand was correlated to computed ligand(s)-to-metal charge transfer values (Varadwaj, Varadwaj, & Jin, 2015). Ligand-to-metal charge transfer is known to stabilize metal complexes and describes the transfer of negative charges in the form of electrons from the ligand to the metal ion. By using these calculated values, it was possible to correctly predict the high stability of copper complexes independent of the type of ligand, the identity of the donor atom (N or O), and the number of interaction sites per ligand (mono-, di-, or polydentate). This correlation was validated for multiple first-row transition metal ions (including $Cu^{2+}$) in combination with various small-molecule ligands (Varadwaj et al., 2015).

## 2.2 Redox Cycling of Copper

The Fenton chemistry of copper is similar to that of iron in principle, describing the metal catalysed, nonenzymatic disproportionation of hydrogen peroxide ($H_2O_2$) into hydroxyl radicals ($^{\bullet}OH$), leaving behind the catalytic metal in an oxidized state (Fig. 1, reaction a). $H_2O_2$ is a natural metabolic byproduct of aerobic growth, both in prokaryotes and in eukaryotes, often by oxidoreductases using oxygen as an electron acceptor. For instance, in *Escherichia coli*, the L-aspartate oxidase NadB, critical for the biogenesis of $NAD^+$, and fumarate reductase are the largest sources of adventitious $H_2O_2$ production under aerobic conditions (Korshunov & Imlay, 2010). In eukaryotic cells, $H_2O_2$ is a byproduct primarily of mitochondrial respiratory chain complexes I, II, and III (Giorgio, Trinei, Migliaccio, & Pelicci, 2007; Kehrer, 2000; Quinlan et al., 2012). Notably, $H_2O_2$ production increases under copper stress in both prokaryotic (Macomber & Imlay, 2009) and eukaryotic cells (Oe, Miyagawa, Honma, & Harada, 2016).

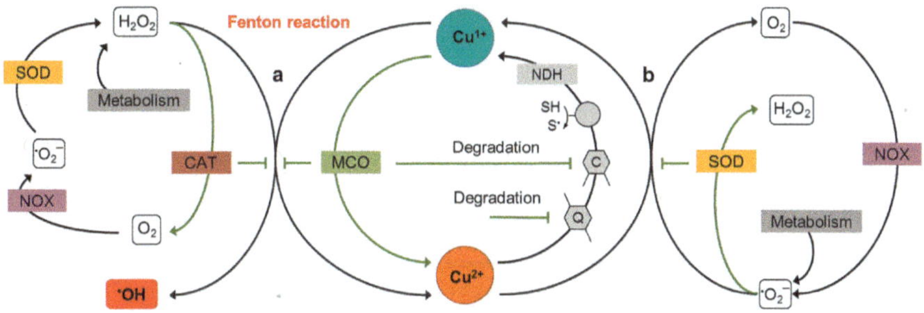


**Fig. 1** Enzymatic and nonenzymatic redox cycling of copper. Through the Fenton reaction (reaction a), copper ($Cu^{1+}$) can act as an inorganic catalyst for the production of hydroxyl radicals ($^{•}OH$) from hydrogen peroxide ($H_2O_2$). To prevent excessive and uncontrolled formation of $^{•}OH$, cells may deplete $Cu^{1+}$ through oxidation into $Cu^{2+}$ by multicopper oxidases (MCO) or reduce $H_2O_2$ levels by catalases (CAT). $H_2O_2$ arises as a metabolic byproduct or generated from superoxide radicals ($^{•}O_2^{-}$) by superoxide dismutases (SOD). NADPH oxidase (NOX) produce $^{•}O_2^{-}$ from oxygen ($O_2$). $^{•}O_2^{-}$ is a principal reductant of $Cu^{2+}$ (reaction b). In the presence of $H_2O_2$ and $^{•}O_2^{-}$, copper ions redox cycle thereby continuously generating $^{•}OH$. While SOD activity can prevent the reduction of $Cu^{2+}$ by depleting $^{•}O_2^{-}$, alternative reductants such as quinones (Q), catechols (C), thiols (SH), and NADH dehydrogenase (NDH) from *E. coli* may reduce $Cu^{2+}$ instead. Oxidation of catechol siderophores by MCO and/or activation of quinone degradation pathways may reduce excessive $Cu^{2+}$ reduction by these metabolites. *Green arrows* represent pathways that prevent or alleviate toxic radical formation or indicate which reactions in the copper cycling are being subverted by host defence proteins (SOD, MCO, and CAT).

Copper alone is not sufficient for sustaining the reaction, though; a reducing agent capable of returning $Cu^{2+}$ back into $Cu^{1+}$ is needed to complete the redox cycle of copper (Fig. 1, reaction b) and continue Fenton chemistry production of reactive $^{•}OH$ (Fig. 1, reaction a). Several potential reducing agents are known, the most common being superoxide radical ($^{•}O_2^{-}$) (Kehrer, 2000). The same as $H_2O_2$, $^{•}O_2^{-}$ is a metabolic byproduct of respiration. However, it can also be generated in large quantities by professional phagocytes expressing NADPH oxidase (Fig. 1) in the form of an oxidative burst when they encounter invading pathogens (Slauch, 2011). Intracellular thiols may also participate in the reduction of copper ions, thereby generating thiyl radicals ($RS^{•}$) (Luc & Vergely, 2008). Other in vivo-relevant copper-reducing agents are the quinones, including menaquinone or ubiquinone, which are components of the electron transport chain. Their copper-reducing action was demonstrated in *Lactococcus lactis* and *E. coli* (Abicht, Gonskikh, Gerber, & Solioz, 2013; Volentini, Farias, Rodriguez-Montelongo, & Rapisarda, 2011). In addition, catechol siderophores and NADH dehydrogenase may also participate in the reduction of $Cu^{2+}$ in *E. coli* (Grass et al., 2004; Rodriguez-Montelongo, Volentini, Farias, Massa, & Rapisarda, 2006; Volentini et al., 2011).

Irrespective of generation method, the toxic end-product of the Fenton reaction, $^{\bullet}OH$, is highly reactive and has an extremely short diffusion-dependent half-life ($\sim 10^{-9}$ s). It is exclusively generated by enzyme-independent processes and, unlike $^{\bullet}O_2^-$ or $H_2O_2$, cannot be neutralized through enzymatic reactions (Freinbichler et al., 2011). Therefore, the generation of hydroxyl radicals inevitably leads to oxidative damage directly at or in close proximity to the site of its appearance (Powers & Jackson, 2008). A hydroxyl radical's reaction with biomolecules can ignite a radical-forming chain reaction, leading to secondary radicals (Powers & Jackson, 2008) and substantial oxidative damage to proteins, DNA, and lipids. The excessive formation of hydroxyl radicals is therefore toxic for any cell.

In normal physiology, endogenous copper ions or sources of $H_2O_2$ and/or $^{\bullet}O_2^-$ production are of little concern, given that most prokaryotes and eukaryotic cells express genes that neutralize these otherwise toxic reactive oxygen species (ROS) (e.g. superoxide dismutases (SOD), catalases) (Imlay, 2013). However, as illustrated in Fig. 1, SOD activity can be a double-edged sword. In one instance, SOD prevents the reduction of $Cu^{2+}$ by depleting its major reductant ($^{\bullet}O_2^-$), yet the generation of $H_2O_2$ can drive Fenton chemistry. In many bacteria (and eukaryotic cells), oxidative stress responses, induced by oxidative stress, are also upregulated in response to a copper challenge. This can indirectly suggest an increased production of damaging ROS under copper stress, and/or a greater need for a more effective $H_2O_2$/superoxide sink to prevent Fenton chemistry when copper is abundant. Such a relation was observed in *Staphylococcus aureus* (Baker et al., 2010), *M. tuberculosis* (Ward, Hoye, & Talaat, 2008), and *Pseudomonas aeruginosa* (Teitzel et al., 2006), but not in *Bacillus subtilis* (Chillappagari et al., 2010) or *E. coli* (Kershaw, Brown, Constantinidou, Patel, & Hobman, 2005), although several members of the superoxide stress response (SoxRS) regulon were induced in the latter.

## 2.3 Copper's Specific Interaction With Proteins

Traditionally, prokaryotic copper toxicity was assumed to occur predominantly through generation of ROS, discussed earlier, resulting in DNA damage. However, in many systems, recent investigations have shown intracellular proteins as the most vulnerable targets of direct copper toxicity (Macomber & Imlay, 2009). Copper ions have a natural affinity for several nitrogen-, oxygen-, and sulphur-based donor groups commonly found in most biomolecules. In particular, the protein backbone, N- and C-termini, and most amino acid side-chain functional groups are potential interaction sites for metals (Table 1), which is influenced by various factors including accessibility, proximity to other binding sites, and pH (Holm, Kennepohl, & Solomon, 1996). Given the sheer amount of protein binding possibilities (Lu, 2010) and copper's strong affinity with reference to the Irving–Williams series, it is likely that readily available copper ions would lead to occupation of most proteinaceous metal binding sites (Foster et al., 2014). In consequence, damage may occur through multiple mechanisms, including mismetallation of catalytic sites (Johnson, Kehl-Fie, & Rosch, 2015), destruction of catalytic sites reliant on other metals (Djoko & McEwan, 2013), structural alteration of exposed sulphur groups (Hiniker, Collet, & Bardwell, 2005), or ROS-induced damage (Johnson et al., 2015) (Fig. 2A).

### *2.3.1 Copper-Mediated Oxidative Damage on Proteins*

Due to the possible affinities of the protein backbone or its side chains for copper ions (Holm et al., 1996) and the increased $H_2O_2$ levels in copper-stressed cells (Macomber & Imlay, 2009; Oe et al., 2016), copper-mediated oxidative protein damage (as per Fenton chemistry generated hydroxyl radicals) could readily contribute to the proteotoxicity of copper. Metal-catalysed oxidation of proteins forms carbonyl residues (Fig. 2), which are general markers for oxidative protein damage. Amino acid residues most prone to carbonyl formation are the side chains of arginine, proline (Fig. 2B), lysine (Fig. 2C), and threonine (Fig. 2D) (Berlett & Stadtman, 1997; Dalle-Donne, Rossi, Giustarini, Milzani, & Colombo, 2003). In particular, the side chains of arginine, lysine, and threonine also exhibit an affinity for transition metal ions (Holm et al., 1996), which may aid in the formation of carbonyl residues as a consequence of copper's ability to generate the prooxidant radical $^{\bullet}OH$ (Fig. 1). The degree to which these effects actually occur seems heavily dependent on specific systems, with contrasting examples in the literature. In one example, Johnson et al. detected an increased presence of protein carbonyls in copper-stressed *Streptococcus pneumoniae*, which was even more pronounced in a copper efflux-deficient mutant (Johnson et al., 2015). Likewise, a twofold higher protein carbonyl content was detected in copper-stressed *E. coli* mutants lacking the

**Table 1** Potential Copper Binding Sites Found in Amino Acids

| | **Backbone** | **Side Chain** | | |
|---|---|---|---|---|
| **Amino Acids** | **N-Terminal, C-Terminal, Peptide Bond** | **N** | **S** | **O** |
| Ala | ✓ | — | — | — |
| Arg | ✓ | R-$CH_2$(**N**H)–C(**N**H) **N**$H_2$ | — | — |
| Asn | ✓ | R-C(O)**N**$H_2$ | — | R-C(**O**)$NH_2$ |
| Asp | ✓ | — | — | R-$CH_2$–C(**O**)**O**H |
| Cys | ✓ | — | R-$CH_2$**S**H | — |
| Glu | ✓ | — | — | R-$CH_2$–C(**O**)**O**H |
| Gln | ✓ | R-C(O)**N**$H_2$ | — | R-C(**O**)$NH_2$ |
| Gly | ✓ | — | — | — |
| His | ✓ | $CH_2$–(CH–**N**H–CH–**N**–C) | — | — |
| Ile | ✓ | — | — | — |
| Leu | ✓ | — | — | — |
| Lys | ✓ | R-**N**$H_2$ | — | — |
| Met | ✓ | — | R-**S**–$CH_3$ | — |
| Phe | ✓ | — | — | — |
| Pro | ✓ | — | — | — |
| Ser | ✓ | — | — | R-C**O**H |
| Thr | ✓ | — | — | R-C(**O**H)–$CH_3$ |
| Trp | ✓ | — | — | — |
| Tyr | ✓ | — | — | R-C(**O**H)–CH |
| Val | ✓ | — | — | — |

All amino acids provide potential binding sites for copper ions at the N-terminus or C-terminus of proteins/peptides or through the amide groups of the peptide backbone. Of those 20, 12 amino acids contain potential binding sites for copper within their side chains. The sites are divided based on their donor atoms (nitrogen (N), sulphur (S), oxygen (O)).

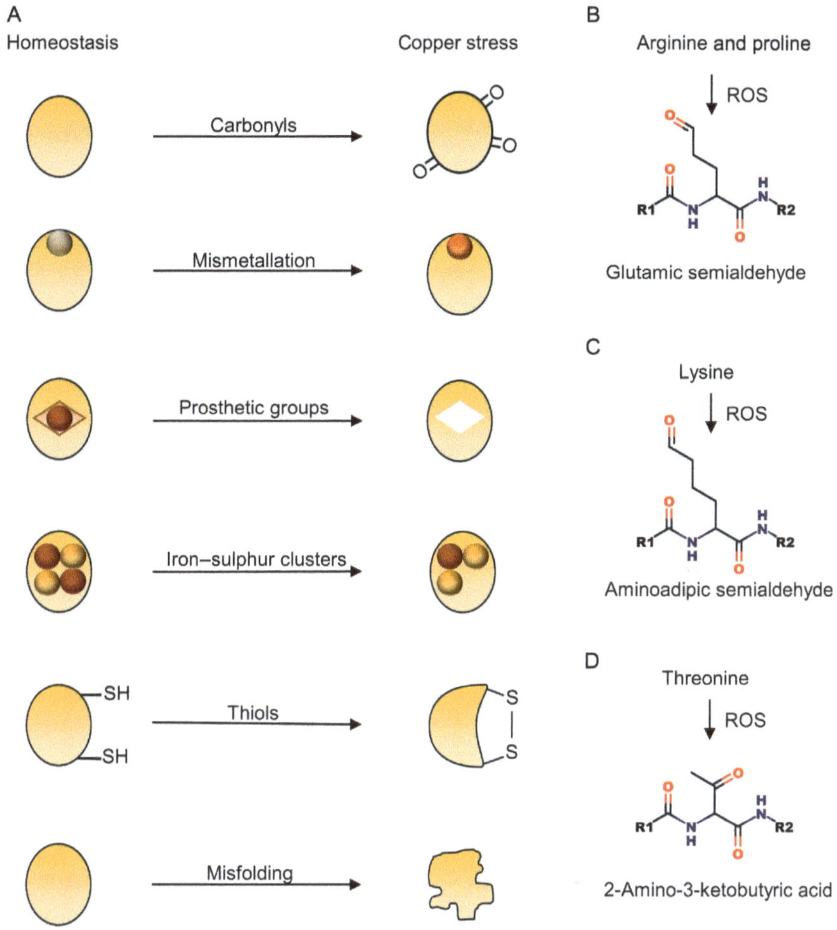


**Fig. 2** Copper-induced protein damage. (A) Copper mediates protein carbonyl formation, causes mis-/transmetallation of metals, interferes with the biogenesis of prosthetic groups (e.g. haem), destroys iron–sulphur clusters, oxidizes thiols, and induces protein misfolding. Examples of nonnative carbonyls derived from (B) arginine, proline, (C) lysine, or (D) threonine due to oxidative damage from reactive oxygen species (ROS).

chaperones IbpA/B, which protect cellular proteins from metal-mediated oxidative protein damage (Matuszewska, Kwiatkowska, Kuczyńska-Wiśnik, & Laskowska, 2008). However, Chillappagari et al. did not find evidence for copper-dependent oxidative stress in *B. subtilis*, as shown by the lack of significant induction of the PerR regulon (Chillappagari et al., 2010), which is a transcriptional marker in *B. subtilis* for hydrogen peroxide and superoxide stress (Mostertz, Scharf, Hecker, & Homuth, 2004). Similarly,

Macomber et al. did not find any significant increase in $^{\bullet}OH$ production in copper-stressed *E. coli* cells relative to their untreated controls (Macomber, Rensing, & Imlay, 2007). It also seems unclear whether carbonylation itself damages proteins and reduces their function, or rather if damaged and misfolded proteins are simply more susceptible to carbonylation, which could serve as a marker for degradation (Dalle-Donne et al., 2006). Thus, while Fenton reaction-driven oxidative damage to proteins may occur in some organisms during copper stress, in most bacterial systems, its contribution to cellular toxicity is likely outweighed by other more detrimental forms of damage.

### 2.3.2 Mismetallation of Proteins

By keeping the free copper content in the cytoplasm vanishingly low (Changela et al., 2003; Rae, Schmidt, Pufahl, Culotta, & O'Halloran, 1999), proteins may be selectively loaded with more weakly binding metals such as $Mn^{2+}$ and $Fe^{2+}$ (Foster et al., 2014). On the other hand, copper proteins acquire copper mostly from periplasmic sources where it is more available as demonstrated for the cupin CucA, multicopper oxidase CueO, Cu, Zn superoxide dismutase SodC, and cytochrome *c* oxidase complex (Ekici, Pawlik, Lohmeyer, Koch, & Daldal, 2012; Osman et al., 2013; Stolle, Hou, & Bruser, 2016; Tottey et al., 2008; Waldron et al., 2010; Waldron, Rutherford, Ford, & Robinson, 2009). Disturbance of cellular metal homeostasis can lead to mismetallation of superoxide dismutases, which is known to occur in prokaryotic (Beyer & Fridovich, 1991) and eukaryotic cells (as reviewed in Culotta, Yang, & O'Halloran, 2006). This applies to most superoxide dismutases (FeSodA, MnSodB, Cu,ZnSodC), although the Fe- and Mn-dependent enzymes seem more vulnerable to inactivation by mismetallation, due to their unusual propensity to bind non-native metals (Culotta et al., 2006). Outside of enzymes directly related to oxidative stress, copper toxicity in *S. pneumoniae* was hypothesized to result from copper outcompeting $Mn^{2+}$ in NrdF, a $Mn^{2+}$-dependent ribonucleotide reductase (Johnson et al., 2015). Addition of $Mn^{2+}$ rescued this copper toxicity, possibly by rebalancing the equilibrium such that $Mn^{2+}$ could again preferentially bind the active site in the enzyme.

### 2.3.3 Degeneration of Iron–Sulphur Clusters

Beyond simple mismetallation of catalytic sites, outright destruction of other catalytic centres is a well-supported mechanism of ultimate copper toxicity. The Imlay group demonstrated that iron–sulphur cluster proteins, primarily those with solvent accessible clusters, are particularly vulnerable targets within *E. coli* (Macomber & Imlay, 2009; Macomber et al., 2007) (Fig. 2A). Sensitive metabolic key enzymes with iron–sulphur clusters prone to copper-mediated attacks in *E. coli* include core metabolic enzymes such as isopropylmalate isomerase, fumarase A, and 6-phosphogluconate dehydratase (Macomber & Imlay, 2009). In *Neisseria gonorrhoeae*, Djoko et al. identified the iron–sulphur cluster protein HemN, an enzyme involved in the biosynthesis of haem, as a direct target of copper toxicity (Djoko & McEwan, 2013). Additionally, microarray studies on copper-stressed *B. subtilis* also implicated iron–sulphur cluster-related metabolic pathways (e.g. SufU, an enzyme participating in the biogenesis of iron–sulphur clusters) as a major target of copper toxicity (Chillappagari et al., 2010). Interestingly, the copper-mediated destruction of these iron–sulphur clusters is an oxygen-independent process, suggesting that Fenton chemistry is not involved (Macomber & Imlay, 2009). However, iron–sulphur clusters of cellular dehydratases are also prone to oxygen radical-mediated degradation (Imlay, 2013). The destruction of iron–sulphur clusters liberates excess iron within the cell, and this liberation could lead to additional Fenton chemistry in the presence of $H_2O_2$.

### 2.3.4 Disulphide or Thioether Bond Linkages

Protein damage may also occur outside of enzymatic active sites through the disruption of the protein structure. One example is the copper-mediated inactivation of RNAse A via formation of nonnative disulphide bonds within the protein (Hiniker et al., 2005) (Fig. 2A). Kachur et al. established in vitro that the copper-catalysed dimerization of cysteines is attributable to a copper-dependent Fenton-type reaction (Kachur, Koch, & Biaglow, 1999). This is relevant, as it was shown that, in *E. coli*, deletion of *dsbC*, coding for a periplasmic disulphide isomerase that can repair nonnative disulphide bonds, induces a copper sensitivity phenotype. Using the activity of periplasmic alkaline phosphatase (AP) as a reporter, they then demonstrated that copper-mediated, nonnative disulphide bond formation can take place in vivo. Restoration of DsbC expression also restored AP activity, suggesting that redox-active copper ions in the periplasm catalyse the formation of nonnative disulphide bonds (Hiniker et al., 2005). Conversely, copper ions may inhibit reduction of existing disulphide bonds, halting the maturation of apoproteins. For example, Duran et al. identified impaired cytochrome *c* assembly as a major mechanism of periplasmic copper toxicity in the β-proteobacterium *Rubrivivax gelatinosus* (Durand et al., 2015). Cytochrome *c* serves as electron carrier in respiration (between Complex III and Complex IV) and its assembly requires the enzymatic attachment of a haem's vinyl groups to a conserved double cysteine motif (–CXXCH–) of the cytochrome apoprotein (Stevens et al., 2011). Because the apoprotein is secreted in its unfolded state, its cysteines are prone to disulphide bond formation in the oxidizing environment of the periplasm (Daltrop, Stevens, Higham, & Ferguson, 2002). Prior to the haem's attachment, the disulphide bond must be reduced to the double thiol state, which is likely accomplished by a cytochrome *c* maturation protein (CcmH). Increased presence of copper is believed to shift the balance in favour to disulphide bond formation, blocking haem attachment and reducing cytochrome *c* levels (Durand et al., 2015).

### 2.3.5 Protein Folding

No matter the cause, copper dyshomeostasis readily leads to protein damage, which must be repaired or removed by the cell. The accumulation of abnormal proteins in response to copper stress (Fig. 2A) is indicated by the upregulation of the protein-misfolding regulon in *S. aureus*, which includes the *clpC* and *clpP* protease genes, the chaperon *clpB*, and the negative transcriptional regulator *ctsR* (Baker et al., 2010). Additional evidence suggests that the regulon may be copper responsive due to direct interaction of copper ions with the *ctsR* modulator protein McsA (Sitthisak et al., 2012). Oxidation and intramolecular disulphide formation in the CXXC motif of McsA subsequently result in the release of McsB, which in turn deactivates the negative transcriptional regulator CtsR and activation of the *clpC*-operon (Frees, Savijoki, Varmanen, & Ingmer, 2007). Interestingly, the CXXC motif of McsA can also act as binding site for copper ions and other metals (Sitthisak et al., 2012), suggesting that McsA may act as intracellular copper sensor capable of activating cellular defence mechanism to mitigate toxicity from misfolded proteins. In addition, a similar transcriptional response to copper stress was observed in *E. coli* where the upregulation of the *cpx*-regulon indicates accumulation of misfolded proteins (Kershaw et al., 2005). In eukaryotes, it was observed that copper treatment leads to an accumulation of abnormal proteins as evidenced by an increase in ubiquitinated proteins, which are subject to degradation by the ubiquitin–proteasome system (Oe et al., 2016). Some eubacteria of the order *Actinomycetales* and *Nitrospirales* are also known to utilize proteasomes that are functionally and architecturally similar to their eukaryotic counterparts (Jastrab & Darwin, 2015). One important member of *Actinomycetales* is the obligate human pathogen *M. tuberculosis*, and defects in its proteasomal activity lead to a copper sensitivity phenotype, suggesting that proteasomal degradation pathways may be critical under copper stress (Samanovic & Darwin, 2016; Shi et al., 2014).

## 2.4 Copper's Interactions With Membranes

### *2.4.1 Copper's Coordination to Select Phospholipids*

Membrane phospholipids are complex molecules that, like proteins, harbour functional groups known to coordinate copper ions. Phosphate, carboxyl, and amine moieties of the surface-exposed hydrophilic head on phospholipids constitute potential binding sites. Copper ions can bind with micro- and nanomolar affinities to the phospholipids phosphatidylethanolamine (PE) and phosphatidylserine (PS), respectively (Fig. 3A) (Cong, Poyton, Baxter, Pullanchery, & Cremer, 2015; Monson et al., 2012; Poyton, Sendecki, Cong, & Cremer, 2016). PE is a major lipid component of prokaryotic (e.g. ~75% in *E. coli* (Sohlenkamp & Geiger, 2016)) and fungal membranes (20%–32% (Rattray, Schibeci, & Kidby, 1975)). PS is more commonly found in yeasts (4%–18% of membrane lipids) than in bacteria, although in some bacteria (e.g. *Bdellovibrio bacteriovorans*) the PS content can be significant (Rattray et al., 1975; Sohlenkamp & Geiger, 2016). Current models implicate the amino and/or carboxyl groups of the hydrophilic head portion in the binding of copper ions (Fig. 3A), and the resulting copper/lipid complexes then consist of two lipid molecules coordinating one copper ion (Monson et al., 2012; Poyton et al., 2016). In general, the coordination of copper ions to phospholipids may change the physicochemical properties of the membrane, perhaps most simply by restricting the movement of phospholipids, thereby decreasing membrane fluidity (Gehman, O'Brien, Shabanpoor, Wade, & Separovic, 2008; Lau et al., 2006). Alternatively, adsorption of copper by phospholipids could immobilize free copper ions and therefore attenuate copper toxicity (Cong et al., 2015). This may be of relevance in the bacterial periplasm, which is a major site of redox-active copper in *E. coli* (Macomber et al., 2007). However, these coordinated copper ions may also potentiate oxidative stress on lipid compartments, perhaps due to an increase in hydroxyl radicals at the membrane surface (Poyton et al., 2016), or perturb the electron transport chain in bacteria via interaction with the quinone pool (Abicht et al., 2013). In addition, copper binding to membranes and altering normal biofunctions may impact cell division and growth in general. In *B. subtilis*, domains of PE lipids were concentrated in the septal region during cell division (Nishibori, Kusaka, Hara, Umeda, &

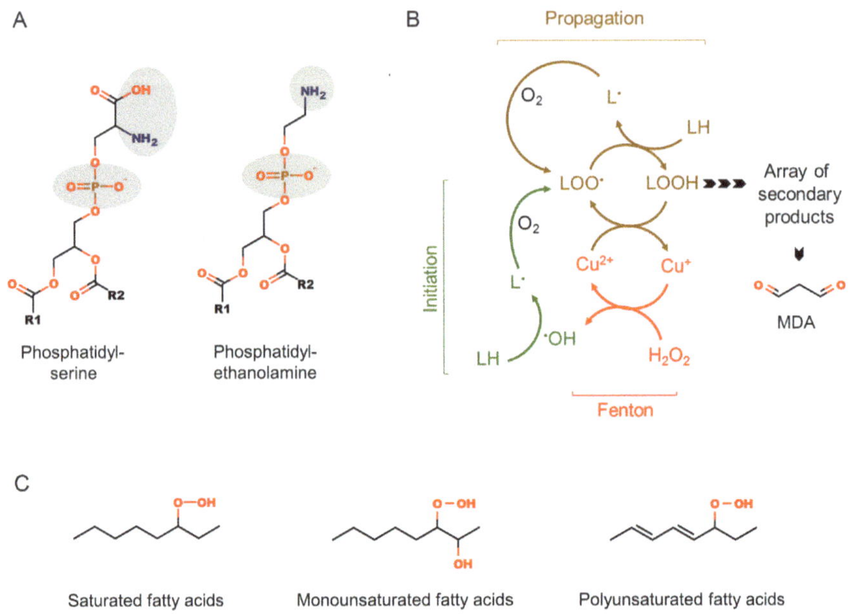


**Fig. 3** Copper-mediated damage of lipids. (A) Potential binding sites for copper of phosphatidylserine (PS) and phosphatidylethanolamine (PE) are highlighted with a *grey background*. R1 and R2 represent variable fatty acid groups. (B) Lipid peroxidation is initiated by hydroxyl radicals ($^{\bullet}OH$), which are formed through copper-catalysed Fenton reaction with hydrogen peroxide ($H_2O_2$). The $^{\bullet}OH$ then react with lipids (LH) forming a carbon-centred lipid radical ($L^{\bullet}$). The resulting radical can then quickly react with oxygen to form a lipid peroxyl radical ($LOO^{\bullet}$). This can then abstract a hydrogen from a close by lipid, propagating the radical reaction. Lipid peroxides (LOOH) can break down into aldehydes, with malondialdehyde (MDA) being the most used marker of lipid-based oxidative damage. (C) Examples of lipid peroxides derived from saturated, monounsaturated, and polyunsaturated fatty acids.

Matsumoto, 2005). In these contexts, copper may interfere with the dispersal of these domains or distribution of PE from these domains to other sites. This may impact the morphology of cells and their ability to replicate, which is consistent with the morphological abnormalities (e.g. wrinkles, clefts, and smaller size) and generally slower growth rate of copper-stressed cells as documented for *E. coli* (Gyawali, Ibrahim, Abu Hasfa, Smqadri, & Haik, 2011).

### 2.4.2 Lipid Oxidation

At least in eukaryotes, hydroxyl radicals are well known to aid in membrane damage, commonly targeting polyunsaturated fatty acid moieties

(Catalá, 2010) and resulting in lipid hydroperoxide (LOOH) formation (Fig. 3B). The process is initiated by $^{\bullet}$OH irreversibly abstracting a hydrogen from a polyunsaturated lipid (LH), leaving a carbon-centred radical (L$^{\bullet}$). This lipid radical rapidly reacts with oxygen, forming a peroxyl radical (LOO$^{\bullet}$) that itself can abstract a hydrogen from another unsaturated lipid, which propagates a radical chain reaction. The newly formed LOOH undergoes complex transformations including isomerization and chain cleavage, leading to an array of secondary products. The chain reaction is only terminated through an antioxidant such as Vitamin E intercepting the lipid peroxyl radical (Ayala, Munoz, & Arguelles, 2014; Niki, 2014; Schneider, 2009). Copper ions can enhance the propagation of the radical chain reaction regenerating peroxyl radicals from lipid peroxides. In prokaryotes, though, copper's interactions with membrane lipids and the subsequent contributions to copper's general toxicity are not well defined. One principle difference between prokaryotes and eukaryotes is that most Gram-positive and Gram-negative bacteria utilize only saturated and monounsaturated fatty acids (Cho & Salton, 1964; Makula, Lockwood, & Finnerty, 1975; Sohlenkamp & Geiger, 2016). Copper-induced oxidative lipid damage in bacteria may therefore diverge from the principle model of polyunsaturated fatty acid peroxidation known to occur in eukaryotic cells. That said, extreme copper stress, such as leaching of copper ions through direct contact with metallic surfaces, can lead to disastrous membrane damage and degradation of lipids in *E. coli* (Calvano et al., 2016; Hong, Kang, Michels, & Gadura, 2012), suggesting that copper ions do at least indirectly damage prokaryotic membranes and membrane lipids. Notably, an increase in monounsaturated fatty acids in the membrane of *E. coli* was associated with an increase in sensitivity towards copper, which was subsequently attributed to lipid peroxidation (Hong et al., 2012). Of course, nonenzymatic oxidation of saturated and monounsaturated fatty acids in controlled model systems is possible, but rate constants are slower than for polyunsaturated chains and forming lipid peroxides differ slightly (Nah et al., 2013) (Fig. 3C). Hydroxyl radicals can abstract a hydrogen from saturated carbon chains, creating an alkyl radical that forms peroxides when oxygen is available. In model systems, the reaction with monounsaturated acyl chains leads primarily to the formation of a hydroxy alkyl radical, which is also likely to form peroxides (Nah et al., 2013).

In eukaryotic cells, the decomposition of lipid peroxides almost always produces aldehydes, and one of the most widely used biomarkers of lipid peroxidation is malondialdehyde (MDA; Fig. 3B). The compound is readily detectable and quantifiable through a condensation reaction with thiobarbituric acid, which produces a pink pigment that is specific for the reaction with MDA (Kosugi & Kikugawa, 1985). The assay is frequently used to evaluate lipid peroxidation in bacteria exposed to a variety of stresses including antibiotics (Ajiboye & Haliru, 2016) and copper (Meyer, Ramlall, Thu, & Gadura, 2015). However, the chemical processes by which MDA forms under copper stress in bacteria remain elusive.

## 2.5 Deoxyribonucleic Acid

Very little is known about copper's interaction with DNA inside a bacterial cell. This topic is somewhat neglected, and aside from binding to some copper-sensing transcriptional repressors such as CsoR and RicR (Festa et al., 2011; Liu et al., 2007), the amount of copper associated with microbial DNA is unknown. In *E. coli*, it was convincingly demonstrated that DNA is not the prime target of copper stress as no significant DNA damage was apparent, even in the presence of $H_2O_2$. Some presented evidence even suggests that copper might even have protective properties by preventing iron-mediated oxidative DNA damage under high iron conditions (Macomber et al., 2007). Nevertheless, DNA is a decent metal chelator (Sagripanti, Goering, & Lamanna, 1991). While intracellular DNA may be sufficiently shielded, extracellular DNA (eDNA) can be in direct contact with environmental copper. The release of eDNA is thought to be a normal function of bacterial physiology. It is liberated by bacterial pathogens while residing inside their eukaryotic host cells (Manzanillo, Shiloh, Portnoy, & Cox, 2012), which is also a well-established environment for bacterial copper exposure (Wagner et al., 2005; White, Lee, Kambe, Fritsche, & Petris, 2009). In a different context, eDNA is also a major matrix component of bacterial biofilms (Whitchurch, Tolker-Nielsen, Ragas, & Mattick, 2002). For example, *P. aeruginosa* cells use DNA within their biofilm matrix (Allesen-Holm et al., 2006) and are up to 600-fold more resistant towards copper than their planktonic counterparts (Teitzel & Parsek, 2003). Given that earlier studies revealed that DNA holds an average of one copper binding site per nucleotide (Sagripanti, 1999; Sagripanti et al., 1991), it is readily conceivable that eDNA may contribute to copper tolerance by serving as copper scavenger protecting cells from free copper ions.

## 2.6 Metabolites

Beyond the dominant macromolecules discussed earlier (proteins, lipids, and DNA), copper ions can interact with many other smaller metabolites within the cell, and these potential interactions are recognized as an alternative mechanism of bacterial copper toxicity. Siderophores such as enterobactin and other catechols are able to reduce $Cu^{2+}$ into $Cu^{1+}$ which may lead to $Cu^{1+}$-related toxicity (Grass et al., 2004; Kamau & Jordan, 2002). As a countermeasure, catechol siderophores can be oxidized by multicopper oxidases, removing the ability to act as a $Cu^{2+}$-reducing agent (Fig. 1) (Grass et al., 2004). One of the most intriguing examples is the reduction of $Cu^{2+}$ to the more toxic $Cu^{1+}$ by menaquinone (Vitamin K) and ubiquinone (coenzyme Q) on the bacterial membrane. Both types of quinones are lipid soluble and serve as internal electron carriers in processes such as respiration (aerobic) and fermentation (anaerobic). This ability of bacteria to reduce $Cu^{2+}$ into $Cu^{1+}$ was noted decades ago (Beswick et al., 1976). In *Enterococcus hirae*, this activity was originally attributed to a putative enzyme believed to be located at the cell envelope (Solioz, 2002; Solioz & Stoyanov, 2003; Wunderli-Ye & Solioz, 1999), though the gene was never identified (Abicht et al., 2013). Instead, the copper reductase activities of *E. coli* (Volentini et al., 2011), *L. lactis* (Abicht et al., 2013), and *E. hirae* (Abicht et al., 2013) were finally linked to quinone-mediated nonenzymatic processes, which also may explain the copper reductase activity observed in *Streptomyces* spp. (Albarracín, Ávila, Amoroso, & Abate, 2008). Mutants of *E. coli* deprived of menaquinone (Δ*menA*) and ubiquinone (Δ*ubiCA*) synthesis displayed deficiencies in copper reductase activity (Volentini et al., 2011). Interestingly, induction of cytochrome *bd* oxidase in *E. hirae*, which is expressed when haem is present in the growth medium, reduced copper reductase activity, suggesting that sensitivity to copper depends not only on menaquinones but also on the metabolic state of the bacterial cells (Abicht et al., 2013). It is also important to note that the ubiquinone-mediated reductase activity was copper specific, as reduction of $Fe^{3+}$ to $Fe^{2+}$ was not seen in *E. coli* (Volentini et al., 2011).

However, some questions remain. It is currently unknown whether quinones reduce $Cu^{2+}$ on both leaflets of the cytoplasmic membrane, as proposed by Volentini et al. (2011), or if this activity is restricted to the periplasmic compartment in Gram-negative or the cell surface of Gram-positive organisms. If this process would occur at the cytoplasmic side, copper ions would have to cross the plasma membrane as $Cu^{2+}$, which is considered unlikely, or form through oxidative processes from $Cu^{1+}$ near the inner leaflet of the plasma membrane. Copper's reaction and interaction with abundant intracellular thiols (e.g. glutathione) would stabilize copper ions in their monovalent state making redox cycling in the reducing environment of the cytoplasm difficult (Pederson, Steinkuhler, Weser, & Rotilio, 1996). In support of this, redox labile copper ions are absent in the cytoplasmic compartment of copper-stressed *E. coli* cells (Macomber et al., 2007).

# 3. MICROBIAL COPPER ACQUISITION AND TOLERANCE MECHANISMS

## 3.1 Copper Acquisition

Oddly, for all that is known about copper, its intracellular effects, and its metabolic handling, how these ions gain cytoplasmic access in prokaryotes is still largely an open question. An early, likely player in the diffusion of copper across the outer membrane of Gram-negative bacteria or mycobacteria is porins, which is supported by studies on *Mycobacterium smegmatis* (Haeili, Speer, Rowland, Niederweis, & Wolschendorf, 2015; Speer, Rowland, Haeili, Niederweis, & Wolschendorf, 2013) and *E. coli* (Lutkenhaus, 1977). However, such porin channels would only grant access to the periplasm, and little is known about systems that mediate uptake of copper ions across the inner membrane. The probably best described bacterial inner membrane copper uptake protein to date is CcoA, which belongs to the major facilitator superfamily of membrane transporters (Ekici, Yang, Koch, & Daldal, 2012; Khalfaoui-Hassani, Verissimo, Koch, & Daldal, 2016; Trasnea et al., 2016). Outside of this relatively recent discovery, in *Pseudomonas syringae*, CopC (a periplasmic copper chaperone) and CopD (a probable plasma membrane protein) were identified as possible mediators of copper uptake (Cha & Cooksey, 1993). PcoD of the *pco* operon in *E. coli* and other Gram-negative bacteria was also assigned a tentative function as an inner membrane copper importer (Lee et al., 2002) but not yet confirmed. Interestingly, the genome of *B. subtilis* encodes for a single protein, YcnJ, with homology to CopCD for which a similar function was proposed (Chillappagari, Miethke, Trip, Kuipers, & Marahiel, 2009). Copper might also bind to promiscuous siderophores as was demonstrated for the staphylococcal metallophore staphylopin, which was shown to mediate uptake of a variety of metal ions, including copper, across the plasma membrane (Ghssein et al., 2016). Whether copper is able to piggyback on free amino acids to gain access to the interior of cells, as shown for $Ni^{2+}$ using free L-histidine (Lebrette et al., 2015), remains to be investigated.

There is, though, strong evidence that suggests uptake of copper across the inner membrane is tightly controlled and limited. In *E. coli* and most other bacteria, most copper-containing enzymes are integral to the plasma membrane or localized in the periplasm (Changela et al., 2003; Stolle et al., 2016), yet metallation of these enzymes often depends on copper efflux across the plasma membrane. Examples include the metallation of Cupin A (CupA), a periplasmic copper binding protein in cyanobacteria (Raimunda, Gonzalez-Guerrero, Leeber, & Arguello, 2011; Waldron et al., 2010), Cu,Zn superoxide dismutase of *Salmonella enterica* serovar Typhimurium (Osman et al., 2013), and cytochrome *c* oxidase in *P. aeruginosa* (Gonzalez-Guerrero, Raimunda, Cheng, & Arguello, 2010) or *Rhodobacter capsulatus* (Ekici, Yang, et al., 2012). Even the multicopper oxidase CueO, which is secreted by the twin-arginine translocation machinery, was recently shown to acquire copper in the periplasm, but in this case, most likely without the help of periplasmic copper chaperons (Stolle et al., 2016).

## 3.2 Copper Defence Strategies

Depending on a bacterium's native environment, lifestyle, and physiology, evolution has led to a variety of copper defence strategies and resistance mechanisms, many of which are summarized in Fig. 4. The most fundamental is that of the plasma membrane, forming a barrier to charged ions such as copper. Even before encountering the plasma membrane, though, extracellular defences are often employed. Previous studies revealed that the biosynthetic pathways for siderophores are induced in copper-stressed *E. coli* (Kershaw et al., 2005) and *P. aeruginosa* (Braud, Geoffroy, Hoegy, Mislin, & Schalk, 2010; Teitzel et al., 2006). It was subsequently shown that the presence of two pseudomonal siderophores, pyoverdine and pyocheline, in the growth medium leads to a lower cell associated copper content in copper-rich conditions, suggesting that these copper-loaded siderophores prevent copper uptake (Braud et al., 2010). Supporting this, pyochelin was found to form stable copper complexes (Brandel et al., 2012). The protective function of another siderophore, virulence-associated yersiniabactin of uropathogenic *E. coli*, was first discovered by Chaturverdi et al. The authors demonstrated that a yersiniabactin–copper complex can be formed in infected animals and humans, remains stable even in the presence of iron, and is associated with a copper resistance phenotype in vitro (Chaturvedi,

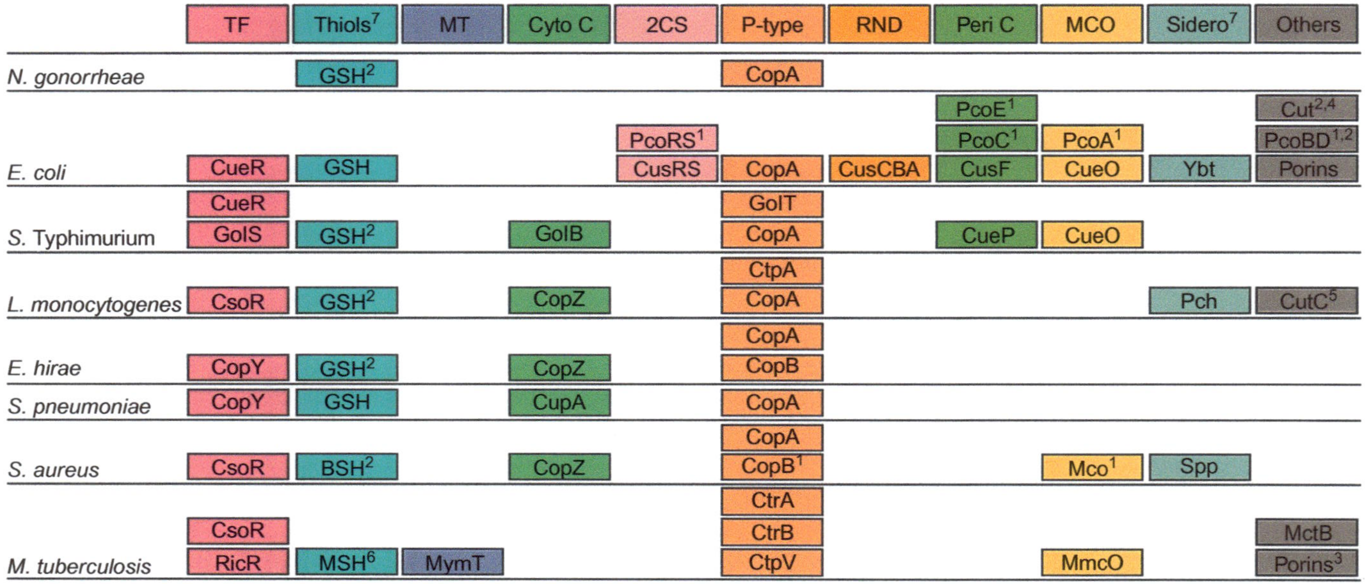


**Fig. 4** Copper homeostasis determinants in select human pathogens. Eight human pathogens and their respective copper resistance machinery are listed to demonstrate the diversity and complexity that vary between different bacteria. The systems start inside the cell and pass through the membrane(s) to reach the extracellular space. *TF*, transcription factors; *Thiols*, intracellular antioxidants associated with copper detoxification; *MT*, metallothionein; *CytoC*, cytoplasmic copper chaperones; *2CS*, copper-sensing two-component systems; *P-type*, P-type copper ATPase; *RND*, resistance-nodulation-division type transmembrane efflux pump; *PeriC*, periplasmic copper chaperons; *MCO*, multicopper oxidase; *Sidero*, siderophores; *Others*, components not classifiable that are implicated in copper homeostasis. [1]Plasmid borne systems that are not located in every strain. [2]Implicated but direct role in copper resistance has not been demonstrated in this organism. [3]Demonstrated for *M. smegmatis*, but there is only indirect evidence in *M. tuberculosis*. [4]Cut represents the Cut operon in *E. coli* that consists of CutABCDEF. [5]CutC of *L. monocytogenes* is not homologous to the Cut operon genes of *E. coli*. [6]Mycothiol (MSH) is less sensitive to copper-mediated oxidation than glutathione (GSH) (Newton et al., 1995). [7]Small-molecule abbreviations: bacillithiol (BSH), glutathione (GSH), yersiniabactin (Ybt), pyochelin (Pch), and staphylopine (Spp).

Hung, Crowley, Stapleton, & Henderson, 2012). Moreover, the yersiniabactin–copper complex was found to have superoxide dismutase-like activity, thereby protecting cells from both copper and superoxide-related exposure (Chaturvedi et al., 2014). Such siderophores likely form extracellular copper sinks, chelating copper ions and preventing interaction with cells.

Should ions bypass siderophores and reach cells themselves, multiple studies have implicated cell surface carboxyl and phosphoryl groups as likely binding sites for metal ions (Fang et al., 2011; González et al., 2010; Tourney, Ngwenya, Mosselmans, & Magennis, 2009). Exopolysaccharides were found to protect functional groups of the cell surface from copper exposure as in the case of the rhizospheric soil bacterium *Pseudomonas aureofaciens*, which produces a fairly inert exopolysaccharide layer, relatively resistant to copper binding (González et al., 2010). Conversely, *B. subtilis* or *Pseudomonas putida* produces exopolysaccharides capable of actively sequestering copper (Fang et al., 2011). Furthermore, when secreted into the extracellular environment, polysaccharides can shield the bacteria from copper insults (González et al., 2010; Nwodo, Green, & Okoh, 2012). The shielding or buffer capacity of exopolysaccharides may therefore serve as an initial defence mechanism towards sudden encounter of potentially toxic copper concentrations. The production of exopolysaccharides either in the form of a capsule or in the context of biofilms is well known to promote pathogenicity of *S. pneumoniae* (Mitchell & Mitchell, 2010), *Klebsiella pneumoniae* (Broberg, Palacios, & Miller, 2014; Paczosa & Mecsas, 2016), *Haemophilus influenzae* (Ulanova & Tsang, 2014), *P. aeruginosa* (Leid et al., 2005), *Neisseria meningitidis* (Hill, Griffiths, Borodina, & Virji, 2010), and *Cryptococcus neoformans* (O'Meara & Alspaugh, 2012) in their mammalian hosts. However, given that the exposure of pathogenic microbes to copper during infection was only recently discovered to be part of the host's innate immune defence strategy (see later), specific studies examining a direct role of bacterial or fungal exopolysaccharide capsules in coping with copper stress during virulence are still lacking.

Once inside, copper must be effluxed, sequestered, or detoxified, leading to all sequenced bacteria possessing some manner of copper defence mechanisms (Ladomersky & Petris, 2015; Solioz et al., 2010). The simplest defence is that of a single efflux pump, such as in *N. gonorrhoeae*, driving copper ions from the cytosol to the periplasm (Djoko et al., 2012). Interestingly, this *copA* gene is within a regulon responding to oxidative and nitrosative stress (Kidd, Potter, Apicella, Jennings, & McEwan, 2005), instead of regulated by a metal-sensing transcription factor, highlighting the link between copper ions and oxidative damage. More generally, two types of efflux pumps are commonly used. In Gram-positive bacteria, P-type ATPases pump copper from the cytoplasm to the extracellular milieu, whereas in Gram-negative and mycobacteria, P-type ATPases transfer copper into the periplasm which requires additional detoxification pathways. Transenvelope efflux of copper in Gram-negative bacteria is usually accomplished by another transporter belonging to the RND superfamily of membrane transporters. The best-characterized tripartite copper efflux system is the *E. coli cusABC* complex, which consists of the inner membrane RND transport protein, a periplasmic linker protein, and an outer membrane component through which copper is released to the outside (Hernandez-Montes, Arguello, & Valderrama, 2012). *E. coli*'s inner membrane CopA delivers copper ions to the periplasmic copper chaperone CusF, which then passes them to the CusABC efflux pump complex. Regardless of the type, many such efflux pumps operate in conjunction with chaperone proteins needed to efficiently deliver ions to their efflux partners. These are generally cytoplasmic proteins that bind with high affinity to copper ions, such as CopZ in *E. hirae* (Solioz & Stoyanov, 2003). In *E. hirae*, CopZ is able to traffic its cargo to both the CopA efflux pump and a CopY transcription factor regulating *copY*, *copZ*, and *copA*. Other chaperones are membrane associated, such as *S. pneumoniae*'s CupA. Interestingly, removal of CupA's membrane binding domain led to an extreme copper susceptibility as severe as deletion of CupA entirely (Fung, Lau, Chan, & Yan, 2013).

If ions cannot be prevented from entering the cell and cannot be effluxed quickly enough, bacteria may attempt to detoxify or sequester copper ions. Multicopper oxidases are present within the periplasm of many Gram negatives and mycobacteria, including *E. coli* (Grass & Rensing, 2001), *S.* Typhimurium (Achard et al., 2010), and *M. tuberculosis* (Rowland & Niederweis, 2013). Gram positives may also contain multicopper oxidases, as in the case of *S. aureus*'s plasmid-encoded *mco* gene (Baker, Sengupta, Jayaswal, & Morrissey, 2011; Sitthisak, Howieson, Amezola, & Jayaswal, 2005). These enzymes oxidize $Cu^{1+}$ to $Cu^{2+}$, reducing the toxicity of uncontrolled ions. Alternatively, intracellular copper binding molecules such as metallothioneins (e.g. *mymT* of *M. tuberculosis* (Gold et al., 2008) or *smtA* from *Synechococcus* sp. (Huckle, Morby, Turner, & Robinson, 1993)), glutathione (Helbig, Bleuel, Krauss, & Nies, 2008), or cytoplasmic copper chaperones (e.g. CupA in *S. pneumoniae* (Fu et al., 2013), CopZ in *Listeria monocytogenes*

(Corbett et al., 2011)) serve as copper sinks in the cytoplasm and buffer the free copper content to undetectable levels (Changela et al., 2003). Accordingly, copper-sensing transcription factors (e.g. CsoR, RicR, CueR) have extremely high affinities for free copper ions (Changela et al., 2003; Festa et al., 2011; Liu et al., 2007) and are therefore able to sense subtle homeostatic imbalances. Binding of copper to these transcription factors will initiate expression of copper defence genes, which include the already mentioned metallothioneins and cytoplasmic copper chaperons, various copper efflux pumps, and multicopper oxidases (Fu, Chang, & Giedroc, 2014; Shi & Darwin, 2015).

## 3.3 Nutritional Immunity

The immunological function of copper as an antiinfective places it within the context of nutritional immunity, the selective modulation of certain nutrients in response to infection. The term was likely established in the mid-1970s and originally only described host strategies to withhold essential iron from microbial invaders (Weinberg, 1975). Only recently has this concept expanded to zinc and manganese (Kehl-Fie & Skaar, 2010), and eventually copper (Festa & Thiele, 2012; Hodgkinson & Petris, 2012; Hood & Skaar, 2012; Samanovic et al., 2012). Though most evidence implicating copper's role in nutritional immunity is related to the macrophage phagosome, this concept may also be applicable to whole tissues. It has long been recognized that copper levels in serum and tissues increase in response to bacterial infection (Chaturvedi & Henderson, 2014). Uropathogenic bacteria are seemingly exposed to copper as evidenced by the release of copper sequestering siderophores (Chaturvedi et al., 2012), and *M. tuberculosis*-infected granulomatous lesions had 70% more copper than uninfected tissue (Wolschendorf et al., 2011). Furthermore, copper homeostasis and resistance proteins are known to contribute to virulence of a broad collection of bacterial pathogens including *M. tuberculosis* (Shi et al., 2014; Ward, Abomoelak, Hoye, Steinberg, & Talaat, 2010; Wolschendorf et al., 2011), *P. aeruginosa* (Schwan, Warrener, Keunz, Stover, & Folger, 2005), *L. monocytogenes* (Corbett et al., 2011; Francis & Thomas, 1997), *S. pneumoniae* (Shafeeq et al., 2011), *S.* Typhimurium (Osman et al., 2010), *N. gonorrhoeae* (Djoko et al., 2012), and *E. coli* (Subashchandrabose et al., 2014; White, Lee, et al., 2009).

Much more is known about the role of copper within phagocytic cells, though, than its role and regulation extracellularly. As early as the mid-1990s a role for metal ions within the phagosomal oxidative burst was hypothesized (Elzanowska, Wolcott, Hannum, & Hurst, 1995), although it was another decade before the first direct evidence (Wagner et al., 2005). The authors measured copper levels in infected macrophage phagosomes using a hard X-ray microprobe. One hour after infection copper concentrations had rocketed from $17.3 \pm 10.3$ to $426 \pm 393$ μM. However, judging by the standard error of the mean values, the experimental procedure seems prone to large discrepancies which the authors partially attributed to the natural heterogeneity of phagosomes (Wagner et al., 2005). A mechanistic explanation was identified shortly after, when White et al. identified the eukaryotic protein ATP7A as a determinant of copper influx into the phagosome using an *E. coli*/RAW264.7 macrophage infection model (White, Kambe, et al., 2009). In response to proinflammatory mediators with key roles in infectious diseases (IFN-γ or bacterial lipopolysaccharide (LPS)), ATP7A expression increased notably within 24 h and in a concentration-dependent manner as determined by Western blot analysis. RNAi-mediated depletion of ATP7A enhanced survival of intraphagosomal *E. coli* by almost threefold (White, Lee, et al., 2009). Likewise, infections of bone marrow-derived macrophages with *S.* Typhimurium induced the expression of copper acquisition genes in macrophages (e.g. *ATP7A*, *SLC31A1* (*CTR1*)), suggesting that macrophages have an increased demand for this metal when facing bacterial pathogens (Achard et al., 2012). These results are in line with long-standing observations that Menkes disease patients, suffering copper deficiency due to a functionally impaired ATP7A protein, are prone to microbial infections (Samanovic et al., 2012).

Like most eukaryotic cells, macrophages must acquire copper ions from their environment primarily through the high-affinity copper uptake transporter CTR1. Within the cells, metallothioneins are involved in limiting copper toxicity by sequestering surplus copper ions, likely restricting their redox-cycling capabilities, and by acting as a (sacrificial) scavenger of oxygen-based radicals (Ogra, Aoyama, & Suzuki, 2006). Immediately after uptake and in concert with glutathione (Maryon, Molloy, & Kaplan, 2013), CTR1 transfers copper ions onto designated copper chaperons (e.g. COX17, CCS) which deliver their cargo to various intracellular points of use (e.g. cytochrome *c* oxidase, superoxide dismutase). The chaperone ATOX1 handles delivery of copper to the secretory pathway in conjunction with the endosomal copper transport proteins ATP7A and ATP7B (Wilson's protein). ATP7A is instrumental in feeding cytoplasmic copper ions into the *trans*-Golgi network, where copper is incorporated as a cofactor into various enzymes such as tyrosinase or lysyl oxidase (Balamurugan & Schaffner, 2006; Linder, 2012; Lutsenko, 2010). Importantly, ATP7A has been implicated in controlling cytoplasmic copper levels by translocating to endosomal vesicles in which surplus copper can be either stored or subsequently released into the extracellular space via fusion with the plasma membrane (Barnes et al., 2009; Petris & Mercer, 1999). Copper uptake and acquisition were found to increase in response to proinflammatory molecules such as IFN-$\gamma$ and LPS, as well as hypoxia, which is also a hallmark of infection and inflammation (White, Kambe, et al., 2009; Zimnicka et al., 2014). Unsurprisingly, both the principal cellular copper uptake protein CTR1 and the intracellular transporter ATP7A are under the control of hypoxia-inducible factors (HIF-1$\alpha$ and HIF-2$\alpha$, respectively) (Xie & Collins, 2011; Zimnicka et al., 2014), again linking cellular copper uptake and its endosomal distribution to microbial infections.

The biochemistry of the actively destructive phagosome is highly complex. What is often regarded as an "hostile environment" is the result of multiple (bio)chemical processes that may occur in parallel or sequentially. In general, the current view of phagosomal functions suggests that engulfed bacteria are subjected to various attacks, including chemical (e.g. low pH, ROS and RNS production, copper), enzymatic (e.g. lipolytic, proteolytic, and carbohydrate-degrading enzymes), and physicochemical (e.g. impairment of membrane integrity by host defensins and cathelicidins) sources (Flannagan, Cosio, & Grinstein, 2009; Guilhelmelli et al., 2013). In this context, it was proposed that copper's primary function in the phagosome is to accelerate the phagosomal respiratory burst. This assumption is supported by in vitro studies which found that, in the absence of copper, *E. coli* cells are relatively resistant to an oxidative burst comparable in intensity to that of macrophages and neutrophils (Elzanowska et al., 1995). It is therefore imperative that copper ions remain relatively mobile and in a redox-active state. Given that the phagosomal compartment may be low on other metal ions due to the action of phagosomal metal-efflux transporters (e.g. NRAMP1 for manganese or iron, ZIP8 for zinc; reviewed in Kehl-Fie & Skaar, 2010), copper ions may encounter a wealth of native binding sites which either need to be saturated or avoided in order for copper ions to reach the bacterial membrane. In addition to natural binding sites, bacteria may release eDNA (Manzanillo et al., 2012) and/or metallophores (Chaturvedi & Henderson, 2014; Chaturvedi et al., 2012) to capture and prevent copper ions from interacting with vulnerable targets.

Interestingly, the low pH in the phagosome may synergistically aid in keeping copper ions relatively mobile. Under acidic conditions, copper ions are less likely to occupy cysteine and histidine residues, as the functional groups (thiol and imidazole, respectively) remain protonated (Rubino, Chenkin, Keller, Riggs-Gelasco, & Franz, 2011). Further, histidine residues, unlike cysteine and methionine, have reasonable affinity not only for $Cu^{1+}$ but also for $Cu^{2+}$ ions. In this regard, oxidative damage was observed in the presence of $H_2O_2$, suggesting the formation of damaging radicals through copper's Fenton chemistry (Rubino et al., 2011). Since these radicals are relatively short lived and therefore incapable of travelling great distances, Fenton chemistry would need to occur either at the surface of the bacteria or inside them for damage to occur. Copper's interaction with and oxidation of phospholipids (see Section 2.4 and Fig. 3B) may work towards sensitizing the bacterial cells to other phagosome functions. Critically, though, the relatively high concentrations of copper within infected tissues and infected phagocytes could offer an opportunity for chemical mimicry. As discussed in the next section, small metal-chelating compounds may be able to harness these in vivo pools of copper for significant antibacterial effects.

# 4. COPPER-DEPENDENT INHIBITORS AS ANTIBACTERIAL AGENTS

The paucity of new antibiotics means that new approaches are urgently needed. Given the unique properties of copper, discussed earlier, it has been readily explored as an antimicrobial agent. Unfortunately, therapeutic use of pure copper is unreasonable outside of niche applications such as self-sterilizing surfaces, due to its potential toxicity to host cells as well as the targeted pathogens. However, selective modulation of copper's intrinsic antibacterial properties may be possible through the use of small-molecule ligands, termed copper-dependent inhibitors (CDIs). The following section discusses a brief history of drug discovery, many of the foundational CDIs, and the future of targeted discovery of additional CDIs.

## 4.1 Brief History of the Successes and Failure of Antibiotic Discovery

The discovery of penicillin in 1928 heralded an end to perhaps the biggest scourge of mankind, that of bacterial infections. In 1900, pneumonia and tuberculosis comprised the two most common sources of mortality in the

United States. In combination with other infectious diseases such as typhoid fever and diphtheria, infections caused 540.9 deaths per 100,000, nearly one-third of all mortality (Linder & Grove, 1947). In contrast, over 100 years later, only the "influenza and pneumonia" designation makes the top 10 causes of death, at a drastically improved 17.3 deaths per 100,000 (Kochanek, Murphy, Xu, & Tejada-Vera, 2016). Many aspects of modern medicine depend on the miracle that is antibiotic therapy, from routine surgeries to immunosuppressive treatments and a life with HIV.

Following penicillin's wild success during World War II, an explosion of antibacterial discovery took place, focused on natural sources of antibiotics. The so-called Waksman platform (Schatz, Bugle, & Waksman, 1944) exploited millions of years of evolution, systematically screening soil microbes' abilities to impair growth of other bacteria. These early efforts relied on phenotypic screening, activity against whole cells without a preselected target. Upon successful inhibition, crude culture filtrate would be purified to obtain the active compound of interest. Pharmaceutical companies quickly followed suit, with the large majority of clinically critical antibiotic classes, including but not limited to macrolides, tetracyclines, rifamycins, and quinolones, found during the "Golden Age" of discovery from ~1940 to ~1960 (Lewis, 2013).

Unfortunately, for as quickly as antibiotic discovery expanded, those abundant sources of natural compounds were rapidly depleted. By the mid-1960s, most promising natural hits were found to be repeated discoveries, and further screens were largely fruitless. The Waksman platform fell into disuse as companies instead focused on synthetically improved version of existing antibiotics. Further compounding the problem, antibiotic resistance began developing quickly after clinical introduction for most drugs, sometimes even prior to medical use (Lewis, 2013). For decades, little progress was made outside of iterative improvements on existing antibiotic classes (Brown & Wright, 2016).

However, the sequencing of *H. influenzae* in 1995 (Fleischmann et al., 1995) prompted a renaissance in antimicrobial discovery (Sams-Dodd, 2005). Powerful genomic tools allowed, for the first time, explorations of preselected target pathways. Coupled with advances in computational power and high-throughput screening (HTS) capabilities, the appeal of satisfying Paul Ehrlich's century-old "one drug, one target" ideal (Strebhardt & Ullrich, 2008) drove significant investment into target-centric drug discovery (Brown & Wright, 2016; Swinney & Anthony, 2011). In contrast with traditional whole-cell screens, target-focused approaches sought to identify conserved pathways or enzymes present only in the organism of interest, and absent in human hosts. After target identification and validation, hundreds of thousands of small molecules might be screened in biochemical assays, looking for outputs such as binding affinity of tested molecules to the target, or altered enzymatic function (Hughes, Rees, Kalindjian, & Philpott, 2011; Livermore, 2011).

Such a focus on target-based "magic bullet" discovery is not without faults, though, and many have called for a return to multitarget or phenotypic screens (Medina-Franco, Giulianotti, Welmaker, & Houghten, 2013; Morphy, Kay, & Rankovic, 2004). In vivo metabolic pathways are highly complex, with numerous interactions not replicable in biochemical assays. Exclusion of other pathways might produce false positives (i.e. hits unable to produce a therapeutic effect when subjected to a whole cell's biology) or false negatives (i.e. compounds with potentially strong activity, but appearing as silent in the limited assay system at hand). An even greater challenge, easily ignored by mammalian drug screening systems yet inherently crucial to the success of antibacterials, is cell permeability (Fair & Tor, 2014). Except for agents targeting cellular membrane components (e.g. daptomycin) or cell walls (e.g. β-lactams), bacterial inhibitors must gain access to their intracellular targets, with Gram-negative bacteria posing particular difficulties. Entrance of hydrophilic compounds is dependent upon porin expression, which are readily downregulated during antibiotic stress. Hydrophobic compounds can diffuse directly into cellular membranes, though poor solubility can heavily limit their effectiveness, and they tend to get trapped in membranes rather than diffusing to cytosolic targets (Blair, Webber, Baylay, Ogbolu, & Piddock, 2015; Valade, Davin-Regli, Bolla, & Pagès, 2013).

Consequently, it is not completely unexpected that modern antibiotic discovery campaigns have largely failed, illustrated by two industrial case studies painting a dismal reality. A highly cited report by GlaxoSmithKline (GSK) (nearly 400 citations by the time of this writing), released in 2007, detailed a decade of unsuccessful antibacterial discovery programmes, despite tens of millions spent in material costs alone (Payne, Gwynn, Holmes, & Pompliano, 2007). Out of 67 target-based HTS campaigns, less than a quarter produced acceptable hits, and only five resulted in lead compounds, all of which attrited prior to even Phase I clinical trials. Such an experience cannot be attributed to simply the inexperience of yesteryear: while GSK's screens ran from 1995 to 2001, early in the heyday of antibacterial HTS, a recent report from AstraZeneca describes screens from 2001 to 2010, with results equally as disappointing (Tommasi, Brown,

Walkup, Manchester, & Miller, 2015). Ultimately, since the Golden Age of the 40s, 50s, and 60s, only three new classes of antibiotics have successfully moved from discovery to full clinical use, the oxazolidinones (i.e. linezolid), the lipopeptides (i.e. daptomycin), and the diarylquinolines (i.e. bedaquiline) (Lewis, 2013). Perhaps somewhat tellingly, none of these classes were discovered through a target-based screening platform: the oxazolidinones arose during an early random screening programme (Slee et al., 1987), the lipopeptides are derived from a product of *Streptomyces roseosporus* (Eliopoulos, Thauvin, Gerson, & Moellering, 1985), and the diarylquinolines resulted from a small shotgun whole-cell screen of selected chemical backbones (Andries et al., 2005). Thus, it seems inescapable that, more than simply new drugs, we desperately need new drug discovery methods, as traditional expertise has completely failed to generate results.

## 4.2 Repurposing Copper as a Directed Antimicrobial Therapeutic

The inability of modern screening techniques to satisfy the need for new antibiotics has driven exploration into alternative discovery platforms. Perhaps most simply, many screeners are rethinking the assays themselves, with a growing call for decreased reliance on biochemical cell-free screens and instead preferring phenotypic assays (Mullard, 2015). Though such screens require eventual target identification and deconvolution (Medina-Franco et al., 2013), they preselect for compounds able to penetrate the notoriously difficult bacterial membrane. Tackling the opposite side of the issue, medicinal chemists are rethinking chemical libraries: most commercial collections were optimized for eukaryotic activity, not for targeting prokaryotes (Fair & Tor, 2014). Other approaches include a significant return to natural inspirations and products (Clardy, Fischbach, & Walsh, 2006; Demain, 2002; Fischbach & Walsh, 2009), as in the case of the exciting new compound teixobactin (Ling et al., 2015), or a focus on nontraditional screening environments and conditions (Farha & Brown, 2015), such as mimicking in vivo situations.

One promising natural inspiration receiving increased attention is that of metal-mediated innate immunity, specifically the antibacterial properties of copper (Becker & Skaar, 2014; Djoko, Goytia, et al., 2015; Ladomersky & Petris, 2015; Neyrolles et al., 2015), discussed in depth in an earlier section. As the immune system has evolved a controlled application of ionic copper in fighting infection, so too might ionic copper be utilized medicinally. Copper ions hold a privileged physiological role, being both essential in eukaryotic and many prokaryotic systems, while simultaneously disastrous if not properly controlled by the cell. Though direct and internal therapeutic application of copper is infeasible outside of surface sterilization, small organic ligands can allow manipulation and targeted application of copper ions. Not confined to copper alone, transition metal ions possess complex chemistry allowing for intricate coordination schemes with various organic and inorganic ligands (Haas & Franz, 2009).

Practical applications reliant on the interaction between ligand molecules and copper ions have been studied, albeit not fully appreciated, since the introduction of traditional antibiotics. In copper's case, its complexes have been used as antiinflammatory agents (Sorenson, 1982), anticancer treatments, and radiological probes (Szymanski et al., 2012). Many of the most extensively characterized scaffolds with copper-dependent activities, such as dithiocarbamates, bipyridyls, phenanthrolines, or 8-hydroxyquinolines, have a strong history as analytical probes to quantify copper content in solution (Smith & McCurdy, 1952) or as pesticides (Russell, 2006; Short, Vargas, Thomas, O'Hanlon, & Enright, 2006). Therapeutic uses were readily found, too: disulfiram, a dithiocarbamate, is primarily employed today as a sobriety aid, but its first clinical appearance was as a scabicide and vermicide (Gordon & Seaton, 1942), and activity was generally attributed to its affinity for copper and other metallic cofactors in metalloenzymes (Eneanya, Bianchine, Duran, & Andresen, 1981; Kragh, 2008; Maresca, Carta, Vullo, & Supuran, 2013). Thiosemicarbazones (TSCs) were noted for metal-related antitumour activity as early as 1958 (French & Freedlander, 1958). GTSM (glyoxal-bis($N^4$-methylthiosemicarbazone)) and several of its derivatives have found numerous potential therapeutic applications today, including against Alzheimer's and as a hypoxia marker (Paterson & Donnelly, 2011).

Many of these compounds have had noted antibacterial properties as well, warranting characterization as CDIs. Such compounds display remarkable inhibitory abilities in the presence of copper, yet are less active or even impotent in its absence. For instance, TSCs, along with diethyldithiocarbamate, were discovered in 1950 to have copper-dependent antituberculosis activity (Liebermeister, 1950). Recently, GTSM has been found to exert profound copper-dependent antibacterial effects on a range of microbes, including *M. tuberculosis* (Speer, Shrestha, et al., 2013), *N. gonorrhoeae* (Djoko, Goytia, et al., 2015), and *S. aureus* (Haeili et al., 2014). Finally, the phenanthroline neocuproine was found in the late 1970s to hold significant antimycoplasmal activity in a copper-dependent fashion, along with other phenanthroline analogues (Smit et al., 1980, 1981, 1982). Examples of these foundational compounds are covered in detail later.

## 4.3 Foundational Compounds

### *4.3.1 Disulfiram*

Disulfiram (Fig. 5A) is a prodrug for the treatment of alcohol dependence, clinically used since the 1940s (Bell & Smith, 1949). Its application as sobriety aid is based upon the irreversible inhibition of human aldehyde dehydrogenase enzyme, leading to a buildup of the acetylaldehyde byproduct. The inability to properly metabolize alcohol causes extreme discomfort within minutes of alcohol intake, discouraging continued consumption (Koppaka et al., 2012). Its absorption in the stomach and intestine is most likely associated with conversion into diethyldithiocarbamate (DETC; Fig. 5B) and other metabolites, which are the active forms of disulfiram (Koppaka et al., 2012). This initial breakdown can occur enzymatically or through interaction with copper ions, thereby forming an acid stable, highly hydrophobic, and absorbable copper complex (Johansson, 1992). Disulfiram itself is only able to complex with copper ions (Fig. 5C), following copper-catalysed reduction of the molecule into two DETC molecules (Lewis, Deshmukh, Tedstone, Tuna, & O'Brien, 2014), but once split these DETC molecules can complex with a variety of transition metals including manganese, iron, and zinc (Dalecki et al., 2015).

Disulfiram's use as a sobriety aid was a secondary use, though; its original clinical appearance was as a scabicide and vermicide (Gordon & Seaton, 1942). Soon after it was recognized that many biological effects were due to the aforementioned ability to complex copper or zinc ions, often stripping the metals from enzyme active sites (Eneanya et al., 1981; Kragh, 2008). As an antibacterial disulfiram has some limited activity against *S. aureus*, but not Enterobacteriaceae spp. or *Pseudomonas* spp. (Phillips, Malloy, Nedunchezian, Lukrec, & Howard, 1991), and a relationship between

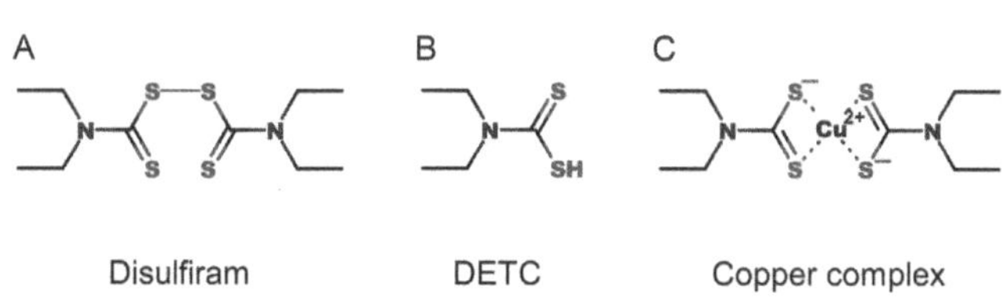


**Fig. 5** (A) Disulfiram; (B) diethyldithiocarbamate (DETC); (C) copper complex of disulfiram or DETC ligands.

activity and copper was not explored. Inhibition against *M. tuberculosis* was equivalent in traditional, rich media with an evident copper dependency, as exclusion (Dalecki et al., 2015) or chelation (Horita et al., 2012) of the metal greatly attenuated disulfiram's effectiveness. Further, despite the ability to form complexes with other transition metals, DETC complexes with zinc, manganese, and iron had no antibacterial effectiveness (Dalecki et al., 2015), highlighting that activity does not simply arise from the formation of a generic metal complex. Interestingly, after disulfiram ingestion, murine and human blood samples possess enhanced ex vivo antimycobacterial profiles (Horita et al., 2012; Hubner, Ernst, Von Laer, Schwander, & Flad, 1991). While tracking the in vivo state of disulfiram is notoriously difficult given the extensive metabolic breakdown profile (Faiman, Jensen, & Lacoursiere, 1984; Johansson, 1992), such activity heavily indicates the formation of a copper complex within the blood, given the established dependency upon copper for antibacterial effects. This activity extends to killing *M. tuberculosis* in infected mice either alone or in combination with other drugs (Horita et al., 2012). The original explanation proposed for disulfiram's antibacterial properties was a mechanism mediated by tumour necrosis factor alpha (TNF-α) or Vitamin $D_3$, leading to macrophage activation (Hubner et al., 1991), or intercalation of the complex into plasma membranes, causing cell lysis (Agar, Mahoney, & Eaton, 1991). However, these conjectures were made well before the discovery of the macrophage copper burst (Wagner et al., 2005; White, Lee, et al., 2009). Given the accumulation of copper and enhanced derepression of the copper stress-induced RicR-regulon in disulfiram/copper relative to copper-only treated *M. tuberculosis* cells (Dalecki et al., 2015), a mode of action relating to general copper toxicity derived from macrophage stores seems more plausible.

### *4.3.2 8-Hydroxyquinoline*

8-Hydroxyquinoline (8HQ; Fig. 6A) and its derivatives have a wide spectrum of biologically relevant activities, and the 8HQ scaffold is regarded as a privileged structure due to its chemical accessibility and broad scope of potential medicinal applications which include cancer, neurodegenerative diseases, and microbial infections (Song, Xu, Chen, Zhan, & Liu, 2015). It is one of the oldest antibacterial agents with documented antiseptic uses dating back to 1895 (Albert & Magrath, 1947), and antiinfective uses in humans dating to before the age of modern antibiotics (Mcelroy, 1910). In these contexts, 8HQ and several of its derivatives (i.e. nitroxoline, clioquinol, 8-hydroxyquinoline-2-carboxylic acid; Fig. 6D–F) are

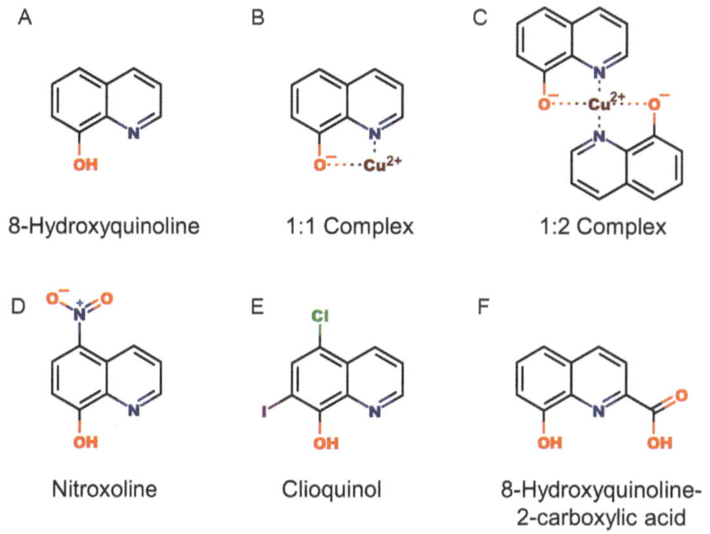


**Fig. 6** (A) 8-Hydroxyquinoline (8HQ); copper complex of 8HQ with a metal:ligand ratio of 1:1 (B) and 1:2 (C); (D) nitroxoline; (E) clioquinol; (F) 8-hydroxyquinoline-2-carboxylic acid (HCA).

considered for drug repositioning or drug development (Musiol, Serda, Hensel-Bielowka, & Polanski, 2010; Oliveri & Vecchio, 2016; Shim & Liu, 2014; Song et al., 2015; Zhang et al., 2016). As early as the 1940s, 8HQ's strong affinity for metal ions was recognized to bestow the molecule with potent antimicrobial properties against a variety of bacterial pathogens at micromolar concentrations (Albert, Gibson, & Rubbo, 1953; Albert, Hampton, Selbie, & Simon, 1954; Albert & Rubro, 1947; Rubbo, Albert, & Gibson, 1950; Shah et al., 2016). 8HQ is a classical ligand that closely follows the Irving–Williams series in terms of affinity for bivalent transition metal ions. When in a ligand/metal complex, two coordination geometries are possible: at low ligand concentrations (relative to the metal ion) a 1:1 complex (Fig. 6B) will form, but when the ligand is in excess an additional molecule joins, forming a dihydroxyquinoline metal complex (1:2 complex, Fig. 6C) with square planar coordination geometry (Albert et al., 1953). It has not yet been established with certainty which of the two complexes is responsible for 8HQ's different bioactivities. Arguments in favour of the 1:1 complex or the 1:2 complex as the bioactive form are currently discussed in the literature (Helsel, White, Razvi, Alies, & Franz, 2017; Shah et al., 2016).

In the context specifically of its antimicrobial properties, 8HQ's most well-characterized targets are the prokaryote *M. tuberculosis* and eukaryote *C. neoformans* (Festa et al., 2014; Helsel et al., 2017; Hongmanee, Rukseree, Buabut, Somsri, & Palittapongarnpim, 2007; Urbanski, Slopek, & Venulet, 1951) and were even identified as a hit cluster during a recent high-throughput physiological screen against *M. tuberculosis* (Ananthan et al., 2009). As the copper concentration of 7H9 growth media is in excess of 6 μM (before accounting for copper bound to BSA), it is perhaps unsurprising that a copper dependency in standard medium was only recently elucidated (Shah et al., 2016). Intriguingly, 8HQ treatment in a cell culture infection model decreased intracellular *M. tuberculosis* by approximately 50%. Given the established copper dependency of 8HQ, it was concluded that this effect may be the result of 8HQ synergizing with phagosomal copper (Shah et al., 2016), which is known to increase within the first 24 h after engulfment by activated macrophages (Wagner et al., 2005). A similar result was seen in a *C. neoformans* infection model, in which treatment with a prochelator analogue of 8HQ resulted in a 60% reduction in fungal CFUs within macrophages (Festa et al., 2014). In addition to *M. tuberculosis* and *C. neoformans*, the 8HQ scaffold may also be considered for targeting neuroinvasive pathogens. For example, 8HQ is active against *L. monocytogenes* in vitro, presumably by acting as an iron scavenger (Simon, Coulanges, Andre, & Vidon, 1995). Although copper-related activities of the 8HQ scaffold against *Listeria* were not considered or investigated, an intact copper homeostasis system was found critical for full virulence of *Listeria* (Francis & Thomas, 1997), suggesting that copper exposure, at least in liver-related infection sites, takes place.

One mechanistic theory considered early on was that 8HQ displaces or interacts with catalytic metal ions of metalloenzymes. This proposed mode of action is supported by target-based in vitro drug discovery screening for metalloenzyme inhibitors and applies to some purified enzymes in chemically defined assays (Olaleye et al., 2011). For instance, 8-hydroxyquinoline-2-carboxylic acid (HCA; Fig. 6F) and clioquinol (Fig. 6E) were found to inhibit methionine aminopeptidase of *M. tuberculosis*, a confirmed drug target and metalloenzyme possibly functionalized by manganese, cobalt, or zinc (Capodagli, Sedhom, Jackson, Ahrendt, & Pegan, 2014; Olaleye et al., 2010, 2011). For HCA, interaction with and displacement of the enzyme's catalytic $Zn^{2+}$ ion was demonstrated in crystal-soaking experiments (Capodagli et al., 2014), though activity on intact *M. tuberculosis* cells was rather weak

(MIC = 500 μM), suggesting that this compound may not reach or target this enzyme in vivo (Capodagli et al., 2014; Shah et al., 2016). Without copper or elevated zinc concentrations, clioquinol also appears inactive towards *M. tuberculosis* at concentrations that would be attractive for therapy, suggesting that only the copper complex and, to some extent, the zinc complex are endowed with antibacterial properties (Shah et al., 2016). Dissipation of the copper complex inside the cell would therefore be necessary so that the free ligand could interact with other metals (Anderson & Swaby, 1951). However, given the much stronger binding affinities for copper than other ions (Oliveri & Vecchio, 2016), in vivo interactions with enzyme-trapped metal ions seem thermodynamically unlikely.

### 4.3.3 Thiosemicarbazones

TSCs have found potential uses in a variety of disease contexts including cancer, neurodegeneration, and infection (Hung et al., 2012; Pahontu et al., 2015; Park et al., 2016). The antitubercular potential of this compound class was recognized more than 70 years ago with clinical trials dating back to the 1950s (Cunningham, Hurford, Erwin, Nagley, & Yell, 1951; Liebermeister, 1950; Mertens & Bunge, 1950; Shane, Riley, Laurie, & Boutilier, 1951). TSCs demonstrated an acceptable toxicity spectrum and had some clinical efficacy, but lost importance as antiinfectives with the introduction of the much more potent antibiotic isoniazid in 1952, which is still one of the most important antituberculosis drugs today. Recently, TSCs were rediscovered for their potent copper-dependent activity against multiple microbes, including *M. tuberculosis*, *N. gonorrhoeae*, *S. aureus*, and *H. influenzae*, while *E. coli* and *S.* Typhimurium are relatively resistant (Dalecki et al., 2016; Djoko, Goytia, et al., 2015; Djoko et al., 2014; Speer, Shrestha, et al., 2013). Some of the best-characterized TSCs are diacetyl-bis($N^4$-methylthiosemicarbazone) (ATSM; Fig. 7A), pyruvaldehyde-bis($N^4$-methylthiosemicarbazone) (PTSM; Fig. 7B), and glyoxal-bis($N^4$-methylthiosemicarbazone) (GTSM; Fig. 7C), which form lipophilic, uncharged, and stable complexes with copper. Unlike 8HQs, TSCs contain rotatable bonds, affording the scaffold conformational flexibility. GTSM, PTSM, and ATSM harbour two TSC motifs, and an individual molecule is able to act as a single tetradentate ligand for one copper ion. The primary target of the GTSM-copper complex (Fig. 7D) is the respiratory chain where Djoko et al. identified succinate and NADH dehydrogenases as most vulnerable targets specifically of the copper complexes, and not the free ligands or other unrelated copper-coordinating scaffolds (Djoko et al., 2014). Direct

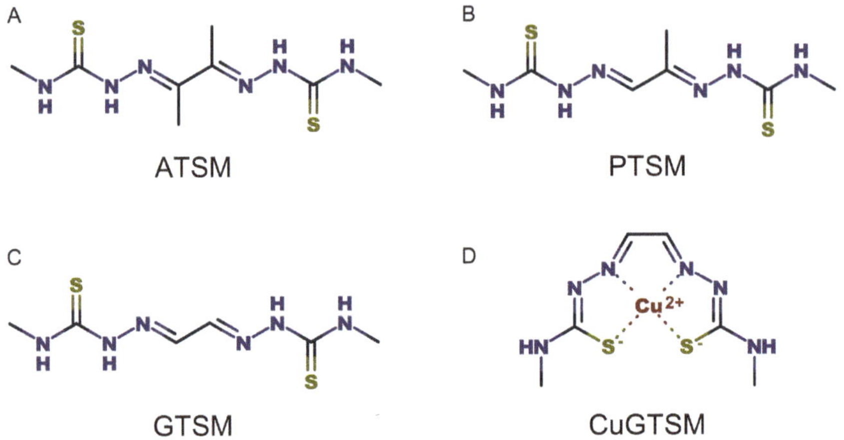


**Fig. 7** (A) Diacetyl-bis($N^4$-methylthiosemicarbazone) (ATSM); (B) pyruvaldehyde-bis($N^4$-methylthiosemicarbazone) (PTSM); (C) glyoxal-bis($N^4$-methylthiosemicarbazone) (GTSM); (D) copper complex of GTSM.

interaction between the copper complex and its targets is required for inhibition, acting as a traditional antimicrobial. In addition, the copper complex is likely able to redox cycle upon its interaction with NADH dehydrogenase. This process consumes molecular oxygen and may therefore induce oxidative stress by generating superoxide anions (Djoko et al., 2014).

However, there is also evidence that impairment of respiratory enzymes is not the only mechanism by which GTSM exerts its antibacterial properties. *S. pneumoniae*, an organism that does not respire, is also highly susceptible to the GTSM-copper complex. Interestingly, the antimicrobial activity of GTSM, ATSM, and PTSM (an intermediate compound) was readily correlated with their reduction potential (Djoko et al., 2014; Haeili et al., 2014; Speer, Shrestha, et al., 2013). The higher reduction potential of GTSM allows redox cycling of the copper ion within the cytoplasm, thought to destabilize the complex and lead to its dissociation (Xiao, Donnelly, Zimmermann, & Wedd, 2008). Consequently, GTSM may partially function through acting as a copper ionophore. Such an effect could not be measured in *N. gonorrhoeae* and was hypothesized to be due to extreme susceptibility to low concentrations, putting any increase in copper content below detectable limits. Instead, *E. coli*, not normally susceptible to GTSM treatment, was used as a model for this cellular increase. Dual treatment with both GTSM and copper led to a twofold increase in cellular copper compared to copper alone (Djoko, Goytia, et al., 2015). Further, GTSM and

ATSM were both able to induce a copper sensitive *lacZ* reporter system in *E. coli*, which is consistent with the release and accumulation of copper in the bacterial cytoplasm. Loss of the inner membrane copper efflux pump CopA also increased sensitivity of *E. coli* to GTSM and copper, but not to copper itself (Djoko, Goytia, et al., 2015).

One caveat of GTSM is that its aforementioned respiratory interference extends to mitochondrial respiration. Though actually explored in some chemotherapy regimens, its use as an antimicrobial is likely heavily limited due to host toxicity. Further improvement on the molecule is needed to generate selectivity between mammalian and prokaryotic respiratory chains. However, use as topical antibiotic may be reasonable and was proposed in context of antiinfective *N. gonorrhoeae* therapy (Djoko, Goytia, et al., 2015).

### 4.3.4 Phenanthroline

Investigations on the antimicrobial activities of phenanthrolines date back to the 1950s and suggested a copper-dependent mode of action (Feeney, Petersen, & Sahinkaya, 1957). Unfortunately, early efficacy studies on mice infected with *M. tuberculosis*, *S. aureus*, or *S. pneumoniae* demonstrated no therapeutic benefits against these particular pathogens in vivo (Dwyer, Reid, Shulman, Laycock, & Dixson, 1969). It was later shown that the copper complex of 1,10-phenanthroline (Fig. 8A) displays in vitro nuclease activity in an oxygen-dependent manner with a strong preference for double over single-stranded DNA (Jessee, Gargiulo, Razvi, & Worcel, 1982; Marshall, Graham, Reich, & Sigman, 1981; Pope & Sigman, 1984). Subsequently, phenanthrolines were exploited to engineer molecules or artificial enzymes with site-specific nuclease activity mostly for analytical applications (Chen et al., 1998; Chen & Sigman, 1986; Papavassiliou, 1994; Soultanas, Dillingham, Wiley, Webb, & Wigley, 2000). Some members of this compound class, namely neocuproine (Fig. 8B) and bathocuproinedisulphonic acid (Fig. 8C), have become valuable probes for studying copper homeostasis, cellular accumulation, and toxicity of copper in vitro (Almeida, Galhardo, Felício, Cabral-Neto, & Leitão, 2000; Asahi et al., 2014; Mohindru, Fisher, & Rabinovitz, 1983a, 1983b). A major advantage of the two molecules comes from their similar metal binding characteristics, yet selective permeabilities, as neocuproine is membrane permeable, while bathocuproinedisulphonic acid is membrane impermeable; the latter is often used as a copper-specific chelating agent in growth medium. In addition, the antibacterial properties

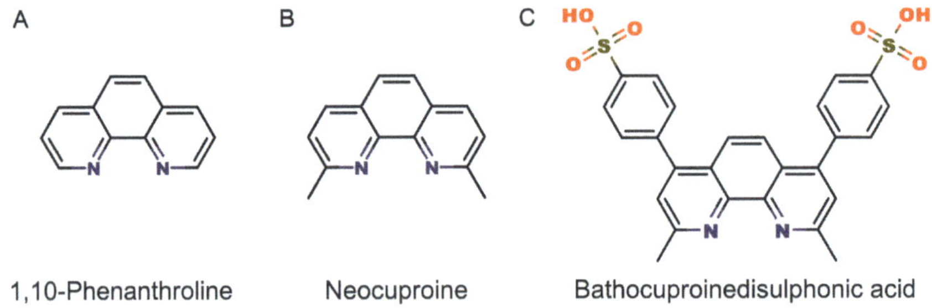


**Fig. 8** (A) 1,10-Phenanthroline; (B) neocuproine; (C) bathocuproinedisulphonic acid.

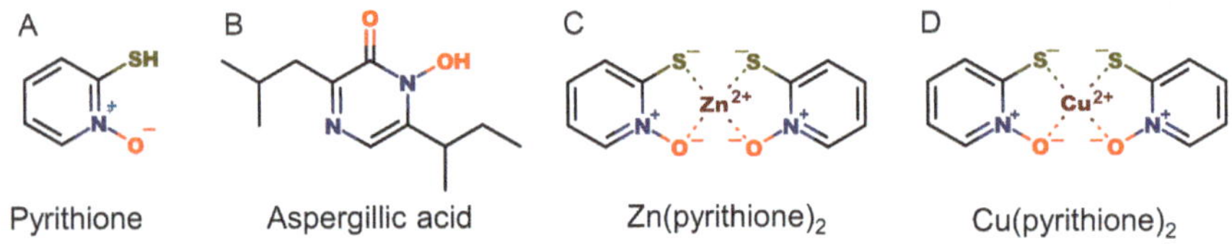


**Fig. 9** (A) Pyrithione; (B) aspergillic acid; (C) zinc complex of pyrithione; (D) copper complex of pyrithione.

of neocuproine were further investigated in various microbial contexts linking the antibacterial action of the neocuproine metal complex to interference with respiration in *Mycoplasma gallisepticum* and *Paracoccus denitrificans* (Smit et al., 1980, 1981, 1982), as well as interference with DNA integrity and inhibition of RNA polymerase in *E. coli* (Perrin, Hoang, Xu, Mazumder, & Sigman, 1996). More recently, neocuproine served as a proof-of-concept compound during HTS for novel compounds in chemical libraries with unknown copper-dependent antibacterial potential (Haeili et al., 2014; Speer, Shrestha, et al., 2013).

### 4.3.5 Pyrithione

Like many of the other foundational CDIs, the pyrithiones (2-mercaptopyridine-*N*-oxide; Fig. 9A) were intensely studied in the 1950s for their fungistatic and bacteriostatic properties (Pansy, Stander, Koerber, & Donovick, 1953; Shaw, Bernstein, Losee, & Lott, 1950), and the metal dependency of this compound's activity was first described shortly after (Albert, Rees, & Tomlinson, 1956). It was found four times as potent on *S. aureus* and *Streptococcus pyogenes* than 8HQ (Albert et al., 1956) and far more active than its naturally occurring sulphur-free analogue aspergillic acid (Fig. 9B) (Shaw et al., 1950). Today, pyrithione is known for its uses in antifouling paints and as a topical antibiotic for the treatment of some mild forms of dermatitis (e.g. seborrhoea, dandruff) (Schwartz, 2016). Therapeutic use is generally accomplished through administration of a zinc-pyrithione complex (Fig. 9C), believed to act as a zinc ionophore and widely accepted as a potential mode of action for its antiviral and anticancer activity (Andersson, Gentry, Moss, & Bevan, 2009; Krenn et al., 2009; Magda et al., 2008). In addition, membrane depolarization was observed in the fungal model organism *Neurospora crassa*, and ablation of several independent membrane transport processes in the *Penicillium* model was attributed to a collapsing proton motive force (Chandler & Segel, 1978; Ermolayeva & Sanders, 1995). However, genetic screening in yeast (*Saccharomyces cerevisiae*) and gene expression profiling on *Malassezia globosa*, one of the fungal pathogens causing dandruff, recently revealed that the activity of the zinc-pyrithione complex is most likely related to copper influx and copper's known toxicity towards iron–sulphur cluster proteins (Reeder et al., 2011). Whole cell-based assays further supported the copper-related mode of action on microbial targets as zinc-pyrithione activity was nearly eliminated in the presence of a membrane impermeable copper-specific chelator (bathocuproinedisulphonic acid), while the addition of copper enhanced the efficacy of zinc-pyrithione treatment by ~10-fold. The formation of a copper-pyrithione complex (Fig. 9D) from zinc-pyrithione through transmetallation was proposed to explain the copper-dependent boost of activity (Reeder et al., 2011).

One noteworthy activity of metal-complexed pyrithione is its ability to restore aminoglycoside susceptibility of amikacin-resistant *K. pneumoniae* isolates (Chiem et al., 2015). The study found that this restoration was due to inhibition of a mostly plasmid-encoded resistance gene, aminoglycoside 6′-*N*-acetyltransferase type Ib. Interestingly, the enzyme's activity is also sensitive to copper alone (Chiem et al., 2015).

## 4.4 Interactions With Clinical Antibiotics

In addition to the canonical CDIs discussed earlier, some clinically utilized antibiotics are known to interact with copper ions, often modulating their therapeutic effects. In the following, three examples are highlighted.

Tetracycline (Fig. 10A), like most of its clinical derivatives, is densely decorated with functional groups (five hydroxyls, two carbonyls, and an amide) creating four possible metal interaction sites; however, only one is typically occupied (Abosede et al., 2016; Zhao et al., 2013). Its interaction with $Mg^{2+}$ is pivotal for porin-mediated uptake across the outer membrane

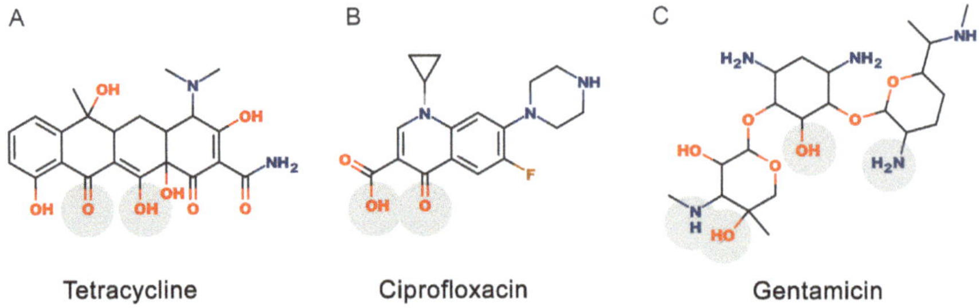


**Fig. 10** (A) Tetracycline; (B) ciprofloxacin; (C) gentamicin. Metal ion binding sites are highlighted in *grey*.

of Gram-negative bacteria, though this complex dissociates in the periplasm enabling tetracycline to diffuse across the inner membrane as uncharged molecule. Reformation of the $Mg^{2+}$ complex in the cytoplasm is a prerequisite for its interaction with and inhibition of ribosomal activity (Schnappinger & Hillen, 1996; Zakeri & Wright, 2008). Unsurprisingly, copper ions are also able to coordinate with the antibiotic, but in a *Vibrio fischeri* model, tetracycline actually acted as a copper sink, ablating copper toxicity (Tong, Zhao, Gu, Gu, & Lee, 2015). Given the high levels of copper used by the innate immune system, and its presence at sites of infection, it would be interesting to learn if such interactions may reduce immune effectiveness of the copper burst or have an impact on the antibacterial properties of the antibiotic itself.

Another example is that of fluoroquinolones (Fig. 10B), which act primarily as bidendate ligands involving the deprotonated carboxylic group and the ring carbonyl oxygen atom. These readily bind physiologically relevant bivalent cations including $Mg^{2+}$, $Ca^{2+}$, $Cu^{2+}$, $Zn^{2+}$, $Fe^{2+}$, and $Co^{2+}$, forming complexes of 1:1 or 1:2 (metal:ligand) stoichiometry. The interaction of copper with quinolones inconsistently alters their antibacterial properties depending on the actual ligand, type of metal ion, and test organism (Ma, Chiu, & Li, 1997; Uivarosi, 2013). As a general trend, copper complexation by fluoroquinolones has little effect on their activity on *S. aureus*, while activity against *E. coli* can be enhanced with MIC values decreasing by up to eightfold. Increased potency against Gram-negative bacteria may be related to enhanced hydrophobic diffusion across cell membranes, which is higher for the copper complexes relative to the free ligands (Sousa, Ferreira, Abreu, Medforth, & Gameiro, 2015).

Finally, aminoglycosides such as gentamicin (Fig. 10C), kanamycin, and amikacin also strongly coordinate copper ions. As for most ligands, their affinity for transition metal ions is in good agreement with the Irving–Williams series. Aminoglycosides exert nephro- and/or ototoxicity during prolonged therapy. Coadministration of metal chelators or antioxidants suppressed these toxicities (Sha & Schacht, 1999, 2000; Song, Sha, & Schacht, 1998), which was taken as evidence that redox-active transition metal complexes, in particular with copper and iron, form in vivo and cause these toxicities potentially in conjunction with DNA damage (Jeżowska-Bojczuk, Bal, & Kozłowski, 1998; Jezowska-Bojczuk et al., 2002; Priuska & Schacht, 1995; Sha & Schacht, 2000). Although aminoglycoside copper complexes are stable under controlled experimental conditions at physiological pH (Szczepanik, Kaczmarek, & Jezowska-Bojczuk, 2004), it was argued that these complexes actually do not form in vivo considering that most copper ions in the blood are tightly bound to carrier proteins (e.g. albumin, ceruloplasmin) or small ligands (e.g. histidine) which would make a ligand exchange reaction with aminoglycosides thermodynamically unfavourable (Lesniak, Harris, Kravitz, Schacht, & Pecoraro, 2003). No significant enhancement of antibacterial activity was observed for the kanamycin A copper complex relative to its free ligand (Szczepanik et al., 2003).

## 4.5 High-Throughput Discovery of CDIs

Despite existence of the numerous CDIs detailed earlier, repurposing existing molecules has largely remained a piecemeal endeavour. Antimicrobial activity might be discovered in established chelators or probes by accident, or batch synthesis efforts might assay antibacterial properties as an afterthought. In the latter case, these efforts are bounded by current knowledge of coordination chemistry, as well as synthetic limitations in what can be batch produced. Further, purposefully designed ligands often have high affinities and may not disassociate in vivo, thus failing to fully potentiate copper itself, with activity likely only resulting from traditional one-drug-one-target interactions. Most glaringly, regardless of the source, the tepid exploration of a few compounds' antibacterial effects divorces copper-dependent drug discovery from the power of HTS. Assaying chemical collections, be they synthetic libraries (as has proven successful for many noninfectious diseases) or natural product extracts (the source of nearly all antibiotics), requires the ability to screen for ligand/ion interaction (and subsequent inhibitory effect) without prior knowledge of how each complex may form.

To accomplish this, copper ions and the ligand must be considered two separate components of the assay system, instead of singular consideration of the expected complex. This is therapeutically justified by the potential large pools of copper in vivo, both in whole tissues and within individual immune cells. On a tissue level, copper concentrations often rise in response to infection, in direct contrast to iron, which is heavily sequestered from pathogens (Palmer & Skaar, 2016). For instance, this was directly observed in a guinea pig model of tuberculosis infection, with copper in infected granulomas increasing 70% over that in uninfected tissue (Wolschendorf et al., 2011). Individual cells may provide a likely source for these heightened copper levels, as macrophages employ a "copper burst" within phagolysosomes (Section 3.3). Measured at up to 400 μM, this appears a rich opportunity for environmental selectivity of antibiotics. If compounds are only active in the presence of copper, and copper is in high concentrations at sites of infection, activity can be directed to pathogens with minimal off-target effects, either on the eukaryotic host or on the microbiome. Utilization of these site-specific sources of copper would require inhibitors able to dynamically coordinate ions, rather than be administered as a prebound complex. Consequently, in a screening assay, identification requires focusing on media modulation, through copper supplementation, instead of focusing on compound modulation, through synthesis. These types of ligands are readily exposed by traditional MIC assays. Dynamic CDIs have been found against a wide variety of pathogens, including Gram positives such as *S. aureus* (Haeili et al., 2014) and *S. pneumoniae* (Djoko, Goytia, et al., 2015), Gram negatives such as *N. gonorrhoeae* (Djoko et al., 2014), mycobacteria (Dalecki et al., 2015; Speer, Shrestha, et al., 2013), mycoplasma (Smit et al., 1982), and eukaryotes such as *C. neoformans* (Festa et al., 2014).

By treating copper supplementation as media modulation and separate from the compound itself, a variety of information can be obtained through parallel MIC measurements, one in the presence of copper and one in the absence. Conceptually, four possible cases arise from the comparison of two results. Most commonly, a compound may have no inhibitory activity either in the presence or in the absence of copper. Conversely, inhibition may result irrespective of copper concentration, producing a copper-independent hit; these are the bulk of hits from traditional, copper-blind drug screens. The third case, of copper-dependent activity, is the primary goal of such an assay: high levels of activity in the presence of copper, but significant or complete impotence in its absence. The final, fourth case may be somewhat unexpected: inverse hits, in which activity is eliminated by the presence of copper. This activity appears rarely (Dalecki et al., 2016), though it can readily obfuscate a copper-blind screen, as in vitro activity could readily disappear in vivo upon encountering environmental copper. The mechanisms of this inactivation are not known, though some possibilities include a phenotypic change in response to copper stress that engenders resistance to the ligand, modification/destruction of the compound by labile copper ions, or conformational change due to complexing with a copper ion that inhibits penetration into the cell or binding to the drug target. Overall, this single-assay approach is readily scaled up and allows the full application of modern screening technology, needing only supplementation of assay media with physiological levels of copper to reveal copper-dependent inhibitory effects. To date, pilot screening campaigns for new scaffolds have been conducted against *M. tuberculosis* (Dalecki & Wolschendorf, 2016) and *S. aureus* (Dalecki et al., 2016).

## 4.6 The NNSNs as a New CDI Scaffold

When deploying a methodological advancement, a crucial early question is whether information learned is new, or at least superior to existing techniques. As applied to large-scale screening for CDIs, this remained unanswered until the characterization of a new chemical motif, the NNSNs. This substructure, composed of a heterocyclic nitrogen-containing ring linked by a carbon to a noncyclic thiourea (Fig. 11A and B), was discovered during a screen for CDIs against *S. aureus* (Dalecki et al., 2016). Though present in many chemical collections, a meta-analysis using the ChEMBL database (Bento et al., 2014) revealed the NNSNs as unexplored antistaphylococcal agents. Curiously, the NNSNs were not completely unknown as antiinfectives, and first appeared in the early 1990s as the phenylethylthioureathiazole family of HIV-1 nonnucleoside reverse transcription inhibitors (Ahgren et al., 1995). The first-generation compound, trovirdine (Fig. 11A), entered Phase I clinical trials, but was eventually withdrawn (Pedersen & Pedersen, 1999). Subsequent generations had improved activity and, despite exiting clinical trials due to poor systemic availability, are under investigation as inhibitors in a vaginal explant model (Barnable et al., 2015; Singer et al., 2011).

The NNSNs displayed a range of MICs against *S. aureus*, with the most active (APT-6i; Fig. 11B) as low as 0.3 μM. This activity was strikingly copper dependent and copper specific, while retaining low eukaryotic toxicity. UV–Vis spectroscopy and $^{1}$H NMR analysis revealed that the NNSNs do

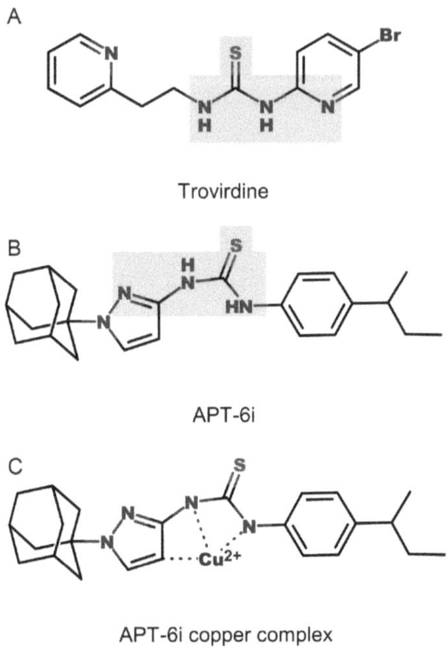


**Fig. 11** (A) Trovirdine; (B) APT-6i; (C) copper complex of APT-6i. Water as the fourth ligand coordinating the copper ion has been omitted. The NNSN motif is highlighted in *grey*.

chelate a copper ion (Fig. 11C), in accordance with other known CDIs such as GTSM and 8HQ. Activity likely results from the formation of this complex, though whether the ligand simply serves as a carrier for the copper ion through a "Trojan Horse"-type mechanism (Dalecki et al., 2016) is unknown. Most intriguingly, the complex forms with a significantly different geometry from canonical thiourea/copper complexes, coordinating a carbon atom in the nearby pyrazole ring along with both thioamides. Synthetic efforts would be highly unlikely to purposefully generate such a geometry. Thus, screens for CDIs can reveal chemical discoveries as well as biological, if one frames bacterial viability as a reporter for the formation of a bioactive ligand/ion complex.

## 5. CONCLUSIONS

As the old cliché states, necessity is the mother of invention, and the looming antibiotic crisis has encouraged exploration into evermore sources of antimicrobials. Nature has long been the primary inspiration for combating infectious agents, from the isolation of penicillin in the 1920s to the modern discovery of teixobactin from a previously unculturable soil microbe. Over the last decade, research on the biology of copper has shown that exploitation of its chemistry in biological systems goes beyond simple use as an enzymatic cofactor. In particular, the discovery of the controlled use of copper within phagolysosomes has created exciting opportunities for new antiinfective therapies. Through modern engineering and screening techniques, this previously intractable toxin can be redirected into purposed therapeutics. Small organic ligands often can coordinate with copper ions into complexes with emergent properties present in neither alone. Dutiful consideration of both components is essential, and failure to do so leaves most antimicrobial screens blind to a host of metal-dependent inhibitions. Consequently, the serendipitous and sporadic discovery that has yielded nearly all of the known CDIs must pivot towards a more systematic exploration.

Recent copper-focused pilot screens have readily established the relative abundance of CDIs within existing libraries. Modulation of this singular variable may thus effectively double the existing accessible chemical space. Though future work requires evaluating whether these compounds retain activity in vivo, they possess great potential to comprise an entirely new source of needed antibiotics.

## ACKNOWLEDGEMENTS

Our thanks goes to Dr Jim Sun for insightful discussions conceptualizing this chapter. A.G.D. was supported by the Carmichael Fund and F.W. by National Institute of Health (NIH) grants R01AI104952 and R01AI121364. C.L.C. was supported by NIH R01AI121364.

## REFERENCES

Abicht, H. K., Gonskikh, Y., Gerber, S. D., & Solioz, M. (2013). Non-enzymic copper reduction by menaquinone enhances copper toxicity in *Lactococcus lactis* Il1403. *Microbiology, 159*(Pt. 6), 1190–1197. http://dx.doi.org/10.1099/Mic.0.066928-0.

Abosede, O. O., Vyas, N. A., Singh, S. B., Kumbhar, A. S., Kate, A., Kumbhar, A. A., et al. (2016). Copper(Ii) mixed-ligand polypyridyl complexes with doxycycline—Structures and biological evaluation. *Dalton Transactions, 45*(7), 3003–3012. http://dx.doi.org/10.1039/C5dt04405g.

Achard, M. E., Stafford, S. L., Bokil, N. J., Chartres, J., Bernhardt, P. V., Schembri, M. A., et al. (2012). Copper redistribution in murine macrophages in response to *Salmonella infection. The Biochemical Journal, 444*(1), 51–57. http://dx.doi.org/10.1042/Bj20112180.

Achard, M. E., Tree, J. J., Holden, J. A., Simpfendorfer, K. R., Wijburg, O. L., Strugnell, R. A., et al. (2010). The multi-copper-ion oxidase CueO of *Salmonella enterica* serovar Typhimurium is required for systemic virulence. *Infection and Immunity, 78*(5), 2312–2319. http://dx.doi.org/10.1128/Iai.01208-09.

Agar, N. S., Mahoney, J. R., Jr., & Eaton, J. W. (1991). Hemolytic and microbicidal actions of diethyldithiocarbamic acid. *Biochemical Pharmacology*, *41*(6–7), 985–993.

Ahgren, C., Backro, K., Bell, F. W., Cantrell, A. S., Clemens, M., Colacino, J. M., et al. (1995). The PETT series, a new class of potent nonnucleoside inhibitors of human immunodeficiency virus type 1 reverse transcriptase. *Antimicrobial Agents and Chemotherapy*, *39*(6), 1329–1335.

Ajiboye, T. O., & Haliru, F. Z. (2016). Redox and respiratory chain related alterations in the lophirones B and C-mediated bacterial lethality. *Microbial Pathogenesis*, *100*, 95–111. http://dx.doi.org/10.1016/j.micpath.2016.08.027.

Albarracín, V. H., Ávila, A. L., Amoroso, M. J., & Abate, C. M. (2008). Copper removal ability by *Streptomyces* strains with dissimilar growth patterns and endowed with cupric reductase activity. *FEMS Microbiology Letters*, *288*(2), 141–148. http://dx.doi.org/10.1111/J.1574-6968.2008.01335.X.

Albert, A., Gibson, M. I., & Rubbo, S. D. (1953). The influence of chemical constitution on antibacterial activity. VI. The bactericidal action of 8-hydroxyquinoline (oxine). *British Journal of Experimental Pathology*, *34*(2), 119–130.

Albert, A., Hampton, A., Selbie, F. R., & Simon, R. D. (1954). The influence of chemical constitution on anti-bacterial activity. Part VII. The site of action of 8-hydroxy-quinoline (oxine). *British Journal of Experimental Pathology*, *35*(1), 75–84.

Albert, A., & Magrath, D. (1947). The choice of a chelating agent for inactivating trace metals: II. Derivatives of oxine (8-hydroxyquinoline). *The Biochemical Journal*, *41*(4), 534–545.

Albert, A., Rees, C. W., & Tomlinson, A. J. (1956). The influence of chemical constitution on antibacterial activity. VIII. 2-mercaptopyridine-N-oxide, and some general observations on metal-binding agents. *British Journal of Experimental Pathology*, *37*(5), 500–511.

Albert, A., & Rubro, S. D. (1947). The influence of chemical constitution of antibacterial activity: A study of 8-hydroxyquinolin (oxine) and related compounds. *British Journal of Experimental Pathology*, *28*(2), 69–87.

Allesen-Holm, M., Barken, K. B., Yang, L., Klausen, M., Webb, J. S., Kjelleberg, S., et al. (2006). A characterization of DNA release in *Pseudomonas aeruginosa* cultures and biofilms. *Molecular Microbiology*, *59*(4), 1114–1128. http://dx.doi.org/10.1111/J.1365-2958.2005.05008.X.

Almeida, C. E. B., Galhardo, R. S., Felício, D. L., Cabral-Neto, J. B., & Leitão, A. C. (2000). Copper ions mediate the lethality induced by hydrogen peroxide in low iron conditions in *Escherichia coli*. *Mutation Research*, *460*(1), 61–67. http://dx.doi.org/10.1016/S0921-8777(00)00014-8.

Ananthan, S., Faaleolea, E. R., Goldman, R. C., Hobrath, J. V., Kwong, C. D., Laughon, B. E., et al. (2009). High-throughput screening for inhibitors of *Mycobacterium tuberculosis* H37rv. *Tuberculosis (Edinburgh, Scotland)*, *89*(5), 334–353. http://dx.doi.org/10.1016/J.Tube.2009.05.008.

Anderson, B. I., & Swaby, R. J. (1951). Factors influencing the fungistatic action of 8-hydroxyquinoline (oxine) and its metal complexes. *Australian Journal of Scientific Research Series B*, *4*(3), 275–282.

Andersson, D. A., Gentry, C., Moss, S., & Bevan, S. (2009). Clioquinol and pyrithione activate Trpa1 by increasing intracellular Zn2+. *Proceedings of the National Academy of Sciences of the United States of America*, *106*(20), 8374–8379. http://dx.doi.org/10.1073/Pnas.0812675106.

Andries, K., Verhasselt, P., Guillemont, J., Göhlmann, H. W. H., Neefs, J. M., Winkler, H., et al. (2005). A diarylquinoline drug active on the ATP synthase of *Mycobacterium tuberculosis*. *Science*, *307*(5707), 223–227. http://dx.doi.org/10.1126/Science.1106753.

Argüello, J. M., Raimunda, D., & Padilla-Benavides, T. (2013). Mechanisms of copper homeostasis in bacteria. *Frontiers in Cellular and Infection Microbiology*, *3*, 73. http://dx.doi.org/10.3389/Fcimb.2013.00073.

Asahi, H., Tolba, M. E. M., Tanabe, M., Sugano, S., Abe, K., & Kawamoto, F. (2014). Perturbation of copper homeostasis is instrumental in early developmental arrest of intraerythrocytic *Plasmodium falciparum*. *BMC Microbiology*, *14*, 167. http://dx.doi.org/10.1186/1471-2180-14-167.

Ayala, A., Munoz, M. F., & Arguelles, S. (2014). Lipid peroxidation: Production, metabolism, and signaling mechanisms of malondialdehyde and 4-hydroxy-2-nonenal. *Oxidative Medicine and Cellular Longevity*, *2014*, 360438. http://dx.doi.org/10.1155/2014/360438.

Baker, J., Sengupta, M., Jayaswal, R. K., & Morrissey, J. A. (2011). The *Staphylococcus aureus* CsoR regulates both chromosomal and plasmid-encoded copper resistance mechanisms. *Environmental Microbiology*, *13*(9), 2495–2507. http://dx.doi.org/10.1111/J.1462-2920.2011.02522.X.

Baker, J., Sitthisak, S., Sengupta, M., Johnson, M., Jayaswal, R. K., & Morrissey, J. A. (2010). Copper stress induces a global stress response in *Staphylococcus aureus* and represses Sae and Agr expression and biofilm formation. *Applied and Environmental Microbiology*, *76*(1), 150–160. http://dx.doi.org/10.1128/aem.02268-09.

Balamurugan, K., & Schaffner, W. (2006). Copper homeostasis in eukaryotes: Teetering on a tightrope. *Biochimica et Biophysica Acta*, *1763*(7), 737–746. http://dx.doi.org/10.1016/j.bbamcr.2006.05.001.

Barnable, P., Calenda, G., Bonnaire, T., Menon, R., Levendosky, K., Gettie, A., et al. (2015). Miv-150/zinc acetate gel inhibits cell-associated simian-human immunodeficiency virus reverse transcriptase infection in a macaque vaginal explant model. *Antimicrobial Agents and Chemotherapy*, *59*(7), 3829–3837. http://dx.doi.org/10.1128/Aac.00073-15.

Barnes, N. L., Bartee, M. Y., Braiterman, L., Gupta, A., Ustiyan, V., Zuzel, V., et al. (2009). Cell-specific trafficking suggests a new role for renal Atp7b in the intracellular copper storage. *Traffic (Copenhagen, Denmark)*, *10*(6), 767–779. http://dx.doi.org/10.1111/J.1600-0854.2009.00901.X.

Battistoni, A. (2003). Role of prokaryotic Cu, Zn superoxide dismutase in pathogenesis. *Biochemical Society Transactions*, *31*(Pt. 6), 1326–1329. http://dx.doi.org/10.1042/.

Becker, K. W., & Skaar, E. P. (2014). Metal limitation and toxicity at the interface between host and pathogen. *FEMS Microbiology Reviews*, *38*, 1235–1249. http://dx.doi.org/10.1111/1574-6976.12087.

Bell, R. G., & Smith, H. W. (1949). Preliminary report on clinical trials of antabuse. *Canadian Medical Association Journal*, *60*(3), 286–288.

Bento, A. P., Gaulton, A., Hersey, A., Bellis, L. J., Chambers, J., Davies, M., et al. (2014). The ChEMBL bioactivity database: An update. *Nucleic Acids Research*, *42*(Database Issue), D1083–D1090. http://dx.doi.org/10.1093/Nar/Gkt1031.

Berlett, B. S., & Stadtman, E. R. (1997). Protein oxidation in aging, disease, and oxidative stress. *The Journal of Biological Chemistry*, *272*(33), 20313–20316. http://dx.doi.org/10.1074/Jbc.272.33.20313.

Beswick, P. H., Hall, G. H., Hook, A. J., Little, K., McBrien, D. C., & Lott, K. A. (1976). Copper toxicity: Evidence for the conversion of cupric to cuprous copper in vivo under anaerobic conditions. *Chemico-Biological Interactions*, *14*(3–4), 347–356.

Beyer, W. F., Jr., & Fridovich, I. (1991). In vivo competition between iron and manganese for occupancy of the active site region of the manganese-superoxide dismutase of *Escherichia coli*. *The Journal of Biological Chemistry*, *266*(1), 303–308.

Blair, J. M. A., Webber, M. A., Baylay, A. J., Ogbolu, D. O., & Piddock, L. J. V. (2015). Molecular mechanisms of antibiotic resistance. *Nature Reviews Microbiology*, *13*(1), 42–51. http://dx.doi.org/10.1038/Nrmicro3380.

Brandel, J., Humbert, N., Elhabiri, M., Schalk, I. J., Mislin, G. L., & Albrecht-Gary, A. M. (2012). Pyochelin, a siderophore of *Pseudomonas aeruginosa*: Physicochemical characterization of the iron(III), copper(II) and zinc(II) complexes. *Dalton Transactions*, *41*(9), 2820–2834. http://dx.doi.org/10.1039/C1dt11804h.

Braud, A., Geoffroy, V., Hoegy, F., Mislin, G. L., & Schalk, I. J. (2010). Presence of the siderophores pyoverdine and pyochelin in the extracellular medium reduces toxic metal accumulation in *Pseudomonas aeruginosa* and increases bacterial metal tolerance. *Environmental Microbiology Reports*, *2*(3), 419–425. http://dx.doi.org/10.1111/J.1758-2229.2009.00126.X.

Broberg, C. A., Palacios, M., & Miller, V. L. (2014). *Klebsiella*: A long way to go towards understanding this enigmatic jet-setter. *F1000prime Reports*, *6*, 64. http://dx.doi.org/10.12703/P6-64.

Brown, E. D., & Wright, G. D. (2016). Antibacterial drug discovery in the resistance Era. *Nature*, *529*(7586), 336–343. http://dx.doi.org/10.1038/Nature17042.

Broxton, C. N., & Culotta, V. C. (2016). Sod enzymes and microbial pathogens: Surviving the oxidative storm of infection. *PLoS Pathogens*, *12*(1), e1005295. http://dx.doi.org/10.1371/journal.ppat.1005295

Calvano, C. D., Picca, R. A., Bonerba, E., Tantillo, G., Cioffi, N., & Palmisano, F. (2016). MALDI-TOF mass spectrometry analysis of proteins and lipids in *Escherichia coli* exposed to copper ions and nanoparticles. *Journal of Mass Spectrometry*, *51*, 828–840. http://dx.doi.org/10.1002/jms.3823.

Capodagli, G. C., Sedhom, W. G., Jackson, M., Ahrendt, K. A., & Pegan, S. D. (2014). A noncompetitive inhibitor for *Mycobacterium tuberculosis*'s class IIa fructose 1,6-bisphosphate aldolase. *Biochemistry*, *53*(1), 202–213. http://dx.doi.org/10.1021/bi401022b.

Catalá, A. (2010). A synopsis of the process of lipid peroxidation since the discovery of the essential fatty acids. *Biochemical and Biophysical Research Communications*, *399*(3), 318–323. http://dx.doi.org/10.1016/j.bbrc.2010.07.087.

Cha, J. S., & Cooksey, D. A. (1993). Copper hypersensitivity and uptake in *Pseudomonas syringae* containing cloned components of the copper resistance operon. *Applied and Environmental Microbiology*, *59*(5), 1671–1674.

Chandler, C. J., & Segel, I. H. (1978). Mechanism of the antimicrobial action of pyrithione: Effects on membrane transport, Atp levels, and protein synthesis. *Antimicrobial Agents and Chemotherapy*, *14*(1), 60–68.

Changela, A., Chen, K., Xue, Y., Holschen, J., Outten, C. E., O'Halloran, T. V., et al. (2003). Molecular basis of metal-ion selectivity and zeptomolar sensitivity by CueR. *Science*, *301*(5638), 1383–1387. http://dx.doi.org/10.1126/science.1085950.

Chaturvedi, K. S., & Henderson, J. P. (2014). Pathogenic adaptations to host-derived antibacterial copper. *Frontiers in Cellular and Infection Microbiology*, *4*, 3. http://dx.doi.org/10.3389/fcimb.2014.00003.

Chaturvedi, K. S., Hung, C. S., Crowley, J. R., Stapleton, A. E., & Henderson, J. P. (2012). The siderophore yersiniabactin binds copper to protect pathogens during infection. *Nature Chemical Biology*, *8*(8), 731–736. http://dx.doi.org/10.1038/nchembio.1020.

Chaturvedi, K. S., Hung, C. S., Giblin, D. E., Urushidani, S., Austin, A. M., Dinauer, M. C., et al. (2014). Cupric yersiniabactin is a virulence-associated superoxide dismutase mimic. *ACS Chemical Biology*, *9*(2), 551–561. http://dx.doi.org/10.1021/cb400658k.

Chen, C. H., Landgraf, R., Walts, A. D., Chan, L., Schlonk, P. M., Terwilliger, T. C., et al. (1998). Scission of DNA at a preselected sequence using a single-strand-specific chemical nuclease. *Chemistry & Biology*, *5*(5), 283–292.

Chen, C. H., & Sigman, D. S. (1986). Nuclease activity of 1,10-phenanthroline-copper: Sequence-specific targeting. *Proceedings of the National Academy of Sciences of the United States of America*, *83*(19), 7147–7151.

Chiem, K., Fuentes, B. A., Lin, D. L., Tran, T., Jackson, A., Ramirez, M. S., et al. (2015). Inhibition of aminoglycoside 6'-N-acetyltransferase type Ib-mediated amikacin resistance in *Klebsiella pneumoniae* by zinc and copper pyrithione. *Antimicrobial Agents and Chemotherapy*, *59*(9), 5851–5853. http://dx.doi.org/10.1128/aac.01106-15.

Chillappagari, S., Miethke, M., Trip, H., Kuipers, O. P., & Marahiel, M. A. (2009). Copper acquisition is mediated by YcnJ and regulated by YcnK and CsoR in *Bacillus subtilis*. *Journal of Bacteriology*, *191*(7), 2362–2370. http://dx.doi.org/10.1128/jb.01616-08.

Chillappagari, S., Seubert, A., Trip, H., Kuipers, O. P., Marahiel, M. A., & Miethke, M. (2010). Copper stress affects iron homeostasis by destabilizing iron-sulfur cluster formation in *Bacillus subtilis*. *Journal of Bacteriology*, *192*(10), 2512–2524. http://dx.doi.org/10.1128/JB.00058-10.

Cho, K. Y., & Salton, M. R. J. (1964). Fatty acid composition of the lipids of membranes of gram-positive bacteria and "walls" of gram-negative bacteria. *Biochimica et Biophysica Acta*, *84*(6), 773–775. http://dx.doi.org/10.1016/0926-6542(64)90043-5.

Clardy, J., Fischbach, M. A., & Walsh, C. T. (2006). New antibiotics from bacterial natural products. *Nature Biotechnology*, *24*(12), 1541–1550. http://dx.doi.org/10.1038/nbt1266.

Cong, X., Poyton, M. F., Baxter, A. J., Pullanchery, S., & Cremer, P. S. (2015). Unquenchable surface potential dramatically enhances Cu(2+) binding to phosphatidylserine lipids. *Journal of the American Chemical Society*, *137*(24), 7785–7792. http://dx.doi.org/10.1021/jacs.5b03313.

Corbett, D., Schuler, S., Glenn, S., Andrew, P. W., Cavet, J. S., & Roberts, I. S. (2011). The combined actions of the copper-responsive repressor CsoR and copper- metallochaperone CopZ modulate CopA-mediated copper efflux in the intracellular pathogen *Listeria monocytogenes*. *Molecular Microbiology*, *81*(2), 457–472. http://dx.doi.org/10.1111/j.1365-2958.2011.07705.x.

Crichton, R. R., & Pierre, J. L. (2001). Old iron, young copper: From Mars to Venus. *Biometals*, *14*(2), 99–112.

Culotta, V. C., Yang, M., & O'Halloran, T. V. (2006). Activation of superoxide dismutases: Putting the metal to the pedal. *Biochimica et Biophysica Acta*, *1763*(7), 747–758. http://dx.doi.org/10.1016/j.bbamcr.2006.05.003.

Cunningham, R., Hurford, J. V., Erwin, G. S., Nagley, M. M., & Yell, W. L. (1951). Thiosemicarbazone in chemotherapy of tuberculosis. *Tubercle*, *32*(10), 217–222. http://dx.doi.org/10.1016/S0041-3879(51)80018-7.

Dalecki, A. G., Haeili, M., Shah, S., Speer, A., Niederweis, M., Kutsch, O., et al. (2015). Disulfiram and copper ions kill *Mycobacterium tuberculosis* in a synergistic manner. *Antimicrobial Agents and Chemotherapy*, *59*, 4835–4844. http://dx.doi.org/10.1128/aac.00692-15.

Dalecki, A. G., Malalasekera, A. P., Schaaf, K., Kutsch, O., Bossmann, S. H., & Wolschendorf, F. (2016). Combinatorial phenotypic screen uncovers unrecognized family of extended thiourea inhibitors with copper-dependent anti-staphylococcal activity. *Metallomics*, *8*, 412–421. http://dx.doi.org/10.1039/c6mt00003g.

Dalecki, A. G., & Wolschendorf, F. (2016). Development of a web-based tool for automated processing and cataloging of a unique combinatorial drug screen. *Journal of Microbiological Methods*, *126*, 30–34. http://dx.doi.org/10.1016/j.mimet.2016.04.013.

Dalle-Donne, I., Aldini, G., Carini, M., Colombo, R., Rossi, R., & Milzani, A. (2006). Protein carbonylation, cellular dysfunction, and disease progression. *Journal of Cellular and Molecular Medicine*, *10*(2), 389–406. http://dx.doi.org/10.1111/j.1582-4934.2006.tb00407.x.

Dalle-Donne, I., Rossi, R., Giustarini, D., Milzani, A., & Colombo, R. (2003). Protein carbonyl groups as biomarkers of oxidative stress. *Clinica Chimica Acta*, *329*(1–2), 23–38. http://dx.doi.org/10.1016/S0009-8981(03)00003-2.

Daltrop, O., Stevens, J. M., Higham, C. W., & Ferguson, S. J. (2002). The CcmE protein of the C-type cytochrome biogenesis system: Unusual in vitro heme incorporation into Apo-CcmE and transfer from holo-CcmE to apocytochrome. *Proceedings of the National Academy of Sciences of the United States of America*, *99*(15), 9703–9708. http://dx.doi.org/10.1073/pnas.152120699.

Darwin, K. H. (2015). *Mycobacterium tuberculosis* and copper: A newly appreciated defense against an old foe? *The Journal of Biological Chemistry, 290*, 18962–18966. http://dx.doi.org/10.1074/jbc.R115.640193.

Demain, A. L. (2002). Prescription for an ailing pharmaceutical industry. *Nature Biotechnology, 20*(4), 331. http://dx.doi.org/10.1038/nbt0402-331.

Denoyer, D., Masaldan, S., La Fontaine, S., & Cater, M. A. (2015). Targeting copper in cancer therapy: 'Copper that cancer'. *Metallomics, 7*, 1459–1476. http://dx.doi.org/10.1039/c5mt00149h.

Djoko, K. Y., Franiek, J. A., Edwards, J. L., Falsetta, M. L., Kidd, S. P., Potter, A. J., et al. (2012). Phenotypic characterization of a CopA mutant of *Neisseria gonorrhoeae* identifies a link between copper and nitrosative stress. *Infection and Immunity, 80*(3), 1065–1071. http://dx.doi.org/10.1128/IAI.06163-11.

Djoko, K. Y., Goytia, M. M., Donnelly, P. S., Schembri, M. A., Shafer, W. M., & McEwan, A. G. (2015). Copper(II)-bis(thiosemicarbazonato) complexes as antibacterial agents: Insights into their mode of action and potential as therapeutics. *Antimicrobial Agents and Chemotherapy, 59*, 6444–6453. http://dx.doi.org/10.1128/AAC.01289-15.

Djoko, K. Y., & McEwan, A. G. (2013). Antimicrobial action of copper is amplified via inhibition of heme biosynthesis. *ACS Chemical Biology, 8*(10), 2217–2223. http://dx.doi.org/10.1021/cb4002443.

Djoko, K. Y., Ong, C. L., Walker, M. J., & McEwan, A. G. (2015). The role of copper and zinc toxicity in innate immune defense against bacterial pathogens. *The Journal of Biological Chemistry, 290*(31), 18954–18961. http://dx.doi.org/10.1074/jbc.R115.647099.

Djoko, K. Y., Paterson, B. M., Donnelly, P. S., & McEwan, A. G. (2014). Antimicrobial effects of copper(II) bis(thiosemicarbazonato) complexes provide new insight into their biochemical mode of action. *Metallomics, 6*, 854–863. http://dx.doi.org/10.1039/c3mt00348e.

Dupont, C. L., Grass, G., & Rensing, C. (2011). Copper toxicity and the origin of bacterial resistance—New insights and applications. *Metallomics, 3*(11), 1109–1118. http://dx.doi.org/10.1039/c1mt00107h.

Durand, A., Azzouzi, A., Bourbon, M. L., Steunou, A. S., Liotenberg, S., Maeshima, A., et al. (2015). C-type cytochrome assembly is a key target of copper toxicity within the bacterial periplasm. *mBio, 6*(5), E01007–E01015. http://dx.doi.org/10.1128/mBio.01007-15.

Dwyer, F. P., Reid, I. K., Shulman, A., Laycock, G. M., & Dixson, S. (1969). The biological actions of 1,10-phenanthroline and 2,2'-bipyridine hydrochlorides, quaternary salts and metal chelates and related compounds. 1. Bacteriostatic action on selected gram-positive, gram-negative and acid-fast bacteria. *The Australian Journal of Experimental Biology and Medical Science, 47*(2), 203–218.

Ekici, S., Pawlik, G., Lohmeyer, E., Koch, H. G., & Daldal, F. (2012). Biogenesis of cbb(3)-type cytochrome C oxidase in *Rhodobacter capsulatus*. *Biochimica et Biophysica Acta, 1817*(6), 898–910. http://dx.doi.org/10.1016/j.bbabio.2011.10.011.

Ekici, S., Yang, H., Koch, H. G., & Daldal, F. (2012). Novel transporter required for biogenesis of Cbb3-type cytochrome C oxidase in *Rhodobacter capsulatus*. *mBio, 3*(1), e00293-11. http://dx.doi.org/10.1128/mBio.00293-11.

Eliopoulos, G. M., Thauvin, C., Gerson, B., & Moellering, R. C. (1985). In vitro activity and mechanism of action of A21978c1, a novel cyclic lipopeptide antibiotic. *Antimicrobial Agents and Chemotherapy, 27*(3), 357–362. http://dx.doi.org/10.1128/AAC.27.3.357.

Elzanowska, H., Wolcott, R. G., Hannum, D. M., & Hurst, J. K. (1995). Bactericidal properties of hydrogen peroxide and copper or iron-containing complex ions in relation to leukocyte function. *Free Radical Biology & Medicine, 18*(3), 437–449. http://dx.doi.org/10.1016/0891-5849(94)00150-I.

Eneanya, D. I., Bianchine, J. R., Duran, D. O., & Andresen, B. D. (1981). The actions and metabolic fate of disulfiram. *Annual Review of Pharmacology and Toxicology*, *21*(1), 575–596. http://dx.doi.org/10.1146/annurev.pa.21.040181.003043.

Ermolayeva, E., & Sanders, D. (1995). Mechanism of pyrithione-induced membrane depolarization in *Neurospora crassa*. *Applied and Environmental Microbiology*, *61*(9), 3385–3390.

Faiman, M. D., Jensen, J. C., & Lacoursiere, R. B. (1984). Elimination kinetics of disulfiram in alcoholics after single and repeated doses. *Clinical Pharmacology and Therapeutics*, *36*(4), 520–526.

Fair, R. J., & Tor, Y. (2014). Antibiotics and bacterial resistance in the 21st century. *Perspectives in Medicinal Chemistry*, *6*, 25–64. http://dx.doi.org/10.4137/PMC.S14459.

Fang, L., Wei, X., Cai, P., Huang, Q., Chen, H., Liang, W., et al. (2011). Role of extracellular polymeric substances in Cu(Ii) adsorption on *Bacillus subtilis* and *Pseudomonas putida*. *Bioresource Technology*, *102*(2), 1137–1141. http://dx.doi.org/10.1016/j.biortech.2010.09.006.

Farha, M. A., & Brown, E. D. (2015). Unconventional screening approaches for antibiotic discovery. *Annals of the New York Academy of Sciences*, *1354*(1), 54–66. http://dx.doi.org/10.1111/nyas.12803.

Fatfat, M., Merhi, R. A., Rahal, O., Stoyanovsky, D. A., Zaki, A., Haidar, H., et al. (2014). Copper chelation selectively kills colon cancer cells through redox cycling and generation of reactive oxygen species. *BMC Cancer*, *14*, 527. http://dx.doi.org/10.1186/1471-2407-14-527.

Feeney, R. E., Petersen, I. M., & Sahinkaya, H. (1957). Liesegang-like rings of growth and inhibition of bacteria in agar caused by metal ions and chelating agents. *Journal of Bacteriology*, *73*(2), 284–290.

Festa, R. A., Helsel, M. E., Franz, K. J., & Thiele, D. J. (2014). Exploiting innate immune cell activation of a copper-dependent antimicrobial agent during infection. *Chemistry & Biology*, *21*(8), 977–987. http://dx.doi.org/10.1016/j.chembiol.2014.06.009.

Festa, R. A., Jones, M. B., Butler-Wu, S., Sinsimer, D., Gerads, R., Bishai, W. R., et al. (2011). A novel copper-responsive regulon in *Mycobacterium tuberculosis*. *Molecular Microbiology*, *79*(1), 133–148. http://dx.doi.org/10.1111/j.1365-2958.2010.07431.x.

Festa, R. A., & Thiele, D. J. (2012). Copper at the front line of the host-pathogen battle. *PLoS Pathogens*, *8*(9), e1002887. http://dx.doi.org/10.1371/journal.ppat.1002887.

Finney, J., Moon, H. J., Ronnebaum, T., Lantz, M., & Mure, M. (2014). Human copper-dependent amine oxidases. *Archives of Biochemistry and Biophysics*, *546*, 19–32. http://dx.doi.org/10.1016/j.abb.2013.12.022.

Fischbach, M. A., & Walsh, C. T. (2009). Antibiotics for emerging pathogens. *Science*, *325*(5944), 1089–1093. http://dx.doi.org/10.1126/science.1176667.

Flannagan, R. S., Cosio, G., & Grinstein, S. (2009). Antimicrobial mechanisms of phagocytes and bacterial evasion strategies. *Nature Reviews Microbiology*, 7(5), 355–366.

Fleischmann, R. D., Adams, M. D., White, O., Clayton, R. A., Kirkness, E. F., Kerlavage, A. R., et al. (1995). Whole-genome random sequencing and assembly of *Haemophilus influenzae* Rd. *Science*, *269*(5223), 496–512. http://dx.doi.org/10.1126/science.7542800.

Foster, A. W., Osman, D., & Robinson, N. J. (2014). Metal preferences and metallation. *The Journal of Biological Chemistry*, *289*(41), 28095–28103. http://dx.doi.org/10.1074/jbc.R114.588145.

Foster, A. W., & Robinson, N. J. (2011). Promiscuity and preferences of metallothioneins: The cell rules. *BMC Biology*, *9*, 25. http://dx.doi.org/10.1186/1741-7007-9-25.

Francis, M. S., & Thomas, C. J. (1997). Mutants in the CtpA copper transporting P-type ATPase reduce virulence of *Listeria monocytogenes*. *Microbial Pathogenesis*, *22*(2), 67–78. http://dx.doi.org/10.1006/mpat.1996.0092.

Fraústo Da Silva, J. J. R., & Williams, R. J. P. (2001). *The biological chemistry of the elements: The inorganic chemistry of life* (2nd ed.). New York, USA: Oxford University Press.

Frees, D., Savijoki, K., Varmanen, P., & Ingmer, H. (2007). Clp ATPases and ClpP proteolytic complexes regulate vital biological processes in low GC, gram-positive bacteria. *Molecular Microbiology, 63*(5), 1285–1295. http://dx.doi.org/10.1111/j.1365-2958.2007.05598.x.

Freinbichler, W., Colivicchi, M. A., Stefanini, C., Bianchi, L., Ballini, C., Misini, B., et al. (2011). Highly reactive oxygen species: Detection, formation, and possible functions. *Cellular and Molecular Life Sciences, 68*(12), 2067–2079. http://dx.doi.org/10.1007/s00018-011-0682-x.

French, F. A., & Freedlander, B. L. (1958). Carcinostatic action of polycarbonyl compounds and their derivatives. IV. Glyoxal bis (thiosemicarbazone) and derivatives. *Cancer Research, 18*(11), 1290–1300.

Fu, Y., Chang, F. M., & Giedroc, D. P. (2014). Copper transport and trafficking at the host-bacterial pathogen interface. *Accounts of Chemical Research, 47*(12), 3605–3613. http://dx.doi.org/10.1021/ar500300n.

Fu, S., Naing, A., Fu, C., Kuo, M. T., & Kurzrock, R. (2012). Overcoming platinum resistance through the use of a copper-lowering agent. *Molecular Cancer Therapeutics, 11*(6), 1221–1225. http://dx.doi.org/10.1158/1535-7163.MCT-11-0864.

Fu, Y., Tsui, H. T., Bruce, K. E., Sham, L.-T., Higgins, K. A., Lisher, J. P., et al. (2013). A new structural paradigm in copper resistance in *Streptococcus pneumoniae*. *Nature Chemical Biology, 9*(3), 177–183. http://dx.doi.org/10.1038/nchembio.1168.

Fung, D. K., Lau, W. Y., Chan, W. T., & Yan, A. (2013). Copper efflux is induced during anaerobic amino acid limitation in *Escherichia coli* to protect iron-sulfur cluster enzymes and biogenesis. *Journal of Bacteriology, 195*(20), 4556–4568. http://dx.doi.org/10.1128/JB.00543-13.

Garber, K. (2015a). Cancer's copper connections. *Science, 349*(6244), 129. http://dx.doi.org/10.1126/science.349.6244.129.

Garber, K. (2015b). Targeting copper to treat breast cancer. *Science Progress, 349*, 128–129.

Gehman, J. D., O'Brien, C. C., Shabanpoor, F., Wade, J. D., & Separovic, F. (2008). Metal effects on the membrane interactions of amyloid-beta peptides. *European Biophysics Journal, 37*(3), 333–344. http://dx.doi.org/10.1007/s00249-007-0251-2.

Ghssein, G., Brutesco, C., Ouerdane, L., Fojcik, C., Izaute, A., Wang, S., et al. (2016). Biosynthesis of a broad-spectrum nicotianamine-like metallophore in *Staphylococcus aureus*. *Science, 352*(6289), 1105–1109. http://dx.doi.org/10.1126/science.aaf1018.

Giorgio, M., Trinei, M., Migliaccio, E., & Pelicci, P. G. (2007). Hydrogen peroxide: A metabolic by-product or a common mediator of ageing signals? *Nature Reviews Molecular Cell Biology, 8*(9), 722–728.

Gold, B., Deng, H., Bryk, R., Vargas, D., Eliezer, D., Roberts, J., et al. (2008). Identification of a copper-binding metallothionein in pathogenic mycobacteria. *Nature Chemical Biology, 4*(10), 609–616. http://dx.doi.org/10.1038/nchembio.109.

González, A. G., Shirokova, L. S., Pokrovsky, O. S., Emnova, E. E., Martínez, R. E., Santana-Casiano, J. M., et al. (2010). Adsorption of copper on *Pseudomonas aureofaciens*: Protective role of surface exopolysaccharides. *Journal of Colloid and Interface Science, 350*(1), 305–314. http://dx.doi.org/10.1016/j.jcis.2010.06.020.

Gonzalez-Guerrero, M., Raimunda, D., Cheng, X., & Arguello, J. M. (2010). Distinct functional roles of homologous Cu+ efflux ATPases in *Pseudomonas aeruginosa*. *Molecular Microbiology, 78*(5), 1246–1258. http://dx.doi.org/10.1111/j.1365-2958.2010.07402.x.

Gordon, R. M., & Seaton, D. R. (1942). Observations on the treatment of scabies. *British Medical Journal, 1*(4248), 685–687.

Grass, G., & Rensing, C. (2001). CueO is a multi-copper oxidase that confers copper tolerance in *Escherichia coli*. *Biochemical and Biophysical Research Communications*, *286*(5), 902–908. http://dx.doi.org/10.1006/bbrc.2001.5474.

Grass, G., Thakali, K., Klebba, P. E., Thieme, D., Muller, A., Wildner, G. F., et al. (2004). Linkage between catecholate siderophores and the multicopper oxidase CueO in *Escherichia coli*. *Journal of Bacteriology*, *186*(17), 5826–5833. http://dx.doi.org/10.1128/JB.186.17.5826-5833.2004.

Guilhelmelli, F., Vilela, N., Albuquerque, P., Derengowski, L. D. S., Silva-Pereira, I., & Kyaw, C. M. (2013). Antibiotic development challenges: The various mechanisms of action of antimicrobial peptides and of bacterial resistance. *Frontiers in Microbiology*, *4*, 353. http://dx.doi.org/10.3389/fmicb.2013.00353.

Gyawali, R., Ibrahim, S. A., Abu Hasfa, S. H., Smqadri, S. Q., & Haik, Y. (2011). Antimicrobial activity of copper alone and in combination with lactic acid against *Escherichia coli* O157:H7 in laboratory medium and on the surface of lettuce and tomatoes. *Journal of Pathogens*, *2011*, 650968. http://dx.doi.org/10.4061/2011/650968.

Haas, K. L., & Franz, K. J. (2009). Application of metal coordination chemistry to explore and manipulate cell biology. *Chemical Reviews*, *109*(10), 4921–4960. http://dx.doi.org/10.1021/cr900134a.

Haeili, M., Moore, C., Davis, C. J. C., Cochran, J. B., Shah, S., Shrestha, T. B., et al. (2014). Copper complexation screen reveals compounds with potent antibiotic properties against methicillin-resistant *Staphylococcus aureus*. *Antimicrobial Agents and Chemotherapy*, *58*(7), 3727–3736. http://dx.doi.org/10.1128/AAC.02316-13.

Haeili, M., Speer, A., Rowland, J. L., Niederweis, M., & Wolschendorf, F. (2015). The role of porins in copper acquisition by mycobacteria. *International Journal Of Mycobacteriology*, *4*, 91–92. http://dx.doi.org/10.1016/j.ijmyco.2014.11.052.

Helbig, K., Bleuel, C., Krauss, G. J., & Nies, D. H. (2008). Glutathione and transition-metal homeostasis in *Escherichia coli*. *Journal of Bacteriology*, *190*(15), 5431–5438. http://dx.doi.org/10.1128/JB.00271-08.

Helsel, M. E., White, E. J., Razvi, S. Z., Alies, B., & Franz, K. J. (2017). Chemical and functional properties of metal chelators that mobilize copper to elicit fungal killing of *Cryptococcus neoformans*. *Metallomics*, *9*, 69–81. http://dx.doi.org/10.1039/c6mt00172f.

Hernandez-Montes, G., Arguello, J. M., & Valderrama, B. (2012). Evolution and diversity of periplasmic proteins involved in copper homeostasis in gamma proteobacteria. *BMC Microbiology*, *12*, 249. http://dx.doi.org/10.1186/1471-2180-12-249.

Hill, D. J., Griffiths, N. J., Borodina, E., & Virji, M. (2010). Cellular and molecular biology of *Neisseria meningitidis* colonization and invasive disease. *Clinical Science (London, England)*, *118*(9), 547–564. http://dx.doi.org/10.1042/CS20090513.

Hiniker, A., Collet, J. F., & Bardwell, J. C. (2005). Copper stress causes an in vivo requirement for the *Escherichia coli* disulfide isomerase DsbC. *The Journal of Biological Chemistry*, *280*(40), 33785–33791. http://dx.doi.org/10.1074/jbc.m505742200.

Hodgkinson, V., & Petris, M. J. (2012). Copper homeostasis at the host-pathogen interface. *The Journal of Biological Chemistry*, *287*(17), 13549–13555. http://dx.doi.org/10.1074/jbc.R111.316406.

Holm, R. H., Kennepohl, P., & Solomon, E. I. (1996). Structural and functional aspects of metal sites in biology. *Chemical Reviews*, *96*(7), 2239–2314.

Hong, R., Kang, T. Y., Michels, C. A., & Gadura, N. (2012). Membrane lipid peroxidation in copper alloy-mediated contact killing of *Escherichia coli*. *Applied and Environmental Microbiology*, *78*(6), 1776–1784. http://dx.doi.org/10.1128/AEM.07068-11.

Hongmanee, P., Rukseree, K., Buabut, B., Somsri, B., & Palittapongarnpim, P. (2007). *In vitro* activities of cloxyquin (5-chloroquinolin-8-Ol) against *Mycobacterium tuberculosis*.

*Antimicrobial Agents and Chemotherapy*, *51*(3), 1105–1106. http://dx.doi.org/10.1128/AAC.01310-06.
Hood, M. I., & Skaar, E. P. (2012). Nutritional immunity: Transition metals at the pathogen-host interface. *Nature Reviews Microbiology*, *10*(8), 525–537. http://dx.doi.org/10.1038/nrmicro2836.
Horita, Y., Takii, T., Yagi, T., Ogawa, K., Fujiwara, N., Inagaki, E., et al. (2012). Antitubercular activity of disulfiram, an antialcoholism drug, against multidrug- and extensively drug-resistant *Mycobacterium tuberculosis* isolates. *Antimicrobial Agents and Chemotherapy*, *56*(8), 4140–4145. http://dx.doi.org/10.1128/AAC.06445-11.
Hubner, L., Ernst, M., Von Laer, D., Schwander, S., & Flad, H. D. (1991). Enhancement of monocyte antimycobacterial activity by diethyldithiocarbamate (DTC). *International Journal of Immunopharmacology*, *13*(8), 1067–1072.
Huckle, J. W., Morby, A. P., Turner, J. S., & Robinson, N. J. (1993). Isolation of a prokaryotic metallothionein locus and analysis of transcriptional control by trace metal ions. *Molecular Microbiology*, *7*(2), 177–187.
Hughes, J. P., Rees, S., Kalindjian, S. B., & Philpott, K. L. (2011). Principles of early drug discovery. *British Journal of Pharmacology*, *162*(6), 1239–1249. http://dx.doi.org/10.1111/j.1476-5381.2010.01127.x.
Hung, L. W., Villemagne, V. L., Cheng, L., Sherratt, N. A., Ayton, S., White, A. R., et al. (2012). The hypoxia imaging agent CuII(atsm) is neuroprotective and improves motor and cognitive functions in multiple animal models of Parkinson's disease. *The Journal of Experimental Medicine*, *209*(4), 837–854. http://dx.doi.org/10.1084/jem.20112285.
Imlay, J. A. (2013). The molecular mechanisms and physiological consequences of oxidative stress: Lessons from a model bacterium. *Nature Reviews Microbiology*, *11*(7), 443–454. http://dx.doi.org/10.1038/nrmicro3032.
Irving, H., & Williams, R. J. P. (1953). 637. The stability of transition-metal complexes. *Journal of the Chemical Society*, 3192–3210. http://dx.doi.org/10.1039/JR9530003192.
Jahn, H. A., & Teller, E. (1937). Stability of polyatomic molecules in degenerate electronic states. I. Orbital degeneracy. *Proceedings of the Royal Society A*, *161*(905), 220–235. http://dx.doi.org/10.1098/rspa.1937.0142.
Jastrab, J. B., & Darwin, K. H. (2015). Bacterial proteasomes. *Annual Review of Microbiology*, *69*, 109–127. http://dx.doi.org/10.1146/annurev-micro-091014-104201.
Jessee, B., Gargiulo, G., Razvi, F., & Worcel, A. (1982). Analogous cleavage of DNA by micrococcal nuclease and a 1-10-phenanthroline-cuprous complex. *Nucleic Acids Research*, *10*(19), 5823–5834.
Jeżowska-Bojczuk, M., Bal, W., & Kozłowski, H. (1998). Kanamycin revisited: A combined potentiometric and spectroscopic study of copper(Ii) binding to kanamycin B. *Inorganica Chimica Acta*, *275*, 541–545. http://dx.doi.org/10.1016/S0020-1693(97)06113-6.
Jezowska-Bojczuk, M., Szczepanik, W., Lesniak, W., Ciesiolka, J., Wrzesinski, J., & Bal, W. (2002). DNA and RNA damage by Cu(II)-amikacin complex. *European Journal of Biochemistry*, *269*(22), 5547–5556.
Johansson, B. (1992). A review of the pharmacokinetics and pharmacodynamics of disulfiram and its metabolites. *Acta Psychiatrica Scandinavica Supplementum*, *369*, 15–26.
Johnson, M. D. L., Kehl-Fie, T. E., & Rosch, J. W. (2015). Copper intoxication inhibits aerobic nucleotide synthesis in *Streptococcus pneumoniae*. *Metallomics*, 7, 786–794. http://dx.doi.org/10.1039/c5mt00011d.
Kachur, A. V., Koch, C. J., & Biaglow, J. E. (1999). Mechanism of copper-catalyzed autoxidation of cysteine. *Free Radical Research*, *31*(1), 23–34.
Kamau, P., & Jordan, R. B. (2002). Kinetic study of the oxidation of catechol by aqueous copper(II). *Inorganic Chemistry*, *41*(12), 3076–3083.

Kehl-Fie, T. E., & Skaar, E. P. (2010). Nutritional immunity beyond iron: A role for manganese and zinc. *Current Opinion in Chemical Biology*, *14*(2), 218–224. http://dx.doi.org/10.1016/j.cbpa.2009.11.008.

Kehrer, J. P. (2000). The Haber-Weiss reaction and mechanisms of toxicity. *Toxicology*, *149*(1), 43–50.

Kershaw, C. J., Brown, N. L., Constantinidou, C., Patel, M. D., & Hobman, J. L. (2005). The expression profile of *Escherichia coli* K-12 in response to minimal, optimal and excess copper concentrations. *Microbiology*, *151*(Pt. 4), 1187–1198. http://dx.doi.org/10.1099/mic.0.27650-0.

Khalfaoui-Hassani, B., Verissimo, A. F., Koch, H. G., & Daldal, F. (2016). Uncovering the transmembrane metal binding site of the novel bacterial major facilitator superfamily-type copper importer CcoA. *mBio*, 7(1), e01981-01915. http://dx.doi.org/10.1128/mBio.01981-15.

Kidd, S. P., Potter, A. J., Apicella, M. A., Jennings, M. P., & McEwan, A. G. (2005). NmlR of *Neisseria gonorrhoeae*: A novel redox responsive transcription factor from the MerR family. *Molecular Microbiology*, *57*(6), 1676–1689. http://dx.doi.org/10.1111/j.1365-2958.2005.04773.x.

Kochanek, K. D., Murphy, S. L., Xu, J., & Tejada-Vera, B. (2016). Deaths: Final data for 2014. *National Vital Statistics Reports*, *65*(4), 1–122.

Koppaka, V., Thompson, D. C., Chen, Y., Ellermann, M., Nicolaou, K. C., Juvonen, R. O., et al. (2012). Aldehyde dehydrogenase inhibitors: A comprehensive review of the pharmacology, mechanism of action, substrate specificity, and clinical application. *Pharmacological Reviews*, *64*(3), 520–539. http://dx.doi.org/10.1124/pr.111.005538.

Korshunov, S., & Imlay, J. A. (2010). Two sources of endogenous hydrogen peroxide in *Escherichia coli*. *Molecular Microbiology*, *75*(6), 1389–1401. http://dx.doi.org/10.1111/j.1365-2958.2010.07059.x.

Kosugi, H., & Kikugawa, K. (1985). Thiobarbituric acid reaction of aldehydes and oxidized lipids in glacial acetic acid. *Lipids*, *20*(12), 915–921. http://dx.doi.org/10.1007/BF02534777.

Kragh, H. (2008). From disulfiram to antabuse: The invention of a drug. *Bulletin for the History of Chemistry*, *33*(2), 82–88.

Krenn, B. M., Gaudernak, E., Holzer, B., Lanke, K., Van Kuppeveld, F. J., & Seipelt, J. (2009). Antiviral activity of the zinc ionophores pyrithione and hinokitiol against picornavirus infections. *Journal of Virology*, *83*(1), 58–64. http://dx.doi.org/10.1128/JVI.01543-08.

Ladomersky, E., & Petris, M. J. (2015). Copper tolerance and virulence in bacteria. *Metallomics*, 7(6), 957–964. http://dx.doi.org/10.1039/c4mt00327f.

Lau, T. L., Ambroggio, E. E., Tew, D. J., Cappai, R., Masters, C. L., Fidelio, G. D., et al. (2006). Amyloid-beta peptide disruption of lipid membranes and the effect of metal ions. *Journal of Molecular Biology*, *356*(3), 759–770. http://dx.doi.org/10.1016/j.jmb.2005.11.091.

Lebrette, H., Borezee-Durant, E., Martin, L., Richaud, P., Boeri Erba, E., & Cavazza, C. (2015). Novel insights into nickel import in *Staphylococcus aureus*: The positive role of free histidine and structural characterization of a new thiazolidine-type nickel chelator. *Metallomics*, 7(4), 613–621. http://dx.doi.org/10.1039/c4mt00295d.

Lee, S. M., Grass, G., Rensing, C., Barrett, S. R., Yates, C. J. D., Stoyanov, J. V., et al. (2002). The Pco proteins are involved in periplasmic copper handling in *Escherichia coli*. *Biochemical and Biophysical Research Communications*, *295*(3), 616–620. http://dx.doi.org/10.1016/S0006-291X(02)00726-X.

Lee, H. J., Reimann, J., Huang, Y., & Adelroth, P. (2012). Functional proton transfer pathways in the heme-copper oxidase superfamily. *Biochimica et Biophysica Acta*, *1817*(4), 537–544. http://dx.doi.org/10.1016/j.bbabio.2011.10.007.

Leid, J. G., Willson, C. J., Shirtliff, M. E., Hassett, D. J., Parsek, M. R., & Jeffers, A. K. (2005). The exopolysaccharide alginate protects *Pseudomonas aeruginosa* biofilm bacteria from IFN-Γ-mediated macrophage killing. *Journal of Immunology*, *175*(11), 7512–7518. http://dx.doi.org/10.4049/jimmunol.175.11.7512.
Lesniak, W., Harris, W. R., Kravitz, J. Y., Schacht, J., & Pecoraro, V. L. (2003). Solution chemistry of copper(II)-gentamicin complexes: Relevance to metal-related aminoglycoside toxicity. *Inorganic Chemistry*, *42*(5), 1420–1429. http://dx.doi.org/10.1021/ic025965t.
Lewis, K. (2013). Platforms for antibiotic discovery. *Nature Reviews Drug Discovery*, *12*(5), 371–387. http://dx.doi.org/10.1038/nrd3975.
Lewis, D. J., Deshmukh, P., Tedstone, A. A., Tuna, F., & O'Brien, P. (2014). On the interaction of copper(II) with disulfiram. *Chemical Communications*, *50*(87), 13334–13337. http://dx.doi.org/10.1039/c4cc04767b.
Liebermeister, V. K. (1950). Zum Wirkungsprinzip schwefelhaltiger tuberkulostatischer chemotherapeutica. *Zeitschrift für Naturforschung Teil B*, *5*(2), 79–86. http://dx.doi.org/10.1515/znb-1950-0203.
Linder, M. C. (2012). The relationship of copper to Dna damage and damage prevention in humans. *Mutation Research*, *733*(1-2), 83–91.
Linder, F., & Grove, R. (1947). *Vital statistics rates in the united states 1900–1940*. Washington, DC, USA: United States Government Printing Office.
Ling, L. L., Schneider, T., Peoples, A. J., Spoering, A. L., Engels, I., Conlon, B. P., et al. (2015). A new antibiotic kills pathogens without detectable resistance. *Nature*, *517*(7535), 455–459. http://dx.doi.org/10.1038/nature14098.
Lisher, J. P., & Giedroc, D. P. (2013). Manganese acquisition and homeostasis at the host-pathogen interface. *Frontiers in Cellular and Infection Microbiology*, *3*, 91. http://dx.doi.org/10.3389/fcimb.2013.00091.
Liu, T., Ramesh, A., Ma, Z., Ward, S. K., Zhang, L., George, G. N., et al. (2007). CsoR is a novel *Mycobacterium tuberculosis* copper-sensing transcriptional regulator. *Nature Chemical Biology*, *3*(1), 60–68. http://dx.doi.org/10.1038/nchembio844.
Livermore, D. M. (2011). Discovery research: The scientific challenge of finding new antibiotics. *The Journal of Antimicrobial Chemotherapy*, *66*(9), 1941–1944. http://dx.doi.org/10.1093/jac/dkr262.
Lu, Y. (2010). Metal ions as matchmakers for proteins. *Proceedings of the National Academy of Sciences of the United States of America*, *107*(5), 1811–1812. http://dx.doi.org/10.1073/pnas.0914008107.
Luc, R., & Vergely, C. (2008). Forgotten radicals in biology. *International Journal of Biomedical Sciences*, *4*(4), 255–259.
Lutkenhaus, J. F. (1977). Role of a major outer membrane protein in *Escherichia coli*. *Journal of Bacteriology*, *131*(2), 631–637.
Lutsenko, S. (2010). Human copper homeostasis: A network of interconnected pathways. *Current Opinion in Chemical Biology*, *14*(2), 211–217. http://dx.doi.org/10.1016/j.cbpa.2010.01.003.
Ma, H. H., Chiu, F. C., & Li, R. C. (1997). Mechanistic investigation of the reduction in antimicrobial activity of ciprofloxacin by metal cations. *Pharmaceutical Research*, *14*(3), 366–370.
Macomber, L., & Imlay, J. A. (2009). The iron-sulfur clusters of dehydratases are primary intracellular targets of copper toxicity. *Proceedings of the National Academy of Sciences of the United States of America*, *106*(20), 8344–8349. http://dx.doi.org/10.1073/pnas.0812808106.
Macomber, L., Rensing, C., & Imlay, J. A. (2007). Intracellular copper does not catalyze the formation of oxidative DNA damage in *Escherichia coli*. *Journal of Bacteriology*, *189*(5), 1616–1626. http://dx.doi.org/10.1128/JB.01357-06.

Magda, D., Lecane, P., Wang, Z., Hu, W., Thiemann, P., Ma, X., et al. (2008). Synthesis and anticancer properties of water-soluble zinc ionophores. *Cancer Research*, *68*(13), 5318–5325. http://dx.doi.org/10.1158/0008-5472.can-08-0601.

Makula, R. A., Lockwood, P. J., & Finnerty, W. R. (1975). Comparative analysis of the lipids of *Acinetobacter* species grown on hexadecane. *Journal of Bacteriology*, *121*(1), 250–258.

Manzanillo, P. S., Shiloh, M. U., Portnoy, D. A., & Cox, J. S. (2012). *Mycobacterium tuberculosis* activates the DNA-dependent cytosolic surveillance pathway within macrophages. *Cell Host & Microbe*, *11*(5), 469–480. http://dx.doi.org/10.1016/j.chom.2012.03.007.

Maresca, A., Carta, F., Vullo, D., & Supuran, C. T. (2013). Dithiocarbamates strongly inhibit the beta-class carbonic anhydrases from *Mycobacterium tuberculosis*. *Journal of Enzyme Inhibition and Medicinal Chemistry*, *28*(2), 407–411. http://dx.doi.org/10.3109/14756366.2011.641015.

Marshall, L. E., Graham, D. R., Reich, K. A., & Sigman, D. S. (1981). Cleavage of deoxyribonucleic acid by the 1,10-phenanthroline-cuprous complex. Hydrogen peroxide requirement and primary and secondary structure specificity. *Biochemistry*, *20*(2), 244–250.

Maryon, E. B., Molloy, S. A., & Kaplan, J. H. (2013). Cellular glutathione plays a key role in copper uptake mediated by human copper transporter 1. *American Journal of Physiology Cell Physiology*, *304*(8), C768–C779. http://dx.doi.org/10.1152/ajpcell.00417.2012.

Marzano, C., Pellei, M., Tisato, F., & Santini, C. (2009). Copper complexes as anticancer agents. *Anti-Cancer Agents in Medicinal Chemistry*, *9*(2), 185–211.

Matuszewska, E., Kwiatkowska, J., Kuczyńska-Wiśnik, D., & Laskowska, E. (2008). *Escherichia coli* heat-shock proteins IbpA/B are involved in resistance to oxidative stress induced by copper. *Microbiology*, *154*(6), 1739–1747. http://dx.doi.org/10.1099/mic.0.2007/014696-0.

Mcelroy, J. (1910). The treatment of pulmonary tuberculosis by intravenous injections of chinosol with formaldehyde. *The Lancet*, *176*(4550), 1408–1409. http://dx.doi.org/10.1016/s0140-6736(01)08447-1.

McPherson, M. J., Parsons, M. R., & Wilmot, C. M. (2006). Prokaryotic copper amine oxidases. *Handbook of Metalloproteins, 6*. http://dx.doi.org/10.1002/0470028637.met191.

Medina-Franco, J. L., Giulianotti, M. A., Welmaker, G. S., & Houghten, R. A. (2013). Shifting from the single- to the multitarget paradigm in drug discovery. *Drug Discovery Today*, *18*, 495–501. http://dx.doi.org/10.1016/j.drudis.2013.01.008.

Mertens, A., & Bunge, R. (1950). The present status of the chemotherapy of tuberculosis with conteben a substance of the thiosemicarbazone series: A review. *American Review of Tuberculosis*, *61*(1), 20–38.

Meyer, T. J., Ramlall, J., Thu, P., & Gadura, N. (2015). Antimicrobial properties of copper in gram-negative and gram-positive bacteria. *International Journal of Biological, Biomolecular, Agricultural, Food and Biotechnological Engineering*, *9*, 273–278.

Mitchell, A. M., & Mitchell, T. J. (2010). Streptococcus pneumoniae: Virulence factors and variation. *Clinical Microbiology and Infection*, *16*(5), 411–418. http://dx.doi.org/10.1111/j.1469-0691.2010.03183.x.

Mohindru, A., Fisher, J. M., & Rabinovitz, M. (1983a). 2,9-Dimethyl-1,10-phenanthroline (neocuproine): A potent, copper-dependent cytotoxin with anti-tumor activity. *Biochemical Pharmacology*, *32*(23), 3627–3632.

Mohindru, A., Fisher, J. M., & Rabinovitz, M. (1983b). Bathocuproine sulphonate: A tissue culture-compatible indicator of copper-mediated toxicity. *Nature*, *303*(5912), 64–65.

Monson, C. F., Cong, X., Robison, A., Pace, H. P., Liu, C., Poyton, M. F., et al. (2012). Phosphatidylserine reversibly binds Cu(2+) with extremely high affinity. *Journal of the American Chemical Society*, *134*(18), 7773–7779. http://dx.doi.org/10.1021/ja212138e.

Morphy, R., Kay, C., & Rankovic, Z. (2004). From magic bullets to designed multiple ligands. *Drug Discovery Today*, *9*(15), 641–651. http://dx.doi.org/10.1016/S1359-6446(04)03163-0.

Mostertz, J., Scharf, C., Hecker, M., & Homuth, G. (2004). Transcriptome and proteome analysis of *Bacillus subtilis* gene expression in response to superoxide and peroxide stress. *Microbiology*, *150*(Pt. 2), 497–512. http://dx.doi.org/10.1099/mic.0.26665-0.

Mullard, A. (2015). The phenotypic screening pendulum swings. *Nature Reviews Drug Discovery*, *14*(12), 807–809. http://dx.doi.org/10.1038/nrd4783.

Musiol, R., Serda, M., Hensel-Bielowka, S., & Polanski, J. (2010). Quinoline-based antifungals. *Current Medicinal Chemistry*, *17*(18), 1960–1973.

Nah, T., Kessler, S. H., Daumit, K. E., Kroll, J. H., Leone, S. R., & Wilson, K. R. (2013). Oh-initiated oxidation of sub-micron unsaturated fatty acid particles. *Physical Chemistry Chemical Physics*, *15*(42), 18649–18663. http://dx.doi.org/10.1039/c3cp52655k.

Newton, G. L., Bewley, C. A., Dwyer, T. J., Horn, R., Aharonowitz, Y., Cohen, G., et al. (1995). The structure of U17 isolated from *Streptomyces clavuligerus* and its properties as an antioxidant thiol. *European Journal of Biochemistry*, *230*(2), 821–825. http://dx.doi.org/10.1111/j.1432-1033.1995.0821h.x.

Neyrolles, O., Wolschendorf, F., Mitra, A., & Niederweis, M. (2015). Mycobacteria, metals, and the macrophage. *Immunological Reviews*, *264*(1), 249–263. http://dx.doi.org/10.1111/imr.12265.

Niki, E. (2014). Antioxidants: Basic principles, emerging concepts, and problems. *Biomedical Journal*, *37*(3), 106–111. http://dx.doi.org/10.4103/2319-4170.128727.

Nishibori, A., Kusaka, J., Hara, H., Umeda, M., & Matsumoto, K. (2005). Phosphatidylethanolamine domains and localization of phospholipid synthases in *Bacillus subtilis* membranes. *Journal of Bacteriology*, *187*(6), 2163–2174. http://dx.doi.org/10.1128/JB.187.6.2163-2174.2005.

Nwodo, U. U., Green, E., & Okoh, A. I. (2012). Bacterial exopolysaccharides: Functionality and prospects. *International Journal of Molecular Sciences*, *13*(11), 14002–14015. http://dx.doi.org/10.3390/ijms131114002.

Oe, S., Miyagawa, K., Honma, Y., & Harada, M. (2016). Copper induces hepatocyte injury due to the endoplasmic reticulum stress in cultured cells and patients with Wilson disease. *Experimental Cell Research*, *347*(1), 192–200. http://dx.doi.org/10.1016/j.yexcr.2016.08.003.

Ogra, Y., Aoyama, M., & Suzuki, K. T. (2006). Protective role of metallothionein against copper depletion. *Archives of Biochemistry and Biophysics*, *451*(2), 112–118. http://dx.doi.org/10.1016/j.abb.2006.04.017.

Olaleye, O., Raghunand, T. R., Bhat, S., Chong, C., Gu, P., Zhou, J., et al. (2011). Characterization of clioquinol and analogues as novel inhibitors of methionine aminopeptidases from *Mycobacterium tuberculosis*. *Tuberculosis (Edinburgh, Scotland)*, *91*(Suppl. 1), S61–S65. http://dx.doi.org/10.1016/j.tube.2011.10.012.

Olaleye, O., Raghunand, T. R., Bhat, S., He, J., Tyagi, S., Lamichhane, G., et al. (2010). Methionine aminopeptidases from *Mycobacterium tuberculosis* as novel antimycobacterial targets. *Chemistry & Biology*, *17*(1), 86–97. http://dx.doi.org/10.1016/j.chembiol.2009.12.014.

Oliveri, V., & Vecchio, G. (2016). 8-Hydroxyquinolines in medicinal chemistry: A structural perspective. *European Journal of Medicinal Chemistry*, *120*, 252–274. http://dx.doi.org/10.1016/j.ejmech.2016.05.007.

O' Meara, T. R., & Alspaugh, J. A. (2012). The *Cryptococcus neoformans* capsule: A sword and a shield. *Clinical Microbiology Reviews*, *25*(3), 387–408. http://dx.doi.org/10.1128/CMR.00001-12.

Osman, D., Patterson, C. J., Bailey, K., Fisher, K., Robinson, N. J., Rigby, S. E., et al. (2013). The copper supply pathway to a *Salmonella* Cu,Zn-superoxide dismutase (SodCII) involves P(1B)-type ATPase copper efflux and periplasmic CueP. *Molecular Microbiology*, *87*(3), 466–477. http://dx.doi.org/10.1111/mmi.12107.

Osman, D., Waldron, K. J., Denton, H., Taylor, C. M., Grant, A. J., Mastroeni, P., et al. (2010). Copper homeostasis in *Salmonella* is atypical and copper-CueP is a major periplasmic metal complex. *The Journal of Biological Chemistry, 285*(33), 25259–25268. http://dx.doi.org/10.1074/jbc.M110.145953.

Paczosa, M. K., & Mecsas, J. (2016). Klebsiella pneumoniae: Going on the offense with a strong defense. *Microbiology and Molecular Biology Reviews, 80*(3), 629–661. http://dx.doi.org/10.1128/MMBR.00078-15.

Pahontu, E., Julea, F., Rosu, T., Purcarea, V., Chumakov, Y., Petrenco, P., et al. (2015). Antibacterial, antifungal and in vitro antileukaemia activity of metal complexes with thiosemicarbazones. *Journal of Cellular and Molecular Medicine, 19*(4), 865–878. http://dx.doi.org/10.1111/jcmm.12508.

Palmer, L. D., & Skaar, E. P. (2016). Transition metals and virulence in bacteria. *Annual Review of Genetics, 50*, 67–91. http://dx.doi.org/10.1146/annurev-genet-120215-035146.

Pansy, F. E., Stander, H., Koerber, W. L., & Donovick, R. (1953). In vitro studies with 1-hydroxy-2(1h) pyridinethione. *Proceedings of the Society for Experimental Biology and Medicine, 82*(1), 122–124.

Papavassiliou, A. G. (1994). 1,10-Phenanthroline-copper ion nuclease footprinting of DNA-protein complexes in situ following mobility-shift electrophoresis assays. *Methods in Molecular Biology, 30*, 43–78. http://dx.doi.org/10.1385/0-89603-256-6:43.

Paranjpe, A., Zhang, R., Ali-Osman, F., Bobustuc, G. C., & Srivenugopal, K. S. (2014). Disulfiram is a direct and potent inhibitor of human O6-methylguanine-DNA methyltransferase (MGMT) in brain tumor cells and mouse brain and markedly increases the alkylating DNA damage. *Carcinogenesis, 35*, 692–702. http://dx.doi.org/10.1093/carcin/bgt366.

Park, K. C., Fouani, L., Jansson, P. J., Wooi, D., Sahni, S., Lane, D. J., et al. (2016). Copper and conquer: Copper complexes of di-2-pyridylketone thiosemicarbazones as novel anti-cancer therapeutics. *Metallomics, 8*(9), 874–886. http://dx.doi.org/10.1039/c6mt00105j.

Paterson, B. M., & Donnelly, P. S. (2011). Copper complexes of bis(thiosemicarbazones): From chemotherapeutics to diagnostic and therapeutic radiopharmaceuticals. *Chemical Society Reviews, 40*(5), 3005–3018. http://dx.doi.org/10.1039/c0cs00215a.

Payne, D. J., Gwynn, M. N., Holmes, D. J., & Pompliano, D. L. (2007). Drugs for bad bugs: Confronting the challenges of antibacterial discovery. *Nature Reviews Drug Discovery, 6*(1), 29–40.

Pedersen, O. S., & Pedersen, E. B. (1999). Non-nucleoside reverse transcriptase inhibitors: The NNRTI boom. *Antiviral Chemistry & Chemotherapy, 10*(6), 285–314. http://dx.doi.org/10.1177/095632029901000601.

Pederson, J. Z., Steinkuhler, C., Weser, U., & Rotilio, G. (1996). Copper-glutathione complexes under physiological conditions: Structures in solution different from the solid state coordination. *Biometals, 9*(1), 3–9.

Perrin, D. M., Hoang, V. M., Xu, Y., Mazumder, A., & Sigman, D. S. (1996). Inhibitors of *Escherichia coli* RNA polymerase specific for the single-stranded DNA of transcription intermediates. Tetrahedral cuprous chelates of 1,10-phenanthrolines. *Biochemistry, 35*(16), 5318–5326. http://dx.doi.org/10.1021/bi952197r.

Petris, M. J., & Mercer, J. F. (1999). The Menkes protein (ATP7A; MNK) cycles via the plasma membrane both in basal and elevated extracellular copper using a C-terminal di-leucine endocytic signal. *Human Molecular Genetics, 8*(11), 2107–2115.

Phillips, M., Malloy, G., Nedunchezian, D., Lukrec, A., & Howard, R. G. (1991). Disulfiram inhibits the in vitro growth of methicillin-resistant *Staphylococcus aureus*. *Antimicrobial Agents and Chemotherapy, 35*(4), 785–787.

Pope, L. E., & Sigman, D. S. (1984). Secondary structure specificity of the nuclease activity of the 1,10-phenanthroline-copper complex. *Proceedings of the National Academy of Sciences of the United States of America*, *81*(1), 3–7.

Potrykus, J., Ballou, E. R., Childers, D. S., & Brown, A. J. P. (2014). Conflicting interests in the pathogen–host tug of war: Fungal micronutrient scavenging versus mammalian nutritional immunity. *PLoS Pathogens*, *10*(3), e1003910. http://dx.doi.org/10.1371/journal.ppat.1003910

Powers, S. K., & Jackson, M. J. (2008). Exercise-induced oxidative stress: Cellular mechanisms and impact on muscle force production. *Physiological Reviews*, *88*(4), 1243–1276. http://dx.doi.org/10.1152/physrev.00031.2007.

Poyton, M. F., Sendecki, A. M., Cong, X., & Cremer, P. S. (2016). Cu(2+) binds to phosphatidylethanolamine and increases oxidation in lipid membranes. *Journal of the American Chemical Society*, *138*(5), 1584–1590. http://dx.doi.org/10.1021/jacs.5b11561.

Priuska, E. M., & Schacht, J. (1995). Formation of free radicals by gentamicin and iron and evidence for an iron/gentamicin complex. *Biochemical Pharmacology*, *50*(11), 1749–1752.

Quinlan, C. L., Orr, A. L., Perevoshchikova, I. V., Treberg, J. R., Ackrell, B. A., & Brand, M. D. (2012). Mitochondrial complex II can generate reactive oxygen species at high rates in both the forward and reverse reactions. *The Journal of Biological Chemistry*, *287*(32), 27255–27264. http://dx.doi.org/10.1074/jbc.M112.374629.

Rae, T. D., Schmidt, P. J., Pufahl, R. A., Culotta, V. C., & O'Halloran, T. V. (1999). Undetectable intracellular free copper: The requirement of a copper chaperone for superoxide dismutase. *Science*, *284*(5415), 805–808.

Rae, C., Tesson, M., Babich, J. W., Boyd, M., Sorensen, A., & Mairs, R. J. (2013). The role of copper in disulfiram-induced toxicity and radiosensitization of cancer cells. *Journal of Nuclear Medicine*, *54*(6), 953–960. http://dx.doi.org/10.2967/jnumed.112.113324.

Raimunda, D., Gonzalez-Guerrero, M., Leeber, B. W., 3rd, & Arguello, J. M. (2011). The transport mechanism of bacterial Cu+-ATPases: Distinct efflux rates adapted to different function. *Biometals*, *24*(3), 467–475. http://dx.doi.org/10.1007/s10534-010-9404-3.

Rattray, J. B., Schibeci, A., & Kidby, D. K. (1975). Lipids of yeasts. *Bacteriological Reviews*, *39*(3), 197–231.

Reeder, N. L., Kaplan, J., Xu, J., Youngquist, R. S., Wallace, J., Hu, P., et al. (2011). Zinc pyrithione inhibits yeast growth through copper influx and inactivation of iron-sulfur proteins. *Antimicrobial Agents and Chemotherapy*, *55*(12), 5753–5760. http://dx.doi.org/10.1128/AAC.00724-11.

Robinson, N. J., & Winge, D. R. (2010). Copper metallochaperones. *Annual Review of Biochemistry*, *79*, 537–562. http://dx.doi.org/10.1146/annurev-biochem-030409-143539.

Rodriguez-Montelongo, L., Volentini, S. I., Farias, R. N., Massa, E. M., & Rapisarda, V. A. (2006). The Cu(II)-reductase NADH dehydrogenase-2 of *Escherichia coli* improves the bacterial growth in extreme copper concentrations and increases the resistance to the damage caused by copper and hydroperoxide. *Archives of Biochemistry and Biophysics*, *451*(1), 1–7. http://dx.doi.org/10.1016/j.abb.2006.04.019.

Rowland, J. L., & Niederweis, M. (2013). A multicopper oxidase is required for copper resistance in *Mycobacterium tuberculosis*. *Journal of Bacteriology*, *195*(16), 3724–3733. http://dx.doi.org/10.1128/JB.00546-13.

Rubbo, S. D., Albert, A., & Gibson, M. I. (1950). The influence of chemical constitution on antibacterial activity. V. The antibacterial action of 8-hydroxyquinoline (oxine). *British Journal of Experimental Pathology*, *31*(3), 425–441.

Rubino, J. T., Chenkin, M. P., Keller, M., Riggs-Gelasco, P., & Franz, K. J. (2011). A comparison of methionine, histidine and cysteine in copper(I)-binding peptides reveals differences relevant to copper uptake by organisms in diverse environments. *Metallomics*, *3*(1), 61–73.

Russell, P. E. (2006). The development of commercial disease control. *Plant Pathology, 55*(5), 585–594. http://dx.doi.org/10.1111/j.1365-3059.2006.01440.x.

Sagripanti, J. L. (1999). Dna damage mediated by metal ions with special reference to copper and iron. In A. Sigel & H. Sigel (Eds.), *Metal ions in biological systems: Vol 36* (pp. 180–203). New York, USA: Marcel Decker, Inc.

Sagripanti, J. L., Goering, P. L., & Lamanna, A. (1991). Interaction of copper with Dna and antagonism by other metals. *Toxicology and Applied Pharmacology, 110*(3), 477–485.

Samanovic, M. I., & Darwin, K. H. (2016). Game of 'Somes: Protein destruction for *Mycobacterium tuberculosis* pathogenesis. *Trends in Microbiology, 24*(1), 26–34. http://dx.doi.org/10.1016/j.tim.2015.10.001.

Samanovic, M. I., Ding, C., Thiele, D. J., & Darwin, K. H. (2012). Copper in microbial pathogenesis: Meddling with the metal. *Cell Host & Microbe, 11*(2), 106–115. http://dx.doi.org/10.1016/j.chom.2012.01.009.

Sams-Dodd, F. (2005). Target-based drug discovery: Is something wrong? *Drug Discovery Today, 10*(2), 139–147. http://dx.doi.org/10.1016/S1359-6446(04)03316-1.

Schatz, A., Bugle, E., & Waksman, S. A. (1944). Streptomycin, a substance exhibiting antibiotic activity against gram-positive and gram-negative bacteria. *Experimental Biology and Medicine, 55*(1), 66–69. http://dx.doi.org/10.3181/00379727-55-14461.

Schimmer, A. D. (2011). Clioquinol—A novel copper-dependent and independent proteasome inhibitor. *Current Cancer Drug Targets, 11*(3), 325–331.

Schnappinger, D., & Hillen, W. (1996). Tetracyclines: Antibiotic action, uptake, and resistance mechanisms. *Archives of Microbiology, 165*(6), 359–369.

Schneider, C. (2009). An update on products and mechanisms of lipid peroxidation. *Molecular Nutrition & Food Research, 53*(3), 315–321. http://dx.doi.org/10.1002/mnfr.200800131.

Schwan, W. R., Warrener, P., Keunz, E., Stover, C. K., & Folger, K. R. (2005). Mutations in the *cueA* gene encoding a copper homeostasis P-type ATPase reduce the pathogenicity of *Pseudomonas aeruginosa* in mice. *International Journal of Medical Microbiology, 295*(4), 237–242.

Schwartz, J. R. (2016). Zinc pyrithione: A topical antimicrobial with complex pharmaceutics. *Journal of Drugs in Dermatology, 15*(2), 140–144.

Sessions, A. L., Doughty, D. M., Welander, P. V., Summons, R. E., & Newman, D. K. (2009). The continuing puzzle of the great oxidation event. *Current Biology, 19*(14), R567–R574. http://dx.doi.org/10.1016/j.cub.2009.05.054.

Sha, S. H., & Schacht, J. (1999). Salicylate attenuates gentamicin-induced ototoxicity. *Laboratory Investigation, 79*(7), 807–813.

Sha, S. H., & Schacht, J. (2000). Antioxidants attenuate gentamicin-induced free radical formation in vitro and ototoxicity in vivo: D-Methionine is a potential protectant. *Hearing Research, 142*(1-2), 34–40.

Shafeeq, S., Yesilkaya, H., Kloosterman, T. G., Narayanan, G., Wandel, M., Andrew, P. W., et al. (2011). The cop operon is required for copper homeostasis and contributes to virulence in *Streptococcus pneumoniae*. *Molecular Microbiology, 81*(5), 1255–1270. http://dx.doi.org/10.1111/j.1365-2958.2011.07758.x.

Shah, S., Dalecki, A. G., Malalasekera, A. P., Crawford, C. L., Michalek, S. M., Kutsch, O., et al. (2016). 8-Hydroxyquinolines are boosting agents of copper-related toxicity in *Mycobacterium tuberculosis*. *Antimicrobial Agents and Chemotherapy, 60*(10), 5765–5776. http://dx.doi.org/10.1128/AAC.00325-16.

Shane, S. J., Riley, C., Laurie, J. H., & Boutilier, M. (1951). Thiosemicarbazones in tuberculosis. *Canadian Medical Association Journal, 64*(4), 289–293.

Shaw, E., Bernstein, J., Losee, K., & Lott, W. A. (1950). Analogs of aspergillic acid. IV. Substituted 2-bromopyridine-N-oxides and their conversion to cyclic thiohydroxamic acids. *Journal of the American Chemical Society, 72*(10), 4362–4364. http://dx.doi.org/10.1021/ja01166a008.

Shi, X., & Darwin, K. H. (2015). Copper homeostasis in *Mycobacterium tuberculosis*. *Metallomics*, 7, 929–934. http://dx.doi.org/10.1039/c4mt00305e.

Shi, X., Festa, R. A., Ioerger, T. R., Butler-Wu, S., Sacchettini, J. C., Darwin, K. H., et al. (2014). The copper-responsive RicR regulon contributes to *Mycobacterium tuberculosis* virulence. *mBio*, *5*(1), e00876-13. http://dx.doi.org/10.1128/mBio.00876-13.

Shim, J. S., & Liu, J. O. (2014). Recent advances in drug repositioning for the discovery of new anticancer drugs. *International Journal of Biological Sciences*, *10*(7), 654–663. http://dx.doi.org/10.7150/ijbs.9224.

Short, B. R., Vargas, M. A., Thomas, J. C., O'Hanlon, S., & Enright, M. C. (2006). In vitro activity of a novel compound, the metal ion chelating agent Aq+, against clinical isolates of *Staphylococcus aureus*. *The Journal of Antimicrobial Chemotherapy*, *57*(1), 104–109. http://dx.doi.org/10.1093/jac/dki428.

Simon, N., Coulanges, V., Andre, P., & Vidon, D. J. (1995). Utilization of exogenous siderophores and natural catechols by *Listeria monocytogenes*. *Applied and Environmental Microbiology*, *61*(4), 1643–1645.

Singer, R., Derby, N., Rodriguez, A., Kizima, L., Kenney, J., Aravantinou, M., et al. (2011). The nonnucleoside reverse transcriptase inhibitor Miv-150 in carrageenan gel prevents rectal transmission of simian/human immunodeficiency virus infection in macaques. *Journal of Virology*, *85*(11), 5504–5512. http://dx.doi.org/10.1128/JVI.02422-10.

Sitthisak, S., Howieson, K., Amezola, C., & Jayaswal, R. K. (2005). Characterization of a multicopper oxidase gene from *Staphylococcus aureus*. *Applied and Environmental Microbiology*, *71*(9), 5650–5653. http://dx.doi.org/10.1128/AEM.71.9.5650-5653.2005.

Sitthisak, S., Kitti, T., Boonyonying, K., Wozniak, D., Mongkolsuk, S., & Jayaswal, R. K. (2012). Mcsa and the roles of metal-binding motif in *Staphylococcus aureus*. *FEMS Microbiology Letters*, *327*(2), 126–133. http://dx.doi.org/10.1111/j.1574-6968.2011.02468.x.

Slauch, J. M. (2011). How does the oxidative burst of macrophages kill bacteria? Still an open question. *Molecular Microbiology*, *80*(3), 580–583. http://dx.doi.org/10.1111/j.1365-2958.2011.07612.x.

Slee, A. M., Wuonola, M. A., McRipley, R. J., Zajac, I., Zawada, M. J., Bartholomew, P. T., et al. (1987). Oxazolidinones, a new class of synthetic antibacterial agents: In vitro and in vivo activities of Dup 105 and Dup 721. *Antimicrobial Agents and Chemotherapy*, *31*(11), 1791–1797. http://dx.doi.org/10.1128/AAC.31.11.1791.

Smit, H., Van Der Goot, H., Nauta, W. T., Pijper, P. J., Balt, S., De Bolster, M. W., et al. (1980). Mode of action of copper complexes of some 2,2'-bipyridyl analogs on *Paracoccus denitrificans*. *Antimicrobial Agents and Chemotherapy*, *18*(2), 249–256. http://dx.doi.org/10.1128/AAC.18.2.249.

Smit, H., Van Der Goot, H., Nauta, W. T., Timmerman, H., De Bolster, M. W., Jochemsen, A. G., et al. (1981). Mode of action of the copper(I) complex of 2,9-dimethyl-1,10-phenanthroline on *Mycoplasma gallisepticum*. *Antimicrobial Agents and Chemotherapy*, *20*(4), 455–462. http://dx.doi.org/10.1128/AAC.20.4.455.

Smit, H., Van Der Goot, H., Nauta, W. T., Timmerman, H., De Bolster, M. W., Stouthamer, A. H., et al. (1982). Mechanism of action of the copper(I) complex of 2,9-dimethyl-1,10-phenanthroline on *Mycoplasma gallisepticum*. *Antimicrobial Agents and Chemotherapy*, *21*(6), 881–886.

Smith, G. F., & McCurdy, W. H. (1952). 2,9-Dimethyl-1,10-phenanthroline. *Analytical Chemistry*, *24*(2), 371–373. http://dx.doi.org/10.1021/ac60062a029.

Sohlenkamp, C., & Geiger, O. (2016). Bacterial membrane lipids: Diversity in structures and pathways. *FEMS Microbiology Reviews*, *40*(1), 133–159. http://dx.doi.org/10.1093/femsre/fuv008.

Solioz, M. (2002). Role of proteolysis in copper homoeostasis. *Biochemical Society Transactions*, *30*(4), 688–691. http://dx.doi.org/10.1042/bst0300688.

Solioz, M., Abicht, H. K., Mermod, M., & Mancini, S. (2010). Response of gram-positive bacteria to copper stress. *Journal of Biological Inorganic Chemistry*, *15*(1), 3–14. http://dx.doi.org/10.1007/s00775-009-0588-3.

Solioz, M., & Stoyanov, J. V. (2003). Copper homeostasis in *Enterococcus hirae*. *FEMS Microbiology Reviews*, *27*(2-3), 183–195.

Song, B. B., Sha, S. H., & Schacht, J. (1998). Iron chelators protect from aminoglycoside-induced cochleo- and vestibulo-toxicity. *Free Radical Biology & Medicine*, *25*(2), 189–195.

Song, Y., Xu, H., Chen, W., Zhan, P., & Liu, X. (2015). 8-hydroxyquinoline: A privileged structure with a broad-ranging pharmacological potential. *Medicinal Chemical Communications*, *6*(1), 61–74. http://dx.doi.org/10.1039/c4md00284a.

Soon, C. P., Donnelly, P. S., Turner, B. J., Hung, L. W., Crouch, P. J., Sherratt, N. A., et al. (2011). Diacetylbis(N(4)-methylthiosemicarbazonato) copper(II) ($Cu^{II}$(atsm)) protects against peroxynitrite-induced nitrosative damage and prolongs survival in amyotrophic lateral sclerosis mouse model. *The Journal of Biological Chemistry*, *286*(51), 44035–44044. http://dx.doi.org/10.1074/jbc.M111.274407.

Sorenson, J. R. J. (1982). Copper complexes as the active metabolites of antiinflammatory agents. In J. R. J. Sorenson (Ed.), *Inflammatory diseases and copper: The metabolic and therapeutic roles of copper and other essential metalloelements in humans* (pp. 289–301). Totowa, NJ: Humana Press.

Soultanas, P., Dillingham, M. S., Wiley, P., Webb, M. R., & Wigley, D. B. (2000). Uncoupling DNA translocation and helicase activity in PcrA: Direct evidence for an active mechanism. *The EMBO Journal*, *19*(14), 3799–3810. http://dx.doi.org/10.1093/emboj/19.14.3799.

Sousa, C. F., Ferreira, M., Abreu, B., Medforth, C. J., & Gameiro, P. (2015). Interactions of a non-fluorescent fluoroquinolone with biological membrane models: A multi-technique approach. *International Journal of Pharmaceutics*, *495*(2), 761–770. http://dx.doi.org/10.1016/j.ijpharm.2015.09.037.

Speer, A., Rowland, J. L., Haeili, M., Niederweis, M., & Wolschendorf, F. (2013). Porins increase copper susceptibility of *Mycobacterium tuberculosis*. *Journal of Bacteriology*, *195*(22), 5133–5140. http://dx.doi.org/10.1128/JB.00763-13.

Speer, A., Shrestha, T. B., Bossmann, S. H., Basaraba, R. J., Harber, G. J., Michalek, S. M., et al. (2013). Copper-boosting compounds: A novel concept for antimycobacterial drug discovery. *Antimicrobial Agents and Chemotherapy*, *57*(2), 1089–1091. http://dx.doi.org/10.1128/AAC.01781-12.

Stevens, J. M., Mavridou, D. A., Hamer, R., Kritsiligkou, P., Goddard, A. D., & Ferguson, S. J. (2011). Cytochrome C biogenesis system I. *The FEBS Journal*, *278*(22), 4170–4178. http://dx.doi.org/10.1111/j.1742-4658.2011.08376.x.

Stolle, P., Hou, B., & Bruser, T. (2016). The Tat substrate CueO is transported in an incomplete folding state. *The Journal of Biological Chemistry*, *291*, 13520–13528. http://dx.doi.org/10.1074/jbc.M116.729103.

Strebhardt, K., & Ullrich, A. (2008). Paul Ehrlich's magic bullet concept: 100 years of progress. *Nature Reviews Cancer*, *8*(6), 473–480. http://dx.doi.org/10.1038/nrc2394.

Subashchandrabose, S., Hazen, T. H., Brumbaugh, A. R., Himpsl, S. D., Smith, S. N., Ernst, R. D., et al. (2014). Host-specific induction of *Escherichia coli* fitness genes during human urinary tract infection. *Proceedings of the National Academy of Sciences of the United States of America*, *111*(51), 18327–18332. http://dx.doi.org/10.1073/pnas.1415959112.

Swinney, D. C., & Anthony, J. (2011). How were new medicines discovered? *Nature Reviews Drug Discovery*, *10*(7), 507–519. http://dx.doi.org/10.1038/nrd3480.

Szczepanik, W., Dworniczek, E., Ciesiołka, J., Wrzesiński, J., Skała, J., & Jeżowska-Bojczuk, M. (2003). In vitro oxidative activity of cupric complexes of kanamycin

A in comparison to in vivo bactericidal efficacy. *Journal of Inorganic Biochemistry, 94*(4), 355–364. http://dx.doi.org/10.1016/S0162-0134(03)00029-1.

Szczepanik, W., Kaczmarek, P., & Jezowska-Bojczuk, M. (2004). Oxidative activity of copper(II) complexes with aminoglycoside antibiotics as implication to the toxicity of these drugs. *Bioinorganic Chemistry and Applications, 2*(1–2), 55–68. http://dx.doi.org/10.1155/S1565363304000056.

Szymanski, P., Fraczek, T., Markowicz, M., & Mikiciuk-Olasik, E. (2012). Development of copper based drugs, radiopharmaceuticals and medical materials. *Biometals, 25*(6), 1089–1112. http://dx.doi.org/10.1007/s10534-012-9578-y.

Teitzel, G. M., Geddie, A., De Long, S. K., Kirisits, M. J., Whiteley, M., & Parsek, M. R. (2006). Survival and growth in the presence of elevated copper: Transcriptional profiling of copper-stressed *Pseudomonas aeruginosa*. *Journal of Bacteriology, 188*(20), 7242–7256. http://dx.doi.org/10.1128/JB.00837-06.

Teitzel, G. M., & Parsek, M. R. (2003). Heavy metal resistance of biofilm and planktonic *Pseudomonas aeruginosa*. *Applied and Environmental Microbiology, 69*(4), 2313–2320. http://dx.doi.org/10.1128/AEM.69.4.2313-2320.2003.

Tommasi, R., Brown, D. G., Walkup, G. K., Manchester, J. I., & Miller, A. A. (2015). Eskapeing the labyrinth of antibacterial discovery. *Nature Reviews Drug Discovery, 14*(8), 529–542. http://dx.doi.org/10.1038/nrd4572.

Tong, F., Zhao, Y., Gu, X., Gu, C., & Lee, C. C. (2015). Joint toxicity of tetracycline with copper(II) and cadmium(II) to Vibrio fischeri: Effect of complexation reaction. *Ecotoxicology, 24*(2), 346–355. http://dx.doi.org/10.1007/s10646-014-1383-7.

Tottey, S., Waldron, K. J., Firbank, S. J., Reale, B., Bessant, C., Sato, K., et al. (2008). Protein-folding location can regulate manganese-binding versus copper- or zinc-binding. *Nature, 455*(7216), 1138–1142. http://dx.doi.org/10.1038/nature07340.

Tourney, J., Ngwenya, B., Mosselmans, J. W. F., & Magennis, M. (2009). Physical and chemical effects of extracellular polymers (EPS) on Zn adsorption to *Bacillus licheniformis* S-86. *Journal of Colloid and Interface Science, 337*(2), 381–389. http://dx.doi.org/10.1016/j.jcis.2009.05.067.

Trasnea, P. I., Utz, M., Khalfaoui-Hassani, B., Lagies, S., Daldal, F., & Koch, H. G. (2016). Cooperation between two periplasmic copper chaperones is required for full activity of the Cbb3-type cytochrome c oxidase and copper homeostasis in *Rhodobacter capsulatus*. *Molecular Microbiology, 100*(2), 345–361. http://dx.doi.org/10.1111/mmi.13321.

Uivarosi, V. (2013). Metal complexes of quinolone antibiotics and their applications: An update. *Molecules, 18*(9), 11153–11197. http://dx.doi.org/10.3390/molecules180911153.

Ulanova, M., & Tsang, R. S. (2014). *Haemophilus influenzae* serotype a as a cause of serious invasive infections. *The Lancet Infectious Diseases, 14*(1), 70–82. http://dx.doi.org/10.1016/S1473-3099(13)70170-1.

Urbanski, T., Slopek, S., & Venulet, J. (1951). Antitubercular activity of some 8-hydroxyquinoline derivatives. *Nature, 168*(4262), 29.

Urso, E., & Maffia, M. (2015). Behind the link between copper and angiogenesis: Established mechanisms and an overview on the role of vascular copper transport systems. *Journal of Vascular Research, 52*(3), 172–196. http://dx.doi.org/10.1159/000438485.

Valade, E., Davin-Regli, A., Bolla, J.-M., & Pagès, J.-M. (2013). *Bacterial membrane, a key for controlling drug influx and efflux,* Antibiotics. Weinheim, Germany: Wiley-Vch Verlag Gmbh & Co. KGaA, pp. 217–240. http://dx.doi.org/10.1002/9783527659685.ch9.

Varadwaj, P. R., Varadwaj, A., & Jin, B. Y. (2015). Ligand(S)-to-metal charge transfer as a factor controlling the equilibrium constants of late first-row transition metal complexes: Revealing the Irving-Williams thermodynamical series. *Physical Chemistry Chemical Physics, 17*(2), 805–811. http://dx.doi.org/10.1039/c4cp03953j.

Volentini, S. I., Farias, R. N., Rodriguez-Montelongo, L., & Rapisarda, V. A. (2011). Cu(II)-reduction by *Escherichia coli* cells is dependent on respiratory chain components. *Biometals*, *24*(5), 827–835. http://dx.doi.org/10.1007/s10534-011-9436-3.

Wagner, D., Maser, J., Lai, B., Cai, Z., Barry, C. E., 3rd, Honer Zu Bentrup, K., et al. (2005). Elemental analysis of *Mycobacterium avium*-, *Mycobacterium tuberculosis*-, and *Mycobacterium smegmatis*-containing phagosomes indicates pathogen-induced microenvironments within the host cell's endosomal system. *Journal of Immunology*, *174*(3), 1491–1500.

Waldron, K. J., Firbank, S. J., Dainty, S. J., Perez-Rama, M., Tottey, S., & Robinson, N. J. (2010). Structure and metal loading of a soluble periplasm cuproprotein. *The Journal of Biological Chemistry*, *285*(42), 32504–32511. http://dx.doi.org/10.1074/jbc.M110.153080.

Waldron, K. J., Rutherford, J. C., Ford, D., & Robinson, N. J. (2009). Metalloproteins and metal sensing. *Nature*, *460*(7257), 823–830. http://dx.doi.org/10.1038/nature08300.

Ward, S. K., Abomoelak, B., Hoye, E. A., Steinberg, H., & Talaat, A. M. (2010). CtpV: A putative copper exporter required for full virulence of *Mycobacterium tuberculosis*. *Molecular Microbiology*, 77(5), 1096–1110. http://dx.doi.org/10.1111/j.1365-2958.2010.07273.x.

Ward, S. K., Hoye, E. A., & Talaat, A. M. (2008). The global responses of *Mycobacterium tuberculosis* to physiological levels of copper. *Journal of Bacteriology*, *190*(8), 2939–2946. http://dx.doi.org/10.1128/JB.01847-07.

Weinberg, E. D. (1975). Nutritional immunity. Host's attempt to withhold iron from microbial invaders. *JAMA*, *231*(1), 39–41.

Whitchurch, C. B., Tolker-Nielsen, T., Ragas, P. C., & Mattick, J. S. (2002). Extracellular DNA required for bacterial biofilm formation. *Science*, *295*(5559), 1487.

White, C., Kambe, T., Fulcher, Y. G., Sachdev, S. W., Bush, A. I., Fritsche, K., et al. (2009). Copper transport into the secretory pathway is regulated by oxygen in macrophages. *Journal of Cell Science*, *122*(Pt. 9), 1315–1321. http://dx.doi.org/10.1242/jcs.043216.

White, A. R., Kanninen, K. M., & Crouch, P. J. (2015). Editorial: Metals and neurodegeneration: Restoring the balance. *Frontiers in Aging Neuroscience*, 7, 127. http://dx.doi.org/10.3389/fnagi.2015.00127.

White, C., Lee, J., Kambe, T., Fritsche, K., & Petris, M. J. (2009). A role for the ATP7A copper-transporting ATPase in macrophage bactericidal activity. *The Journal of Biological Chemistry*, *284*(49), 33949–33956. http://dx.doi.org/10.1074/jbc.m109.070201.

Whittaker, J. W. (2005). The radical chemistry of galactose oxidase. *Archives of Biochemistry and Biophysics*, *433*(1), 227–239. http://dx.doi.org/10.1016/j.abb.2004.08.034.

Wolschendorf, F., Ackart, D., Shrestha, T. B., Hascall-Dove, L., Nolan, S., Lamichhane, G., et al. (2011). Copper resistance is essential for virulence of *Mycobacterium tuberculosis*. *Proceedings of the National Academy of Sciences of the United States of America*, *108*(4), 1621–1626. http://dx.doi.org/10.1073/pnas.1009261108.

Wunderli-Ye, H., & Solioz, M. (1999). Copper homeostasis in *Enterococcus hirae*. *Advances in Experimental Medicine and Biology*, *448*, 255–264.

Xiao, Z., Donnelly, P. S., Zimmermann, M., & Wedd, A. G. (2008). Transfer of copper between bis(thiosemicarbazone) ligands and intracellular copper-binding proteins. Insights into mechanisms of copper uptake and hypoxia selectivity. *Inorganic Chemistry*, *47*(10), 4338–4347. http://dx.doi.org/10.1021/ic702440e.

Xie, L., & Collins, J. F. (2011). Transcriptional regulation of the Menkes copper ATPase (Atp7a) gene by hypoxia-inducible factor (HIF2A) in intestinal epithelial cells. *American Journal of Physiology Cell Physiology*, *300*(6), C1298–C1305. http://dx.doi.org/10.1152/ajpcell.00023.2011.

Yin, D., Urresti, S., Lafond, M., Johnston, E. M., Derikvand, F., Ciano, L., et al. (2015). Structure–function characterization reveals new catalytic diversity in the galactose oxidase and glyoxal oxidase family. *Nature Communications*, *6*, 10197. http://dx.doi.org/10.1038/ncomms10197.

Zakeri, B., & Wright, G. D. (2008). Chemical biology of tetracycline antibiotics. *Biochemistry and Cell Biology*, *86*(2), 124–136. http://dx.doi.org/10.1139/O08-002.

Zhang, Q. I., Wang, S., Yang, D., Pan, K., Li, L., & Yuan, S. (2016). Preclinical pharmacodynamic evaluation of antibiotic nitroxoline for anticancer drug repurposing. *Oncology Letters*, *11*(5), 3265–3272. http://dx.doi.org/10.3892/ol.2016.4380.

Zhao, Y., Tan, Y., Guo, Y., Gu, X., Wang, X., & Zhang, Y. (2013). Interactions of tetracycline with Cd (II), Cu (II) and Pb (II) and their cosorption behavior in soils. *Environmental Pollution*, *180*, 206–213. http://dx.doi.org/10.1016/j.envpol.2013.05.043.

Zimnicka, A. M., Tang, H., Guo, Q., Kuhr, F. K., Oh, M. J., Wan, J., et al. (2014). Upregulated copper transporters in hypoxia-induced pulmonary hypertension. *Plos One*, *9*(3), e90544. http://dx.doi.org/10.1371/journal.pone.0090544.

CHAPTER SEVEN

# Metal Resistance and Its Association With Antibiotic Resistance

**Chandan Pal*[,†], Karishma Asiani[‡], Sankalp Arya[‡], Christopher Rensing[§,¶], Dov J. Stekel[‡], D.G. Joakim Larsson*[,†], Jon L. Hobman[‡,1]**

*Institute of Biomedicine, Sahlgrenska Academy, University of Gothenburg, Gothenburg, Sweden
[†]Centre for Antibiotic Resistance Research (CARe) at University of Gothenburg, Gothenburg, Sweden
[‡]School of Biosciences, the University of Nottingham, Sutton Bonington Campus, Sutton Bonington, Leicestershire, United Kingdom
[§]College of Resources and Environment, Fujian Agriculture and Forestry University, Fuzhou, Fujian, China
[¶]J. Craig Venter Institute, La Jolla, CA, United States
[1]Corresponding author: e-mail address: jon.hobman@nottingham.ac.uk

## Contents



*Advances in Microbial Physiology*, Volume 70
ISSN 0065-2911
http://dx.doi.org/10.1016/bs.ampbs.2017.02.001

## Abstract

Antibiotic resistance is recognised as a major global threat to public health by the World Health Organization. Currently, several hundred thousand deaths yearly can be attributed to infections with antibiotic-resistant bacteria. The major driver for the development of antibiotic resistance is considered to be the use, misuse and overuse of antibiotics in humans and animals. Nonantibiotic compounds, such as antibacterial biocides and metals, may also contribute to the promotion of antibiotic resistance through co-selection. This may occur when resistance genes to both antibiotics and metals/biocides are co-located together in the same cell (co-resistance), or a single resistance mechanism (e.g. an efflux pump) confers resistance to both antibiotics and biocides/metals (cross-resistance), leading to co-selection of bacterial strains, or mobile genetic elements that they carry. Here, we review antimicrobial metal resistance in the context of the antibiotic resistance problem, discuss co-selection, and highlight critical knowledge gaps in our understanding.

## ABBREVIATIONS

**ABC** ATP-binding cassettes
**AMR** antimicrobial resistance
**ARGs** antibiotic resistance genes
**HGT** horizontal gene transfer
**MATE** multidrug and toxic compound extrusion
**MDR** multidrug resistance
**MGEs** mobile genetic elements
**MIC** minimal inhibitory concentration
**MRGs** metal resistance genes
**QACs** quaternary ammonium compounds
**RND** resistance–nodulation–division
**SCENIHR** Scientific Committee on Emerging and Newly Identified Health Risks
**SMR** small multidrug resistance
**WHO** World Health Organization
**WWTP** wastewater treatment plant

## 1. INTRODUCTION

Antibiotics, antibacterial biocides and metals have many uses in medicine, agriculture, aquaculture, and in consumer products. Antibiotics and biocides are produced, and antimicrobial metals are mined and refined, on industrial scales, eventually resulting in widespread opportunities for selection of resistant bacteria. Bacterial communities are exposed to these compounds at selective levels for example through human and animal medicine or through biocidal products, and in unintentional environmental contexts, for example through waste streams, where antimicrobials are often present at subinhibitory concentrations. Many metals and other antimicrobials are also used for other purposes (e.g. as in-feed growth promoters, in water sanitation systems) but still retain their antimicrobial properties. Resistance genes to metals and antibacterial biocides are found in all types of environments (Pal, Bengtsson-Palme, Kristiansson, & Larsson, 2015, 2016), including pristine environments not influenced by large-scale human use of antimicrobials (Wardwell et al., 2009). Sometimes, these genes are found together with antibiotic resistance genes (ARGs) on mobile genetic elements (MGEs) such as plasmids, transposable elements and genomic islands. Additionally, some resistance mechanisms protect bacteria from toxic effects of both antibiotics and other antibacterial compounds. Therefore, an important developing concern is that biocides and antimicrobial metals could be selecting for antibiotic resistance (SCENIHR, 2009). In this review, we give a short history of antimicrobial resistance (AMR) studies in bacteria, with particular reference to the discovery of metal resistance in human pathogens. In addition, we highlight mechanisms for metal resistance, studies reporting co-selection between antibiotics and metals in bacteria, discuss the mechanisms of co-selection for antibiotic and metal resistance, and identify and discuss critical knowledge gaps.

## 2. ANTIMICROBIALS

### 2.1 Antibiotics

Antibiotics target specific bacterial structures or processes, and may act by inhibiting growth (bacteriostatic), by directly killing bacteria (bactericidal) or by a combination of both mechanisms. The introduction of antibiotics revolutionised the treatment of bacterial diseases, and antibiotics have saved millions of lives since their introduction into clinical practise. They are primarily used to treat or prevent bacterial infections. Since they come into direct contact with human and animal tissues and cells, selective toxicity to bacteria over human cells is crucial to avoid harmful side effects. This is less critical for other antibacterial compounds that are not ingested or injected (see below).

Originally the term antibiotic (defined in its current sense by Selman Waksman, the discoverer of streptomycin in 1941; Pramer, 1988) meant a product from a living organism that kills another microorganism. That definition has now widened to include semisynthetic antibiotics (e.g. amoxicillin) and most often also fully synthetic compounds, such as sulphonamides, trimethoprim and fluoroquinolones. The main classes of antibiotic compounds include sulphonamides, β-lactams (which include semisynthetic penicillins cephalosporins and carbapenems), aminoglycosides, tetracyclines, phenicols, macrolides, lincosamides, streptogramins, glycopeptides, ansamycins, quinolones, oxazolidinones, and polypeptides as well as some smaller classes of compounds such as nitrofurans, trimethoprim and mupirocin. β-Lactam antibiotics are sometimes used in conjunction with β-lactamase inhibitors, e.g. clavulanate, sulbactam and tazobactam. Resistance to each new class of antibiotic has in most cases been detected in pathogens that the antibiotic was intended to treat within a few years of the introduction of the antibiotic in clinical use (Schmieder & Edwards, 2012). Consequently, resistance has developed to every class of antibiotics introduced into the clinic. The discovery and development of fundamentally new classes of antibiotics has been exceptionally slow during the past decades. Thus, the World Health Organization (WHO) has declared the gradual loss of the effective antibiotic arsenal as a serious threat to human health in all parts of the globe and warned that a postantibiotic era could be near (WHO, 2014).

## 2.2 Antibacterial Biocides

Biocides usually have a broad spectrum of antimicrobial activity. Many different biocides are currently used, with diverse activities and cellular target sites, including alcohols, acids and alkalis, aldehydes, anilides and biguanides, diamides, halogen releasing compounds, oxidising agents, organic acids, peroxygens, phenolics (phenols, bisphenols and halophenols), quaternary ammonium compounds (QACs) and vapour phase sterilants (McDonnell & Russell, 1999; Pal, Bengtsson-Palme, Rensing, Kristiansson, & Larsson, 2014). Biocidal compounds are widely used in antiseptics, disinfectants, preservatives, antifouling compounds and anti-infectives in healthcare, agriculture and livestock farming, industry, food preparation and in consumer goods (e.g. toothpastes and cosmetics). Biocides are not the prime focus of this review, as their potential for co-selection of antibiotic resistance has been discussed in detail recently by Wales and Davies (2015). However, the challenges in relation to antibiotic resistance very much resemble those of antibacterial metals.

## 2.3 Antimicrobial Metals

Antimicrobial metals also have a broad spectrum of antimicrobial activity. The modes of action of these metals are not always fully understood, but they affect multiple cellular targets through their chemical properties (thiophilicity, ability to redox cycle, the function of the metal as a soft Lewis acid) (Lemire, Harrison, & Turner, 2013). Antimicrobial metal compounds used prior to the widespread use of antibiotics included inorganic and organic mercury compounds, silver, copper, gold, tellurium, potassium, magnesium and zinc salts, and inorganic and organic arsenic and antimony compounds (Hobman & Crossman, 2015). The current WHO list of essential medicines includes many metal-containing medicines: sodium stibogluconate or meglumine antimoniate (pentavalent antimonial compounds for the treatment of Leishmanniasis), melarsoprol (arsenic, for the treatment of trypanosomiasis), cisplatin, carboplatin and oxaliplatin (platinum, cytotoxic drugs with antimicrobial properties), silver sulfadiazine (burns—an antiinfective), calamine (zinc oxide and ferric oxide; antiinflammatory), selenium sulphide (antifungal) and zinc sulphate (acute diarrhoea treatment) (WHO, 2015). Antimicrobial metal compounds containing copper, zinc, cadmium and arsenic are used in agriculture and animal husbandry as growth promoters, fungicides, herbicides and antimicrobials (Castillo, Martín-Orúe, Taylor-Pickard, Pérez, & Gasa, 2008; Li et al., 2010; Lucas, Livingstone, & McDonald, 1961; Nachman et al., 2013). In addition, metal-containing products are regularly used in clinical surgery and in medicines. For example, mercury-based amalgams in dental fillings, nickel in dental bridges, copper in intrauterine contraceptive devices, aluminium and bismuth in antacids, and zinc and titanium in nappy/diaper rash ointments. Antimicrobial metal(loid)s such as arsenic, mercury, copper and zinc have been used for many other purposes beyond healthcare and have been released into the environment through anthropogenic activities for more than 2000 years by human activities (Nriagu, 1996).

# 3. AMR: ANCIENT AND MODERN

## 3.1 Antibiotic Resistance

Antibiotic resistance and multidrug resistance (MDR) are modern phenomena that have arisen following the large-scale production and use of antibiotics in medicine and agriculture. Specifically, clinical antibiotic resistance has developed into a full-blown internationally recognised crisis within just over 70 years; the MDR *Shigella* sp. isolated in Japan in the 1950s were harbingers of MDR enterobacteria and were first isolated within 7 years of antibiotic use (Falkow, 1975). However, there is accumulating evidence that ARGs themselves are ancient (Aminov & Mackie, 2007; Aminov, 2009; D'Costa et al., 2011; Bhullar et al., 2012; and reviewed in Perry, Waglechner, & Wright, 2016). Antibiotics are often naturally produced by soil-borne bacteria and fungi, which carry resistance genes to the antibiotics they produce. Thus, antibiotic producers act as potential sources of resistance genes. Moreover, other environmental bacteria have evolved resistance due to their exposure to antibiotics produced by soil organisms (Forsberg et al., 2012). Therefore, the preexistence of the ARGs and MGEs that carry them is likely to have contributed to the very short evolutionary timeframe for the development of MDR.

Antibiotic resistance can be acquired via mutations of target proteins or porins and/or regulatory genes or motifs (e.g. of AmpC) or through acquisition of resistance genes from other bacteria. The mobilisation of ARGs on plasmids, transposons and integrons has enabled their ability to spread horizontally across strains and species, greatly facilitating the development of resistance in numerous pathogens. Other contributions to AMR, and particularly to antibiotic and biocide resistance, are intrinsic resistance (e.g. cell membrane permeability barriers, the LPS in Gram-negative bacteria and cell surface charge properties) or broad-spectrum efflux pumps belonging to the ABC (ATP-binding cassette), MFS (major facilitator), MATE (multidrug and toxic efflux), RND (resistance, nodulation and division) or SMR (small multidrug resistance) families (Webber & Piddock, 2003). Efflux systems, such as AcrAB/TolC appear to significantly contribute to multiple antibiotic resistance in Enterobacteria (Webber & Piddock, 2003). The broad specificity and large number of these pumps suggest that they have a generalised role in protecting the cell from toxic compounds (including metals) encountered in the environment or contribute to virulence within hosts (Blanco et al., 2016; Nies, 2003), and their wide distribution is consistent with their ancient nature (Alcalde-Rico, Hernando-Amado, Blanco, & Martínez, 2016).

## 3.2 Metal Resistance

Metals have been biologically available since the Great Oxidation Event 2.4 billion years ago (Chi Fru et al., 2016). Resistance genes to toxic metals and metalloids are believed to be ancient (Boyd & Barkay, 2012; Jackson &

Dugas, 2003; Staehlin, Gibbons, Rokas, O'Halloran, & Slot, 2016); in silico evolution studies would suggest that metal resistance genes (MRGs) should be as ancient as metal toxicity (Jenkins & Stekel, 2010), but this is difficult to prove with sequence analysis due to the timescales involved.

Metal resistance is a common phenotype in many microorganisms that are exposed to metals in their habitats. Pal et al. (2015) showed that resistance genes to biocides and metals are present in the great majority of genomes isolated from different environments ranging from humans, animals and insect symbionts, to extreme environments, such as hydrothermal vents and environments polluted by discharges from antibiotic manufacture. In contrast, the presence of such genes on plasmids was considerably less common in all types of environments.

Antimicrobial metals have been released into the biosphere in huge quantities through geological events for billions of years, and used by humans in medicine, agriculture and manufacturing, for thousands of years. There is data that associate metal contamination in different environments (due to either geological or anthropogenic activities) and the presence of MRGs (Farias et al., 2015; Poulain et al., 2016; Staehlin et al., 2016). Recent work analysing the occurrence of bacterial MRGs in dated permafrost cores and deep subterranean bacterial isolates links increases in the numbers of mercury (Poulain et al., 2016) and divergence of copper (Staehlin et al., 2016) resistance genes to global deposition of toxic metals due to industrial activity. This not only includes the rapid increase in metal production during the industrial revolution, but also dissemination of these metals due to inefficient methods of smelting in preindustrial revolution eras.

Mercury resistance is especially important because some *mer* transposons can accumulate other resistances, and are therefore a vector for co-resistance (Liebert, Hall, & Summers, 1999; Summers, 2004). Mercury resistance transposons, which are closely related to modern Tn*21*-family transposons, but lacking ARGs, or the integron carrying them, have been detected from 'preantibiotic era' bacteria (Essa, Julian, Kidd, Brown, & Hobman, 2003), and from permafrost isolates from ice cores that are over 8000 years old (Kholodii, Mindlin, Petrova, & Minakhina, 2003). Therefore, the preexistence of these MGEs may also have contributed to the rapid evolution of resistance in the modern era.

Predation by protists might also act as a driver for the presence of MRGs in bacteria (Hao et al., 2017, 2015, 2016). Metal poisoning is employed by protists to first inactivate and then kill bacteria (Hao et al., 2015, 2016). In response, bacteria have evolved metal detoxification strategies including copper/zinc resistance determinants and thus have selected for metal (copper/zinc) resistance to avoid killing by metal poisoning (Hao et al., 2016). Since these resistance genes would aid survival in protists, one could expect a higher occurrence of additional copper, zinc and arsenic resistance determinants. Thus, this could be an important factor/driver to select metal resistance or co-select antibiotic resistance. These mechanisms would also predate the antibiotic era.

Thus, the ancient nature and broad distribution of metal ion resistance and homeostasis genes, efflux pumps, MGEs, and ARGs, suggests that the 'tool-kit' of genes and other elements required for the evolution of multi-resistant bacteria already existed before the modern antibiotic era. This leads to the question: Has the development and spread of resistance to antibiotics in pathogens been further promoted by the exposure to metals? In the next section, we will provide a short historical reflection on bacterial resistance with a particular emphasis on the detection of metal resistance occurrence, in conjunction with antibiotic resistance.

## 3.3 AMR and Antimicrobial Metal Resistance in the 20th Century

Humans have been using metals for 4000–5000 years, although more rarely for the treatment of infectious diseases. The most widely used antimicrobial metals and metalloids have been mercury, arsenic, antimony, copper, silver and zinc, mainly in inorganic water soluble compounds. As organic chemistry developed during the 19th century, organometallic compounds became used in treatments (and still are) (Hobman & Crossman, 2015). Metal ion and other inorganic compounds were used alongside plant extracts to treat disease, although there is some sporadic evidence that crude preparations of antibiotics may have been used in some human cultures (reviewed in Aminov, 2009). The expansion of environmental pollution by metals due to human activity has been directly correlated with the presence of bacterial MRGs in the same dated samples going back to the time of the Roman Empire (Poulain et al., 2016; Staehlin et al., 2016). The empirical use of antimicrobial metals in medicine for preventing or treating infections declined when antibiotics were introduced, but medical, agricultural and public health uses of antimicrobial metals continue today. Exposure of bacteria to antimicrobial metals in the environmental, human and animal microbiomes has also occurred due to the use of these metal(loid)s in manufactured products, and release from mining and fossil fuel use.

The first descriptions of acquired antimicrobial metal resistance in bacterial pathogens came in the 1960s with mercury-resistant *Staphylococcus aureus* isolated from wounds (Moore, 1960), and linkage of mercury resistance to penicillin resistance plasmids in *S. aureus* (Richmond & John, 1964), and multiple metal ion resistance linked to *S. aureus* penicillinase plasmid carriage (arsenic/antimony, mercury, lead/zinc, cadmium and bismuth) (Novick, 1967; Novick & Roth, 1968). Smith (1967) found that R-plasmids from clinical *Escherichia coli* and *Salmonella* species carried combinations of antibiotic resistances as well resistance to mercury, nickel, cadmium, cobalt and arsenate. Retrospective studies have shown that metal and antibiotic resistances co-occurred in isolates from early in the age of wide-scale antibiotic use (see Hobman & Crossman, 2015).

Many of the original clinical isolate plasmids (then known as resistance transfer factors and subsequently as R-plasmids) were given R or NR designations such as R100 (NR1) (Watanabe, 1963). Plasmid R100 was first isolated in the late 1950s in Japan and confers resistance to streptomycin, tetracycline, sulphonamide and chloramphenicol, and latterly it was also found to confer mercury resistance (Davies & Rownd, 1972; Nakaya, Nakamura, & Murata, 1960). Other R-plasmids isolated in the 1960s/1970s were found to confer resistance to antimicrobial metals (Smith, 1967) as well as antibiotics: for example plasmids such as R773 conferred resistance to antibiotics and arsenic (Hedges & Bamberg, 1973). Silver resistance was reported to be linked with plasmid-mediated MDR (McHugh, Moellering, Hopkins, & Swartz, 1975) and sulphonamide resistance (Bridges & Lowbury, 1977), and other studies using transformation, plasmid capturing and sequencing demonstrated that resistance genes to metals and antibiotics could be linked on plasmids, although the detailed mechanisms involved were at the time not determined (Nakahara, Ishikawa, Sarai, Kondo, & Mitsuhashi, 1977). The linkage between metal and antibiotic resistances on plasmids is now well established, and, despite a reduction in use of antimicrobial metals in medicine and agriculture, antimicrobial metal resistances are still often found on the same MGEs as antibiotic and other AMRs (Pal et al., 2015).

It is still largely accepted that R-plasmids evolved by the insertion of ARGs carried by mobile elements into a preexisting pool of Enterobacterial plasmids, rather than by the spread of preexisting R-plasmids. However, there is now more evidence that ancient MDR plasmids did exist, at least in permafrost. Plasmid pKLH 80 from 15,000 to 35,000-year-old Siberian permafrost *Psychrobacter maritimus* strain carries resistance to tetracycline, streptomycin and penicillin (Petrova, Kurakov, Shcherbatova, & Mindlin, 2014), and Tn*5045* from a *Pseudomonas* sp. isolate from 15,000 to 40,000-year-old permafrost carries streptomycin, sulphonamide and chromate resistance (Petrova, Gorlenko, & Mindlin, 2011). There are other mercury-resistant and antibiotic-resistant bacterial isolates of similar antiquity (Mindlin et al., 2009).

Evidence from the 'preantibiotic' era has also been obtained by resuscitation of (pathogenic and nonpathogenic) bacterial strains from long-term storage. The complete genome sequence of the original *E. coli* (*Bacterium coli commune*) NCTC 86 has recently been determined. This strain was isolated around 1885 by Theodor Escherich from a child with no signs of diarrhoea, and carries silver resistance genes (Dunne et al., 2017). *Enterobacter cloacae* ATCC 13047, the type strain isolated in 1890, contains multiple mercury, copper, silver and arsenic resistance genes, including resistances carried on plasmid pECL-A, and efflux pumps (Ren et al., 2010). Further evidence from the Murray collection made in the UK between 1919 and 1949, shows that although metal resistances (to mercury, copper, arsenic, tellurite) were present, antibiotic resistance was much rarer (Baker et al., 2015; Essa et al., 2003; Hughes & Datta, 1983; Jones & Stanley, 1992). Nevertheless, recent work has shown that 30% of *Klebsiella pneumoniae* isolates from the Murray collection were penicillin resistant and carried the $bla_{SHV}$ genes, but showed lower levels of biocide resistance (Wand et al., 2015).

## 3.4 Antimicrobial Metals and Metal Resistances

Metals, such as mercury, are persistent in the environment. They can be globally distributed, e.g. within the mercury cycle, or can cause localised high levels of pollution. Toxic metals are particularly problematic as they are immutable, but their valency and bioavailability in the environment can be altered through physical or biological processes.

### *3.4.1 Mercury Resistances and Integrons*

Mercury and mercury salts have been used as antimicrobials since at least the time of Pliny the elder (in the first century AD), and also independently in China. The use of mercury in medicine is believed to have been developed by Arab scholars and doctors who passed on this knowledge into Europe around the 12th century. Mercury (and guaiacum—holy wood) was used to treat syphilis in Europe from the 15th century, but the heyday of mercury treatment, specifically for syphilis, was probably from the 16th to late 19th centuries, where various applications of mercury vapour, ointments, pills or injections of mercurials were used in treatment. Initially these treatments were mercury metal (Hg) in ointments or through inunction (exposure to mercury vapour), but the less toxic mercurous chloride ($Hg_2Cl_2$) (calomel) or more highly toxic mercuric chloride ($HgCl_2$) (corrosive sublimate) were used in later syphilis treatments, as well as other inorganic mercury compounds such as ammoniated and salicylated mercury ointments (Frith, 2012). Other mercurials have been used as antimicrobials; particularly mercury nitrate $Hg(NO_3)_2$ and mercury oxide (HgO) compounds (Hobman & Crossman, 2015; Silver & Hobman, 2007). Although mercury treatments for infections were superseded by Salvarsan (albeit in combination with other antimicrobial metals) in the early 20th century, and then by sulphonamides and antibiotics, mercury compounds were still used in other medical treatments (for treating constipation, in teething powders, for the elimination of parasitic worms, as diuretics) and also as disinfectants and antiseptics. Mercurials such as phenylmercuric nitrate, merbromin (mercurochrome), nitromersol (metaphen), thimerosal (merthiolate-sodium ethylmercurithiosalicylate) and others have been used in these roles in hospitals and some were used until the 1970s. Thimerosal and phenylmercuric acetate have also been used as preservatives in pharmaceuticals and eyedrops—and are still used. There has been a widespread controversy over the use of thiomersal as a vaccine preservative. Mercury amalgam dental fillings containing mercury, silver, tin, copper and zinc, have been widely used since the 1840s (although it is likely that amalgam has been in use in China since the seventh century) and are still used in some countries. Amalgam fillings are probably the major contribution to mercury body burden in humans that have them (WHO, 2003).

Mercury compounds have also been widely used in agriculture, up until the middle/late 20th century where organomercurial antifungal seed dressings (methyl-, ethyl-mercury derivatives) were widely used, and prior to that as timber preservatives, pesticides and fungicides (Huisingh, 1974). Consumption of organomercury treated cereal grains, or consumption of animals that have consumed these cereals has led to outbreaks of organomercury poisoning affecting many hundreds of people in Guatemala, Iraq and Pakistan during the 1950–1970s (reviewed in Silver & Hobman, 2007). Large-scale pollution of the environment by mercury in the past has been the result of coal burning, losses of mercury during industrial processes (such as sodium hydroxide production), and metal mining or refining either of mercury, or when using mercury to extract precious metals as amalgams. Evidence for the mobilisation and deposition of mercury during the industrial revolution has been found in dated ice cores from the arctic (Poulain et al., 2016). Gold-mining activities where gold is extracted from ore as an amalgam with mercury have led to serious pollution in Africa and South America. Pollution effects can be widely distributed, as in the case of mercury release into the atmosphere from coal burning. In the case of Minamata disease, a localised environmental release of mercury has had effects on marine life, and on humans who consumed seafood caught in the polluted waters, as well as selection for mercury-resistant marine bacteria (Harada, 1995). The EU is reducing the demand for and use of mercury and has banned mercury exports. Norway, Sweden and Denmark have banned all mercury products including the use of mercury in amalgam fillings.

#### 3.4.1.1 The Mer Mercury Resistance System

Mercury resistance was the first metal resistance discovered in bacteria and is arguably the most well known. Although there are differences in detail of the mechanism of mercury resistance in Gram-negative and Gram-positive bacteria, the underlying resistance mechanism is the same, and absolutely requires the reduction of $Hg^{2+}$ to $Hg^0$ in the bacterial cytoplasm by the enzyme mercuric reductase encoded by the *merA* gene. The reduced (metallic) mercury is volatile at room temperature and diffuses out of bacterial cells as a vapour. The volatilisation of mercury by bacteria was reported in 1964 (Magos, Tuffery, & Clarkson, 1964), and researchers started to elucidate the mechanism of resistance to inorganic and organic mercury compounds in the late 1960s–early 1970s (Komura, Funaba, & Izaki, 1971; Komura & Izaki, 1971; Komura, Izaki, & Takahashi, 1970; Summers & Silver, 1972; Tonomura & Kanzaki, 1969; Tonomura, Maeda, Futai, Nakagami, & Yamada, 1968). Essentially the mercury resistance volatilisation mechanism converts toxic, soluble and bioavailable $Hg^{2+}$ to insoluble metallic mercury, which in its reduced metallic state is essentially nontoxic. Mercury methylation by anaerobic bacteria was also demonstrated, which increased the bioavailability and toxicity of mercury to higher organisms.

#### 3.4.1.2 Tn21 Mercury Resistance

The role of other genes involved in mercury resistance began to be understood from the mid-1970s on (Morby, Hobman, & Brown, 1995; Nakahara, Silver, Miki, & Rownd, 1979; Schottel, Mandal, Clark, Silver, & Hedges, 1974) and the detailed mechanism of mercury resistance is now well known. The Gram-negative mercury resistance transposon Tn*21*, which was originally isolated on plasmid R100 in Japan in the mid-late 1950s will be used here as the model to describe the mechanism of mercury resistance and the association of integrons with Tn*21* subgroup mercury resistance transposons (Liebert et al., 1999). The resistance mechanism appears counterintuitive, as $Hg^{2+}$ is transported into the bacterial cytoplasm before it is reduced to $Hg^{0}$ by mercuric reductase. This is achieved by a periplasmic protein, MerP sequestering $Hg^{2+}$ in the cytoplasm and passing it to inner membrane spanning proteins MerT (and MerC in Tn*21*). These proteins pass mercuric ions through the membrane via cysteine coordination, to the N-terminal of mercuric reductase. Mercuric reductase reduces $Hg^{2+}$ to $Hg^{0}$ in an NADPH dependent reaction. Expression of the resistance genes is regulated by the mercuric ion-responsive activator MerR, with an ancillary role in regulation of expression for MerD (Fig. 1). Organomercurial resistance is encoded by an additional enzyme, MerB, organomercurial lyase, which cleaves the organo-moiety from the mercury.

#### 3.4.1.3 Integrons Carried on Mercury Transposons

Tn*21* is a Tn*3* subfamily transposable element, which also carries the In*2* integron, inserted between the mercury resistance genes and the transposition genes (Fig. 2). The integron carries the *qacEΔ*1 gene (which is a truncated version of the *qacE* gene) that effluxes biocides and confers low-level resistance to QACs and other disinfectants (Gillings, Xuejun, Hardwick, Holley, & Stokes, 2009; Kazama, Hamashimam, Sasatsu, & Arai, 1999; Küchen et al., 2000; Paulsen et al., 1993), and streptomycin and sulphonamide resistances, but also carries the integrase gene, *intI1* which can acquire, express and reassort the antibiotic resistance cassettes carried by the integrons. There are multiple reports of mercury-resistant Tn*21* or Tn*1696*-related transposons which carry integrons, which themselves carry antibiotic and biocide resistance genes (Mindlin & Petrova, 2013; Partridge, Brown, Stokes, & Hall, 2001; Rosewarne, Pettigrove, Stokes, & Parsons, 2010). Integrons were originally located on chromosomes, but are also found on MGEs. Integrons found in mercury transposons are no longer mobile by themselves but are mobile through their carriage on transposons and conjugative plasmids (Brown, Stokes, & Hall, 1996; Escudero, Loot, Nivina, & Mazel, 2015). Integrons from the In2 family have been of particular interest as vehicles for antimicrobial gene acquisition (Escudero et al., 2015; Gaze, Abdouslam, Hawkey, & Wellington, 2005; Gillings et al., 2015; Wright et al., 2008), but require the transposition capability of the Tn*3* family transposon for mobility, and appear to be closely associated with mercury resistance transposons such as Tn*21*.

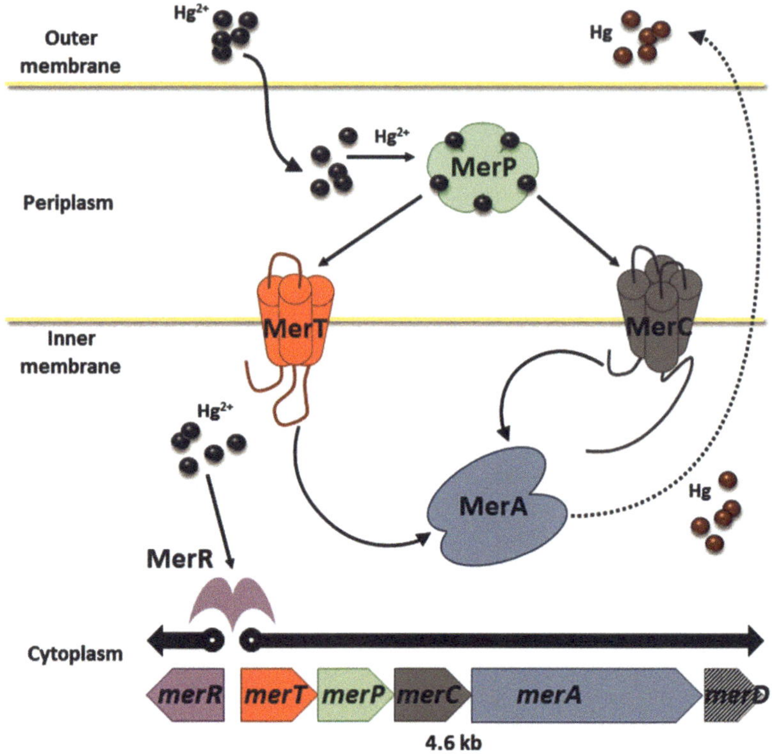


**Fig. 1** The Tn*21* mercuric ion resistance mechanism in Gram-negative bacteria. Divalent mercuric ions ($Hg^{2+}$) cross into the periplasm via outer membrane porins and bind to cysteine residues in the periplasmic mercuric ion transport protein MerP. The ions are then passed on to the transmembrane (inner membrane) MerT or MerC proteins which transport ionic mercury into the cytosol. In the cytoplasm the mercuric ions ($Hg^{2+}$) are reduced by mercuric reductase, MerA, to $Hg^0$, which is volatile at room temperature and leaves the cell as mercury vapour. MerR, is the key regulatory protein of the operon, activating transcription in response to ionic mercury with secondary regulation of the operon by MerD. Additionally, MerB (not shown) acts as an organomercurical lyase that catalyses the cleavage of the organic moiety from organomercury compounds. *Adapted from Hobman, J. L., & Crossman, L.C. (2015). Bacterial antimicrobial metal ion resistance.* Journal of Medical Microbiology, 64, *471–497.*

### *3.4.2 Copper/Silver Resistance*

Both silver and copper are increasingly being utilised because of their toxicity and broad-spectrum activity against bacteria, viruses, yeasts and fungi (Finley, Peterson, & Finley, 2013; Lemire et al., 2013; Silver, Phung, & Silver, 2006). Today, both copper and silver are used in medicine (Durán et al., 2016; Page, Wilson, & Parkin, 2009). Copper and its alloys are often

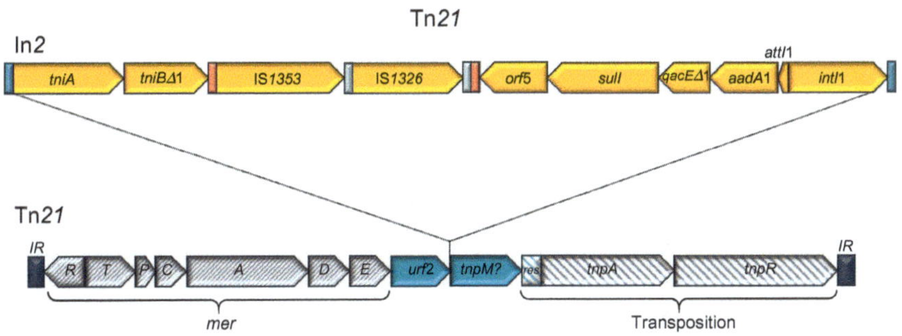


**Fig. 2** Genetic map of Tn*21*. The mercury resistance (*merRTPCADE*) and transposition (*tnpAR*) genes are shown. *Urf2* and *tnpM* are the remnants of a single gene (*urf2M*) which was interrupted by the insertion of the 11.0 kb In2 integron (Liebert et al., 1999). *Vertical bars* represent the Tn*21* flanking inverted repeat sequences (IR), or IS element inverted repeats. The In2 integron carries an unknown open reading frame (orf5) *sulI* sulphonamide resistance, the *qacEΔ1*, quaternary ammonium efflux protein and *aadA1* aminoglycoside resistances. The *intI*1 (integrase) and *attI*1 insertion site. IS*1326* and IS*1353* have inserted into In2. The integron transposition genes are incomplete in Tn*21* (*tniA* and *tniBΔ1* remain). *Adapted from Liebert, C. A., Hall, R. M., & Summers, A. O. (1999). Tn21-flagship of the floating genome.* Microbiology and Molecular Biology Reviews, 63*(3), 507–522.*

used in hospital surfaces (Schmidt et al., 2016), while silver is used in indwelling devices, wound (burn) dressings and silver catheters (Finley et al., 2015; Page et al., 2009; Sandegren, Linkevicius, Lytsy, Melhus, & Andersson, 2012). Outside of the clinical setting, copper is used in aquaculture, horticulture and livestock production as a pesticide or antimicrobial. Where copper is used in animal husbandry, it is added to animal feed to promote growth stimulation, by influencing the gut microbiota, or is used in footbaths to prevent/treat footrot (Bondarczuk & Piotrowska-Seget, 2013; Hobman & Crossman, 2015). High copper supplementation (125 ppm) in swine feed showed increased levels of copper-resistant isolates (21.1% vs 2.8%) compared to those in swine fed normal levels of copper (16.5 ppm), which suggests that supplementation of copper in animal feed can select for copper resistance (Amachawadi et al., 2011). More recently, silver has established its commercial use in household water purification systems, personal care applications, nonprescription wound dressings, clothing and in paints to prevent fouling (Gupta, Matsui, Lo, & Silver, 1999; Hobman & Crossman, 2015; Randall, Gupta, Jackson, Busse, & O'Neill, 2015; Silver, 2003). Silver nanoparticles are increasingly being used as an antimicrobial agent in antimicrobial coatings in water purification filters, computer keyboards and biomedical devices that continuously release a low level of silver ions to provide protection against bacteria, and in textiles (Mijnendonckx, Leys, Mahillon, Silver, & Van Houdt, 2013).

#### 3.4.2.1 Copper Resistance

The transition metal copper has the ability to alternate between oxidation states; from cuprous Cu(I) to cupric Cu(II). This redox potential has allowed copper to serve as an ideal biological co-factor, particularly in aerobic organisms (Outten, Huffman, Hale, & O'Halloran, 2001; Rensing & Grass, 2003). Copper appears to be able to enter bacterial cells without the need for specific copper uptake systems, but high concentrations of copper are extremely toxic to the cell (Rensing, Fan, Sharma, Mitra, & Rosen, 2000). Precise copper homeostasis has to be achieved to avoid any excess copper-mediated toxicity, while retaining an adequate supply of copper for cellular processes. Bacterial copper-homeostatic protein networks in enterobacteria are made up of four systems: *cue*, *cus*, *pco* and *cop*. Most gram-negative bacteria use the chromosomal *cue* and *cus* systems (with the exception of *Salmonella* which use the *gol* system instead of *cus*). These systems detoxify and correctly compartmentalise copper through charge modification and efflux (Hobman & Crossman, 2015).

**3.4.2.1.1 The *Cue* System** The *cue* (for Cu efflux) system is considered to be the primary mechanism responsible for copper resistance in *E. coli* in both aerobic and anaerobic conditions (Bondarczuk & Piotrowska-Seget, 2013; Rensing & Grass, 2003). The *cue* system is comprised of the inner membrane Cu(I)-translocating $P_{1B}$-type ATPase effluxer CopA and the periplasmic multicopper oxidase CueO; predicted to oxidise Cu(I) to Cu(II) (Fig. 3). An additional periplasmic copper-binding protein CueP has been discovered but only in *Salmonella* as yet. Expression of the Cue proteins is regulated by the MerR homologue protein, CueR. CueR is the cytoplasmic copper sensor protein that can be induced by elevated Cu(I) levels (Franke, Grass, Rensing, & Nies, 2003; Hao et al., 2015; Hobman & Crossman, 2015; Osman et al., 2013; Zimmermann et al., 2012). Alongside copper-mediated induction, the expression of *copA* can also be induced by silver or gold (Bondarczuk & Piotrowska-Seget, 2013; Rensing et al., 2000; Stoyanov & Brown, 2003).

**3.4.2.1.2 The *Cus* System** While the *cue* system provides copper-tolerance under low to moderate copper levels and aerobic conditions, due to oxygen limitations periplasmic copper detoxification by CueO, is inactive in anaerobic conditions (Singh et al., 2011). Therefore, under

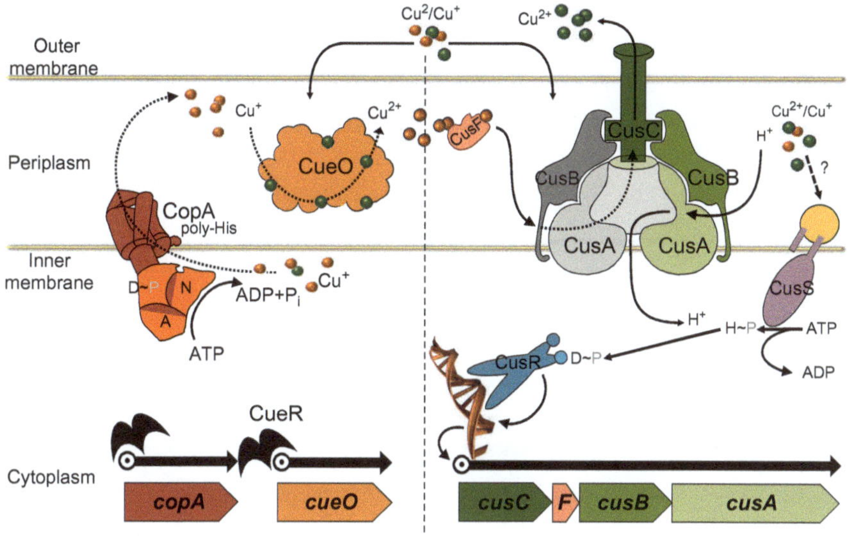


**Fig. 3** Copper homeostatic mechanisms in *E. coli* detoxify the cell of copper, using the chromosomal *cue-cus*—copper efflux systems. Copper probably enters the cell through porins, possibly OmpC and OmpF, and enters the cytoplasm. CopA is a $Cu^+$ translocating P-type ATPase and CueO a multicopper oxidase. Expression of *copA* and *cueO* is activated by CueR, a MerR family regulator. CusCFBA form a four-component copper efflux pump in which CusF is a $Cu^+$-binding metallochaperone, and CusCBA form an RND tripartite efflux complex. Transcription of these proteins is regulated by CusS, a histidine kinase, which senses copper and subsequently phosphorylates CusR, a response regulator. *Modified from Rensing, C., & Grass, G. (2003). Escherichia coli mechanisms of copper homeostasis in a changing environment.* FEMS Microbiology Reviews, *27(2–3),197–213; Hobman, J. L., & Crossman, L.C. (2015). Bacterial antimicrobial metal ion resistance.* Journal of Medical Microbiology, 64, *471–497.*

anaerobic and extreme copper stress conditions, *E. coli* uses the CusCFBA efflux mechanism found within the *cus* system (Franke et al., 2003; Munson, Lam, Outten, & O'Halloran, 2000; Outten et al., 2001; Zimmermann et al., 2012). The *cusCFBA* operon is transcriptionally regulated, in response to stress and elevated Cu(I) levels at the cell envelope, by the divergently transcribed two-component sensor-regulatory system CusRS (Munson et al., 2000; Outten et al., 2001) (Fig. 3). CusS, which is also induced by silver ions (Ag(I)) (Franke, Grass, & Nies, 2001), senses the periplasmic concentration of copper and phosphorylates CusR, thereby activating the expression of *cusCFBA*. The protein products of the *cusC*, *cusB* and *cusA* genes form CusCBA: a multiunit copper efflux pump that spans both membranes and the periplasm as a proton-substrate antiporter. The CusCBA complex,

(CusC, the outer membrane protein; CusB, the membrane fusion protein; and CusA, the substrate-binding inner membrane transporter protein from the RND family of proteins) is proposed to expel copper directly from the cytoplasm into the extracellular environment (Franke et al., 2001, 2003). CusF is a small periplasmic metallochaperone (~10 kDa) that recruits a single Cu(I) or Ag(I) to CusB and/or CusC of the efflux CusCBA machinery (Franke et al., 2003).

**3.4.2.1.3 The *Pco* System** Plasmid-mediated copper resistance was first found on plasmid pRJ1004 isolated from pigs fed on copper sulphate supplemented feed (Tetaz & Luke, 1983). The *pco* system (Fig. 4) is plasmid-encoded and provides further copper resistance in *E. coli* (Brown, Barrett, Camakaris, Lee, & Rouch, 1995). The *pco* system is encoded by a cluster of nine (or occasionally 10) *pco* genes: *pcoGFE*, *pcoABCDRS* and *pcoE*, which are arranged as two operons and a separate gene (Hao et al., 2015; Zimmermann et al., 2012). Although less well studied, an analogy-based proposed model of the system involves transcriptional regulation of the *pcoABCDEFG* genes by a CopRS-like two-component regulatory system: PcoRS (Brown et al., 1995). PcoA is a multicopper oxidase that is homologous to CueO that can detoxify copper, alongside oxidised catechol siderophore sequestration, by oxidising Cu(I) to the less toxic Cu(II) (Bondarczuk & Piotrowska-Seget, 2013; Rensing & Grass, 2003). Together with PcoA, PcoC is postulated to be a crucial component of the *pco* system. PcoC is a small periplasmic chaperone that can bind to Cu(I) and Cu(II) and has been suggested to perform several roles. PcoC could either deliver Cu(I) to PcoA for further oxidation (Huffman et al., 2002), chaperone copper to PcoD (Hao et al., 2015; Rensing & Grass, 2003) or be involved in the transport of electrons (Staehlin et al., 2016). PcoD is a copper pump which sits in the inner membrane, and although its exact function is unknown, it is predicted to function as a single unit with PcoC to deliver copper to PcoA (Huffman et al., 2002). PcoB is a predicted outer membrane transporter (Hao et al., 2015; Rensing & Grass, 2003) that potentially interacts with PcoA (Bondarczuk & Piotrowska-Seget, 2013; Hobman & Crossman, 2015). Another protein possibly interacting with and shuttling copper to PcoA, is PcoE, which is co-regulated by CusRS (Munson et al., 2000). Studies by Zimmermann et al. (2012) suggest that PcoE is an additional copper chaperone that may act as a first line of defence 'molecular sponge' protein, also conferring limited resistance to silver.

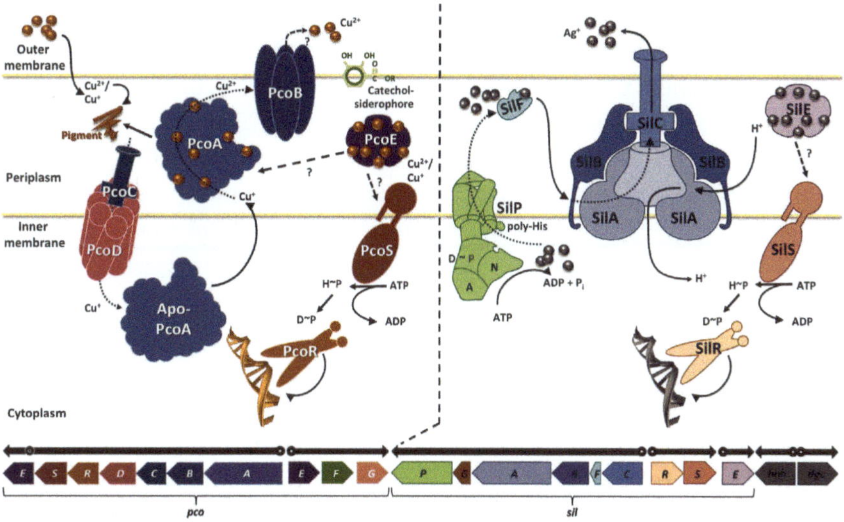


**Fig. 4** The proposed bacterial *pco* copper resistance and *sil* silver resistance systems are together known as the 'copper pathogenicity' or 'CHASRI' island. The different genes, the direction of transcription of the ORFs and the different transcriptional start sites (*circle*) are shown. The predicted function of each of the genes based on homology studies is shown in the diagram. The mechanism by which the *pco* system, encoded by *pcoABCDRSEGF* (*bottom line*), contributes to copper resistance is not fully understood. However, it is proposed to involve several periplasmic components thought to be involved in copper detoxification. Once copper has entered the cell, it is believed to be sequestered on oxidised catechol siderophores, or $Cu^{+}$ may be oxidised to the less toxic $Cu^{2+}$ by the copper oxidase PcoA. PcoC and PcoD may transport copper across the cytoplasmic membrane and PcoE bind to copper in the periplasm, where all three are involved in the shuttling of copper to PcoA. PcoB is a probable outer membrane transporter. PcoR and PcoS appear to regulate the expression of *pcoABCD*. The exact functions of PcoG and PcoF are still unknown. For the *sil* system, SilP is a P-type ATPase pump that transports $Ag^{+}$ from the cytoplasm to the periplasm. The SilCBA complex is a cation/proton antiporter, which transports $Ag^{+}$ from both the cytoplasm and periplasm out of the cell. SilF transports silver in the periplasm from SilP to the SilCBA complex. SilG is also a proposed periplasmic $Ag^{+}$-chaperone. SilS is a histidine kinase that phosphorylates SilR in the presence of $Ag^{+}$. Once phosphorylated SilR acts as a transcriptional activator for expression of SilP, G, A, B, F and C. SilE is a periplasmic metal-binding protein controlled by its own promoter and is strongly expressed in the presence of $Ag^{+}$. It is thought to act as a first line of defence to $Ag^{+}$ while the other proteins are synthesised. *This diagram has been reproduced and used with permissions from the publisher from Hao, X., Lüthje, F. L., Qin, Y., McDevitt, S. F., Lutay, N., Hobman, J. L., et al. (2015). Survival in amoeba- a major selective pressure on the presence of bacterial copper and zinc resistance determinants? Identification of a "copper pathogenicity island".* Applied Microbiology and Biotechnology, *99(14), 5817–5824.*

**3.4.2.1.4 The *Cop* System** Homologous to the *pco* system, the *cop* system is another plasmid-mediated copper resistance system that is also found on chromosomes. Composed of *copABCDRS*, the *cop* determinants have been found in *Pseudomonas syringae* pv. *tomato* PT23 isolated from plants exposed to high levels of copper, *Cupriavidus metallidurans* CH34 isolated from sediments of zinc decantation basins in Belgium, in addition to *Xanthomonas axonopodis* pv. *vesicatoria* E3C5 and *Pseudomonas aeruginosa* PAO1 (Monchy et al., 2006; Staehlin et al., 2016; Zimmermann et al., 2012). The operon structure of the *cop* genes and the protein sequences are similar to their *pco* gene counterparts (Monchy et al., 2006). Monchy et al. (2006) discovered the *copABCDRS* gene cluster was part of a larger cluster of *cop* genes, *copVTMKNSRABCDIJGFLQHE*, identified in plasmid pMOL30 from *C. metallidurans* CH34. While no putative function could be assigned for the roles of CopV, CopT, CopM, CopK, CopN, CopG, CopQ and CopE; CopI is thought to play a part in periplasmic copper detoxification/transport mediated by *copABCD* genes; CopJ shares identity to cytochrome c proteins, CopF is a putative P1-type ATPase; CopL may be involved in the regulation of CopF and finally CopH shows similarities to copper-binding CzcE (Monchy et al., 2006).

Confusingly, Gram-positive bacteria are known to use a similarly named *cop* system, recently described by Hobman and Crossman in a review (Hobman & Crossman, 2015). The mechanism has been best explained through the presence of the *copYZAB* operon in *Enterococcus hirae*. A different CopA, an ATPase, imports copper into the cytoplasm, CopZ binds to excess copper and chaperones it either to the copper export ATPase, CopB, or to the copper-responsive repressor for the operon itself, CopY (in turn activating the *cop* resistance mechanism). In *Bacillus subtilis* and *Staphylococcus*, CsoR, the cytosolic copper sensing repressor, triggers the export of copper by CopA, the $P_{1B}$-type ATPase, with CopZ assisting as a copper-binding chaperone, while in *Enterococcus faecium* from livestock a further copper resistance gene *tcrB*, which is associated with the erythromycin resistance (encoded by *ermB gene*), has been found (Hasman & Aarestrup, 2002). Copper resistance may also be linked to antibiotic resistance in *S. aureus*, as methicillin resistance is higher in copper-resistant strains (Cavaco et al., 2010).

#### 3.4.2.2 Silver Resistance

Silver resistance is conferred by the *sil* locus, which was first found in 1973 on the MDR plasmid pMG101, from *Salmonella enterica* serovar

Typhimurium following the death of several septicaemia patients in a burns unit who had been treated with silver nitrate (McHugh et al., 1975). The *sil* cluster of genes is composed of nine open reading frames—*silPGABFCRSE*, and the associated mechanism of silver resistance has been proposed through DNA sequence analysis and homologies to the bacterial *cop*, *cus* and *pco* systems (Hobman & Crossman, 2015; Silver, 2003), and in ancestral operons may have been part of a larger copper resistance operon. The *sil* genes are divided into three transcriptional units: *silCFBAGP*, *silRS* and *silE* (Fig. 4)—each of which is expressed from a different promoter (Silver, 2003; Silver, Gupta, Matsui, & Lo, 1999).

SilP is a putative inner membrane cation $P_{1B}$-type ATPase efflux pump. SilG is a putative periplasmic silver-binding chaperone (Randall et al., 2015). SilCBA is homologous to the CusCBA complex and is predicted to be a cell membrane spanning tripartite RND protein cation/proton antiporter (Gupta et al., 1999). Collectively, the outer membrane located SilC (periplasmic protein), the inner membrane antiporter SilA, and the periplasmic SilB which stabilises the complex, forms the efflux pump which pumps silver ions out of the cell. The periplasmic SilF (homologous to CusF; Osman et al., 2013) is a probable chaperone involved in transferring silver ions from SilP to the SilCBA complex (Mijnendonckx et al., 2013). The *silCFBAGP* gene unit is believed to be transcriptionally regulated by divergently expressed SilRS (homologous to CusRS and PcoRS) (Gupta et al., 1999). Together SilS, the transmembrane histidine kinase sensor and SilR, the transcriptional regulator responder, form the two-component sensor-regulator responder system. Located downstream of *silRS* is *silE*, a well-defined biomarker for silver resistance. Expression of the *silE* gene is under the control of its own promoter (Silver et al., 1999) and has been proposed by Munson et al. (2000) to be co-regulated by CusRS. Although it's precise role in resistance has not been confirmed, SilE is an intrinsically disordered protein (like PcoE), that functions as a periplasmic metal-binding protein—binding up to eight silver ions (Asiani et al., 2016). SilE has been suggested to contribute to the resistance mechanism as a periplasmic 'metal sponge' and/or first responder that can chaperone and deliver silver to other *sil* components for further activation of the resistance network (Asiani et al., 2016). However, many loci termed 'silver resistance' are in reality mainly copper resistance systems that coincidentally also give silver resistance (Hao et al., 2015).

Homologues of the *pco/sil* locus are being repeatedly found on multi-resistance plasmids, particularly isolated from swine where copper is used as a feed additive (Fang et al., 2016; Moreno Switt et al., 2012) but also in clinical isolates (Mourão, Novais, Machado, Peixe, & Antunes, 2015; Sütterlin et al., 2014; Zhai et al., 2016), and recent work has shown positive correlation of long-term copper contamination with ARGs and MGEs; and significant co-occurrence patterns between ARGs and bacterial communities exposed to elevated copper in soils for 4–5 years (Hu et al., 2016). These resistances may also have been selected because of copper use by protists grazing on bacteria (Hao et al., 2016).

The copper and silver resistance genes form a cluster of genes on a MGE (Fig. 4), which has recently been called the 'copper homeostasis and silver resistance island' (CHASRI) by Staehlin and coworkers, and is also referred to as *pco/sil* (Hao et al., 2015; Hobman & Crossman, 2015; Randall et al., 2015; Staehlin et al., 2016).

#### 3.4.2.3 Arsenic and Antimony Resistance

Arsenicals and antimonials have been used as medicinal compounds for at least 2000 years. Inorganic arsenics such as arsenic trioxide ($AS_2O_3$), arsenic sulphide ($As_4S_4$) and arsenic trisulphide ($As_2S_3$) have been widely used up until the 20th century. In addition to Salvarsan, organic arsenicals such as melarsoprol are still included in the WHO list of essential medicines. Agricultural uses of arsenics have been as herbicides, rodenticides, insecticides, fungicides, defoliants and wood preservatives. Organic arsenical compounds, such as carbarsone (4-carbamoylaminophenylarsonic acid), nitarsone (4-nitrophenylarsonic acid) and roxarsone (3-nitro-4-hydroxyphenylarsonic acid) have been used as in-feed additives to control coccidiosis in chickens (Hughes, Beck, Chen, Lewis, & Thomas, 2011).

Antimony compounds such as tartar emetic (antimony potassium tartrate) have been used in medicine to induce vomiting, to treat smallpox and skin diseases and as treatments for schistosomiasis and leishmaniasis. Sodium stibogluconate (Pentostam) and meglumine antimonite (Glucantime) are also in the WHO list of essential medicines as treatments for these parasites (WHO, 2015). Agricultural uses of tartar emetic include use as a pesticide spray and for treating parasitic infections in farm animals.

Arsenic (often with antimony) is widely distributed in soil and rocks, and naturally released into the environment by geological and microbial actions. Particularly well known is arsenic contamination of drinking water in Bangladesh, the eastern Indian state of West Bengal, areas of China and parts of the United States, but arsenic is ubiquitous in the environment (Hughes et al., 2011; Nriagu, 1989). Anthropological environmental contamination by arsenic is also related to mining or metallurgical processing release of arsenic which contaminates many other metal ores (Garelick, Jones, Dybowska, & Valsami-Jones, 2008), but also because arsenic has been used in pesticides, in pigments, and as a food additive (Hughes et al., 2011).

**3.4.2.3.1 Arsenic and Antimony Resistance Conferred by the Ars Operon** Resistance to arsenic (and antimony) was first detected in R-plasmids in the 1960s (Novick & Roth, 1968; Smith, 1967). The *ars* resistance mechanism confers resistance to $As^{3+}$, $As^{5+}$ and $Sb^{3+}$ compounds. The resistance mechanism found in both Gram-positive and Gram-negative systems requires the activity of a minimum of three proteins, ArsR, ArsB and ArsC (Fig. 5). ArsR is a trans-acting transcriptional repressor, which senses $As^{3+}$ in the cytoplasm. ArsC, arsenate reductase, reduces $As^{5+}$ to $As^{3+}$ so that it can be effluxed by the inner membrane-bound ArsB antiporter. Additional arsenic resistance genes found on Gram-negative plasmids such as R773 increase the efficiency of the system. An ArsA dimer binds to ArsB to form an ATP-energised efflux system, which is more efficient than ArsB alone, and ArsD appears to act as a metallochaperone.

In addition to resistance to inorganic arsenite and arsenate, there are now numerous genes conferring resistance to various organic arsenicals: such as *arsP* encoding the transporter ArsP for efflux of methylated arsenic species, and *arsH* encoding ArsH which oxidises the extremely toxic reduced methylated arsenic monomethylarsenous acid, MMA(III). The organo-arsenical biocycle has been reviewed recently (Li, Pawitwar, & Rosen, 2016).

Arsenic resistance in Gram-positive and Gram-negative pathogens, and carried on MDR plasmids is widely reported, with examples of carriage on mobile elements in bacteria such as *S. aureus*, *K. pneumoniae*, *A. baumanii*, *C. jejuni* and *Y. pestis* (Hobman & Crossman, 2015). Resistances carried by animal pathogens have also been linked to the use of arsenic in-feed additives (Sapkota, Lefferts, McKenzie, & Walker, 2007).

#### 3.4.2.4 Other Metals

**3.4.2.4.1 Zinc** Zinc is an essential metal, which in excess can be toxic to cells. Zinc has been known since the sixth century BC and is now widely used in alloys, coatings, batteries, wood preservatives and antifungals. Direct evidence for zinc use as an antimicrobial in antiquity comes from a shipwreck from the second century BC (Giachi et al., 2013) where tablets made from hydrozincite, smithsonite, pine resin, fats and beeswax were found, which are believed to have been used for eye treatments. Current uses

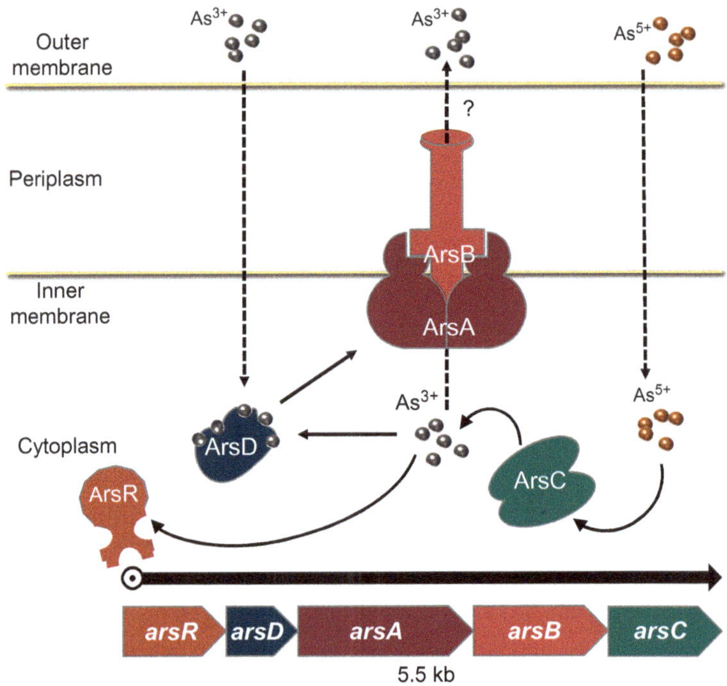


**Fig. 5** The *ars* arsenic transformation, and resistance operon and mechanism of arsenic resistance in bacteria. Arsenic enters the cell through cellular transport systems—the phosphate (arsenate) or aquaglyceroporins (arsenite) channels. In the cytoplasm, the arsenite-responsive trans-acting transcriptional repressor ArsR senses $As^{3+}$ and regulates the expression of the structural arsenic resistance genes. Once in the cell, arsenate is reduced to arsenite by the arsenate reductase ArsC. The arsenite can then be extruded by ArsB, an arsenite antiporter. ArsA binds to ArsB as a dimer and functions as an ATP-energised effluxer adding efficiency to the arsenite efflux process. Alongside its minor role in transcription, ArsD has also recently been found to act as a metallochaperone for arsenite efflux via ArsAB. Not shown here are resistance determinants for methylated arsenic species such as *arsH* and *arsP* (discussed in detail in Lin, Walmsley, & Rosen, 2006). *Adapted from Hobman, J. L., & Crossman, L.C. (2015). Bacterial antimicrobial metal ion resistance.* Journal of Medical Microbiology, 64, *471–497.*

for zinc include dietary supplements, and as a mild antimicrobial in consumer products such as toothpastes and shampoos, as well as in skin treatments such as calamine lotion. Zinc is also used as a feed additive in livestock production as well as in footbaths, at very high levels (2500 ppm) in medicated feed for piglets suffering from postweaning diarrhoea caused by enterotoxigenic *E. coli*. Zinc sulphate is also in the WHO list of essential medicines (2015) for treatment of acute diarrhoea in humans.

**3.4.2.4.2 Zinc Resistance** Zinc homeostasis of physiological levels of zinc is controlled by a combination of zinc importers, and zinc efflux pumps, such as the P-type ATPase ZntA, in *E. coli* (discussed in Takahashi et al., 2015). Expression of ZntA is regulated by the MerR family regulator ZntR. Higher levels of zinc are believed to be effluxed in Gram-negative bacteria by homologues of the Czc (Cadmium, zinc and cobalt) efflux system first found in the multiply metal-resistant *C. metallidurans* CH34 (Mergeay et al., 2003). In strain CH34, CzcCBA is a cation–proton antiporter complex, while CzcD is a cation–diffusion facilitator. CzcRS is a two-component regulator controlling expression of the *czc* operon. The *czcAB* genes have been found in *E. coli*, although there is little published work on their function.

Zinc usage has been linked with increased levels of antibiotic resistance. Studies on the effects on *E. coli* populations of feed supplementation by 2500 ppm zinc oxide by Bednorz et al. (2013) showed a significant increase in numbers of antibiotic-resistant strains, compared to control populations in animals not fed high zinc levels. Other research links zinc and copper use in animal feeds to increased incidence, or linkage to, antibiotic resistance (Medardus et al., 2014; Seiler & Berendonk, 2012; Yazdankhah, Rudi, & Bernhoftm, 2014), and zinc to increased antibiotic resistance in microbial communities from activated sludge bioreactors (Peltier, Vincent, Finn, & Graham, 2010). In December 2016, the Committee for Medicinal Products for Veterinary Use (CVMP) of the European Medicines Agency recommended that new market authorisations should be refused, and existing marketing authorisations withdrawn in the EU, for orally administered veterinary medicinal products containing zinc oxide. The reason for referral was the potential risk to the environment and risk of co-selection of antibiotic-resistant bacteria as a result of use of zinc oxide containing products outweighed treatment benefits for the prevention of diarrhoea in pigs (Veterinary Medicines Directorate, 2016).

## 4. CO-SELECTION, CO-RESISTANCE, AND CROSS-RESISTANCE

### 4.1 Co-Selection

Co-selection of antibiotic and metal resistance in bacteria is important primarily because it can maintain and promote antibiotic resistance in bacterial populations in the absence of antibiotics. Co-selection is mainly caused by cross- or co-resistance mechanisms (Fig. 6). Cross-resistance occurs when a single mechanism (e.g. an efflux pump) provides resistance to different

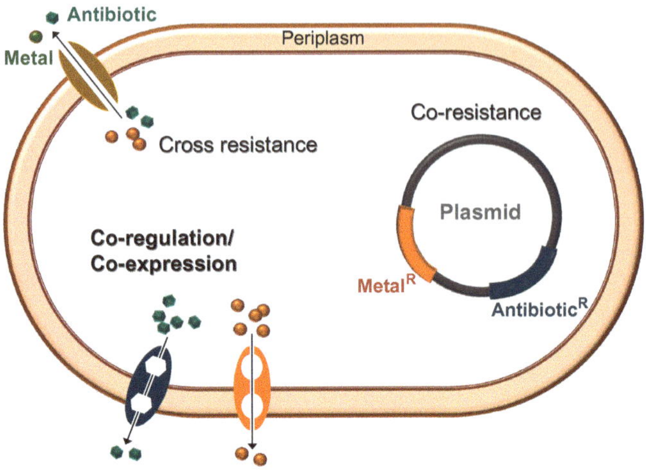


**Fig. 6** Mechanisms of cross-resistance, co-resistance and co-regulation/co-expression of metal and antibiotic resistance. Cross-resistance: where one resistance system confers resistance to both an antibiotic and a metal. Co-resistance: where resistances to antibiotics and metals are physically co-located on the same genetic element; for example on a plasmid. Co-regulation/co-expression: where expression of resistance systems to metals and antibiotics are controlled by a common regulator. *Figure modified from Baker-Austin, C., Wright, M. S., Stepanauskas, R., & McArthur, J. V. (2006). Co-selection of antibiotic and metal resistance.* Trends in Microbiology, 14*(1),176–182; and Sarma, B., Acharya, C., & Joshi, S. R. (2010). Pseudomonads: A versatile bacterial group exhibiting dual resistance to metals and antibiotics.* African Journal of Microbiology Research, 4*(25), 2828–2835.*

compounds simultaneously (Chapman, 2003). In contrast, co-resistance occurs when two or more different resistance genes are physically co-located on the same genetic element, such as a plasmid or a transposon, or are present in the same bacterial strain where each resistance gene provides resistance to different compounds (Baker-Austin, Wright, Stepanauskas, & McArthur, 2006; Pal et al., 2015). In addition to co- and cross-resistance, co-regulatory mechanisms can also promote the co-selection process. This occurs when multiple resistance genes that confer resistance to different toxic compounds are controlled by a single regulatory gene (Baker-Austin et al., 2006). For example, the regulatory protein CzcR regulates the expression of the CzcCBA efflux pump, which confers resistance to zinc, cadmium and cobalt. CzcR also co-regulates resistance to carbapenems, a class of last resort antibiotic, by repressing the expression of the OprD porin, the route of entry for these antibiotics to the bacterial cell (Perron et al., 2004).

## 4.2 Co-Resistance Mechanisms (Genetic Linkage of Resistance Genes) Between Antibiotic and Metal Resistance

Understanding the genetic linkage for resistance genes, including their association with different MGEs, such as integrons and transposons, is critical to fully comprehend the risks for horizontal transfer of resistance genes between bacteria (Bengtsson-Palme & Larsson, 2015; Martinez, Coque, & Baquero, 2015). While antibiotic resistance was already identified before penicillin became widely available, the potential consequences of the links between biocide/metal and antibiotic resistance have only been recognised more recently (SCENIHR, 2009).

Carriage of mercury and antibiotic resistance due to integron containing mercury transposons has been documented in a wide range of bacterial habitats, including oral and faecal microbial flora of primates (Wireman, Liebert, Smith, & Summers, 1997). The copper resistance gene, *tcrB*, is horizontally transferrable and linked to genes encoding macrolide and glycopeptide resistance (Amachawadi et al., 2013; Hasman & Aarestrup, 2002). Co-existence of *oqxAB* (which confers resistance to QACs, chlorhexidine, fluoroquinolones, etc.), beta-lactamases ($bla_{\text{CTX-M}}$), and the copper (*pco*) and silver (*sil*) resistance operons on the same plasmids have also been reported (Fang et al., 2016; Mourão et al., 2015). Cavaco et al. (2010) reported that cadmium and zinc can drive co-selection for methicillin resistance in *S. aureus* through horizontal transfer of plasmids containing genes for both methicillin and metal resistance (*mec* and *czr*). Plasmids carrying copper/silver resistance (*pco*/*sil*) as well as antibiotic resistance ($bla_{CTX\text{-}M}$ and *oqxAB*) have also been recently isolated, from diseased chickens, pigs and ducks in China (Fang et al., 2016); interestingly, they do not carry *ars* or *mer* genes.

The recently discovered *mcr-1* transferrable colistin resistance has also been found in copper-tolerant isolates (Campos, Cristino, Peixe, & Antunes, 2016). Thus, copper could co-select for resistance to last resort antibiotics such as colistin, which is regularly used in animal farming in Asia. This supports restriction of the use of copper as a growth promoter in animals. Pal et al. (2015) showed that although the majority of sequenced bacterial strains carry genes involved in biocide/metal tolerance/resistance mechanisms, only one out of six of the bacterial strains carry both ARGs and biocide/MRGs. Specifically, resistance genes to metals such as copper, silver, arsenic, antimony, cobalt, nickel, cadmium, iron, zinc and mercury frequently co-occur with resistance genes for many classes of antibiotics, e.g. sulphonamides, beta-lactams, amphenicols, tetracyclines and aminoglycosides. Only genes involved in resistance mechanisms to mercury,

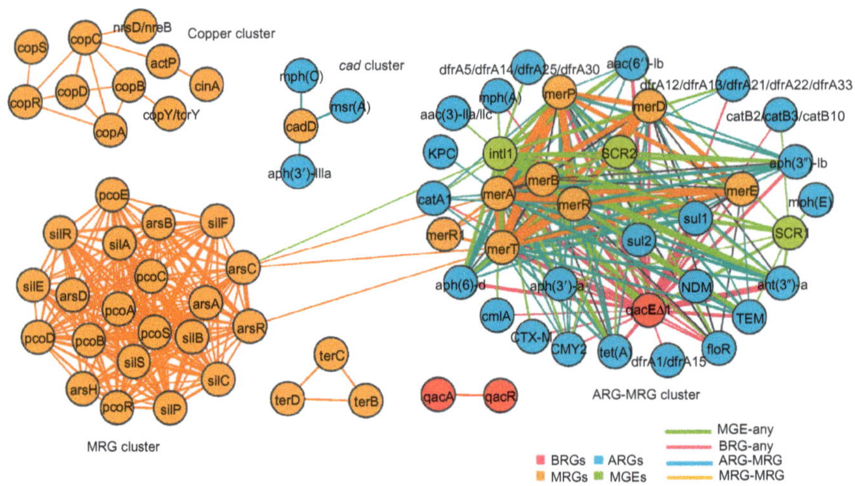


**Fig. 7** Co-occurrence of resistance genes to antibiotics, biocides and metals on plasmids. The network shows the most frequently observed co-occurrences of resistance genes to antibiotics, biocides and metals on plasmids. The network also includes connections between antibiotic/metal resistance genes and markers of mobile genetic elements (integron-associated integrases and ISCR transposases). Connections between the genes are only shown if the genes are found together on at least 10 different plasmids. The thickness of each connection between two genes is proportional to the number of times the genes co-occurred on a plasmid. The network does not show connections between antibiotic resistance genes. *Figure reproduced from Pal, C., Bengtsson-Palme, J., Kristiansson, E., & Larsson, D. G. J. (2015). Co-occurrence of resistance genes to antibiotics, biocides and metals reveals novel insights into their co-selection potential.* BMC Genomics, 16, *964.*

cadmium, zinc, tellurium, arsenic and copper/silver being frequently observed on plasmids, with the other MRGs being chromosomal (Fig. 7). Notably, both mercury and arsenic resistance genes are highly connected with integron-associated integrases but only resistance genes to mercury are highly linked to ARGs (Pal et al., 2015). This frequent linkage between mercury resistance and antibiotic resistance has been documented in bacteria from a wide range of habitats, including oral and faecal microbiota of primates (Wireman et al., 1997), oral microbial flora of patients with amalgam fillings (Summers et al., 1993), fish gastrointestinal tracts (Akinbowale, Peng, Grant, & Barton, 2007), mine sediments (Nemergut, Martin, & Schmidt, 2004) and freshwater microcosms (Stepanauskas et al., 2006). Lloyd, Janssen, Reinfelder, and Barkay (2016) showed that fish from a contaminated site contained mercury and a high abundance of mercury resistance genes and resistance to three or more antibiotics was more common in mercury-resistant bacteria isolated from the environmentally exposed fish compared to the mercury-sensitive bacterial isolates, demonstrating a potential co-occurrence of mercury and antibiotic resistances under the influence of metal pollution. In addition, Pal et al. (2015) showed that cadmium and zinc resistance genes occasionally co-occur with resistance genes to antibiotics such as macrolides and aminoglycosides. Thus, it is not only copper but also cadmium and zinc which have the potential to co-select for antibiotic resistance bacteria. Therefore, unrestricted use of these metals could potentially promote antibiotic resistance towards macrolides and aminoglycosides. Table 1 lists selected studies that have reported genetic linkage between metal and antibiotic resistances.

Whereas metal and biocide resistance genes are commonly encountered in a wide variety of bacteria from all types of environments, ARGs are considerably more common in bacteria isolated from humans (Li et al., 2016; Pal et al., 2016), and humans and domestic animals (Pal et al., 2015). This pattern is particularly clear for plasmid-borne genes (Pal et al., 2015).

**Table 1** Examples of Studies Reporting Co-Resistance (Genetic Linkage of Resistance Genes) Due to Co-Occurrence of Metal and Antibiotic Resistance Genes

| Metal | Antibiotic | References |
|---|---|---|
| Copper | Erythromycin and tetracycline | Amachawadi et al. (2011) |
| Copper | Erythromycin and vancomycin | Hasman and Aarestrup (2002) |
| Zinc | Methicillin | Cavaco et al. (2010) |
| Arsenic, copper, mercury, silver and tellurium | Chloramphenicol, kanamycin and tetracycline | Gilmour, Thomson, Sanders, Parkhill, and Taylor (2004) |
| Cadmium and zinc | Methicillin | Cavaco, Hasman, and Aarestrup (2011) |
| Arsenic, copper, silver and QACs | Trimethoprim, beta-lactam, macrolide, tetracycline and sulfonamide | Sandegren et al. (2012) |
| Copper and silver | Beta-lactam and fluoroquinolone | Fang et al. (2016) |
| Copper, silver and mercury | Colistin, ampicillin, sulfonamide, tetracycline, streptomycin and chloramphenicol | Campos et al. (2016) |

Accordingly, plasmids carrying resistance genes to both antibiotics and metals/biocides are primarily found in these two environments. Hence, the co-occurrence is most likely a consequence of the fact that humans and domestic animals constitute precisely those two microbial niches that are regularly and intentionally exposed to a strong selection pressure from antibiotics. While this does not speak in favour of metal selection as the main driver for the evolution of bacteria to carry both types of genes together, exposure to metals now apparently has the ability to provide selective advantages to such bacteria through co-resistance mechanisms. As many of these co-resistance plasmids are carried by pathogens, co-selection from metals may in some cases have direct clinical implications (Pal et al., 2015).

## 4.3 Cross-Resistance and Co-Regulation Mechanisms for Antibiotic and Metal Resistance

Cross-resistance to $Cd^{2+}$, $Zn^{2+}$, sodium dodecyl sulphate, beta-lactams, kanamycin, erythromycin, novobiocin and ofloxacin due to the DsBA–DsbB disulphide bond formation system in *Burkholderia cepacia* has been described (Hayashi, Abe, Kimoto, Furukawa, & Nakazawa, 2000), and in *Salmonella* Typhimurium, BaeSR regulation of the AcrD and MdtABC drug efflux systems confers resistance to copper, zinc and β-lactam antibiotic resistance (Nishino, Nikaido, & Yamaguchi, 2007). Cross-resistance is also potentially possible via enhanced efflux systems, which can confer cross-resistance to metals and antibiotics, for instance the TetA(L) protein effluxes tetracycline and cobalt (Cheng, Hicks, & Kruwlich, 1996) as well as via alteration in cell membrane structure, outer membrane adaptation or via mutations. Similarly, a multidrug efflux pump encoded by MdrL protein in *Listeria monocytogenes* has been shown to pump out heavy metals such as zinc, cobalt and cadmium, as well antibiotics such as erythromycin, josamycin and clindamycin and biocides, such as benzalkonium chloride (Mata, Baquero, & Pérez-Díaz, 2000). Table 2 lists selected studies that have reported cross-resistance between metal and antibiotics.

Co-regulation of antibiotic and metal resistance has also been reported such as in *P. aeruginosa* where CzcRS, controls (increase) the expression of the *czc*CBA efflux system conferring resistance to cadmium, zinc and cobalt, and decreases expression of the OprD porin, leading to increased resistance to carbapenems (Perron et al., 2004). Similarly, overexpression of the Rob protein (a binding protein) encoded by the *rob*A gene increases resistance to heavy metals, as well as multiple antibiotics in *E. coli* (Nakajima, Kobayashi, Kobayashi, Asako, & Aono, 1995).

**Table 2** Examples of Studies Reporting Cross-Resistance (A Shared Mechanism of Resistance) to Both Antibiotics and Metals

| Metal/Biocide | Antibiotic | References |
|---|---|---|
| Zinc, cobalt, chromium and benzalkonium chloride | Erythromycin, josamycin and clindamycin | Mata et al. (2000) |
| Cobalt | Tetracycline | Cheng et al. (1996) |
| Gold, acriflavine (class: acridine) and alexidine (biguanide class) | Chloramphenicol, cloxacillin, thiamphenicol and nafcillin | Conroy, Kim, McEvoy, and Rensing (2010) |
| Copper and zinc | Tetracycline | Flach et al. (2017) |

## 4.4 Selected Studies Evaluating the Co-Selective Ability of Metals by Metal Exposure

Metals are widespread across environments (Nriagu, 1996), therefore providing ample opportunities for selection given that local concentrations are sufficiently high. Co-resistance plasmids tend to be most common in bacteria of human or domestic animal origin (Pal et al., 2015). The fact that these are the prime environments where pathogenic bacteria reside, the largest potential for metals to promote clinically relevant antibiotic resistance is therefore in and on our bodies, but also environments that serve as potential transmission routes for pathogens.

Many studies reported significant positive correlations between the presence of ARGs and MRGs in different environmental contexts such as in swine (Holman & Chénier, 2015), swine waste lagoons (McKinney, Loftin, Meyer, Davis, & Pruden, 2010) and in soils and faeces from dairy farms (Zhou et al., 2016). Positive correlations between antibiotic and metal resistance, and between total AMR genes and copper was observed in a study of manure, compost and soils from a Chinese swine farm (Zhu et al., 2013). However, correlation does not imply causality. As many factors other than metal exposure differ between groups, the actual causes and mechanism behind the correlations are often not clear. Studies based on controlled addition of metals compared with control groups therefore bear a considerably higher value in determining causal relationships.

Furthermore, many community-based studies investigate overall changes in resistance genes or proportion of resistant bacteria, but care should be taken when interpreting those studies as the results almost without exception are confounded by gross changes in taxonomic composition of the investigated communities. Therefore, the selection (by the metal) could simply be for species that also happen to be resistant to an antibiotic. This is a scenario that is very different, and less serious, from that where metals select for specific strains within species, strains that also have acquired resistance to antibiotics. While the former scenario will change the balance between species in the exposed community, the latter specifically selects for resistant strains, and thereby contributes to the fixation of resistant genotypes, derived by spontaneous mutations or gene transfer events. From the perspective of antibiotic resistance, it is the latter that has significant clinical importance. Finally, if changes in antibiotic resistance (phenotype or genes) and metal resistance (phenotype or genes) are simply measured in parallel, and not within the same types of cells in a complex community, then there is no direct evidence for co-selection in the more meaningful sense, i.e. of bacteria carrying both types of resistances. Taken together, while studies on communities better represent a 'real-life scenario', co-selection measured on a community level is from several points of view often considerably less informative than co-selection studied within a species, unless care is taken with regards to controlled exposure and selection for co-resistance within species.

Antibiotics, biocides and metals are often found at subinhibitory concentrations in sewage effluents. Bengtsson-Palme et al. (2016) studied resistance genes to antibiotic, biocide and MRGs in different steps of three Swedish municipal wastewater treatment plants (WWTPs). However, the study found no evidence of co-selection for antibiotic resistance by metals, although large changes in taxonomic composition could have obscured effects. Another study reports strong correlations between resistance genes to both antibiotics and metal resistance (Di Cesare et al., 2016), however whether this is a result of co-selection is not known.

There are many field-based studies that investigate co-selection of antibiotic resistance by short- or long-term exposure of metals in different environments (Berg et al., 2010; Singer, Ward, & Maldonado, 2006; Stepanauskas et al., 2006; Wright, Loeffler Peltier, Stepanauskas, & McArthur, 2006). Microcosm studies of freshly collected Savannah River (USA) surface water samples found that exposure of bacterial communities to individual toxic metals selected for multiantibiotic and metal-resistant microorganisms, with increases in the frequency of multiple resistance to gentamicin, tetracycline and ampicillin in freshwater bacteria resulting from addition of cadmium. The antibiotic and metal resistance mechanisms were not determined, but ribotypes from the microcosms suggested that *Pseudomonas* sp., *Burkholderia* sp., *Acinetobacter* sp. and *Ralstonia* sp., related to opportunistic human pathogens were present in the resistant communities (Stepanauskas et al., 2006).

In another study, bacteria sampled along a gradient of metal contamination sites were more tolerant to metals, as well as antibiotics, compared to bacterial samples from a reference site (Wright et al., 2006), which suggests co-selection for antibiotic resistance, but quite possibly on the species level. A controlled, field-based study has also shown that 4–5 years of copper addition (0–800 mg/kg) to soils changed the diversity, abundance and mobility potential of a broad spectrum of ARGs in two types of agricultural soils (Hu et al., 2017). Taxonomic changes were large, and isolates were not studied, hence it was not clear if the copper exposure selected for co-tolerant strains within species.

Copper exposure has also been correlated with increased levels of bacterial resistance to various antibiotics in soil environments. In one study, two soils from a Cu-gradient field site contaminated solely with copper over 80 years ago was compared to a soil from an adjacent site just outside the contaminated area. High levels of copper correlated with Cu-resistant microbial communities, and also with higher community-level tolerance to vancomycin and tetracycline. Isolates from the contaminated soils were shown to be more tolerant to chloramphenicol and nalidixic acid (Berg et al., 2010). Similarly, a Scottish study on randomly selected archived agricultural soils collected between 1940 and 1970 reported that the presence of 11 ARGs correlated to soil copper, chromium, lead and iron levels (Knapp et al., 2011). However, these studies too only demonstrate correlation and not causation, and the changes in ARGs could again be due to taxonomic shifts. Recently, long-term soil exposure experiments have shown that nickel has significantly increased the abundance, diversity and horizontal transfer potential of ARGs from soil bacteria (Hu et al., 2017).

Antifouling paint, containing high levels of copper and zinc, selects not only for metal-resistant bacteria in marine biofilms but also for bacteria resistant to tetracycline (Flach et al., 2017). Interestingly, known mobile ARGs, including *tet*-genes, did not increase as a result of the exposure. On the other hand, a range of efflux pumps, capable of providing resistance both to metals and antibiotics, increased consistently in the exposed microbial communities. This suggests that the main mechanisms of co-selection were via cross-resistance mechanisms. Another effect of the exposure to the paint was a clear and significant enrichment of integrases and insertion sequence common regions (ISCRs). Both integrases and ISCRs are genetic elements associated with the mobilisation of genes, and they are often associated with ARGs. This suggests a possible role of heavy metals in promoting the mobilisation of genes, including those involved in resistance.

Co-selection of antibiotic resistance has also been reported in bacterial isolates from the mouse gut, where exposure to iron and arsenic has been shown to increase types and abundance of ARGs, in some cases, but the alteration in ARGs may be due to alteration in the diversity of gut microbiota (Guo et al., 2014).

There are, however, studies that have investigated changes in resistance patterns *within* species, excluding the confounding effects of taxonomic changes. Human populations with higher body burdens of mercury had a larger proportion of antibiotic-resistant *E. coli*, in their faeces, than did control populations, although the populations that were compared differed in aspects other than the mercury exposure (Skurnik et al., 2010). In experiments with apes, the oral and gut microbial flora of test animals was altered after bursts of mercury were released on installation and removal of amalgam fillings, with peaks of mercury release being correlated with increased numbers of antibiotic- and mercury-resistant bacteria (Summers et al., 1993).

Cadmium and zinc have been reported to co-select for methicillin resistance in *S. aureus* through horizontal transfer of plasmids containing methicillin and MRGs (*mec* and *czr*) (Cavaco et al., 2011). The use of copper and zinc as replacements for, or supplements to, antibiotics as growth promoters in animal feeds, or in medicated feeds for treatment of diarrhoea, shows significant correlation to antibiotic resistance in bacteria isolated from these environments (Bednorz et al., 2013). For example, Hölzel et al. (2012) showed a positive correlation between copper and zinc levels in pig manure to the proportion of beta-lactam resistant *E. coli*. While copper levels and resistance followed a clear concentration response, the same was not true for zinc. Notably, heavy metals may be administered together with antibiotics, potentially confounding the results. As the study was performed on stored, rather than fresh faeces, analyses of antibiotic residues (as briefly mentioned but not reported in the study) the differences in stability of heavy metals and organic antibiotics may have obscured results. Experiments with controlled addition of zinc and/or copper and analyses of fresh manure would be valuable to follow up these studies.

## 5. PLASMIDS, FITNESS COSTS AND SELECTION PRESSURE

### 5.1 Plasmids and Other MGEs

MGEs such as plasmids, integron gene cassettes and transposable elements play important roles in co-selection of resistance (Frost, Leplae, Summers, &

Toussaint, 2005). Plasmids belonging to the same incompatibility groups (and host range) as those that today carry resistance genes to antibiotics and metals were present in Enterobacteria before the antibiotic era (Datta & Hughes, 1983). Integrons provide a system for capturing and expressing resistance gene cassettes. Mobile integrons (carried on other MGEs) such as class I integrons were found to carry ARGs, and disinfectant and small drug resistance genes (130 reported), and many genes with unknown functions, but not known MRGs (Escudero et al., 2015; Gillings, 2017; Partridge, Tsafnat, Coiera, & Iredell, 2009; Summers, 2004, 2006). However, horizontal transmission of these elements requires their association with transposable elements and/or conjugative plasmids (Escudero et al., 2015). In the case of class I integrons, which appear to be the main carriers of antibiotic resistance cassettes, they require transposable elements to move them to plasmids, and then into new hosts. The Tn*3* family mercury resistance transposons (often Tn*21* or Tn*1696* family), have been widely associated with carriage of integrons. Conversely, MGEs could act as reservoirs for resistance gene cassettes that integrons acquire and express, with a huge potential for diversity of integron resistance gene assemblies (Gaze et al., 2011; Gillings et al., 2015).

Many of the first studied AMRs were found to be carried on transposable elements: either on composite (nonreplicative) transposons such as Tn*5*, or on complex (replicative) transposons such as Tn*3* family transposons. Transposable elements can use plasmids (or could be transduced by bacteriophage) to move from one host to another, according to their host range, but other genetic elements play roles in acquiring, retaining and moving AMR genes (Stokes & Gillings, 2011). Genomic islands and integrative conjugative elements (ICEs) have roles in acquiring, moving or acting as reservoirs for resistance genes.

The role of horizontal gene transfer (HGT) in the proliferation of ARGs has been widely studied, and generally, conjugation, as opposed to transduction and natural transformation, is often considered as the most common mechanism for HGT of ARGs in various environments (Huddleston, 2014; Norman, Hansen, & Sørensen, 2009). Plasmids also often carry MRGs, although in general and relative terms, these genes are more often chromosomally encoded than are ARGs (Pal et al., 2015). Plasmids frequently carry traits which confer an advantage to the host under certain environmental conditions. Under conditions where the plasmid may not confer an advantage to the host, but rather a metabolic burden, stable maintenance of the plasmid can be achieved by a number of different mechanisms. Systems to retain plasmids include active partition systems, plasmid addiction systems which actively kill cells which have lost plasmids, plasmid carriage of essential metabolic genes (Kroll, Klinter, Schneider, Voß, & Steinbücher, 2010), or complex regulatory mechanisms to minimise cost of carriage (Herman, Thomas, & Stekel, 2012). All plasmid retention systems stabilise plasmid carriage, and will lead to retention of the plasmid as well as the genes it is carrying, irrespective of whether there is a selective pressure. Interestingly, plasmids that carry toxin–antitoxin systems are also particularly prone to carry biocide and MRGs (Pal et al., 2015).

## 5.2 Fitness Costs and Selection Pressures

Acquiring resistance usually comes with a cost. Resistant bacteria often exhibit a slower growth rate, and occasionally lower virulence or lower transmission rate than its wild type, susceptible counterparts in the absence of the antibiotic (Andersson & Hughes, 2010). The fitness cost of developing resistance varies dramatically, primarily depending on the mechanisms involved. Some resistance mechanisms have no cost for bacteria and thus the resistant strain is always as fit as or fitter than the susceptible counterpart. Resistance due to a target site mutation often has high fitness cost but is variable and depends on the host genetic background. However, bacteria tend to evolve over time to compensate for the fitness cost without losing the resistance (Andersson & Hughes, 2010). Plasmids too can maintain complex regulatory mechanisms that can serve to minimise fitness burdens to their hosts (Herman et al., 2012). On average chromosomal resistance mutations carry a larger cost than acquiring resistance via a plasmid (Vogwill & MacLean, 2014). Metal resistance normally requires specific multigene resistances which are normally horizontally acquired, rather than simple target site mutations. These factors may explain why resistance can be gained by plasmid acquisition and also highlight the need to understand why resistance plasmids carry a relatively low cost. The probability of carrying both antibiotic and other AMR genes is higher if the plasmid-encoded genes display low fitness costs in certain bacteria. This may explain the very low levels of antibiotics and antimicrobial metals required to select for bacteria carrying a multiresistance plasmid (Gullberg, Albrecht, Karlsson, Sandegren, & Andersson, 2014).

To understand risks, it is critically important to define the relationship between the exposure level to metals and the ability to select for metal resistance and associated antibiotic resistance. In natural environments, the concentration of metals such as copper, zinc and cadmium can often reach over a few thousands of mg/kg, which may be sufficient to drive the co-selection process (Seiler & Berendonk, 2012). Concentrations lower than the MIC can also be selective, but how much lower largely depends on the cost of carrying the resistance factors in relation to the growth advantage it gains from carrying it, and the bioavailability of the metal or antibiotic. The minimal selective concentration (MSC) can be expressed at the concentration where the costs for carrying the resistance determinant equal the benefits (Andersson & Hughes, 2010). In vitro studies on the selection for carriage of a MDR plasmid, which confers resistance to both antibiotics and metals, suggested an (MSC) of ~140-fold lower than minimal inhibitory concentration (MIC) for arsenite and an MSC 10-fold lower than MIC for copper sulphate can select for retention of a large MDR plasmid (Gullberg et al., 2014).

In many environments, bacteria encounter several metals, antibiotics and other antibacterial compounds simultaneously, at a range of different concentrations. Exposure to multiple metals has been shown to be more effective for co-selection for antibiotic resistance than exposure to a single metal (Guo et al., 2014). Moreover, it has been suggested that genetically identical bacteria could show heterogeneous MRG expression (Takahashi et al., 2015); this could help maintain metal resistance under sublethal exposure. However, further studies are required in this area to understand the risk for co-selection of antibiotic resistance due to combination effects.

## 6. CURRENT MODELS AND KNOWLEDGE GAPS

Though there are numerous studies which report the association between metal exposure and increased antibiotic resistance, in the majority of them, data on dose/concentration–response relation for metal exposure and resistance are lacking. One of the hypotheses that has received experimental support is that exposure to antibiotics has been the major driver behind the overrepresentation of co-occurrence between ARGs and biocide/MRGs in bacteria from humans and domestic animals, rather than exposure to biocides or metals (Pal et al., 2015). This is based on evidence for the preexistence of metal resistances in clinical bacteria before antibiotics were widely used, and that metal ion resistance genes are found widely on plasmids where there is little evidence of antimicrobial metal usage, whereas ARGs are clearly most common on plasmids of human and animal origin. An alternative hypothesis is of course that exposure to metals has contributed to the selection and development of bacteria and MGEs with co-selective potential. However, further research is needed in this area to uncover the link between metal exposure and antibiotic resistance.

Understanding of the mechanisms and selective pressures for co-occurrence and co-selection is not fully established. Fundamentally, it is not known to what extent co-occurrence is driven by properties of MGEs, for example high levels of transposon/integron activity or plasmid transfer, as opposed to being driven directly by selection for varying concentrations of antibiotics in the environment. Nor is it known to what extent different patterns of exposure, i.e. continuous, repeated or occasional to different concentrations of antibiotics or metals are more or less likely to lead to co-occurrence of resistance genes, and thus the potential for co-selection. Mathematical modelling could be a valuable approach to providing answers to these questions.

Mathematical models have been used to study the effect of antibiotic presence in the environment on AMR emergence (Bell, Schellevis, Stobberingh, Goossens, & Pringle, 2014; Gerrish & García-Lerma, 2003; Iwasa, Michor, & Nowak, 2004; Tanaka, Bergstrom, & Levin, 2003), the effect of interaction between drugs on AMR emergence (Michel, Yeh, Chait, Moellering, & Kishony, 2008), the effect of treatment strategy on resistance selection (Ahmad et al., 2016; D'Agata, Dupont-Rouzeyrol, Magal, Olivier, & Ruan, 2008; Murphy, Walshe, & Devocelle, 2008) and have quantified the relationship between HGT and resistance emergence in bacterial populations (Gehring, Schumm, Youssef, & Scoglio, 2010). Population-level studies have been conducted to understand how the resistance is spread from one host or location to another. These models generally study a community where parameters related to resistance spread such as antibiotic frequency, dose and duration can be readily measured. This has helped in a better understanding of resistance emergence and selection in an environment where antibiotic use is high (Alawieh et al., 2015; Austin, Kakehashi, & Anderson, 1997; Bonhoeffer, Lipsitch, & Levin, 1997; Temime, Boëlle, Courvalin, & Guillemot, 2003). Of particular interest is the study combining both population and pharmacological levels in a single model (Opatowski et al., 2010). The study reaffirms the hypothesis that resistance patterns in communities are related to individual doses. Mathematical models for emergence and spread of AMR in the environment (Ayscue, Lanzas, Ivanek, & Grohn, 2009; Dolliver, Gupta, & Noll, 2008) have shown that resistant organisms can persist even after antibiotic treatment has ceased (Ayscue et al., 2009; Sharifi, Murthy, Takacs, &

Massoudieh, 2014; Volkova, Lu, Lanzas, Scott, & Grohn, 2013), can make predictions for how long resistance is likely to remain (Volkova et al., 2013), and can identify factors to which spread of resistance is likely to be most sensitive (Baker, Hobman, Dodd, Ramsden, & Stekel, 2016). However, to our knowledge, mathematical models for co-selection of antibiotic and metal resistance have not yet been developed. Such models will be important both from explanatory and predictive perspectives. From an explanatory perspective, these models will provide a framework to dissect the open questions about co-occurrence of resistance genes mentioned earlier. From a predictive perspective, models will be able to assess questions such as the extent to which antibiotic resistance may persist following the withdrawal of antibiotics but in the continued presence of metals, for example because of agricultural use or land contamination.

The last decade has seen an enormous increase in the number of sequenced bacterial genomes and plasmids deposited in DNA sequence databases. These are a valuable resource for studying the co-selection potential of metals, biocides and antibiotics, using co-occurrence analysis of resistance genes that are found together on bacterial chromosomes or plasmids, or on other MGEs such as integrons and transposons (Li, Xia, & Zhang, 2017; Pal et al., 2015, 2014). In contrast to co-occurrence analysis, using bioinformatics approaches to predict co-selection potential based on cross-resistance is more challenging, as the proteins involved (often membrane pumps) have a broad and largely uncharacterised substrate specificity. Hence, assessing cross-resistance potential accurately is in relative terms more dependent on phenotypic assays. In addition, in silico approaches to studying resistances and their co-selection in individual bacterial isolates and in complex communities have many limitations. One of the major drawbacks with this approach is that DNA sequences provide information about genes that are assumed to be expressed. However, not all genes are expressed in a given bacteria or under a given environmental condition. Another major technical challenge is the difficulty of completely sequencing/assembling the genomes of microorganisms present in a given sample due to insufficient sequence coverage. As a consequence determining, the true genetic context of resistance genes in a certain bacterial isolate within a metagenome becomes a complicated task, thus hampering identification of the co-selection potential.

As known resistance genes are widespread across environments (Pal et al., 2016), it is likely that many unknown resistance genes have yet to be discovered. Novel putative ARGs have been identified in complex microbial communities using either homology-based searches of protein domains or using functional metagenomic approaches (González-Pastor & Mirete, 2010; Hatosy & Martiny, 2015). This suggests that in order to discover novel mechanisms for metal resistance, other approaches to searches for homology across entire proteins would be valuable (Li, Cai, Zhang, & Zhang, 2014). Bioinformatics searches for metal-binding protein sequence motifs in metagenomes, or using functional metagenomics approaches to isolate novel functional genes have been used in the search for novel ARGs in the past, could be valuable.

Other gaps in knowledge include understanding the biological mechanisms that retain resistance genes even under conditions of no selection pressure, and what other factors associated with antimicrobial metal use may drive HGT, gene acquisition and evolution of resistance co-occurrence. The role of protists in selection for bacterial metal resistance, and how that may select for metal resistance in 'unpolluted' sites is just starting to be established. First experiments have been conducted in amoeba but protists were the first eukaryotes and have thus diverged widely. We do not know if all protists use metals in predation and to what extent the availability of metals and metalloids influences this—where copper is available protists might use more copper whereas arsenic may be used in other environments (Hao et al., 2017, 2015, 2016).

## 7. CONCLUSION AND FUTURE PERSPECTIVES

There are many studies supporting the idea of co-selection of antibiotic resistance by metals, but conclusive data showing a widespread clinical impact of biocide or metal use on antibiotic resistance is still lacking. Thus, more studies are needed to evaluate the relationship between exposure level to antimicrobial metals and the selection/maintenance of antibiotic-resistant strains in different environments. Similarly, knowledge of the bioavailable concentrations of metals that may drive HGT events are important to more comprehensively understand risks (Jutkina, Rutgersson, Flach, & Larsson, 2016).

We have recently predicted that there are, in relative terms, more limited opportunities for biocides and metals to drive HGT of antibiotic resistance, whereas there are ample possibilities for these chemicals to select for antibiotic-resistant bacteria through chromosomal biocide/MRGs (Pal et al., 2015). It is not clear what the dominant mechanisms of co-selection for metal- and antibiotic resistance are. However, based on the available studies that have addressed mechanisms behind co-selection, we hypothesise that for co-selection of resistance between antibiotics and biocides, the dominant mechanism is cross-resistance, mostly involving efflux systems. This may also be the case for metals (Flach et al., 2017) although co-resistance (co-occurrence) may in some cases play a larger role.

The genetic context of the resistance genes, including the MGEs carrying them, need to be studied more closely. While studies of individual bacterial isolates involve the risk of missing what is occurring in other bacteria, there are many technical challenges and shortcomings associated with assembly based metagenomics approaches. The recently developed epicPCR technology has potential to identify the possible hosts of specific genes (including resistance genes) (Spencer et al., 2016). In addition, co-occurrences of specific resistance genes in microbial communities may be identified using inverse PCR in combination with long read sequencing technologies (Pärnänen et al., 2016). Future research adopting this and/or similar approaches may enable better understanding of the co-selection process.

## ACKNOWLEDGEMENTS

We thank Jan Kreft for stimulating discussions over mathematical modelling. This work was supported by UK NERC Grant NE/N019881/1, a UK BBSRC studentship to K.A., a University of Nottingham Vice Chancellor's Scholarship to S.A., and support from the Swedish Research Council FORMAS to D.G.J.L.

## REFERENCES

Ahmad, A., Zachariasen, C., Christiansen, L. E., Græsbøll, K., Toft, N., Matthews, L., et al. (2016). Multistrain models predict sequential multidrug treatment strategies to result in less antimicrobial resistance than combination treatment. *BMC Microbiology*, *16*(1), 118.

Akinbowale, O. L., Peng, H., Grant, P., & Barton, M. D. (2007). Antibiotic and heavy metal resistance in motile aeromonads and pseudomonads from rainbow trout (*Oncorhynchus mykiss*) farms in Australia. *International Journal of Antimicrobial Agents*, *30*(2), 177–182.

Alawieh, A., Sabra, Z., Bizri, A. R., Davies, C., White, R., & Zaraket, F. A. (2015). A computational model to monitor and predict trends in bacterial resistance. *Journal of Global Antimicrobial Resistance*, *3*(3), 174–183.

Alcalde-Rico, M., Hernando-Amado, S., Blanco, P., & Martínez, J. L. (2016). Multidrug efflux pumps at the crossroad between antibiotic resistance and bacterial virulence. *Frontiers in Microbiology*, 7, 1483.

Amachawadi, R. G., Scott, H. M., Alvarado, C. A., Mainini, T. R., Vinasco, J., Drouillard, J. S., et al. (2013). Occurrence of the transferable copper resistance gene *tcrB* among fecal enterococci of U.S. feedlot cattle fed copper-supplemented diets. *Applied and Environmental Microbiology*, *79*(14), 4369–4371.

Amachawadi, R. G., Shelton, N. W., Shi, X., Vinasco, J., Dritz, S. S., Tokach, M. D., et al. (2011). Selection of fecal enterococci exhibiting *tcrB*-mediated copper resistance in pigs fed diets supplemented with copper. *Applied and Environmental Microbiology*, 77(16), 5597–5603.

Aminov, R. I. (2009). The role of antibiotics and antibiotic resistance in nature. *Environmental Microbiology*, *11*(12), 2970–2988.

Aminov, R. I., & Mackie, R. I. (2007). Evolution and ecology of antibiotic resistance genes. *FEMS Microbiology Letters*, *271*(2), 147–161.

Andersson, D. I., & Hughes, D. (2010). Antibiotic resistance and its cost: Is it possible to reverse resistance? *Nature Reviews. Microbiology*, *8*(4), 260–271.

Asiani, K. R., Williams, H., Bird, L., Jenner, M., Searle, M. S., Hobman, J. L., et al. (2016). SilE is an intrinsically disordered periplasmic "molecular sponge" involved in bacterial silver resistance. *Molecular Microbiology*, *101*(5), 731–742.

Austin, D. J., Kakehashi, M., & Anderson, R. M. (1997). The transmission dynamics of antibiotic-resistant bacteria: The relationship between resistance in commensal organisms and antibiotic consumption. *Proceedings of the Royal Society of London B*, *264*(1388), 1629–1638.

Ayscue, P., Lanzas, C., Ivanek, R., & Grohn, Y. T. (2009). Modelling on-farm *Escherichia coli* O157:H7 population dynamics. *Foodborne Pathogens and Disease*, *6*(4), 461–470.

Baker, K. S., Burnett, E., McGregor, H., Deheer-Graham, A., Boinett, C., Langridge, G. C., et al. (2015). The Murray collection of pre-antibiotic era Enterobacteriacae: A unique research resource. *Genome Medicine*, 7, 97.

Baker, M., Hobman, J. L., Dodd, C. E. R., Ramsden, S. J., & Stekel, D. J. (2016). Mathematical modelling of antimicrobial resistance in agricultural waste highlights importance of gene transfer rate. *FEMS Microbiology Ecology*, *92*(4), fiw040.

Baker-Austin, C., Wright, M. S., Stepanauskas, R., & McArthur, J. V. (2006). Co-selection of antibiotic and metal resistance. *Trends in Microbiology*, *14*(1), 176–182.

Bednorz, C., Oelgeschläger, K., Kinnemann, B., Hartmann, S., Neumann, K., Pieper, R., et al. (2013). The broader context of antibiotic resistance: Zinc feed supplementation of piglets increases the proportion of multi-resistant *Escherichia coli* in vivo. *International Journal of Medical Microbiology*, *303*(6–7), 396–403.

Bell, B. G., Schellevis, F., Stobberingh, E., Goossens, H., & Pringle, M. (2014). A systematic review and meta-analysis of the effects of antibiotic consumption on antibiotic resistance. *BMC Infectious Diseases*, *14*, 13.

Bengtsson-Palme, J., Hammarén, R., Pal, C., Östman, M., Björlenius, B., Flach, C. F., et al. (2016). Elucidating selection processes for antibiotic resistance in sewage treatment plants using metagenomics. *Science of the Total Environment*, *572*, 697–712.

Bengtsson-Palme, J., & Larsson, D. G. J. (2015). Antibiotic resistance genes in the environment: Prioritizing risks. *Nature Reviews. Microbiology*, *13*(6), 396.

Berg, J., Thorsen, M. K., Holm, P. E., Jensen, J., Nybroe, O., & Brandt, K. K. (2010). Cu exposure under field conditions coselects for antibiotic resistance as determined by a novel cultivation-independent bacterial community tolerance assay. *Environmental Science and Technology*, *44*(22), 8724–8728.

Bhullar, K., Waglechner, N., Pawlowski, A., Koteva, K., Banks, E. D., Johnston, M. D., et al. (2012). Antibiotic resistance is prevalent in an isolated cave microbiome. *PLoS One*, 7(4), e34953.

Blanco, P., Hernando-Amado, S., Reales-Calderon, J. A., Corona, F., Lira, F., Alcalde-Rico, M., et al. (2016). Bacterial multidrug efflux pumps: Much more than antibiotic resistance determinants. *Microorganisms*, *4*(1), 14.

Bondarczuk, K., & Piotrowska-Seget, Z. (2013). Molecular basis of active copper resistance mechanisms in Gram-negative bacteria. *Cell Biology and Toxicology*, *29*(6), 397–405.

Bonhoeffer, S., Lipsitch, M., & Levin, B. R. (1997). Evaluating treatment protocols to prevent antibiotic resistance. *Proceedings of the National Academy of Sciences of the United States of America*, *94*(22), 12106–12111.

Boyd, E. S., & Barkay, T. (2012). The mercury resistance operon: From an origin in a geothermal environment to an efficient detoxification machine. *Frontiers in Microbiology*, *3*, 349.

Bridges, K., & Lowbury, E. J. (1977). Drug resistance in relation to use of silver sulphadiazine cream in a burns unit. *Journal of Clinical Pathology*, *30*(2), 160–164.

Brown, N. L., Barrett, S. R., Camakaris, J., Lee, B. T., & Rouch, D. A. (1995). Molecular genetics and transport analysis of the copper-resistance determinant (*pco*) from *Escherichia coli* plasmid pRJ1004. *Molecular Microbiology*, *17*(6), 1153–1166.

Brown, H. J., Stokes, H. W., & Hall, R. M. (1996). The integrons In*0*, In*2*, and In*5* are defective transposon derivatives. *Journal of Bacteriology*, *178*(15), 4429–4437.

Campos, J., Cristino, L., Peixe, L., & Antunes, P. (2016). MCR-1 in multidrug-resistant and copper-tolerant clinically relevant Salmonella 1,4,[5],12:i:- and S. Rissen clones in Portugal, 2011 to 2015. *Eurosurveillance*, *21*(26). http://dx.doi.org/10.2807/1560-7917.ES.2016.21.26.30270.

Castillo, M., Martín-Orúe, S. M., Taylor-Pickard, J. A., Pérez, J. F., & Gasa, J. (2008). Use of mannanoligosaccharides and zinc chelate as growth promoters and diarrhea preventative in weaning pigs: Effects on microbiota and gut function. *Journal of Animal Science*, *86*(1), 94–101.

Cavaco, L. M., Hasman, H., & Aarestrup, F. M. (2011). Zinc resistance of *Staphylococcus aureus* of animal origin is strongly associated with methicillin resistance. *Veterinary Microbiology*, *150*(3–4), 344–348.

Cavaco, L. M., Hasman, H., Stegger, M., Andersen, P. S., Skov, R., Fluit, A. C., et al. (2010). Cloning and occurrence of *czrC*, a gene conferring cadmium and zinc resistance in methicillin-resistant *Staphylococcus aureus* CC398 isolates. *Antimicrobial Agents and Chemotherapy*, *54*(9), 3605–3608.

Chapman, J. S. (2003). Disinfectant resistance mechanisms, cross-resistance, and co-resistance. *International Biodeterioration & Biodegradation*, *51*(4), 271–276.

Cheng, J., Hicks, D. B., & Kruwlich, T. A. (1996). The purified Bacillus subtilis tetracycline efflux protein TetA(L) reconstitutes both tetracycline–cobalt/H+ and Na+(K+)/H+ exchange. *Proceedings of the National Academy of Sciences of the United States of America*, *93*(25), 14446–14451.

Chi Fru, E., Rodriguez, N. P., Partin, C. A., Lalonde, S. V., Andersson, P., Weiss, D. J., et al. (2016). Cu isotopes in marine black shales record the great oxidation event. *Proceedings of the National Academy of Sciences of the United States of America*, *113*, 4941–4946.

Conroy, O., Kim, E. H., McEvoy, M. M., & Rensing, C. (2010). Differing ability to transport nonmetal substrates by two RND-type metal exporters. *FEMS Microbiology Letters*, *308*(2), 115–122.

D'Agata, E. M. C., Dupont-Rouzeyrol, M., Magal, P., Olivier, D., & Ruan, S. (2008). The impact of different antibiotic regimens on the emergence of antimicrobial-resistant bacteria. *PLoS One*, *3*(12), e4036.

D'Costa, V. M., King, C. E., Kalan, L., Morar, M., Sung, W. W., Schwarz, C., et al. (2011). Antibiotic resistance is ancient. *Nature*, *477*(7365), 457–461.

Datta, N., & Hughes, V. M. (1983). Plasmids of the same Inc groups in enterobacteria before and after the medical use of antibiotics. *Nature*, *306*(5943), 616–617.

Davies, J., & Rownd, R. (1972). Transmissible multiple drug resistance in Enterobacteriaceae. *Science*, *176*(4036), 758–768.

Di Cesare, A., Eckert, E. M., D'Urso, S., Bertoni, R., Gillan, D. C., Wattiez, R., et al. (2016). Co-occurrence of integrase 1, antibiotic and heavy metal resistance genes in municipal wastewater treatment plants. *Water Research*, *94*, 208–214.

Dolliver, H., Gupta, S., & Noll, S. (2008). Antibiotic degradation during manure composting. *Journal of Environmental Quality*, *37*(3), 1245–1253.

Dunne, K. A., Chaudhuri, R. R., Rossiter, A. E., Beriotto, I., Browning, D. F., Squire, D., et al. (2017). Sequencing a piece of history: Complete genome sequence of the original *Escherichia coli* strain. *Microbial Genomics*, in press.

Durán, N., Durán, M., de Jesus, M. B., Seabra, A. B., Fávaro, W. J., & Nakazato, G. (2016). Silver nanoparticles: A new view on mechanistic aspects on antimicrobial activity. *Nanomedicine: Nanotechnology, Biology and Medicine, 12*(3), 789–799.

Escudero, J. A., Loot, C., Nivina, A. A., & Mazel, D. (2015). The integron: Adaption on demand. *Microbiology Spectrum, 3*. MDNA3-0019-2014.

Essa, A. M., Julian, D. J., Kidd, S. P., Brown, N. L., & Hobman, J. L. (2003). Mercury resistance determinants related to Tn*21*, Tn*1696*, and Tn*5053* in enterobacteria from the 'Pre-antibiotic era'. *Antimicrobial Agents and Chemotherapy, 47*(3), 1115–1119.

Falkow, S. (1975). *Infectious multiple drug resistance.* London: Pion Publishing.

Fang, L., Li, X., Li, L., Li, S., Liao, X., Sun, J., et al. (2016). Co-spread of metal and antibiotic resistance within ST3-IncH12 plasmids from *E. coli* isolates of food-producing animals. *Scientific Reports, 6*, 25312.

Farias, P., Santo, C. E., Branco, R., Francisco, R., Santos, S., Hansen, L., et al. (2015). Natural hot spots for gain of multiple resistances: Arsenic and antibiotic resistances in heterotrophic, aerobic bacteria from marine hydrothermal vent fields. *Applied and Environmental Microbiology, 81*(7), 2534–2543.

Finley, P. J., Norton, R., Austin, C., Mitchell, A., Zank, S., & Durham, P. (2015). Unprecedented silver-resistance in clinically isolated Enterobacteriaceae: Major implications for burn and wound management. *Antimicrobial Agents and Chemotherapy, 59*(8), 4734–4741.

Finley, P. J., Peterson, A., & Finley, P. J. (2013). The prevalence of phenotypic silver resistance in clinical isolates. *Wounds, 25*(4), 84–88.

Flach, C. F., Pal, C., Svensson, C. J., Kristiansson, E., Östman, M., Bengtsson-Palme, J., et al. (2017). Does antifouling paint select for antibiotic resistance? *Science of the Total Environment.* http://dx.doi.org/10.1016/j.scitotenv.2017.01.213.

Forsberg, K. J., Reyes, A., Wang, B., Selleck, E. M., Sommer, M. O., & Dantas, G. (2012). The shared antibiotic resistome of soil bacteria and human pathogens. *Science, 337*(6098), 1107–1111.

Franke, S., Grass, G., & Nies, D. H. (2001). The product of the *ybdE* gene of the Escherichia coli chromosome is involved in detoxification of silver ions. *Microbiology, 147*(4), 965–972.

Franke, S., Grass, G., Rensing, C., & Nies, D. H. (2003). Molecular analysis of the copper-transporting efflux system CusCFBA of *Escherichia coli. Journal of Bacteriology, 185*(13), 3804–3812.

Frith, J. (2012). Syphilis—Its early history and treatment until penicillin and the debate on its origins. *Journal of Military and Veterans' Health, 20*, 4.

Frost, L. S., Leplae, R., Summers, A. O., & Toussaint, A. (2005). Mobile genetic elements: The agents of open source evolution. *Nature Reviews. Microbiology, 3*(9), 722–732.

Garelick, H., Jones, H., Dybowska, A., & Valsami-Jones, E. (2008). Arsenic pollution sources. *Reviews of Environmental Contamination and Toxicology, 197*, 17–60.

Gaze, W. H., Abdouslam, N., Hawkey, P. M., & Wellington, E. M. (2005). Incidence of class 1 integrons in a quaternary ammonium compound-polluted environment. *Antimicrobial Agents and Chemotherapy, 49*(5), 1802–1807.

Gaze, W. H., Zhang, L., Abdouslam, N. A., Hawkey, P. M., Calvo-Bado, L., Royle, J., et al. (2011). Impacts on anthropogenic activity on the ecology of class 1 integrons and integron-associated genes in the environment. *The ISME Journal, 5*(8), 1253–1261.

Gehring, R., Schumm, P., Youssef, M., & Scoglio, C. (2010). A network-based approach for resistance transmission in bacterial populations. *Journal of Theoretical Biology, 262*(1), 97–106.

Gerrish, P. J., & García-Lerma, J. G. (2003). Mutation rate and the efficacy of antimicrobial drug treatment. *Lancet Infectious Diseases*, *3*(1), 28–32.

Giachi, G., Pallecchi, P., Romualdi, A., Ribechini, E., Lucejko, J. J., Colombini, M. P., et al. (2013). Ingredients of a 2,000-y-old medicine revealed by chemical, mineralogical, and botanical investigations. *Proceedings of the National Academy of Sciences of the United States of America*, *110*(4), 1193–1196.

Gillings, M. R. (2017). Lateral gene transfer, bacterial genome evolution, and the Anthropocene. *Annals of the New York Academy of Sciences*, *1389*, 20–36. http://dx.doi.org/10.1111/nyas.13213.

Gillings, M. R., Gaze, W. H., Pruden, A., Smalla, K., Tiedje, J. M., & Zhu, Y.-G. (2015). Using the class 1 integron-integrase gene as a proxy for anthropogenic pollution. *The ISME Journal*, *9*(6), 1269–1279.

Gillings, M. R., Xuejun, D., Hardwick, S., Holley, M. P., & Stokes, H. W. (2009). Gene cassettes encoding resistance to quaternary ammonium compounds: A role in the origin of class 1 integrons? *The ISME Journal*, *3*(2), 209–215.

Gilmour, M. W., Thomson, N. R., Sanders, M., Parkhill, J., & Taylor, D. E. (2004). The complete nucleotide sequence of the resistance plasmid R478: Defining the backbone components of incompatibility group H conjugative plasmids through comparative genomics. *Plasmid*, *52*, 182–202.

González-Pastor, J. E., & Mirete, S. (2010). Novel metal resistance genes from microorganisms: A functional metagenomic approach. *Methods in Molecular Biology*, *668*, 273–285.

Gullberg, E., Albrecht, L. M., Karlsson, C., Sandegren, L., & Andersson, D. I. (2014). Selection of a multidrug resistance plasmid by sublethal levels of antibiotics and heavy metals. *MBio*, *5*(5), e01918. 14.

Guo, X., Liu, S., Wang, Z., Zhang, X. X., Li, M., & Wu, B. (2014). Metagenomic profiles and antibiotic resistance genes in gut microbiota of mice exposed to arsenic and iron. *Chemosphere*, *112*, 1–8.

Gupta, A., Matsui, K., Lo, J.-F., & Silver, S. (1999). Molecular basis for resistance to silver cations in *Salmonella*. *Nature Medicine*, *5*(2), 183–188.

Hao, X., Li, X., Pal, C., Hobman, J. L., Rosen, B. P., Larsson, D. G. J., et al. (2017). Widespread resistance to arsenic in bacteria is influenced by its use in Protists to kill bacterial prey. *Biometals*. http://dx.doi.org/10.1007/s10534-017-0003-4.

Hao, X., Lüthje, F. L., Qin, Y., McDevitt, S. F., Lutay, N., Hobman, J. L., et al. (2015). Survival in amoeba—A major selective pressure on the presence of bacterial copper and zinc resistance determinants? Identification of a "copper pathogenicity island" *Applied Microbiology and Biotechnology*, *99*(14), 5817–5824.

Hao, X., Lüthje, F., Rønn, R., German, N. A., Li, X., Huang, F., et al. (2016). A role for copper in protozoan grazing—Two billion years selecting for bacterial copper resistance. *Molecular Microbiology*, *102*(4), 628–641.

Harada, M. (1995). Minamata disease: Methylmercury poisoning in Japan caused by environmental pollution. *Critical Reviews in Toxicology*, *25*(1), 1–24.

Hasman, H., & Aarestrup, F. M. (2002). *TcrB*, a gene conferring transferable copper resistance in *Enterococcus faecium*: Occurrence, transferability, and linkage to macrolide and glycopeptide resistance. *Antimicrobial Agents and Chemotherapy*, *48*(5), 1410–1416.

Hatosy, S. M., & Martiny, A. C. (2015). The ocean as a global reservoir of antibiotic resistance genes. *Applied and Environmental Microbiology*, *81*(21), 7593–7599.

Hayashi, S., Abe, M., Kimoto, M., Furukawa, S., & Nakazawa, T. (2000). The DsbA-DsbB disulfide bond formation system of *Burkholderia cepacia* is involved in the production of protease and alkaline phosphatase, motility, metal resistance, and multi-drug resistance. *Microbiology and Immunology*, *44*(1), 41–50.

Hedges, R. W., & Bamberg, S. (1973). Resistance to arsenic compounds conferred by a plasmid transmissible between strains of *Escherichia coli*. *Journal of Bacteriology*, *115*(1), 459–460.

Herman, D., Thomas, C. M., & Stekel, D. J. (2012). Adaptation for protein synthesis efficiency in a naturally occurring self-regulating operon. *PLoS One*, 7(11), e49678.

Hobman, J. L., & Crossman, L. C. (2015). Bacterial antimicrobial metal ion resistance. *Journal of Medical Microbiology*, *64*, 471–497.

Holman, D. B., & Chénier, M. R. (2015). Antimicrobial use in swine production and its effect on the swine gut microbiota and antimicrobial resistance. *Canadian Journal of Microbiology*, *61*(11), 1–14.

Hölzel, C. S., Müller, C., Harms, K. S., Mikolajewski, S., Schäfer, S., Schwaiger, K., et al. (2012). Heavy metals in liquid pig manure in light of bacterial antimicrobial resistance. *Environmental Research*, *113*, 21–27.

Hu, H.-W., Wang, J.-T., Li, J., Li, J.-J., Ma, Y.-B., Chen, D., et al. (2016). Field-based evidence for copper contamination changes of antibiotic resistance in agricultural soils. *Environmental Microbiology*, *18*(11), 3896–3909.

Hu, H.-W., Wang, J.-T., Li, J., Shi, X.-Z., Ma, Y.-B., Chen, D., et al. (2017). Long-term nickel contamination increases the occurrence of antibiotic resistance genes in agricultural soils. *Environmental Science and Technology*, *51*(2), 790–800.

Huddleston, J. R. (2014). Horizontal gene transfer in the human gastrointestinal tract: Potential spread of antibiotic resistance genes. *Infection and Drug Resistance*, 7, 167–176.

Huffman, D. L., Huyett, J., Outten, F. W., Doan, P. E., Finney, L. A., Hoffman, B. M., et al. (2002). Spectroscopy of Cu(II)-PcoC and the multicopper oxidase function of PcoA, two essential components of *Escherichia coli pco* copper resistance operon. *Biochemistry*, *41*(31), 10046–10055.

Hughes, M. F., Beck, B. D., Chen, Y., Lewis, A. S., & Thomas, D. J. (2011). Arsenic exposure and toxicology: A historical perspective. *Toxicological Sciences*, *123*(2), 305–332.

Hughes, V. M., & Datta, N. (1983). Conjugative plasmids in bacteria of the 'pre-antibiotic' era. *Nature*, *302*(5910), 725–726.

Huisingh, D. (1974). Heavy metals: Implications for agriculture. *Annual Review of Phytopathology*, *12*, 375–388.

Iwasa, Y., Michor, F., & Nowak, M. A. (2004). Evolutionary dynamics of invasion and escape. *Journal of Theoretical Biology*, *226*(2), 205–214.

Jackson, C. R., & Dugas, S. L. (2003). Phylogenetic analysis of bacterial and archaeal arsC gene sequences suggests an ancient, common origin for arsenate reductase. *BMC Evolutionary Biology*, *3*, 18.

Jenkins, D. J., & Stekel, D. J. (2010). De novo evolution of complex, global and hierarchical gene regulatory mechanisms. *Journal of Molecular Evolution*, *71*(2), 128–140.

Jones, C., & Stanley, J. (1992). *Salmonella* plasmids of the pre-antibiotic era. *Journal of General Microbiology*, *138*(1), 189–197.

Jutkina, J., Rutgersson, C., Flach, C. F., & Larsson, D. G. J. (2016). An assay for determining minimal concentrations of antibiotics that drive horizontal transfer of resistance. *Science of the Total Environment*, *548–549*, 131–138.

Kazama, H., Hamashimam, H., Sasatsu, M., & Arai, T. (1999). Characterization of the antiseptic-resistance gene *qacE* delta 1 isolated from clinical and environmental isolates of *Vibrio parahaemolyticus* and *Vibrio cholerae* non-O1. *FEMS Microbiology Letters*, *174*(2), 379–384.

Kholodii, G., Mindlin, S., Petrova, M., & Minakhina, S. (2003). Tn*5060* from the Siberian permafrost is most closely related to the ancestor of Tn*21* prior to integron acquisition. *FEMS Microbiology Letters*, *226*(2), 251–255.

Knapp, C. W., McCluskey, S. M., Singh, B. K., Campbell, C. D., Hudson, G., & Graham, D. W. (2011). Antibiotic resistance gene abundances correlate with metal and geochemical conditions in archived Scottish soils. *PLoS One*, *6*(11), e27300.

Komura, I., Funaba, T., & Izaki, K. (1971). Mechanism of mercuric chloride resistance in microorganisms II: NADPH-dependent reduction of mercuric chloride and vaporization of mercury from mercuric chloride by a multiple drug resistant strain of *Escherichia coli*. *Journal of Biochemistry*, *70*, 895–901.

Komura, I., & Izaki, K. (1971). Mechanism of mercuric chloride resistance in microorganisms I: Vaporization of a mercury compound from mercuric chloride by multiple drug resistance strains of *Escherichia coli*. *Journal of Biochemistry*, *70*, 885–893.

Komura, I., Izaki, K., & Takahashi, H. (1970). Vaporization of inorganic mercury by cell free extracts of drug resistant *Escherichia coli*. *Agricultural and Biological Chemistry*, *34*, 480–482.

Kroll, J., Klinter, S., Schneider, C., Voß, I., & Steinbüchler, A. (2010). Plasmid addiction systems: Perspectives and applications in biotechnology. *Microbial Biotechnology*, *3*(6), 634–657.

Küchen, D., Feucht, H.-H., & Kaulfers, P.-M. (2000). Association of *qacE* and *qacE*△*1* with multiple resistance to antibiotics and antiseptics in clinical isolates of Gram-negative bacteria. *FEMS Microbiology Letters*, *183*, 95–98.

Lemire, J. A., Harrison, J. J., & Turner, R. J. (2013). Antimicrobial activity of metals: Mechanisms, molecular targets and applications. *Nature Reviews. Microbiology*, *11*, 371–384.

Li, L.-G., Cai, L., Zhang, X.-X., & Zhang, T. (2014). Potentially novel copper resistance genes in copper-enriched activated sludge revealed by metagenomic analysis. *Applied Microbiology and Biotechnology*, *98*, 10255–10266.

Li, J., Pawitwar, S. S., & Rosen, B. P. (2016). The organoarsenical biocycle and the primordial antibiotic methylarsenite. *Metallomics*, *8*(10), 1047–1055.

Li, L.-G., Xia, Y., & Zhang, T. (2017). Co-occurrence of antibiotic and metal resistance genes revealed in complete genome collection. *The ISME Journal*, *11*, 651–662. http://dx.doi.org/10.1038/ismej.2016.155.

Li, Y. X., Xiong, X., Chun-ye, L., Feng-song, Z., Wei, L., & Wei, H. (2010). Cadmium in animal production and its potential hazard on Beijing and Fuxin farmlands. *Journal of Hazardous Materials*, *177*(1), 475–480.

Liebert, C. A., Hall, R. M., & Summers, A. O. (1999). Tn*21*-flagship of the floating genome. *Microbiology and Molecular Biology Reviews*, *63*(3), 507–522.

Lin, Y. F., Walmsley, A. R., & Rosen, B. P. (2006). An arsenic metallochaperone for an arsenic detoxification pump. *Proceedings of the National Academy of Sciences of the United States of America*, *103*(42), 15617–15622.

Lloyd, N. A., Janssen, S. E., Reinfelder, J. R., & Barkay, T. (2016). Co-selection of mercury and multiple antibiotic resistances in bacteria exposed to mercury in the *Fundulus heteroclitus* gut microbiome. *Current Microbiology*, *73*(6), 834–842.

Lucas, I. A. M., Livingstone, R. M., & McDonald, I. (1961). Copper sulphate as a growth stimulant for pigs: Effect of level and purity. *Animal Production*, *3*(02), 111–119.

Magos, L., Tuffery, A. A., & Clarkson, T. W. (1964). Volatilization of mercury by bacteria. *British Journal of Industrial Medicine*, *21*, 294–298.

Martinez, J. L., Coque, T. M., & Baquero, F. (2015). What is a resistance gene? Ranking risk in resistomes. *Nature Reviews. Microbiology*, *13*(2), 116–123.

Mata, M. T., Baquero, F., & Pérez-Díaz, J. C. (2000). A multidrug efflux transporter in *Listeria monocytogenes*. *FEMS Microbiology Letters*, *187*(2), 185–188.

McDonnell, G., & Russell, A. D. (1999). Antiseptics and disinfectants: Activity, action and resistance. *Clinical Microbiology Reviews*, *12*, 147–179.

McHugh, G. L., Moellering, R. C., Hopkins, C. C., & Swartz, M. N. (1975). *Salmonella typhimurium* resistant to silver nitrate, chloramphenicol and ampicillin. A new threat in burn units? *Lancet*, *305*, 235–240.

McKinney, C. W., Loftin, K. A., Meyer, M. T., Davis, J. G., & Pruden, A. (2010). *Tet* and *sul* antibiotic resistance genes in livestock lagoons of various operation type, configuration, and antibiotic occurrence. *Environmental Science and Technology*, *44*(16), 6102–6109.

Medardus, J. J., Molla, B. Z., Nicol, M., Morrow, W. M., Rajala-Schultz, P. J., Kazwala, R., et al. (2014). In-feed use of heavy metal micronutrients in U.S. Swine production systems and its role in the persistence of multidrug resistant *Salmonellae*. *Applied and Environmental Microbiology*, *80*(7), 2317–2325.
Mergeay, M., Monchy, S., Vallaeys, T., Auquier, V., Benotmane, A., Bertin, P., et al. (2003). *Ralstonia metallidurans*, a bacterium specifically adapted to toxic metals: Towards a catalogue of metal-responsive genes. *FEMS Microbiology Reviews*, *27*(2–3), 385–410.
Michel, J.-B., Yeh, P. J., Chait, R., Moellering, R. C., & Kishony, R. (2008). Drug interactions modulate the potential for evolution of resistance. *Proceedings of the National Academy of Sciences of the United States of America*, *105*(39), 14918–14923.
Mijnendonckx, K., Leys, N., Mahillon, J., Silver, S., & Van Houdt, R. (2013). Antimicrobial silver: Uses, toxicity and potential for resistance *Biometals*, *26*(4), 609–621.
Mindlin, S., & Petrova, M. (2013). Mercury resistance transposons. In A. P. Roberts & P. Mullany (Eds.), *Bacterial integrative mobile genetic elements* (pp. 33–52). Texas, USA: Landes Bioscience.
Mindlin, S., Petrova, M., Gorlenko, Zh., Soina, V., Khachikian, N., & Karaevskaya, E. (2009). Multidrug resistant bacteria in permafrost: Isolation, biodiversity, phenotypic and genotypic analysis. *New permafrost and glacier research* (pp. 89–105). Hauppauge: Nova Science Publishers, Inc.
Monchy, S., Benotmane, M. A., Wattiez, R., van Aelst, S., Auquier, V., Borremans, B., et al. (2006). Transcriptomic and proteomic analyses of the pMOL30-encoded copper resistance in *Cupriavidus metallidurans* strain CH34. *Microbiology*, *152*(6), 1765–1776.
Moore, B. (1960). A new screen test and selective medium for the rapid detection of epidemic strains of Staph. aureus. *Lancet*, *276*(7148), 453–458.
Morby, A. P., Hobman, J. L., & Brown, N. L. (1995). The role of cysteine residues in the transport of mercuric ions by the Tn*501* MerT and MerP mercury-resistance proteins. *Molecular Microbiology*, *17*(1), 25–35.
Moreno Switt, A. I., den Bakker, H. C., Cummings, C. A., Rodriguez-Rivera, L. D., Govoni, G., Raneiri, M. L., et al. (2012). Identification and characterization of novel Salmonella mobile elements involved in the dissemination of genes linked to virulence and transmission. *PLoS One*, 7(7), e41247.
Mourão, J., Novais, C., Machado, J., Peixe, L., & Antunes, P. (2015). Metal tolerance in emerging clinically relevant multi-drug resistant *Salmonella enterica* serotype 4, [5], 12:i:- clones circulating in Europe. *International Journal of Antimicrobial Agents*, *45*(6), 610–616.
Munson, G. P., Lam, D. L., Outten, F. W., & O'Halloran, T. V. (2000). Identification of a copper-responsive two-component system on the chromosome of *Escherichia coli* K-12. *Journal of Bacteriology*, *182*(20), 5864–5871.
Murphy, J. T., Walshe, R., & Devocelle, M. (2008). A computational model of antibiotic-resistance mechanisms in Methicillin-Resistant *Staphylococcus aureus* (MRSA). *Journal of Theoretical Biology*, *254*(2), 284–293.
Nachman, K. E., Baron, P. A., Raber, G., Francesconi, K. A., Navas-Acien, A., & Love, D. C. (2013). Roxarsone, inorganic arsenic, and other arsenic species in chicken: A U.S.-based market basket sample. *Environmental Health Perspectives*, *121*, 818–824.
Nakahara, H., Ishikawa, T., Sarai, Y., Kondo, I., & Mitsuhashi, S. (1977). Frequency of heavy-metal resistance in bacteria from inpatients in Japan. *Nature*, *266*(5598), 165–167.
Nakahara, H., Silver, S., Miki, T., & Rownd, R. H. (1979). Hypersensitivity to $Hg^{2+}$ and hyperbinding activity associated with cloned fragments of the mercurial resistance operon of plasmid NR1. *Journal of Bacteriology*, *140*(1), 161–166.
Nakajima, H., Kobayashi, K., Kobayashi, M., Asako, H., & Aono, R. (1995). Over-expression of the *robA* gene increases organic solvent tolerance and multiple antibiotic

and heavy metal ion resistance in *Escherichia coli*. *Applied and Environmental Microbiology*, *61*(6), 2302–2307.
Nakaya, R., Nakamura, A., & Murata, Y. (1960). Resistance transfer agents in Shigella. *Biochemical and Biophysical Research Communications*, *3*, 654–659.
Nemergut, D. R., Martin, A. P., & Schmidt, S. K. (2004). Integron diversity in heavy-metal-contaminated mine tailings and inferences about integron evolution. *Applied and Environmental Microbiology*, *70*(2), 1160–1168.
Nies, D. H. (2003). Efflux-mediated heavy metal resistance in prokaryotes. *FEMS Microbiology Reviews*, *27*, 313–339.
Nishino, K., Nikaido, E., & Yamaguchi, A. (2007). Regulation of multidrug efflux systems involved in multidrug and metal resistance of *Salmonella enterica* serovar Typhimurium. *Journal of Bacteriology*, *189*(24), 9066–9075.
Norman, A., Hansen, L. H., & Sørensen, S. J. (2009). Conjugative plasmids: Vessels of the communal gene pool. *Philosophical Transactions of the Royal Society of London. Series B, Biological Sciences*, *364*(1527), 2275–2289.
Novick, R. P. (1967). Penicillinase plasmids of *Staphylococcus aureus*. *Federation Proceedings*, *26*(1), 29–38.
Novick, R. P., & Roth, C. (1968). Plasmid-linked resistance to inorganic salts in *Staphylococcus aureus*. *Journal of Bacteriology*, *95*(4), 1335–1342.
Nriagu, J. O. (1989). A global assessment of natural sources of atmospheric trace metals. *Nature*, *338*, 47–49.
Nriagu, J. O. (1996). A history of global metal pollution. *Science*, *272*(5259), 223–224.
Opatowski, L., Mandel, J., Varon, E., Boëlle, P. Y., Temime, L., & Guillemot, D. (2010). Antibiotic dose impact on resistance selection in the community: A mathematical model of β-lactams and *Streptococcus pneumoniae* dynamics. *Antimicrobial Agents and Chemotherapy*, *54*(6), 2330–2337.
Osman, D., Patterson, C. J., Bailey, K., Fisher, K., Robinson, N. J., Rigby, S. E. J., et al. (2013). The copper supply pathway to a Salmonella Cu, Zn-superoxide dismutase (SodCII) involves P1B-type ATPase copper efflux and periplasmic CueP. *Molecular Microbiology*, *87*(3), 466–477.
Outten, F. W., Huffman, D. L., Hale, J. A., & O'Halloran, T. V. (2001). The independent *cue* and *cus* systems confer copper tolerance during aerobic and anaerobic growth in *Escherichia coli*. *Journal of Biological Chemistry*, *276*(33), 30670–30677.
Page, K., Wilson, M., & Parkin, I. (2009). Antimicrobial surfaces and their potential in reducing the role of the inanimate environment in the incidence of hospital-acquired infections. *Journal of Materials Chemistry*, *19*, 3819–3831.
Pal, C., Bengtsson-Palme, J., Kristiansson, E., & Larsson, D. G. J. (2015). Co-occurrence of resistance genes to antibiotics, biocides and metals reveals novel insights into their co-selection potential. *BMC Genomics*, *16*, 964.
Pal, C., Bengtsson-Palme, J., Kristiansson, E., & Larsson, D. G. J. (2016). The Structure and diversity of human, animal and environmental resistomes. *Microbiome*, *4*, 54.
Pal, C., Bengtsson-Palme, J., Rensing, C., Kristiansson, E., & Larsson, D. G. J. (2014). BacMet: Antibacterial biocide and metal resistance genes database. *Nucleic Acids Research*, *42*(Database issue), D737–D743.
Pärnänen, K., Karkman, A., Tamminen, M., Lyra, C., Hultman, J., Paulin, L., et al. (2016). Evaluating the mobility potential of antibiotic resistance genes in environmental resistomes without metagenomics. *Scientific Reports*, *6*, 635790.
Partridge, S. R., Brown, H. J., Stokes, H. W., & Hall, R. M. (2001). Transposons Tn*1696* and Tn*21* and their integrons In*4* and In*2* have independent origins. *Antimicrobial Agents and Chemotherapy*, *45*(4), 1263–1270.
Partridge, S. R., Tsafnat, G., Coiera, E., & Iredell, J. R. (2009). Gene cassettes and cassette arrays in mobile resistance integrons. *FEMS Microbiology Reviews*, *33*(4), 757–784.

Paulsen, I. T., Littlejohn, T. G., Radström, P., Sundström, L., Sköld, O., Swedburg, G., et al. (1993). The 3′ conserved segment of integrons contains a gene associated with multidrug resistance to antiseptics and disinfectants. *Antimicrobial Agents and Chemotherapy*, *37*(4), 761–768.

Peltier, E., Vincent, J., Finn, C., & Graham, D. W. (2010). Zinc-induced antibiotic resistance in activated sludge bioreactors. *Water Research*, *44*, 3829–3836.

Perron, K., Caille, O., Rossier, C., Delden, C. V., Dumas, J., & Kohler, T. (2004). CzcR-CzcS, a two-component system involved in heavy metal and carbapenem resistance in *Pseudomonas aeruginosa*. *Journal of Biological Chemistry*, *279*(10), 8761–8768.

Perry, J., Waglechner, N., & Wright, G. (2016). The prehistory of antibiotic resistance. *Cold Spring Harbor Perspectives in Medicine*, *6*(6), a025197.

Petrova, M., Gorlenko, Z., & Mindlin, S. (2011). Tn*5045*, a novel integron-containing antibiotic and chromate resistance transposon isolated from a permafrost bacterium. *Research in Microbiology*, *162*(3), 337–345.

Petrova, M., Kurakov, A., Shcherbatova, N., & Mindlin, S. (2014). Genetic structure and biological properties of the first ancient multiresistance plasmid pKLH80 isolated from a permafrost bacterium. *Microbiology*, *160*(10), 2253–2263.

Poulain, A. J., Aris-Brousou, S., Blais, J. M., Brazeau, M., Keller, W., & Paterson, A. M. (2016). Microbial DNA records historical delivery of anthropogenic mercury. *The ISME Journal*, *9*(12), 2541–2550.

Pramer, D. (1988). *Some choice words from Waksman*. The Scientist April 1988 issue http://www.the-scientist.com/?articles.view/articleNo/9408/title/Some-Choice-Words-From-Waksman/.

Randall, C. P., Gupta, A., Jackson, N., Busse, D., & O'Neill, A. J. (2015). Silver resistance in Gram-negative bacteria: A dissection of endogenous and exogenous mechanisms. *Journal of Antimicrobial Chemotherapy*, *70*(4), 1037–1046.

Ren, Y., Ren, Y., Zhou, Z., Guo, X., Li, Y., Feng, L., et al. (2010). Complete genome sequence of *Enterobacter cloacae* subsp. *cloacae* type strain ATCC13047. *Journal of Bacteriology*, *192*(9), 2463–2464.

Rensing, C., Fan, B., Sharma, R., Mitra, B., & Rosen, B. P. (2000). CopA: An *Escherichia coli* Cu(I)-translocating P-type ATPase. *Proceedings of the National Academy of Sciences of the United States of America*, *97*(2), 652–656.

Rensing, C., & Grass, G. (2003). *Escherichia coli* mechanisms of copper homeostasis in a changing environment. *FEMS Microbiology Reviews*, *27*(2–3), 197–213.

Richmond, M. H., & John, M. (1964). Co-transduction by a staphylococcal phage of the genes responsible for penicillinase synthesis and resistance to mercury salts. *Nature*, *202*, 1360–1361.

Rosewarne, C. P., Pettigrove, V., Stokes, H. W., & Parsons, Y. M. (2010). Class 1 integrons in benthic bacterial communities: Abundance, association with Tn*402*-like transposition modules and evidence for coselection with heavy metal resistance. *FEMS Microbiology Ecology*, *72*(1), 35–46.

Sandegren, L., Linkevicius, M., Lytsy, B., Melhus, Å., & Andersson, D. I. (2012). Transfer of an *Escherichia coli* ST131 multiresistance cassette has created a *Klebsiella pneumoniae*-specific plasmid associated with a major nosocomial outbreak. *Journal of Antimicrobial Chemotherapy*, *67*(1), 74–83.

Sapkota, A. R., Lefferts, L. Y., McKenzie, S., & Walker, P. (2007). What do we feed to food-production animals? A review of animal feed ingredients and their potential impacts on human health. *Environmental Health Perspectives*, *115*(5), 663–670.

Schmidt, M. G., Von Dessauer, B., Benavente, C., Benadof, D., Cifuentes, P., Elgueta, A., et al. (2016). Copper surfaces are associated with significantly lower concentrations of

bacteria on selected surfaces within a pediatric intensive care unit. *American Journal of Infection Control, 44*(2), 203–209.
Schmieder, R., & Edwards, R. (2012). Insights into antibiotic resistance through metagenomic approaches. *Future Microbiology*, 7(1), 73–89.
Schottel, J., Mandal, A., Clark, D., Silver, S., & Hedges, R. W. (1974). Volatilisation of mercury and organomercurials determined by inducible R-factor systems in enteric bacteria. *Nature, 251*(5473), 335–337.
Scientific Committee on Emerging and Newly Identified Health Risks SCENIHR. (2009). *Assessment of the antibiotic resistance effects of biocides*. Brussels: European Commission. http://ec.europa.eu/health/ph_risk/committees/04_scenihr/docs/scenihr_o_021.pdf.
Seiler, C., & Berendonk, T. U. (2012). Heavy metal driven co-selection of antibiotic resistance in soil and water bodies impacted by agriculture and aquaculture. *Frontiers in Microbiology, 3*, 339.
Sharifi, S., Murthy, S., Takacs, I., & Massoudieh, A. (2014). Probabilistic parameter estimation of activated sludge processes using Markov Chain Monte Carlo. *Water Research, 50*, 254–266.
Silver, S. (2003). Bacterial silver resistance: Molecular biology and uses and misuses of silver compounds. *FEMS Microbiology Reviews, 27*(2–3), 341–353.
Silver, S., Gupta, A., Matsui, K., & Lo, J.-F. (1999). Resistance to Ag(I) cations in bacteria: Environments, genes and proteins. *Metal-Based Drugs, 6*(4–5), 315–320.
Silver, S., & Hobman, J. L. (2007). Mercury microbiology: Resistance systems, environmental aspects, methylation and human health. In D. H. Nies & S. Silver (Eds.), *Microbial Monographs: 6. Molecular microbiology of heavy metals* (pp. 357–370). Berlin, Heidelberg: Springer-Verlag.
Silver, S., Phung, L. T., & Silver, G. (2006). Silver as biocides in burn and wound dressings and bacterial resistance to silver compounds. *Journal of Industrial Microbiology and Biotechnology, 33*(7), 627–634.
Singer, R. S., Ward, M. P., & Maldonado, G. (2006). Can landscape ecology untangle the complexity of antibiotic resistance? *Nature Reviews. Microbiology, 4*(12), 943–952.
Singh, S. K., Roberts, S. A., McDevitt, S. F., Weichsel, A., Wildner, G. F., Grass, G. B., et al. (2011). Crystal structures of multicopper oxidase CueO bound to copper(I) and silver(I): Functional role of a methionine-rich sequence. *Journal of Biological Chemistry, 286*(43), 37849–37857.
Skurnik, D., Ruimy, R., Ready, D., Ruppe, E., Bernède-Bauduin, C., Guillemot, D., et al. (2010). Is exposure to mercury a driving force for the carriage of antibiotic resistance genes? *Journal of Medical Microbiology, 59*(7), 804–807.
Smith, D. H. (1967). R factors mediate resistance to mercury, nickel, and cobalt. *Science, 156*(3778), 1114–1116.
Spencer, S. J., Tamminen, M. V., Preheim, S. P., Guo, M. T., Briggs, A. W., Brito, I. L., et al. (2016). Massively parallel sequencing of single cells by epicPCR links functional genes with phylogenetic markers. *ISME Journal, 10*(2), 427–436.
Staehlin, B. M., Gibbons, J. G., Rokas, A., O'Halloran, T. V., & Slot, J. C. (2016). Evolution of a heavy metal homeostasis/resistance island reflects increasing copper stress in enterobacteria. *Genome Biology and Evolution, 8*(3), 811–826.
Stepanauskas, R., Glenn, T. C., Jagoe, C. H., Tuckfield, R. C., Lindell, A. H., King, C. J., et al. (2006). Coselection for microbial resistance to metals and antibiotics in freshwater microcosms. *Environmental Microbiology, 8*(9), 1510–1514.
Stokes, H. W., & Gillings, M. R. (2011). Gene flow, mobile genetic elements and the recruitment of antibiotic resistance genes into Gram-negative pathogens. *FEMS Microbiology Reviews, 35*(5), 790–819.
Stoyanov, J. V., & Brown, N. L. (2003). The *Escherichia coli* copper-responsive *copA* promoter is activated by gold. *Journal of Biological Chemistry, 278*(3), 1407–1410.

Summers, A. O. (2004). Generally overlooked fundamentals of microbial ecology. *Clinical Infectious Diseases*, *34*(Suppl. 3), S85–S92.
Summers, A. O. (2006). Genetic linkage and horizontal gene transfer, the roots of the antibiotic multi-resistance problem. *Animal Biotechnology*, *17*(2), 125–135.
Summers, A. O., & Silver, S. (1972). Mercury resistance in a plasmid-bearing strain of *Escherichia coli*. *Journal of Bacteriology*, *112*(3), 1228–1236.
Summers, A. O., Wireman, J., Vimy, M. J., Lorscheider, F. L., Marshall, B., Levy, S. B., et al. (1993). Mercury released from dental "silver" fillings provokes an increase in mercury- and antibiotic-resistant bacteria in oral and intestinal floras of primates. *Antimicrobial Agents and Chemotherapy*, *37*(4), 825–834.
Sütterlin, S., Edquist, P., Sandegren, L., Adler, M., Tängdén, T., Drobni, M., et al. (2014). Silver resistance genes are overrepresented among *Escherichia coli* isolates with CTX-M production. *Applied and Environmental Microbiology*, *80*(22), 6863–6869.
Takahashi, H., Oshima, T., Hobman, J. L., Doherty, N., Clayton, S. R., Iqbal, M., et al. (2015). The dynamic balance of import and export of zinc in *Escherichia coli* suggests a heterogeneous population response to stress. *Journal of the Royal Society, Interface*, *12*(106), 20150069.
Tanaka, M., Bergstrom, C. T., & Levin, B. R. (2003). The evolution of mutator genes in bacterial populations: The roles of environmental change and timing. *Genetics*, *164*(3), 843–854.
Temime, L., Boëlle, P., Courvalin, P., & Guillemot, D. (2003). Bacterial resistance to penicillin G by decreased affinity of penicillin-binding proteins: A mathematical model. *Emerging Infectious Diseases*, *9*(4), 411–417.
Tetaz, T. J., & Luke, R. K. J. (1983). Plasmid-controlled resistance to copper in *Escherichia coli*. *Journal of Bacteriology*, *154*(3), 1263–1268.
Tonomura, K., & Kanzaki, F. (1969). The reductive decomposition of organic mercurial by cell-free extract of a mercury resistant pseudomonad. *Biochimica et Biophysica Acta*, *184*(1), 227–229.
Tonomura, K., Maeda, K., Futai, F., Nakagami, T., & Yamada, M. (1968). Stimulative vaporization of phenyl-mercuric acetate by mercury-resistant bacteria. *Nature*, *217*(5129), 644–646.
Veterinary Medicines Directorate. (2016). https://www.gov.uk/government/news/veterinary-medicines-containing-zinc-oxide-european-referral-outcome.
Vogwill, T., & MacLean, R. C. (2014). The genetic basis of the fitness costs of antimicrobial resistance: A meta-analysis approach. *Evolutionary Applications*, *8*(3), 284–295.
Volkova, V. V., Lu, Z., Lanzas, C., Scott, H. M., & Grohn, Y. T. (2013). Modelling dynamics of plasmid-gene mediated antimicrobial resistance in enteric bacteria using stochastic differential equations. *Scientific Reports*, *3*, 2463.
Wales, A. D., & Davies, R. H. (2015). Co-selection of resistance to antibiotics, biocides and heavy metals, and its relevance to foodborne pathogens. *Antibiotics (Basel, Switzerland)*, *4*(4), 567–604.
Wand, M. E., Baker, K. S., Benthall, G., McGregor, H., McCowen, J. W. I., Deheer-Graham, A., et al. (2015). Characterization of pre-antibiotic era *Klebsiella pneumoniae* isolates with respect to antibiotic/disinfectant susceptibility and virulence in *Galleria mellonella*. *Antimicrobial Agents and Chemotherapy*, *59*(7), 3966–3972.
Wardwell, L. H., Jude, B. A., Moody, J. P., Olcerst, A. I., Gyure, R. A., Nelson, R. E., et al. (2009). Co-selection of mercury and antibiotic resistance in sphagnum core samples dating back 2000 years. *Geomicrobiology Journal*, *26*(5), 238–247.
Watanabe, T. (1963). Infective heredity of multidrug resistance in bacteria. *Bacteriological Reviews*, *27*, 87–115.
Webber, M. A., & Piddock, L. J. V. (2003). The importance of efflux pumps in bacterial antibiotic resistance. *Journal of Antimicrobial Chemotherapy*, *51*(1), 9–11.

WHO. (2003). *Elemental mercury and inorganic mercury compounds: Human health aspects*. Geneva: World Health Organization. http://www.who.int/ipcs/publications/cicad/en/cicad50.pdf.

WHO. (2014). *Antimicrobial resistance: Global report on surveillance*. Geneva: World Health Organization. http://www.who.int/drugresistance/documents/surveillancereport/en/.

WHO. (2015). *Model list of essential medicines April 2015. 19th list*. Geneva: World Health Organization. http://www.who.int/medicines/publications/essentialmedicines/en/.

Wireman, J., Liebert, C. A., Smith, T., & Summers, A. O. (1997). Association of mercury resistance with antibiotic resistance in the gram-negative fecal bacteria of primates. *Applied and Environmental Microbiology*, *63*(11), 4494–4503.

Wright, M. S., Baker-Austin, C., Lindell, A. H., Stepanauskas, R., Stokes, H. W., & McArthur, J. V. (2008). Influence of industrial contamination on mobile genetic elements: Class 1 integron abundance and gene cassette structure in aquatic bacterial communities. *ISME Journal*, *2*(4), 417–428.

Wright, M. S., Loeffler Peltier, G., Stepanauskas, R., & McArthur, J. V. (2006). Bacterial tolerances to metals and antibiotics in metal-contaminated and reference streams. *FEMS Microbiology Ecology*, *58*(2), 293–302.

Yazdankhah, S., Rudi, K., & Bernhoftm, A. (2014). Zinc and copper in animal feed- development of resistance and co-resistance to antimicrobial agents in bacteria of animal origin. *Microbial Ecology in Health and Disease*, *25*, 25862.

Zhai, Y., He, Z., Kang, Y., Yu, H., Wang, J., Du, P., et al. (2016). Complete nucleotide sequence of pH11, an IncHI2 plasmid conferring multi-antibiotic resistance and multi-heavy metal resistance genes in a clinical *Klebsiella pneumoniae* isolate. *Plasmid*, *86*, 26–31.

Zhou, B., Wang, C., Zhao, Q., Wang, Y., Huo, M., Wang, J., et al. (2016). Prevalence and dissemination of antibiotic resistance genes and coselection of heavy metals in Chinese dairy farms. *Journal of Hazardous Materials*, *320*, 10–17.

Zhu, Y. G., Johnson, T. A., Su, J. Q., Qiao, M., Guo, G. X., Stedtfeld, R. D., et al. (2013). Diverse and abundant antibiotic resistance genes in Chinese swine farms. *Proceedings of the National Academy of Sciences of the United States of America*, *110*(9), 3435–3440.

Zimmermann, M., Udagedara, S. R., Sze, C. M., Ryan, T. M., Howlett, G. J., Xiao, Z., et al. (2012). PcoE—A metal sponge expressed to the periplasm of copper resistance *Escherichia coli*. Implication of its function role in copper resistance. *Journal of Inorganic Biochemistry*, *115*(1), 186–197.

CHAPTER EIGHT

# The Role of Intermetal Competition and Mis-Metalation in Metal Toxicity

**Anna Barwinska-Sendra, Kevin J. Waldron**[1]
Institute for Cell & Molecular Biosciences, Faculty of Medical Sciences, Newcastle University, Newcastle upon Tyne, United Kingdom
[1]Corresponding author: e-mail address: kevin.waldron@ncl.ac.uk

## Contents



## Abstract

The metals manganese, iron, cobalt, nickel, copper and zinc are essential for almost all bacteria, but their precise metal requirements vary by species, by ecological niche and by growth condition. Bacteria thus must acquire each of these essential elements in sufficient quantity to satisfy their cellular demand, but in excess these same elements are toxic. Metal toxicity has been exploited by humanity for centuries, and by the mammalian immune system for far longer, yet the mechanisms by which these elements cause toxicity to bacteria are not fully understood. There has been a resurgence of interest in metal toxicity in recent decades due to the problematic spread of antibiotic resistance amongst bacterial pathogens, which has led to an increased research effort to understand these toxicity mechanisms at the molecular level. A recurring theme from these studies is the role of intermetal competition in bacterial metal toxicity.

*Advances in Microbial Physiology*, Volume 70
ISSN 0065-2911
http://dx.doi.org/10.1016/bs.ampbs.2017.01.003

In this review, we first survey biological metal usage and introduce some fundamental chemical concepts that are important for understanding bacterial metal usage and toxicity. Then we introduce a simple model by which to understand bacterial metal homeostasis in terms of the distribution of each essential metal ion within cellular 'pools', and dissect how these pools interact with each other and with key proteins of bacterial metal homeostasis. Finally, using a number of key examples from the recent literature, we look at specific metal toxicity mechanisms in model bacteria, demonstrating the role of metal–metal competition in the toxicity mechanisms of diverse essential metals.

# 1. INTRODUCTION

All life depends on essential metal ions, which enable biological catalysts to achieve chemical transformations that would be unachievable using solely the organic elements that constitute biological macromolecules. Yet when present in excess, these same life-giving essential metals can be highly toxic to cells, whereas some other metals that are not biologically essential can be toxic at almost any dose. As such, each and every living cell on Earth is in a constant battle to optimise its metal supply in order to ensure sufficient delivery of metal cofactors to its nascent metalloproteins, while minimising the harmful effects caused by metal excess, metal deficiency or inappropriate metal localisation. As an example, iron limitation in the oceans has been demonstrated to be a key factor in limiting marine primary productivity (Boyd et al., 2007; Martin & Fitzwater, 1988; Watson, Bakker, Ridgwell, Boyd, & Law, 2000), and inputs of iron from atmospheric deposition (for example, caused by particulates from the Sahara desert being driven by the wind into the Atlantic) are important in promoting plankton growth (Jickells et al., 2005), but can also provide sufficient copper to cause growth-limiting toxicity (Jordi, Basterretxea, Tovar-Sánchez, Alastuey, & Querol, 2012). The processes of metal homeostasis that maintain intracellular metal concentrations have evolved over geological timescales and enable organisms to respond rapidly to ever-changing exogenous conditions, to adjust metal supply to meet demand and/or to adjust metal demand to meet supply.

The toxicity of metals towards microorganisms has been unknowingly exploited by humanity for millennia. Ancient texts from past civilisations dating back to the 3rd millennium BCE recorded the use of copper salts for the treatment of infections and for sterilisation of drinking water, and later for treatment of headaches and burn wounds (Dollwet & Sorenson, 1985). Medical uses of copper were discussed in ancient Greek, Roman, Indian and Aztec texts (Dollwet & Sorenson, 1985). In more recent times, copper compounds have a long and distinguished history as antifungal agents in agriculture. Bordeaux Mixture, a combination of copper sulphate and lime developed by Millardet in the 1880s, was shown to be highly effective against a range of fungal pathogens including vine mildew (*Peronospora viticola*) and potato blight (*Phytophthora infestans*) and was soon adopted worldwide (Ayres, 2004). Copper-containing antifungal agents remain in use today in significant quantities globally.

Zinc compounds are widely used in commercial products for their antimicrobial properties, for example, in toothpastes, whereas zinc pyrithione is the active ingredient in antidandruff shampoos. Interestingly, the antifungal mechanism of zinc pyrithione has been shown to involve it acting as a copper ionophore, which shuttles copper into the cytoplasm of target fungi, leading to copper toxicity (Reeder et al., 2011). Both copper and zinc remain in widespread use as growth promoters in animal husbandry; although it is unclear whether the beneficial effect is due to antimicrobial activity of the metals, there is some evidence that this metal usage increases the spread of metal resistance amongst zoonotic pathogens (Amachawadi et al., 2011; Cavaco, Hasman, & Aarestrup, 2011; Yazdankhah, Rudi, & Bernhoff, 2014). Indeed, metal resistance genes such as those encoding metal exporters are often present on the mobile genetic elements that are responsible for the spread of antibiotic resistance between microorganisms (Baker-Austin, Wright, Stepanauskas, & McArthur, 2006; Di Pilato et al., 2014). One study has even suggested that the release of copper from cuproenzymes plays a role in the bacterial response to some traditional antibiotic compounds (Hao et al., 2014).

In the last couple of decades, there has been a resurgence in interest in the antimicrobial properties of metals, most notably copper and silver. Solid copper surfaces have been demonstrated to display ‘contact killing’ effects on a wide range of bacterial pathogens, including spores, and even viruses (Elguindi, Wagner, & Rensing, 2009; Faúndez, Troncoso, Navarrete, & Figueroa, 2004; Molteni, Abicht, & Solioz, 2010; Noyce, Michels, & Keevil, 2006, 2007; Weaver, Michels, & Keevil, 2008), zinc nanoparticles are used in food packaging and cosmetics (Espitia et al., 2012), whereas silver is routinely incorporated into wound dressings as well as a wide range of commercial products (Mijnendonckx, Leys, Mahillon, Silver, & Van Houdt, 2013; Silver, Phung, & Silver, 2006). In the face of increasingly widespread bacterial resistance to traditional antibiotics amongst pathogens, and the limited progress in developing new classes of antibiotics (Blair, Webber, Baylay, Ogbolu, & Piddock, 2015; Brown & Wright, 2016), there is great appeal in using the natural 'self-sterilising' properties of copper surfaces in healthcare settings to reduce nosocomial disease transmission (Casey et al., 2010; Inkinen, Makinen, Keinanen-Toivola, Nordstrom, & Ahonen, 2016), and clinical trials replacing stainless steel fixtures and fittings with copper-containing alloys have thus far shown promising results (Hinsa-Leasure, Nartey, Vaverka, & Schmidt, 2016; Salgado et al., 2013). The mechanism of toxicity of solid copper is as yet unclear; there is mixed evidence on the role, or lack thereof, of dissolved copper ions in contact killing of microbes (Elguindi et al., 2009; Molteni et al., 2010).

However, the exploitation of metal toxicity has a far longer history than humanity. The same metal ions are essential micronutrients and potential toxicants for both host and pathogen alike, putting mechanisms of metal acquisition and metal detoxification at the front line of the battle at the host-pathogen interface during infection. The iron-withholding response of the mammalian immune system has been established by several decades of research (reviewed in Drakesmith & Prentice, 2012; Weinberg, 1984), but in recent years this metal-restriction process, termed nutritional immunity, has been extended to other metals (Kehl-Fie & Skaar, 2010). For example, the immune effector protein complex calprotectin has been shown to restrict the growth of pathogens through its propensity to tightly sequester manganese, zinc and, to a lesser extent, iron at sites of bacterial infection (Corbin et al., 2008; Damo et al., 2013; Kehl-Fie et al., 2013; Nakashige, Zhang, Krebs, & Nolan, 2015), creating a selection pressure for metal acquisition or substitution (Garcia et al., 2017).

Metal acquisition systems have been shown to be essential for virulence in a wide range of bacterial pathogens (Boyer, Bergevin, Malo, Gros, & Cellier, 2002; Corbett et al., 2012; Horsburgh, Clements, Crossley, Ingham, & Foster, 2001; Remy et al., 2013), whereas bacterial metal detoxification and efflux systems have been shown to be essential for virulence in other systems (Pi, Patel, Argüello, & Helmann, 2016; Rosch, Gao, Ridout, Wang, & Tuomanen, 2009; Stähler et al., 2006; Wolschendorf et al., 2011). Indeed, in at least one case both acquisition and efflux systems are essential (Jin et al., 2006; Pi et al., 2016). The conclusion must be that immune cells try to starve invading pathogens of some of these essential nutrients under certain conditions and in some bodily niches, where as in other cases the immune response uses metal toxicity to its advantage. For example, there is now substantial evidence that both copper and zinc are used by macrophages and neutrophils as a biocidal tool in combination with its other arsenal of antimicrobial weapons (Botella et al., 2011; Ong, Gillen, Barnett, Walker, & McEwan, 2014; Osman et al., 2010; White, Lee, Kambe, Fritsche, & Petris, 2009).

In this review, we first present a simple model for understanding how bacterial metal homeostasis regulates the buffered intracellular levels of each essential metal ion, and how this controls metal binding to target metalloproteins through regulating metal bioavailability (Section 2). Then we use this model to understand how an excess of a metal ion can give rise to intracellular mis-metalation events, before reviewing the emerging molecular mechanistic evidence of how excess concentrations of essential metals are toxic to bacteria (Section 3). The precise mechanisms of toxicity are complex and multifaceted, but significant advances have been made in this research field over the last decade. Although not intended to be an exhaustive view of the current literature, several key examples will be used to show that metal–metal competition plays a key role in the molecular mechanisms that determine metal toxicity.

## 1.1 Biological Metal Usage

The vast majority (>95%) of a cell's dry biomass is organic, composed of biology's major classes of macromolecules: carbohydrates, lipids, proteins and nucleic acids. These are made from just six elements of the Periodic Table: carbon (C), hydrogen (H), nitrogen (N), oxygen (O), phosphorus (P) and sulphur (S). Beyond these 'big six' organic elements, life's chemical inventory is actually somewhat limited (Fig. 1). Of all of the chemical space that is potentially available from combinations of the 94 naturally occurring elements, it seems rather surprising that most life on Earth relies on just a small, select subsample of the Periodic Table (see Fig. 1) to perform most of its biological chemistry.

The Group 1 alkali metals sodium (Na) and potassium (K), as well as the Group 17 halogen chlorine (Cl), are ubiquitous in biology where they are predominantly employed as carriers of electrical charge and as essential regulators of osmotic pressure. The Group 2 alkali Earth metals magnesium (Mg) and calcium (Ca) are also ubiquitous, which similarly act as charge carriers and osmotic regulators, but additionally perform numerous essential biological roles, including as important structural components of biological membranes, as cofactors for essential enzymes, and as secondary messengers

| | 1 | 2 | | 3 | 4 | 5 | 6 | 7 | 8 | 9 | 10 | 11 | 12 | 13 | 14 | 15 | 16 | 17 | 18 |
|---|---|---|---|---|---|---|---|---|---|---|---|---|---|---|---|---|---|---|---|
| 1 | 1 H 1.008 | | | | | | | | | | | | | | | | | | 2 He 4.003 |
| 2 | 3 Li 6.943 | 4 Be 9.012 | | | | | | | | | | | | 5 B 10.811 | 6 C 12.011 | 7 N 14.007 | 8 O 15.999 | 9 F 18.998 | 10 Ne 20.180 |
| 3 | 11 Na 22.990 | 12 Mg 24.305 | | | | | | | | | | | | 13 Al 26.982 | 14 Si 28.086 | 15 P 30.974 | 16 S 32.065 | 17 Cl 35.453 | 18 Ar 39.948 |
| 4 | 19 K 39.098 | 20 Ca 40.078 | | 21 Sc 44.956 | 22 Ti 47.867 | 23 V 50.942 | 24 Cr 51.996 | 25 Mn 54.938 | 26 Fe 55.845 | 27 Co 58.933 | 28 Ni 58.693 | 29 Cu 63.55 | 30 Zn 65.39 | 31 Ga 69.723 | 32 Ge 72,61 | 33 As 74.922 | 34 Se 78.96 | 35 Br 79.904 | 36 Kr 83.80 |
| 5 | 37 Rb 85.468 | 38 Sr 87.62 | | 39 Y 88.906 | 40 Zr 91.224 | 41 Nb 92.906 | 42 Mo 95.94 | 43 Tc [98] | 44 Ru 101.07 | 45 Rh 102.91 | 46 Pd 106.42 | 47 Ag 107.87 | 48 Cd 112.41 | 49 In 114.82 | 50 Sn 118.71 | 51 Sb 121.76 | 52 Te 127.60 | 53 I 126.90 | 54 Xe 131.29 |
| 6 | 55 Cs 132.91 | 56 Ba 137.33 | 57-70 | 71 Sc 174.97 | 72 Hf 178.49 | 73 Ta 180.95 | 74 W 183.84 | 75 Re 186.21 | 76 Os 190.23 | 77 Ir 192.22 | 78 Pt 195.08 | 79 Au 196.97 | 80 Hg 200.59 | 81 Tl 204.38 | 82 Pb 207.2 | 83 Bi 208.98 | 84 Po [209] | 85 At [210] | 86 Rn [222] |
| 7 | 87 Fr [223] | 88 Ra [226] | ↓ | 89 Ac [227] | 90 Th 232.04 | 91 Pa 231.04 | 92 U 238.03 | 93 Np [237] | 94 Pu [244] | | | | | | | | | | |

| | | | | | | | | | | | | | |
|---|---|---|---|---|---|---|---|---|---|---|---|---|---|
| 57 La 138.91 | 58 Ce 140.12 | 59 Pr 140.91 | 60 Nd 144.24 | 61 Pm [145] | 62 Sm 150.36 | 63 Eu 151.96 | 64 Gd 157.25 | 65 Tb 158.93 | 66 Dy 162.50 | 67 Ho 164.93 | 68 Er 167.26 | 69 Tm 168.93 | 70 Yb 170.04 |

**Fig. 1** The Periodic Table, showing only the 94 naturally occurring elements, depicting the essential organic (*red*), inorganic (*green*) and metal (*blue*) elements utilised by all life on Earth. Additional elements used rarely or of currently unknown essential status are shown in *yellow*.

in eukaryotic signalling (Berridge, Bootman, & Roderick, 2003). Magnesium, for example, plays a critical role in the biochemistry and structure of nucleic acids, and the coordination of Mg(II) ions by the nucleotides ATP and GTP is essential for their biological activity. Although undoubtedly fulfilling the criteria of being essential metals, magnesium and calcium will not be considered further in this review, which is focused on the essential metals of the *d*-block of the Periodic Table. The essential roles of magnesium and calcium have been recently reviewed elsewhere (Domínguez, Guragain, & Patrauchan, 2015; Groisman et al., 2013), but their toxicity mechanisms are largely unstudied.

Of the *d*-block, or 'transition' metals, only seven are considered biologically essential (in approximate order of abundance, from most to least—Andreini, Bertini, Cavallaro, Holliday, & Thornton, 2008; Cameron, House, & Brantley, 2012; Outten & O'Halloran, 2001): iron (Fe), zinc (Zn), manganese (Mn), copper (Cu), nickel (Ni), cobalt (Co) and molybdenum (Mo), the latter of which is utilised catalytically exclusively in the form of the molybdenum-containing cofactor known as molybdopterin (reviewed in Schwarz & Mendel, 2006). A few further elements are utilised in certain limited organisms or have been suggested to be important, even essential, in some systems but whose precise biological role is as yet unclear: vanadium (V) is used in the vanadium-dependent nitrogenases and haloperoxidases (reviewed in Rehder, 2015); tungsten (W) can be incorporated in place of the more usual molybdenum within the molybdpterin cofactor in certain (predominantly thermophilic) prokaryotes and used by enzymes such as aldehyde oxidoreductase and formylmethanofuran dehydrogenase (reviewed in Andreesen & Makdessi, 2008); and a single characterised enzyme, a carbonic anhydrase enzyme from the marine diatom *Thalassiosira weissflogii*, has been shown to be able to use a cadmium (Cd) cofactor (Lane & Morel, 2000; Lane et al., 2005), a 'red-list' metal which is highly toxic to most organisms, presumably representing a zinc-sparing adaptation in the zinc-deficient oceanic environment.

Metal usage can be highly dynamic. For example, some methanotrophic bacteria possess two distinct enzymes for methane oxidation, the methane monooxygenases (MMO). In these cases, the switch between the preferred, copper-containing particulate enzyme (pMMO) (Balasubramanian et al., 2010) and the less energetically efficient, iron-containing soluble enzyme (sMMO) (Lee, McCormick, Lippard, & Cho, 2013) is regulated by the availability of the copper cofactor for pMMO (Semrau et al., 2013). Some cyanobacteria can switch between copper-containing plastocyanin and haem iron-containing cytochrome $c_6$ for photosynthetic and respiratory electron transfer, with the switch in expression again regulated by copper availability (Zhang, McSpadden, Pakrasi, & Whitmarsh, 1992). A subset of the superoxide dismutase (SOD) enzymes, which are usually highly specific for their target metal (Beyer & Fridovich, 1991), are cambialistic in vitro (Amano, Shizukuishi, Tsunemitsu, & Tsunasawa, 1992; Martin et al., 1986), and there is some evidence that this ability to use an alternative cofactor is important in vivo (Amano et al., 1992), including during infection (Garcia et al., 2017). It has also been proposed that many cytosolic enzymes in *Escherichia coli* that require a divalent cation for catalysis predominantly utilise Fe(II) under normal conditions (Anjem & Imlay, 2012; Anjem, Varghese, & Imlay, 2009; Taudte, German, Zhu, Grass, & Rensing, 2016), but can switch to using Mn(II) under conditions in which the cells experience oxidative stress, conditions in which iron uptake is repressed and manganese uptake is induced (Anjem et al., 2009).

Metal usage varies, not only between different species, but also between subcellular compartments, between cell types in multicellular organisms, between different ecological niches, and even by circadian rhythm. For example, the diazotrophic (i.e. nitrogen-fixing) marine cyanobacterium *Crocosphaera watsonii* has been shown to undergo a large-scale reorganisation of its metalloenzyme complement during the transition from day to night, degrading daytime metalloenzymes at night and vice versa, and expending a great deal of additional energy in protein synthesis in the process, in order to reduce its iron demand by as much as 40% during the hours of darkness (Saito et al., 2011). Such an iron-sparing mechanism likely provides a strong selective advantage in the nutrient-poor conditions of the ocean; iron availability is considered to be a major limiting factor in oceanic primary productivity (Boyd et al., 2007; Martin & Fitzwater, 1988; Watson et al., 2000). The intracellular concentration of magnesium, which oscillates over the circadian cycle in a cell-autonomous manner in diverse eukaryotes, has been shown to be a key component of the circadian clock itself, potentially playing an important role in the temporal regulation of energy consumption in the form of Mg(II)-ATP (Feeney et al., 2016). Whether this is replicated in bacteria is unknown, but an early report suggested cell cycle-dependent changes in metal abundance in *E. coli* cells (Kung, Raymond, & Glaser, 1976).

Historically, biological metal usage has been heavily influenced by environmental availability, as would be expected in a model of "'the economical utilisation of resources', i.e. choosing those elements less costly in terms of energy required for uptake, given the function for which they are required," as described by Fraústo da Silva and Williams in their seminal book 'The Biological Chemistry of the Elements: The Inorganic Chemistry of Life' (Fraústo da Silva & Williams, 2001). Life evolved early in Earth's history under strictly anaerobic conditions. Although the precise chemical composition of the ancient ocean and atmosphere during the Archean and Proterozoic periods are subjects of intense discussion (Canfield, 1998; Fraústo da Silva & Williams, 2001; Saito, Sigman, & Morel, 2003), it seems likely that iron was highly available in its reduced, ferrous form, Fe(II) (Saito et al., 2003). Under those same conditions, it is proposed that cobalt, nickel and manganese would also have been soluble and readily available in the oceans, whereas copper and zinc would have been largely unavailable due to their insolubility as sulphides (Fraústo da Silva & Williams, 2001; Saito et al., 2003). Therefore, early life forms would have likely made extensive use of these available metals, especially the redox-versatile iron, for their metalloenzymes. However, the invention and subsequent proliferation of oxygenic photosynthesis by the photoautotrophic ancestors of cyanobacteria is thought to have given rise to the 'great oxidation event', whereby the levels of planetary oxygen rose rapidly around 2.3 billion years ago (Schirrmeister, de Vos, Antonelli, & Bagheri, 2013). Oxygen did not initially accumulate in the atmosphere, instead being consumed in the oxidation of oceanic iron (Catling, Zahnle, & McKay, 2001). The oxidised, ferric form of iron is essentially insoluble, leading to the vast geologic deposition of iron in this period, giving rise to the banded iron deposits that we mine to acquire this economically essential element today (Canfield, 1998). This rapid decrease in iron availability would have created strong selection pressure for evolution, initially for organisms to develop mechanisms to acquire the ever-decreasing available iron, for example, by secreting iron chelators that aid iron solubilisation, and then subsequently to find alternative cofactors for their iron enzymes, utilising metals such as copper and zinc instead. Traces of this development from predominantly iron-requiring to a more metal-diverse cellular complement of metalloproteins can be detected in bioinformatic analyses of gene and protein sequences today (Dupont, Butcher, Valas, Bourne, & Caetano-Anollés, 2010).

Some bacteria appear to have taken this process to its logical conclusion, by foregoing iron usage entirely. The streptococci, for example, have a greatly diminished complement of iron proteins, especially those requiring iron sulphur (Fe-S) clusters, probably due to their innate production of high levels of hydrogen peroxide, which readily damages such clusters (Jang & Imlay, 2007). The pathogenic spirochaete, *Borrelia burgdorferi*, appears to have no requirement for iron at all (Posey & Gherardini, 2000). It makes extensive use of manganese for its essential functions (Aguirre et al., 2013), but even this iron-independent bacterium possesses a Dps protein (Li et al., 2007), a mini-ferritin iron storage protein that protects DNA from iron-mediated toxicity during stress conditions (Almirón, Link, Furlong, & Kolter, 1992), which is essential for its virulence (Li et al., 2007).

An oft-quoted estimate is that approximately one-third of all proteins require one or more essential metal ion for function. Detailed bioinformatics analysis of the vast numbers of uncharacterised protein sequences available in public databases is inevitably difficult (Andreini et al., 2008; Dupont et al., 2010; Gladyshev & Zhang, 2013), but some estimates suggest that this number may rise to almost half when only enzymes, proteins with known or predicted catalytic activities, are considered (Andreini et al., 2008; Waldron, Rutherford, Ford, & Robinson, 2009). Regardless of their accuracy, these estimates nonetheless highlight the prevalence of metalloproteins in biology. Such a preponderance of metalloproteins and metalloenzymes has important implications for biotechnology, and particularly for achieving the grand goals of the synthetic biology community. Wherever attempts are made to produce synthetic biological systems to perform useful tasks, there is a high likelihood that one or more critical protein component of the system will require metals, and thus thought must be given to engineering the metal homeostasis system to ensure sufficient cofactor availability, and intracellular metal supply pathways to ensure their correct function within the heterologous chassis of the genetically modified organism or synthetic host. This problem will be even more challenging when trying to engineer metal supply for man-made enzymes with novel activities that utilise nonessential metal cofactors (Jeschek et al., 2016; Key, Dydio, Clark, & Hartwig, 2016).

## 1.2 Metal Chemistry and Biochemistry

The essential metal ions for consideration herein are several of those in the first row of the *d*-block of the Periodic Table (Fig. 1). The *d*-block elements are also known as the 'transition' elements due to their ability to form cations with a partially filled *d* electron orbital. They can generally form a variety of complexes with different oxidation states, a property that formerly excludes zinc from the definition as its Zn(II) ion has a complete $d^{10}$ electron configuration and its Zn(0) state has a complete $d^{10}s^2$ configuration. They often form coloured complexes due to light-induced electronic transitions, either between *d* electron orbitals on the metal (*d*-*d* transitions), or from orbitals on the bound ligands to incomplete metal *d* orbitals (ligand-to-metal charge transfer), a property that can be exploited in the study of metalloproteins through extensive spectroscopic techniques.

There is a vast array of resources for those interested in learning more about the fundamental principles of inorganic and bioinorganic chemistry, but several key principles are crucial to understanding metal toxicity and therefore are relevant to this review.

Metal binding to metalloproteins is dictated by the same principles of coordination chemistry that apply to small-molecule inorganic complexes. The bonding between a metal ion and its ligands can be described using ligand field theory (LFT) (Griffith & Orgel, 1957), which predicts the electronic structure of a complex based on the composition of metal and ligands and the coordination geometry. In the case of metalloproteins the ligands come primarily from the protein, not only from amino acid side chains but also from the polypeptide backbone, but can also come from metabolites, solvent molecules or the substrate/product of the metalloenzyme. In LFT side chains can be considered as individual ligands rather than one polydentate ligand, although energetically the polydentate nature of metal-protein binding will be important due to an entropic benefit known as the chelate effect. The structural properties of the polypeptide chain, especially once folded, can impose some level of steric restraint on the final coordination geometry, although nascent polypeptides are inherently more flexible and thus could potentially misfold around a metal ion delivered inappropriately.

The thermodynamic stability of metal complexes can be generalised by the principles of hard–soft acid–base theory (Parr & Pearson, 1983; Pearson, 1963), where the metal ion is the acid and the ligands are bases. This states that 'soft' metals such as Cu(I) tend to form more stable complexes with 'soft' ligands such as thiolate or thioether group sulphurs, whereas 'hard' metal ions such as Fe(III) tend to form more stable complexes with 'hard' ligands such as hydroxyl or carboxylate oxygens. In proteins, this is observed as a preference of Cu(I) for cysteine or methionine side chains, against a preference of Fe(III) for aspartate, glutamate and tyrosine side chains. Most biologically relevant metal ions are considered 'borderline' and thus are relatively flexible in terms of their ligand preferences, often being associated with histidine nitrogens as well as mixtures of sulphur and oxygen-containing side chains (Andreini et al., 2008). The strong preference for sulphur ligands of the Cu(I) ion can influence copper selectivity in biological systems and can aid bioinformatics searches for putative Cu(I) binding sites.

A fundamental principle of coordination chemistry, determined empirically from thermodynamic stability data of metal complexes of the first row of the *d*-block elements, is the Irving–Williams series (Irving & Williams, 1953). This states that the divalent cations of these metal ions generally form progressively more stable complexes with a given ligand as we go from the elements on the left of the *d*-block to those on the right side of the block, forming the series that defines the relative affinities as Mn < Fe < Co < Ni < Cu > Zn. It should be noted that copper is higher up this series than would be expected from the general trend due to a specific geometric and electronic effect known as the Jahn-Teller distortion. In biological terms, and relevant to this review, we can think of this key property as defining the essential elements to the left of the *d*-block as 'uncompetitive' and those on the right as 'competitive'. That is to say that the weaker binding, uncompetitive essential metal ions such as Mn(II) and Fe(II) are inherently vulnerable to displacement from their biological binding sites by metal ions that are further up the series, such as the competitive metal ions Cu(II) or Zn(II). It is for this reason that the suggestion that the cell is devoid of 'free' transition metal ions, i.e. ions that are coordinated exclusively with water ligands, makes instinctive sense; the Irving–Williams series implies that a cell that contained 'free' ions of all of the essential metals would be unable to accurately populate a target metalloprotein with an uncompetitive metal such as Fe(II).

When considering kinetics, it must be remembered that metal ions are never alone in aqueous environments. 'Free' metal ions are actually hydrated, surrounded by a complete ligand shell where the metal ion is coordinated to a lone pair of electrons from the oxygen atom of each coordinated water molecule. Thus, when we consider the kinetics of metal binding to a metalloprotein, we should more accurately consider this as the kinetics of swapping one ligand set (for example, six coordinated water molecules arranged in an octahedral geometry) for another (the final arrangement of protein side chain-provided ligands coordinated to the central metal ion), and the desolvation of the metal ion is a nonnegligible component, both kinetically and thermodynamically. This situation is made more complicated by the fact that even these 'free' ions are essentially absent in biology (Section 2.1); metal ions are always coordinated to biological ligands, whether proteinaceous or small molecules, and thus a consideration of the kinetics of metal binding to a nascent metalloprotein must consider the swapping of the ligand set present in its cytosolic, bioavailable state (which is generally unknown) for the ligand set provided by the protein.

Reactions of metal complexes that involve a change of the ligand set can occur via one of two distinct mechanisms. The associative mechanism, in some ways comparable to an $S_N2$ nucleophilic substitution reaction mechanism in organic chemistry, involves the incoming ligand (the nucleophile, due its possession of a lone pair of electrons that will ultimately coordinate to the central metal ion, the electrophile) forming a coordinate bond to the metal, temporarily increasing the coordination number (the total number of ligands bound to the metal ion), before subsequent dissociation of the outgoing ligand (the leaving group), which reduces the coordination number back to its original value. Reaction rates of associative reactions depend on both the concentration (i.e. availability) of the incoming ligand and the concentration of the reactant complex. In the alternative, dissociative mechanism, which can be considered to be analogous to an $S_N1$ nucleophilic substitution mechanism in organic chemistry, involves the dissociation of the outgoing ligand, temporarily decreasing the coordination number, before subsequent binding of the incoming ligand to restore the coordination number. Reaction rates of dissociative reactions are independent of the concentration of the incoming ligand as the rate-limiting step is the dissociation of the outgoing ligand. In practise, most reactions occur through an interchange mechanism (either associative or dissociative interchange) in which the transition state contains both the incoming and outgoing ligands coordinated (either strongly or weakly, respectively) to the metal ion. Neither of these mechanisms involves a net change in the oxidation state (total number of electrons relative to the element's atomic number) of the metal ion, but oxidation or reduction of the metal ion can be important in ligand exchange reactions as the oxidised metal ion can have different ligand preferences to those of the reduced ion. Metal redox reactions are important, for example, in the release of iron ions from siderophores (Cooper, McArdle, & Raymond, 1978).

In proteins, the steric shielding of the metal site can greatly influence the kinetics of such substitution mechanisms by preventing both mechanisms; by preventing an incoming nucleophile from approaching close enough to undergo associative substitution or by restricting an outgoing ligand from diffusing away from the metal site and thus inhibiting the dissociative mechanism. Conversely, the protein can also enforce a strained conformation on the metal complex, thereby facilitating ligand exchange reactions that are needed for an enzyme's catalytic cycle.

Metal complexes can be classified kinetically as either inert, i.e. resistant to ligand exchange reactions, or labile, i.e. undergo ligand exchange reactions readily. It is worth noting that this kinetic description is distinct from complex stability, which is a thermodynamic parameter that describes the relative abundance of starting complex and product complex when the system is at equilibrium. Thermodynamically very stable complexes can nonetheless be kinetically labile, and thermodynamically unstable complexes can be kinetically inert. This is because the reaction rate is determined by the relative energy levels, that is, the difference in energy, of the starting materials and the transition state, whereas thermodynamic stability is determined by the relative energy levels of the starting materials and the reaction products.

One final property of metal complexes that is important for their roles and toxicity in biological systems is their ability to undergo reduction and oxidation (redox) reactions by gaining or losing electrons. This property explains why some metal ions have been recruited by biology, for example copper and iron, which facilitate both electron transfer in energy-producing electron transport chains such as those of photosynthesis and respiration, and in redox reactions by metalloenzymes. The ability of a metal ion to undergo redox reactions can be quantified through the measure of reduction potentials (a relative measure in comparison to a redox couple defined as a standard, in this case the standard hydrogen electrode, under standard experimental conditions of temperature and pressure). Importantly, metalloenzymes are able to alter or 'tune' this redox potential of a metal ion by controlling the metal's coordination environment, both its primary coordination sphere (the direct metal ligands) and the secondary coordination sphere (the chemical properties of the residues that are in the immediate vicinity of the metal ion, without actually coordinating the metal). As well as determining the use of redox-active metal ions by metalloenzymes, the redox properties of essential metal ions are also of importance in determining their ability to cause toxicity. In particular, the redox-active metals copper and iron can facilitate the production of the most toxic of the reactive oxygen species (ROS), the hydroxyl radical, via the Haber-Weiss and Fenton reactions:

$$Fe^{3+} + \bullet O_2^- \rightarrow Fe^{2+} + O_2$$

$$Fe^{2+} + H_2O_2 \rightarrow Fe^{3+} + OH^- + \bullet OH$$

Net reaction:

$$\bullet O_2^- + H_2O_2 \rightarrow \bullet OH + OH^- + O_2$$

The hydroxyl radical is extremely reactive towards a range of biological macromolecules, causing oxidative damage that can inhibit essential cellular processes, and ultimately cause cell death. There is much evidence that this iron-catalysed reaction scheme is important in iron toxicity, including under conditions of hydrogen peroxide and superoxide stress, which lead to release of cellular iron from iron binding sites (Section 3.2). Copper is actually an efficient catalyst of the Fenton reaction (Gunther, Hanna, Mason, & Cohen, 1995), but recent data suggest that the biological control of copper that maintains a cytosol devoid of Cu(I) prevents this reaction from occurring in vivo, even under copper toxicity conditions. Its catalysis of this reaction may, however, be an important source of ROS inside the macrophage phagolysosomal compartment (White et al., 2009).

# 2. METAL HOMEOSTASIS

As with all other nutrients, bacterial cells have evolved complex systems to ensure homeostasis, i.e. to maintain metal concentrations within a narrow target range, despite fluctuations in external concentrations and despite every cell division requiring a halving of the cell's metal resources.

## 2.1 What Determines Metal Binding Inside Cells?

Every cell needs to acquire its own specific complement of essential metal ions from the extracellular environment, and then ensure that these are supplied correctly, both temporally and spatially, to newly synthesised metalloproteins. It is essential for cell functioning that nascent manganese proteins are supplied with manganese, and manganese alone, whereas nascent zinc proteins must be supplied with zinc, and only zinc. A failure to supply the required metal ion, or the supply of the incorrect metal to the newly synthesised polypeptide will yield an inactive resulting enzyme, wasting the energy and resources of synthesising the protein in the first place. For example, the incorrect provision of iron to the strictly manganese-dependent SOD enzyme, SodA, in *E. coli* results in a catalytically inactive Fe-SodA under growth conditions in which *E. coli* fails to take up manganese (Anjem et al., 2009).

So what determines which of a cell's proteins bind a given metal? Surprisingly, other than a few select examples, little research has looked into the specific mechanisms of metal acquisition by target metalloenzymes. In a very few cases such as the supply of nickel to hydrogenase and urease (Olson, Mehta, & Maier, 2001), metal cofactors are known to be delivered to a specific protein target by a class of proteins called metallochaperones, ensuring that in these select cases the metal selectivity of the target metalloenzyme is determined by the specificity of the protein–protein interaction between it and the metallochaperone (Banci et al., 2006; Banci, Bertini, Ciofi-Baffoni,

Poggi, et al., 2010). However, very few bacterial proteins have been shown to acquire their cofactor in this way, and thus it is assumed that the vast majority of metalloproteins acquire their metal directly from one or more cellular metal 'pools'.

It is important to note that metal ions within cells are not generally considered to be 'free' ions (where 'free' ions are defined as metal ions that are associated with no ligands except water). The huge overcapacity of the cytosol to chelate transition metal ions should ensure that all soluble ions are coordinated to biological molecules. Instead, these metal ions can be considered as residing in different 'pools' in which they are associated with different classes of biological ligands. Their availability to a nascent metalloprotein will thus be a function of: (a) the abundance of a given metal ion within each given 'pool', (b) the (kinetic) lability of the given metal complex(es) in that pool, and (c) the respective (thermodynamic) stability constants of the two complexes (i.e. that of the metal complex with the pool ligands compared with that of the metal complex with the proteinaceous ligands).

A simple theoretical model (Fig. 2), aspects of which are supported by some empirical evidence (Banci, Bertini, Ciofi-Baffoni, Kozyreva, et al., 2010; Gu & Imlay, 2013; Macomber & Imlay, 2009), would hold that the cell can be considered as a system comprised of three main metal pools (of each metal).

The primary pool is the main bioavailable metal pool, perhaps associated with one or more abundant small molecules such as the reductants glutathione or bacillithiol, amino acids such as cysteine or histidine, or other abundant metabolites such as polyphosphates or lipoamide, or even with as yet unidentified proteins. Alternatively, a primary metal pool may be constituted of a variety of such ligands, especially for the relatively uncompetitive metals at the bottom of the Irving–Williams series, such as magnesium, calcium and manganese, a model which has been described as a 'poly disperse buffer' (Foster, Osman, & Robinson, 2014). The primary pool is the primary metal 'buffer' and would act as the main source of metal for nascent proteins to acquire. It is presumably fed, directly or indirectly, by the diverse metal importers, and is likewise the source of substrate for most metal exporters. Although complexed, the metal ions within this pool would need to be relatively labile, to facilitate transfer to target proteins, and would require an affinity for its metal sufficiently tight to ensure complete sequestration of the metal (i.e. negligible 'free' ions) and to prevent inappropriate association with nontarget proteins and metabolites but also sufficiently

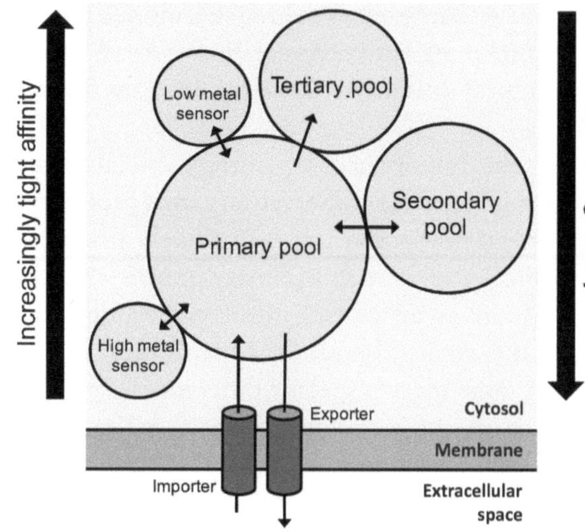


**Fig. 2** Schematic diagram of intracellular metal pools. Imported metals are incorporated into the primary pool. This is the main bioavailable pool, which acts as metal source for metal storage proteins (the secondary pool) and for target metalloproteins (the tertiary pool). It is also the source of metals for detection by the metal sensors and for metal efflux via exporter proteins. The trends in relative metal thermodynamic affinity and kinetic lability of the ligands of each pool are shown with arrows. Note that the role of metallochaperones is not illustrated in this diagram for simplicity.

weak to allow a bona fide metal-requiring protein to outcompete the pool ligand. As such, the affinities of the primary pools' ligands are anticipated to follow the Irving–Williams series.

In some cases, a single cellular molecule may act as the main ligand of the primary pool for more than one metal, especially where those metals have similar ligand preferences and are in similar positions in the Irving–Williams series, for example, manganese and ferrous iron or zinc and cuprous copper ions, whereas it's highly unlikely that metals that differ more significantly, such as manganese and zinc, will share their primary ligand. None of the ligands of these primary pools have as yet been definitively identified in any organism, although a role for cellular reductants glutathione (copper) and bacillithiol (zinc) have been suggested in some cases (Banci, Bertini, Ciofi-Baffoni, Kozyreva, et al., 2010; Ma et al., 2014). One tantalising possibility comes from the fact that the ribosome itself is a significant repository of cellular zinc, containing both tightly bound (i.e. likely structurally important) and weakly bound zinc ions (Gabriel & Helmann, 2009; Hensley, Tierney, & Crowder, 2011; Panina, Mironov, & Gelfand, 2003), and may act as a primary zinc pool. Yet in such a model, how nascent metalloproteins that require metals lower down the Irving–Williams series could avoid acquisition of, and thus inactivation by zinc ions from this ribosomal pool, is unclear.

Given the central importance of correct metal-loading of metalloproteins and the high levels of conservation across prokaryotes of the proteins involved in metal homeostasis, it seems likely (though by no means certain) that some properties of such a primary pool, including the chemical nature of the ligand(s) involved for each metal, will be found to be conserved across bacteria, but this awaits experimental verification. Nonetheless, there are also likely to be some species-specific differences. Species have evolved to make more or less use of each given metal based on the relative metal abundance within their specific ecological niche, and as a result different bacterial species can have very different requirements for the essential metals. They are thus likely to have substantially different primary pools, at least in terms of magnitude. Indeed, this variety explains why metalloproteins often acquire a different metal cofactor when expressed in a heterologous host cell compared with that acquired in their native host organism (Beyer & Fridovich, 1991; Cavet et al., 2002; Cotruvo & Stubbe, 2012; Hohle & O'Brian, 2016; Schäfer & Kardinahl, 2003).

Further pools must also exist in such a model. For example, a metal storage protein, such as ferritins for iron storage (Carrondo, 2003; Le Brun, Crow, Murphy, Mauk, & Moore, 2010), metallothioneins for zinc (Blindauer et al., 2002) or the Csp proteins for copper (Vita et al., 2015, 2016), can be considered to be the secondary pool (Fig. 2). Their metal ions are less bioavailable than those in the primary pool due to their tighter association with proteins either through coordination or through biomineralisation (Blindauer et al., 2001; Le Brun et al., 2010; Vita et al., 2015), but can nonetheless be mobilised under conditions of metal deficiency, presumably by releasing their metal cargo back to the primary pool.

A third pool, the tertiary pool, consists of the metal that is correctly associated with its target proteins (Fig. 2); these metal ions can generally be considered to be kinetically inert, either because they actually have a negligible off-rate (i.e. they are kinetically trapped within the folded target metalloprotein—Tottey et al., 2008), or simply because their affinity for that metal is so much tighter thermodynamically than those of the primary and secondary pools that their metal can be considered to be essentially immune to competition from new metal-requiring proteins. Nonetheless, such tertiary pools may still be affected by high concentrations of other metal ions, especially from more competitive metals higher in the Irving–Williams series (see Section 3.3).

Although this model of cellular metal homeostasis and supply is inevitably primitive, it nonetheless provides a simple theoretical framework for understanding how homeostasis is maintained and the consequences of metal excess.

## 2.2 How Do Cells Detoxify Excess Metal?

A key factor in determining the abundance of these cellular metal pools is the bacterial metal homeostasis system. In the simple model described earlier, we can consider that both the target metalloenzymes and the metal homeostasis proteins themselves, including the metal sensors that sit atop the homeostatic hierarchy (Helmann, Soonsanga, & Gabriel, 2007; Tottey, Harvie, & Robinson, 2007; Waldron & Robinson, 2009; Waldron et al., 2009), must operate primarily through association and disassociation of metal ions at the level of the primary pool as these are the ions that are most thermodynamically accessible and kinetically labile. The activities of metal transporters, both importers and exporters, as well as metal storage proteins, work in concert to maintain the buffered cytosolic level of each metal within a narrow target range (Outten & O'Halloran, 2001) by adding or removing metal to/from the primary pool. Target metalloenzymes remove their required metal from the primary pool at the point of synthesis/folding, and presumably return this cofactor to the primary pool at the point of protein turnover/degradation.

Classes of specific transition metal importers are known for manganese (Dintilhac, Alloing, Granadel, & Claverys, 1997; Kehl-Fie et al., 2013), iron (Braun & Hantke, 2011; Hammer & Skaar, 2011), nickel (Chivers, Benanti, Heil-Chapdelaine, Iwig, & Rowe, 2012; Remy et al., 2013), cobalt (Chivers et al., 2012; Remy et al., 2013) and zinc (Hantke, 2005). The exception here is copper. Copper import has been shown to be mediated by small, soluble, modified peptides called methanobactins in some methanotrophic bacteria (Balasubramanian, Kenney, & Rosenzweig, 2011; El Ghazouni et al., 2012; Kim et al., 2004) that have a high copper demand due to their utilisation of the copper-dependent enzyme particulate methane monooxygenase (pMMO), but despite some bioinformatic evidence that methanobactin production may not be limited solely to these organisms (Kenney & Rosenzweig, 2013; Semrau et al., 2013), this is not anticipated to be a universally conserved copper acquisition mechanism.

Accumulating microbiological and genetic evidence indicates that the CopCD system, encoded by many bacteria, may mediate high-affinity copper uptake (Chillappagari, Miethke, Trip, Kuipers, & Marahiel, 2009), but this awaits rigorous biochemical confirmation.

As the exogenous availability of an essential metal ion rises, rate of influx through these importers will increase and the metal will begin to accumulate in the primary cytosolic metal pool. Once its abundance in this pool increases sufficiently, it will cross the critical concentration threshold of the relevant high-affinity, 'low metal' sensor of the cell (Fig. 3), for example, the iron sensor Fur, the zinc sensors Zur or AdcR, or the nickel sensor NikR (Chivers & Sauer, 2000; Guerra, Dann, & Giedroc, 2011; Lee & Helmann, 2007; Outten & O'Halloran, 2001). In all these cases, metal binding to these 'low metal' sensors leads to an allosteric conformational change in the metal-dependent corepressor protein which leads to increased DNA-binding affinity for the operator sequence upstream of their target genes (Fig. 4), namely those encoding the importer for that metal. Therefore metal binding to the regulator leads to increased DNA binding, and thus inhibition of transcription of the gene encoding the importer protein, making the initial response to the increased abundance of metal in the primary pool, caused by an increased rate of metal influx into the cytosol due to the rising external metal concentration, the repression of expression of the specific metal importer. In the case of iron and zinc, this regulatory event also enables the control of other metal-utilising pathways, for example,

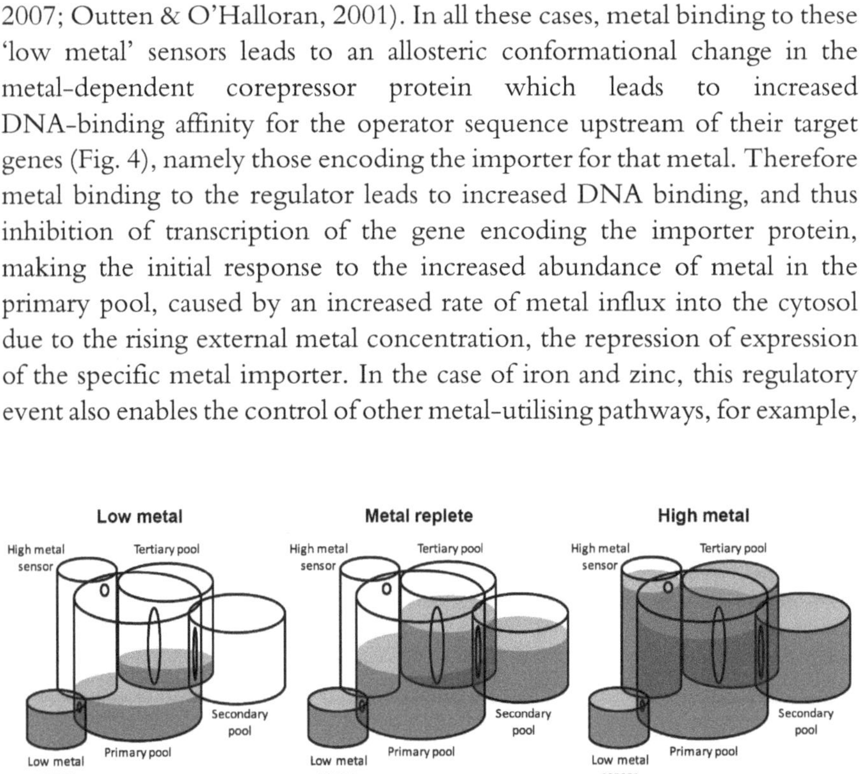


**Fig. 3** Filling of the primary metal pool leads to metal-dependent transcriptional regulatory events. The simple model present in Fig. 2 can be used to visualise the consequences for metal homeostasis as the exogenous metal concentration changes, taking the bacterial cell from metal-deplete conditions (*left*) to metal-replete conditions (*centre*), triggering the low metal sensor, repressing import and activating metal storage, and ultimately to metal excess conditions (*right*), triggering the high metal sensor leading to induction of metal exporters.

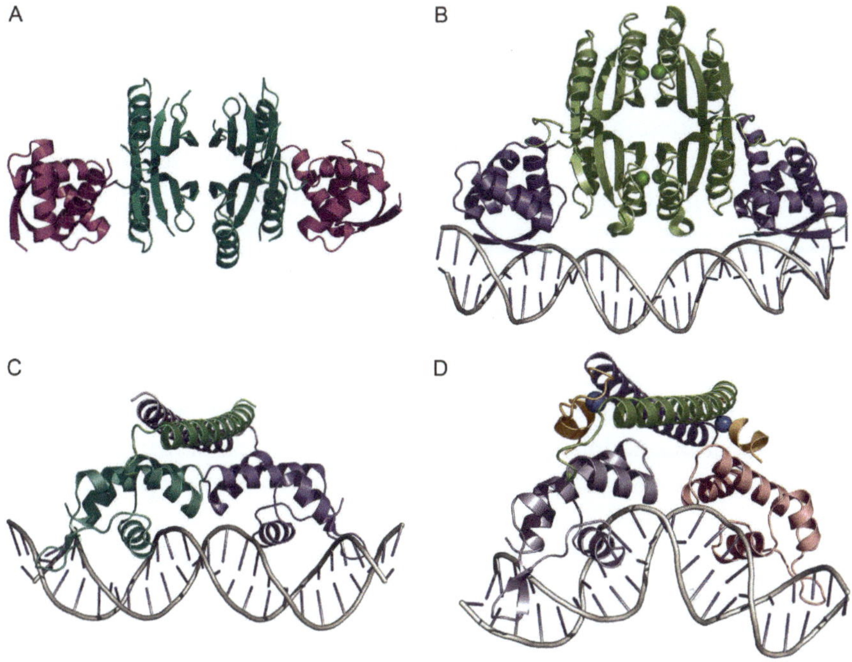


**Fig. 4** Metal-dependent transcriptional regulation by the low metal sensors NikR and CueR. Cartoon representation of the crystal structure of the *E. coli* nickel-responsive transcriptional regulator NikR, (A) in apo form (1Q5V; Schreiter, Wang, Zamble, & Drennan, 2006), with metal-binding domains coloured in *teal* and DNA-binding domains represented in *purple*, and (B) in its operator DNA-bound complex (2HZV; Schreiter et al., 2006), with activating Ni ions represented as *green spheres*. The B-form DNA in the operator complex is represented in *grey*. Cartoon representation of the crystal structure of the *E. coli* copper-responsive transcriptional regulator CueR bound to its operator DNA, showing the N-terminal DNA-binding domain (*teal* and *slate*) and dimerisation helix (*green* and *blue*) (C) in its repressor apo form (PDB ID: 4WLS; Philips et al., 2015), using a mutant lacking its metal-binding residues (C112S, C120S), and (D) in its metal-bound activator form with additional well-ordered metal-binding loop (*yellow*) and a two-turn C-terminal α-helix (*yellow*). Metal binding triggers conformational changes in CueR which in turn modulate the overall shape of the core DNA promoter for control of transcription.

Zur-dependent derepression of Zn-containing ribosomal proteins (Gabriel & Helmann, 2009), and the Fur-dependent induction of the iron storage protein, ferritin (Massé & Gottesman, 2002). The latter suggests that the signal of metal-replete conditions for this critical metal ion, whose extracellular availability is often limiting, is taken as an opportunity for cells to store the excess in anticipation of iron-deprivation conditions in future or in preparation for subsequent rounds of cell division.

Under most normal fluctuations of exogenous metal availability within a bacterium's natural niche, this restriction of metal import may be a sufficient response to prevent further cytosolic metal accumulation. But if the extracellular concentration of the metal continues to increase, eventually the rate of metal influx into the primary pool will continue to rise, for example through nonspecific or adventitious transport events via transporters that function to import other metal ions. As an example of this adventitious transport, the *Bradyrhizobium japonicum* Mg(II) transporter MgtE has been demonstrated to be a competent transporter of Mn(II) (Hohle & O'Brian, 2014) and is regulated by binding of intracellular magnesium under magnesium-sufficient conditions causing closing of the channel (Hattori et al., 2009). This leads to the potential for inappropriate import of Mn(II) under magnesium-deficient conditions resulting in manganese toxicity (Hohle & O'Brian, 2014). If metal input to the primary pool outstrips metal removal by the components of the secondary or tertiary pools, then it will eventually accumulate sufficiently to cross the critical concentration threshold of the low affinity, 'high metal' sensor, triggering a second metal-dependent regulatory event (Fig. 3). These 'high metal' sensors are usually either metal-dependent coactivators (such as the MerR-family regulators, for example, CueR for copper and ZntR for zinc in *E. coli*) that undergo a conformational change upon metal binding while bound to DNA, thus altering the structure of the bound DNA (Philips et al., 2015; Fig. 4) and thereby altering transcription, or metal-dependent corepressors (such as CsoR for copper and SmtB for zinc) whose affinity for DNA is reduced by a conformational change upon metal binding (Busenlehner, Pennella, & Giedroc, 2003; Liu et al., 2007). Regardless of the specific molecular mechanism of transcriptional regulation, metal binding to these regulatory proteins leads to an induction of transcription of the target genes, which are primarily those encoding specific metal export proteins (Baker, Sengupta, Jayaswal, & Morrissey, 2011; Liu et al., 2007; Thelwell, Robinson, & Turner-Cavet, 1998) and those involved in sequestration of the excess metal (Festa et al., 2011; Morby, Turner, Huckle, & Robinson, 1993). Metal exporters are known for manganese, iron, cobalt, nickel, copper and zinc (Botella et al., 2011; Martin, Waters, Storz, & Imlay, 2015; Pi et al., 2016; Rutherford, Cavet, & Robinson, 1999; Thelwell et al., 1998; Wolschendorf et al., 2011).

The effectiveness of this regulatory circuit and its subsequent changes in expression of metal detoxification proteins in response to metal stress is highlighted by a time-dependent expression study by the groups of Nies and Grass (Thieme, Neubauer, Nies, & Grass, 2008). In a temporal study of the expression of the copper efflux machinery in *E. coli*, a diminished transcriptional response was detected after a second 'copper shock' event, relative to that observed to a first copper shock event. This is presumably a result of the continued presence of residual functional copper detoxification machinery in the *E. coli* membrane that was synthesised in response to the first shock event, which enables the cell to detoxify the copper from the second shock without requiring the same high level expression of the relevant copper detoxification genes (Thieme et al., 2008). This difference in response is likely determined by differences in the extent of expansion of the primary pool, and thus the concentration of copper sensed by CueR, in the two copper shock events.

The cellular response that is induced by the action of the 'high metal' sensor is the expression of proteins involved in removal of excess metal ions from the primary pool, either through sequestration and storage in the secondary pool or through efflux out of the cytosol. Presumably, the combination of the two regulatory loops that constitute bacterial metal homeostasis, namely the transcriptional regulation of metal import and storage by the 'low metal' sensors and of metal export and sequestration by the 'high metal' sensors has proven sufficient over evolutionary time to regulate the abundance of metal within the bacterial primary pools within acceptable limits while experiencing fluctuations in extracellular availability.

## 2.3 The Crucial Role of Metal-Dependent Transcriptional Regulators in Bacterial Metal Homeostasis

The metal-dependent transcriptional regulators thus sit at the apex of a hierarchy of metal homeostasis proteins, as the 'arbiters of metal sufficiency' (Helmann et al., 2007); the metal concentrations at which they associate and dissociate their metal cofactor indirectly controls metal availability to all other cellular components through their action on the expression of the metal homeostatic factors that directly regulate cytosolic metal concentration. The molecular mechanisms that dictate their metal selectivity have been reviewed extensively elsewhere (Helmann et al., 2007; Tottey et al., 2007; Waldron & Robinson, 2009), but can be broadly defined as being a function of three terms 'affinity', 'access' and 'allostery' (Tottey et al., 2007; Waldron & Robinson, 2009; Waldron et al., 2009).

Much experimental effort has been expended over recent decades to determine the metal affinities of these sensors (Changela et al., 2003; Golynskiy, Gunderson, Hendrich, & Cohen, 2006; Liu et al., 2007; Outten & O'Halloran, 2001), including in a few cases for metals other than their cognate metal cofactor (Foster, Patterson, Pernil, Hess, & Robinson, 2012; Patterson et al., 2013). In some cases these parameters have been interpreted as having important implications for the cytosolic metal availability of that metal ion. For example, the affinity of the *E. coli* 'high copper' sensor, CueR, for Cu(I) is in the zeptomolar range ($K_d$ $10^{21}$ $M^{-1}$), many orders of magnitude tighter than that which would represent the cellular concentration with a single free Cu(I) ion within the entire *E. coli* cell (which would approximate $10^{-9}$ M) (Changela et al., 2003). This result suggests that the *E. coli* cytosol never contains a single 'free' Cu(I) ion. Other bacterial copper sensors also have extraordinarily tight affinities (Grossoehme et al., 2011; Liu et al., 2007; Ma, Cowart, Scott, & Giedroc, 2009), suggesting that these bacteria are also devoid of cytosolic 'free' Cu(I) ions. For copper specifically, this result might be simply a reflection of the apparent absence of cytosolic copper-requiring enzymes (Changela et al., 2003; Waldron & Robinson, 2009), although the presence of cytosolic Csp proteins in diverse bacteria suggests these organisms may accumulate significant amounts of copper in the cytosol (Vita et al., 2015, 2016). Regardless, this cannot be the explanation in the case of zinc, for which bacteria have a high cytosolic demand (Andreini et al., 2008). The 'low zinc' sensor in *E. coli*, Zur, is estimated to have a Zn(II) affinity of $\sim 5 \times 10^{15}$ $M^{-1}$, whereas the equivalent parameter for the 'high zinc' sensor, ZntR, is estimated at $\sim 1 \times 10^{15}$ $M^{-1}$ (Outten & O'Halloran, 2001). If these affinities do reflect, to some extent, the optimal buffered concentration of zinc in the *E. coli* cell, it's noteworthy that this must be retained within the very narrow range demarcated by the boundary limits set by the affinities of these high and low zinc sensors. But importantly, like Cu(I), this also implies that the *E. coli* cytosol is devoid of 'free' Zn(II) ions. It raises the question of how the cell is able to supply Zn(II) ions to zinc-requiring proteins in the absence of any identified zinc metallochaperones. Nonetheless, these experiments demonstrate that metal partitioning to metalloproteins is not under simple thermodynamic control as has been suggested (Banci, Bertini, Ciofi-Baffoni, Kozyreva, et al., 2010), and that kinetics must play an important role in metal homeostasis and intracellular trafficking to target proteins.

The important contribution of kinetics to metal sensing is demonstrated by experimental evidence of the roles of 'affinity' and 'access' in determining metal binding to the transcriptional regulators. Specifically, when these regulators are expressed heterologously in cells other than that in which they have evolved, they often display a change in their apparent metal selectivity. For example, the nickel- and cobalt-sensing regulator NmtR from *Mycobacterium tuberculosis* shows cobalt-dependent transcriptional regulation when it is expressed heterologously in a cyanobacterial host, but it fails to detect nickel (Cavet et al., 2002). This is presumably a reflection of the stark difference in nickel accumulation of the two organisms (Cavet et al., 2002), with NmtR failing to respond to elevated nickel due to a failure to 'access' nickel pools within the heterologous cyanobacterial host cell. Conversely, the cobalt sensor CoaR from *Synechocystis* PCC 6803 exhibits a weaker affinity for Co(II) in vitro than does either of the Zn(II) sensors present in the same organism, both of which are competent to alter their DNA-binding properties in response to Co(II) in vitro (Patterson et al., 2013). Yet in vivo, only CoaR senses elevated cobalt levels, suggesting that cobalt is in some way supplied, or 'channelled' to the true cobalt sensor, while the zinc sensors have restricted access to cytosolic cobalt pools (Patterson et al., 2013). Similarly, the manganese sensor Mur from *B. japonicum* senses only iron, and not manganese, when heterologously expressed in *E. coli*, and conversely the *E. coli* iron sensor Fur senses only manganese, and not iron, when expressed heterologously in *B. japonicum* (Hohle & O'Brian, 2016). Both of these proteins have similar sequences, conserved metal ligands, bind each metal with similar affinities, and show similar metal-dependent DNA-binding activity, and thus this difference in sensing in vivo appears to be a reflection of differences in intracellular iron availability between the two host species (Hohle & O'Brian, 2016).

A crucial mechanistic step in the action of all of the bacterial metal-dependent transcriptional regulators is the allosteric conformational change that is induced by metal binding, which either enables or prevents DNA binding in order to affect a transcriptional response. In a number of such proteins, nonnative metals have been shown to bind to a sensor in vitro, and occasionally even in vivo (Foster et al., 2012; Golynskiy et al., 2006; Patterson et al., 2013), often with significantly higher affinities than the native metal ion (consistent with the Irving–Williams series), but fail to induce this allosteric change in protein conformation and thus fail to alter transcription. For example, *M. tuberculosis* NmtR binds Zn(II) with a higher affinity than for either of its native effector ions, Ni(II) or Co(II), but zinc binding fails to induce the allosteric conformational change required to reduce the affinity of NmtR for its target operator DNA, probably through its preference for tetrahedral coordination vs the octahedral or square planar coordination of cobalt and nickel (Cavet, Graham, Meng, & Robinson, 2003; Cavet et al., 2002).

In addition to their critical role atop the homeostatic hierarchy, these three organising concepts of affinity, access and allostery (or alternatively, activity) could also be useful in understanding how macromolecules become selective targets of metal toxicity.

# 3. METAL TOXICITY MECHANISMS

Broadly speaking, there are generally considered to be two main mechanisms by which an excess of one or more of the essential transition metal ions can cause metal toxicity, each of which will be discussed in detail here. The first such mechanism is through the generation of toxic ROS, either directly, via the inherent redox activity of some of the essential metal ions themselves, for example, via Fenton and Haber-Weiss reactions (Section 1.2), or indirectly through inhibition or modification of respiratory enzymes, depletion of cellular antioxidant molecules (e.g. glutathione), or through metal-dependent inactivation of enzymes that detoxify the ROS generated naturally by cellular respiration (Section 3.2). This latter effect could also be considered to be consistent with the second main mechanism of metal toxicity, namely the inappropriate or adventitious association of metal ions with biological molecules that do not require them, thereby altering or inactivating their normal function (Section 3.3). This latter mechanism can give rise to diverse downstream effects, depending on the identity of the molecule affected, from small-molecule metabolite to complex macromolecules such as proteins and nucleic acids.

There is evidence for both of these fundamental mechanisms. The extent to which each contributes is anticipated to vary between metals (for example, between redox-active and redox-inactive metals, and between metals in different positions in the Irving–Williams series), between bacterial species (which differ in how they use and handle specific metal ions and also in their ability to produce and detoxify ROS) and between growth conditions (most obviously between aerobic and anaerobic growth). It seems likely that in almost all situations of metal toxicity several specific toxic mechanisms are operating, something that could be considered to be a modern version of that described previously as the 'oligodynamic effect'.

## 3.1 Studying Metal Toxicity

It is inherently challenging to design experiments that definitively demonstrate the mechanism of action of any bactericidal agent, given that the cell is killed in the process of treatment, and sublethal doses do not always affect the same processes as lethal doses. This experimental difficulty is exacerbated in the case of metal toxicity because metal ions appear to affect multiple distinct cellular processes concomitantly. This technical challenge may explain why studies of the molecular mechanisms at work in metal toxicity have, until recent years, made little progress. However, with a constantly improving understanding of bacterial metal homeostasis, we can now use extensive genetic, chemical, biophysical and biochemical tools to facilitate our studies, at least in genetically tractable model systems. The extensive genetic tools explain why the focus so far has been on the study of traditional model systems, such as *E. coli* and *Bacillus subtilis*, but more and more diverse systems are beginning to receive attention and even establish themselves as models for specific bacterial phyla.

The robust activity of the metal detoxification systems can obscure measurements of metal toxicity, and therefore it is no surprise that most of the seminal studies in this field in recent years have taken advantage of mutant bacterial strains that lack key members of their metal homeostasis machinery and therefore experience elevated metal stress (Azzouzi et al., 2013; Macomber, Elsey, & Hausinger, 2011; Macomber & Imlay, 2009; Ranquet, Ollagnier-de-Choudens, Loiseau, Barras, & Fontecave, 2007; Thorgersen & Downs, 2007). To test the hypothesis that a given candidate protein is inhibited by elevated metal concentrations, bacterial strains can be created in which the expression of this protein is modified and the resulting cells assayed (relative to a relevant control strain) for a metal-dependent growth phenotype. A strain that overexpresses the protein would be anticipated to become more metal resistant, whereas one which underexpresses the protein would be anticipated to become more sensitive. Such mutant strains can be produced quickly and cheaply in a modern microbiology laboratory, and with precision control over expression levels that is unprecedented.

In addition, we have increasingly subtle ways to detect excess metal levels within cells. A simple assay to determine whether cells grow or fail to grow is a rather blunt instrument. Similarly, measuring the total cellular metal concentration may not be informative; a change in metal bioavailability to macromolecules, both to target metalloproteins and to adventitious metal toxicity targets, can occur without a change in total metal levels through the alteration of the size of the primary pool relative to the secondary and/or tertiary pools (Chillappagari et al., 2010; Ranquet et al., 2007).

Increasingly complex reporter strains can be constructed, exploiting the regulatory loops of the metal-dependent transcriptional regulators, that give biological read-out of the cell's metal status (Binet & Poole, 2000; Osman et al., 2010; Riether, Dollard, & Billard, 2001), and these reporters can be calibrated by biochemical investigation of the regulator's metal-binding and DNA-binding properties in vitro (Osman et al., 2010). Both chemical and genetically encoded metal probes can be used to detect cellular metal ions (as well as other relevant cellular species such as ROS) in cells which, when coupled with advances in microscopy technology, are enabling an unprecedented cellular-level view of metal homeostasis, as are physical techniques such as X-ray absorption and fluorescence spectroscopies.

All studies of bacterial metal toxicity must take careful consideration of the culture conditions used. It has been demonstrated in a number of systems that the toxic effects of metal excess towards bacteria can vary widely depending on the composition of the growth medium (Macomber & Imlay, 2009; Rathnayake, Megharaj, Krishnamurti, Bolan, & Naidu, 2012; Remy et al., 2013; Wolfram & Bauerfeind, 2009). Complex bacterial growth media that are rich in organic compounds reduce the effective metal concentration experienced by the cultured bacteria through metal chelation, and thus reduce the observed toxicity effect on bacterial growth. Conversely minimal media, which contain both far less diversity and far lower abundance of these organic molecules, have a lower chelation capacity, and thus increase the effective metal concentration experienced by the bacteria. As an example, the $EC_{50}$ (a statistical estimate of the concentration that results in 50% effective reduction in growth within a specified time) for copper towards growth of *Bacillus megaterium* was determined to be $>8\ mg\ L^{-1}$ in the complex medium tryptic soy broth, compared to $0.36\ mg\ L^{-1}$ in a Tris-buffered minimal medium, and as low as $1\ \mu g\ mL^{-1}$ in a MES-buffered minimal medium, which reflects the differences in the available 'free' copper detected in those same media with an ion-specific electrode ($6 \times 10^{-5}$, $3 \times 10^{-3}$ and $4 \times 10^{-2}\ mg\ L^{-1}$, respectively) after supplementation of each medium with $0.05\ mg\ L^{-1}$ copper (Rathnayake et al., 2012). It has been shown that the basal *E. coli* cellular metal content is quite similar whether it is cultured in minimal medium or rich medium, despite the differences in exogenous metal concentration being an order of magnitude or more in some cases (Outten & O'Halloran, 2001), but when the two types of media are supplemented with equal concentrations of additional metal, the resulting total metal content of the cells is higher in cells cultured in the less complex medium.

The observation that metal chelation by media components reduces metal import and inhibits metal toxicity implies that most metal importers, including those used by metal ions to adventitiously enter bacterial cells, primarily access and transport metals as 'free' ions. However, exceptions to this rule do exist; the *E. coli* nickel importer, NikABCDE, has been shown to transport nickel chelates such as $Ni(L\text{-}His)_2$ (Chivers et al., 2012), the methanotrophs are thought to import copper–methanobactin complexes (Balasubramanian et al., 2011; El Ghazouni et al., 2012), and the transport of iron-siderophore chelates has been extensively studied (for example, Ferguson, Hofmann, Coulton, Diederichs, & Welte, 1998; Morrissey, Cockayne, Hill, & Williams, 2000).

Another critical aspect of experimental design relates to cellular adaptation to metal excess. Toxicity will be experienced differently by bacteria after a sudden 'metal shock', where the cells experience a rapid increase in concentration while growing under conditions in which they are expressing minimal metal detoxification proteins, than after long-term culture in a mildly inhibitory concentration of metal, conditions in which the cells will have adapted to the excess over time by maximising their metal detoxification response to achieve a 'steady state' (Thieme et al., 2008).

Clues to the mechanisms of action of metal toxicity can sometimes be gleaned through expression studies, by either proteomic or transcriptomic assays to determine which proteins and/or genes are altered in their expression as the exogenous metal concentration is increased. Knowledge of the effects of excess metal on certain aspects of metabolism can indicate which cellular processes are affected by metal toxicity. However such genome- or proteome-wide studies usually show wide-ranging effects of metal excess on gene/protein expression (Baker et al., 2010; Hu, Brodie, Suzuki, McAdams, & Andersen, 2005; Jacobsen, Kazmierczak, Lisher, Winkler, & Giedroc, 2011; Kimura & Nishioka, 1997; Moore, Gaballa, Hui, Ye, & Helmann, 2005; Nandakumar, Espirito Santo, Madayiputhiya, & Grass, 2011; Ray et al., 2013; Van der Heijden et al., 2016), indicating that a large number of cellular processes are altered, and these collective responses will include both direct and indirect effects. In particular, certain common cellular stress response pathways are often induced under metal stress, especially the oxidative stress response (Baker et al., 2010; Hu et al., 2005; Moore et al., 2005; Rodriguez, Voskuil, Gold, Schoolnik, & Smith, 2002), a recurring observation that has implicated metal-catalysed generation of ROS and/or metal-dependent inhibition of ROS-detoxification systems in the mechanisms of bacterial metal toxicity (Section 3.2). Although occasionally informative, it is impossible to determine which of the cellular responses observed are specific to the metal toxicity molecular mechanism, and which are more indirect effects caused by the resulting cellular stress, using such expression data alone. Nonetheless, proteomic-type approaches have proven useful in identifying new components of the bacterial metal homeostatic repertoire (Almirón et al., 1992; Chillappagari et al., 2009), and both proteomic and transcriptomic analyses are useful in defining the regulon controlled by bacterial metal-dependent transcriptional regulators (Graham et al., 2009; Grifantini et al., 2003; Massé, Vanderpool, & Gottesman, 2005; Thompson et al., 2002). Although application of metabolomic analyses to the study of the mechanisms of metal toxicity to bacteria is still in its infancy (Booth, Weljie, & Turner, 2015; Booth, Workentine, Weljie, & Turner, 2011), such approaches hold great potential for future investigations.

As with the traditional antibiotics, genetic screening techniques have also proven useful in the discovery of mechanisms of bacterial metal resistance. In such experiments, a bacterial population (usually after mutagenesis using chemical means or by irradiation, or using a library of mutant strains defective in individual nonessential genes) is exposed to lethal concentrations of a metal on solid growth medium, and mutant strains that have increased metal resistance are selected and subsequently analysed. The mutated gene can be identified, and the role of the protein encoded investigated for its role in metal resistance. For example, a screen for genetic mutations that suppressed the manganese acquisition deficiency of a mutant strain of *B. japonicum* lacking its high-affinity manganese importer, MntH (Hohle & O'Brian, 2014), identified that the magnesium importer, MgtE, previously defined as highly specific for Mg(II) (Hattori et al., 2009), is prevented from transporting Mn(II) only by an intracellular Mg(II)-mediated gating mechanism (Hohle & O'Brian, 2014). Subsequent study implicated this transporter as a possible cause of cellular manganese toxicity under magnesium limiting conditions. Like the proteomic and transcriptomic approaches, genetic screens have also proven useful in identifying new components of the bacterial metal homeostasis system (Jen et al., 2015).

The diverse methodologies collectively termed metallomics or metalloproteomics have not yet been successfully applied to determining the molecular mechanisms of metal toxicity, but could potentially prove useful in this regard. Such methods aim to detect metal–protein interactions inside cells and cell extracts (Cvetkovic et al., 2010; Ferrer, Golyshina, Beloqui, Golyshin, & Timmis, 2007; Sevcenco et al., 2009; Tottey et al., 2008), and therefore could potentially detect adventitious metal-binding events under toxicity growth conditions. Metalloproteomics approaches have previously been used to identify novel players in bacterial metal homeostasis (Osman et al., 2010; Vita et al., 2015). Indeed, while generally of little value in the study of metal binding by native metalloproteins or of metal homeostasis under normal growth conditions, the study of inappropriate metal binding under metal toxicity conditions may be the one area where the numerous studies employing immobilised metal affinity chromatographic enrichment to detect proteins with potential for adventitious metal binding could prove useful (Barnett, Scanlan, & Blindauer, 2012; Smith, Stauber, Wilson, & Jolley, 2014; Sun et al., 2011).

## 3.2 Oxidative Stress Generation

It is clear from numerous studies in a number of different bacterial systems that an excess of one of the essential metal ions can lead to the cell detecting an increase in oxidative stress, but it is inherently difficult to demonstrate conclusively the role of ROS in a toxic mechanism (Imlay, 2015). As we have seen, transcriptomic, proteomic and biochemical analyses have demonstrated induction of numerous genes whose products are involved in detoxification of ROS (Kimura & Nishioka, 1997; Moore et al., 2005; Ray et al., 2013; Van der Heijden et al., 2016). However, it has been suggested that the additional ROS that are detected by the bacteria are created by direct metal-catalysed reactions occurring within the cells (Kimura & Nishioka, 1997). Whereas in the case of iron toxicity, there is strong evidence in favour of these reactions being important in iron toxicity in vivo (Imlay, Chin, & Linn, 1988; Touati, Jacques, Tardat, Bouchard, & Despied, 1995), in other cases this has been ruled out for other metals, at least in some organisms (Macomber & Imlay, 2009; Macomber, Rensing, & Imlay, 2007). Given the huge overcapacity of the cytosol for chelation of metal ions, it is highly unlikely that 'free' metal ions are present in the bacterial cytosol, even under metal toxicity conditions, rendering most in vitro analyses of metal-catalysed ROS generation redundant (Gunther et al., 1995). Conversely, some biologically relevant metal chelates can also catalyse ROS production or enhance catalysis by other cellular metabolites (Park & Imlay, 2003), and these are more likely to be relevant to the metal toxicity mechanism than free ions. For redox-inactive zinc ions, such direct oxidative reactions are presumably unimportant. Collectively, this implies that in many cases the cell is detecting increased oxidative stress that is caused by some more indirect means other than direct metal-catalysed generation of ROS.

One possible mechanism is that intracellular metal excess somehow stimulates the production of ROS by endogenous sources, primarily by the respiratory enzymes that have been shown to be the primary source of ROS under normal conditions (González-Flecha & Demple, 1995). Metal toxicity events that affect NADH dehydrogenase 2 (Complex I) and the cytochrome *bc* complex (Complex III) have clear potential to increase endogenous ROS production (González-Flecha & Demple, 1995), and several metal ions and metal chelates have been shown to have adverse effects on these respiratory complexes in bacteria (Djoko, Donnelly, & McEwan, 2014; Kleiner, 1978). An alternative possibility is that excess intracellular metal ions can inhibit the cell's ability to detoxify the ROS that is naturally generated endogenously as a by-product of respiration. This can be through metal toxicity-dependent alteration of the activities of the enzymes that catalyse the detoxification of ROS (Djoko & McEwan, 2013), or through alterations to the concentrations of cellular reductants such as glutathione, bacillithiol, cysteine, acetyl-CoA or lipoamide (Thorgersen & Downs, 2007; Van der Heijden et al., 2016). It is noteworthy that increased cytosolic cysteine levels in *E. coli* increase sensitivity to oxidative stress due to its capacity to reduce labile iron for participation in Fenton chemistry (Park & Imlay, 2003), whereas conversely the depletion of cellular glutathione seems to contribute to cobalt toxicity in *Salmonella* (Thorgersen & Downs, 2007) and both cobalt (Thorgersen & Downs, 2007) and nickel (Freeman, Persans, Nieman, & Salt, 2005) stress can be ameliorated by increasing cellular cysteine or glutathione levels.

For example, copper toxicity in *Neisseria gonorrhoeae* leads to inhibition of haem biosynthesis through copper-dependent inactivation of HemN (Section 3.3), which manifests itself as a reduction in the catalytic activity of the enzyme catalase that detoxifies hydrogen peroxide (Djoko & McEwan, 2013). This copper-dependent diminution of catalase activity led to a reduced capacity of copper-treated cells to detoxify endogenous hydrogen peroxide produced naturally through normal gonococcal metabolism, and made copper pretreated cells more sensitive to killing by exogenous peroxide (Djoko & McEwan, 2013). A similar effect was observed during copper toxicity in *E. coli* (Macomber & Imlay, 2009). As we shall see, this is an example of copper-mediated damage to Fe-S clusters, a mechanism that is common to the toxicity of several metal ions (Section 3.3), which in turn is anticipated to increase the cytosolic labile iron concentration (i.e. the abundance of iron in its primary pool). It is anticipated that some or all of the oxidative stress responses observed during metal stress conditions may be related to this effect on iron homeostasis. It has been clearly demonstrated that bacterial growth is inhibited by conditions that increase the labile iron pool, particularly those involving hydrogen peroxide and superoxide poisoning of Fe-S clusters and mononuclear iron enzymes (Anjem & Imlay, 2012; Anjem et al., 2009; Gu & Imlay, 2013; Park & Imlay, 2003).

Another alternative possibility is that the excess metal ion affects the transcriptional regulators that control the oxidative stress response, and it is for this reason that we observe induction of the ROS detoxification regulons under metal stress. In Gram-negative bacteria, the transcriptional responses to hydrogen peroxide and superoxide are mediated by the regulators OxyR and SoxRS, respectively (Christman, Morgan, Jacobson, & Ames, 1985; Wu & Weiss, 1991). Peroxide sensing by OxyR is mediated by cysteine hydroxylation (Jo et al., 2015), which may conceivably be altered by the presence of excess metal ions, whereas SoxR contains a solvent exposed, oxidant-sensing Fe-S cluster (Watanabe, Kita, Kobayashi, & Miki, 2008), which could potentially be affected by excess metal ions. Peroxide sensing in Gram-positive bacteria is mediated by PerR (Bsat, Herbig, Casillas-Martinez, Setlow, & Helmann, 1998), a sensor that binds either Mn(II) or Fe(II) (Herbig & Helmann, 2001) and senses peroxide through an iron-catalysed hydroxylation of one of the coordinating histidine ligands (Lee & Helmann, 2006). Given that PerR is known to be differentially metalated with manganese or iron in vivo under different growth conditions and metal availabilities (Herbig & Helmann, 2001), it seems a reasonable hypothesis that other metals may compete for this site under conditions of metal toxicity and that binding of alternative metals could give rise to transcriptional effects on the peroxide detoxification regulon as binding of other divalent metals alter the DNA-binding properties of PerR in vitro (Herbig & Helmann, 2001). It is also significant that ROS detoxification genes are part of the regulon of some metal-responsive transcriptional regulators such as Fur and MntR (Horsburgh et al., 2002; Niederhoffer, Naranjo, Bradley, & Fee, 1990), whereas the expression of metal homeostasis proteins are modified by the action of ROS-responsive transcriptional regulators such as OxyR and PerR (Anjem et al., 2009; Harrison et al., 2007; Horsburgh et al., 2001).

Finally, it should be emphasised that manganese is something of a special case in terms of oxidative stress. There is some debate about whether excess cytosolic manganese, and resulting manganese complexes, can provide a beneficial effect to bacteria through direct catalysis of the superoxide dismutation reaction; although this reaction occurs in vitro, its rate is very low relative to that catalysed by SOD enzymes (Anjem et al., 2009). Yet a strain of *Streptococcus pneumoniae* in which the manganese efflux protein MntE is absent and which therefore accumulates excess cytosolic manganese has been shown to display increased resistance to ROS-mediated killing without any concurrent increase in enzyme-mediated SOD activity (Rosch et al., 2009). Nonetheless, *E. coli* accumulates high levels of manganese only under conditions in which its iron homeostasis is dysfunctional (Anjem et al., 2009), and the accumulated manganese plays a protective role by enabling activity of iron-utilising enzymes under these conditions (Anjem & Imlay, 2012; Anjem et al., 2009; Taudte et al., 2016).

In conclusion, the data suggest that direct metal-catalysed generation of ROS plays little or no role in the mechanism of metal toxicity for any of the essential metals, with the notable and important exception of iron, for which clear evidence of cytosolic Fenton and Haber–Weiss chemistry is available. However, although direct ROS generation is negligible, an excess of most metal ions can nonetheless give rise to indirect effects that either generate additional ROS or weaken bacterial ROS-detoxification systems. The precise role of this additional oxidative stress is in most cases unclear and warrants further investigation.

## 3.3 Inappropriate Intracellular Metal Binding

Although the potential for inappropriate binding of metal ions that are present in excess to biological molecules in vivo has been recognised for a long time, until recently there have been remarkably few examples of these adventitious metal associations documented in the literature (Anjem et al., 2009; Gu & Imlay, 2013; Macomber et al., 2011). As has been previously suggested, 'this rarity may reflect an exceptional efficiency of metal homeostasis or perhaps the experimental effort necessary to establish which metals are bound to proteins inside cells' (Waldron et al., 2009). Although there have undoubtedly been significant recent improvements in experimental approaches, generally termed metallomics, for detecting such metal-binding events inside cells or within lysed cell extracts, this remains a technical challenge that is dogged by the potential for both false positives and false negatives.

Nonetheless, in the last decade a number of very specific examples of this kind have been identified, in which an excess of a metal ion accumulated in the cytosol has been shown to associate with or cause damage to a specific cellular target. So far, these molecular targets have primarily been enzymes, rather than small-molecule metabolites, but it seems likely that this is again an experimental bias, because the downstream effect of such 'mislocation' to an enzyme can be detected more easily through the subsequent damage to a metabolic pathway and can be tested genetically.

Although inevitably oversimplified, the cellular metal homeostasis model described earlier (see Section 2 and Figs. 2 and 3) gives a simple framework for understanding how inappropriate metal binding occurs intracellularly under the toxic conditions of an excess of one of the essential metals. As the extracellular concentration of the metal initially increases, the rate of its transport (through both specific and nonspecific transporters) across the bacterial cytoplasmic membrane will increase, which in turn will increase the size of the primary pool. This increase may be initially buffered by the metal storage proteins (i.e. by expanding the secondary pool) if they are present and will also be sensed by the metal sensors, which will adjust gene expression in response to this increased metal abundance in the primary pool. The common adjustments made through this sensing event, as we have seen (Section 2), are: (i) downregulation of expression of specific metal importers, thereby decreasing the rate of metal influx into the cell and thus into the primary pool; (ii) upregulation of expression of specific exporters, thereby increasing the rate of metal efflux (at the expense of energy) from the primary pool back into the extracellular milieu; and (iii) upregulation of metal storage proteins, thereby removing excess metal from the primary pool and expanding the secondary pool. The cellular aim of these responses is to return the size of the primary pool back to within its 'normal' buffered range, the ultimate goal of bacterial metal homeostasis.

These genetic responses to metal excess have evolved over millennia under selection pressures of fluctuating exogenous metal availability including those induced by immune responses (White et al., 2009) and used during grazing by other microorganisms (Hao et al., 2016), and are thus sufficient to enable the bacterium to detoxify excess metal under the typical conditions that it has routinely experienced over evolutionary timescales. Some organisms, for example bacterial strains that have been isolated from ecological niches that are extremely metal enriched such as mines and hot springs, have evolved extreme metal detoxification systems through expansion of their complement of specific and nonspecific metal detoxification genes, highlighting the selection pressure that metal excess can exert on bacterial evolution (Kirsten et al., 2011). However, we know that very high metal concentrations can kill bacteria, and therefore larger excesses must be able to overcome even the active metal detoxification systems induced under metal excess conditions.

If we increase the extracellular metal concentration further, we will reach a point where, despite the downregulation of the specific metal importers and induction of exporters and storage systems, the rate of metal entry into the cell and into the primary pool exceeds the cell's ability to remove metal from this pool. In this scenario, the size of the primary pool will inexorably increase. As the primary pool increases, its bioavailability to other molecules will also increase. This may not actually require complete saturation of its primary ligand or accumulation of 'free' metal ions—simply increasing the metal abundance within the labile primary pool beyond its normal range may be sufficient to significantly increase its bioavailability to such a degree that it leads to inappropriate binding of the metal ion to other species that would not normally bind that metal (Fig. 5). The deleterious consequences of such inappropriate metal-binding events depend on the type of molecular species involved.

The best examples in the current literature of this mechanism, by which a metal present in excess in the bacterial cytosol causes toxicity through inappropriate association with biomolecules that do not require that metal, are all enzymes. It is likely that this is an experimental bias, as the deleterious effects of inappropriate metal binding are readily detectable through the metal-dependent inactivation of the enzyme's activity, and potentially even the disruption of an entire metabolic pathway. It is anticipated that binding

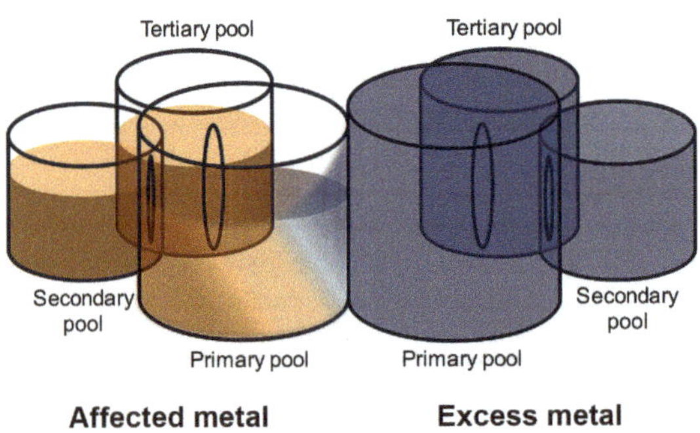


**Fig. 5** Intracellular metal pools can interact to cause metal toxicity. Each of the metal pools, primary, secondary and tertiary, are present in the same cytoplasm simultaneously for each of the essential metal ions. As the intracellular concentration of one metal ion (*blue*) increases and saturates its primary pool, this can have knock-on consequences for homeostasis of another metal ion (*orange*), illustrated schematically here by 'leaking' from the *blue* pool into the *orange* pool. For example, increased copper or cobalt concentrations result in damage to Fe-S cluster-containing proteins, possibly through direct interaction of Cu(I) or Co(II) ions with Fe-S clusters in enzymes.

of excess metal ions to small-molecule metabolites, lipids and nucleic acids will also occur, but such events are yet to be observed.

The key examples of such a mechanism have been identified in recent years in studies of metal toxicity in model organisms, and have a recurring theme; namely the disruption of iron homeostasis by more competitive metal ions that lie above iron in the Irving–Williams series.

Seminal studies in this area from the Imlay lab examined the effects of copper toxicity in the model Gram-negative bacterium, *E. coli* (Macomber & Imlay, 2009; Macomber et al., 2007). Using a mutant strain defective in all three of the known *E. coli* copper detoxification systems, namely the efflux ATPase CopA, the periplasmic multicopper oxidase CueO and the Cus periplasmic efflux system, they showed conclusively that direct copper-catalysed production of ROS via Fenton-type chemistry does not play a significant role in the copper-dependent inhibition of *E. coli* growth (Macomber et al., 2007). In fact, *E. coli* cells were found to be even more sensitive to excess copper when cultured under anaerobic conditions than they were in aerobic culture, and copper supplementation actually reduced, rather than exacerbated, the toxic effects of hydrogen peroxide (the substrate for the metal-catalysed Fenton reaction) on aerobic *E. coli* cells (Macomber et al., 2007), a result that is in contrast to that observed in analogous experiments using excess iron (Imlay et al., 1988).

Having ruled out an important role for direct ROS generation, the same group continued this study to investigate the copper toxicity mechanism in *E. coli*. Based on the hypothesis that copper toxicity may operate via similar mechanisms to ROS toxicity, their critical observation was that copper-dependent growth inhibition could be ameliorated by the addition of specific amino acids to the culture medium. The effect was found to be specific to the addition of the branched-chain amino acids leucine, isoleucine and valine, which reduced copper-mediated growth inhibition in both aerobic and anaerobic culture (Macomber & Imlay, 2009). Subsequent analysis elegantly demonstrated the reason for this protective effect; the excess copper accumulated in the cytosol somehow inactivated critical Fe-S cluster-containing dehydratase enzymes in the branched-chain amino acid biosynthetic pathway, namely isopropylmalate isomerase (IPMI), encoded by the *leuC* gene, and dihydroxy-acid dehydratase, encoded by the *ilvD* gene. This enzyme inactivation was reversible in vivo even when protein synthesis was blocked and could be observed with recombinant Fe-S cluster-containing recombinant proteins in vitro even in the presence of the ROS-detoxifying enzymes SOD and catalase, ruling out a role for copper-generated ROS and instead implicating copper binding in the dehydratase inactivation mechanism. Further in vitro studies showed that the mechanism of damage to these enzymes involved release of iron ions through disintegration of the Fe-S cluster, interestingly without apparent formation of mixed copper/iron clusters and that this damage could be prevented by the presence of saturating concentrations of substrate (presumably through occupancy of the active site protecting the cluster). It is thus hypothesised 'that copper treatment first converts the $[4Fe\text{-}4S]^{2+}$ native enzyme to a $[3Fe\text{-}4S]^{0}$ form and that subsequent events further degrade the enzyme through undefined intermediates ultimately to an apoenzyme' (Macomber & Imlay, 2009).

The copper-dependent inactivation of Fe-S clusters was not restricted only to IPMI. Several other enzymes known to have solvent-exposed Fe-S clusters were also tested and were found to be similarly damaged by the accumulated copper. This included fumarase A (Fig. 6), an essential enzyme in the tricarboxylic acid (TCA) cycle, damage to which had no detectable effect under growth conditions that included abundant glucose, where the TCA cycle is nonessential, but did prevent growth on succinate in the presence of copper (Macomber & Imlay, 2009). Conversely, a range of other Fe-S cluster-containing enzymes, in which the cluster is buried within the protein structure and is not solvent exposed, were found to be immune from copper-dependent inactivation (Macomber & Imlay, 2009).

Importantly, the addition of the branched-chain amino acids to *E. coli* growth medium reduced, but did not completely eliminate the toxic effects of excess copper on *E. coli*, demonstrating that other toxicity mechanisms in addition to the damage to the amino acid biosynthetic pathway must also operate (Macomber & Imlay, 2009). It is worth noting that a similar effect, by which the addition of branched-chain amino acids reduced copper toxicity, was also observed in the cyanobacterium *Synechocystis* PCC 6803 (Tottey et al., 2012), implying that this specific toxicity mechanism is conserved at least between these two distantly related Gram-negative bacteria.

A subsequent study by the Miethke group extended this observed effect of copper toxicity on the stability of Fe-S clusters into Gram-positive bacteria (Chillappagari et al., 2010). A previous microarray study had shown that members of the *B. subtilis* Fur regulon of iron homeostasis genes were induced under conditions of copper excess and suggested that this may be due to direct displacement of Fe(II) from the Fur protein by accumulated Cu(I) (Moore et al., 2005). The Miethke group confirmed this observation, showing by microarray analysis the induction of Fur-regulated high-affinity

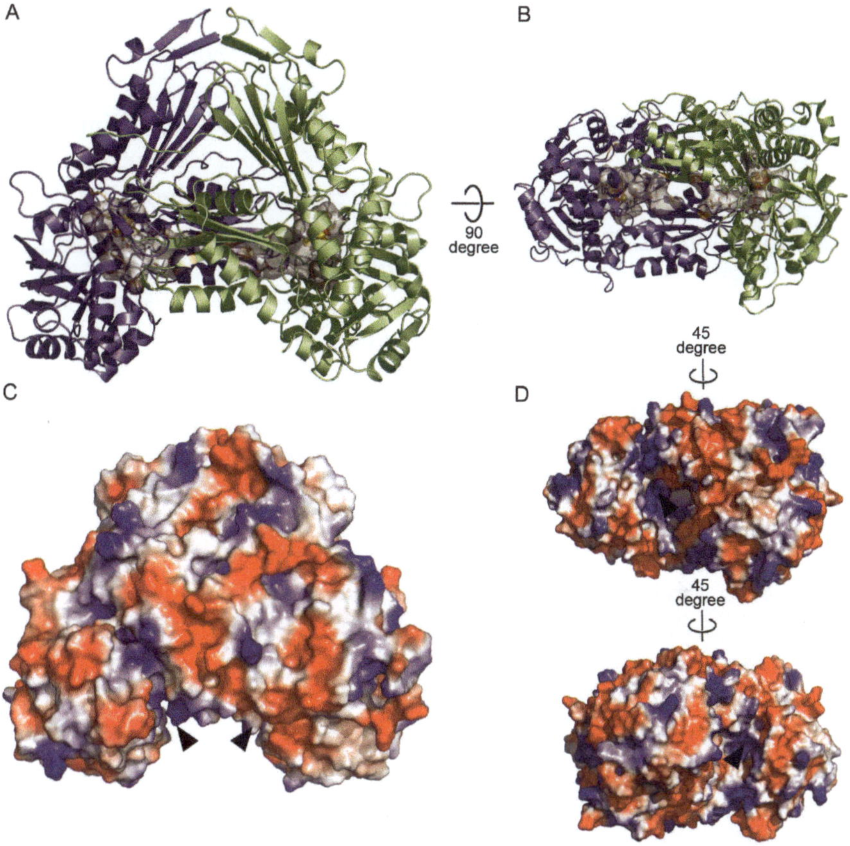


**Fig. 6** The Fe-S cluster of the TCA cycle enzyme fumarase is solvent exposed and vulnerable to attack by excess Cu(I) ions. (A and B) Structural model of *E. coli* fumarase A generated with SWISS-MODEL (Biasini et al., 2014) based on the published 2.05 Å crystal structure of fumarate hydratase from *Leishmania major* (PDB ID: 5L2R; Feliciano, Drennan, & Nonato, 2016), with which it shares 64.7% sequence identity. Based on the template, four molecules of ligand (2*S*)-2-hydroxybutanedioic acid (L-malate) and two [4Fe-4S] clusters were modelled into a homo-dimeric assembly. The Fe-S cluster is represented as a rhombus with iron (*orange sphere*) and sulphur (*yellow sphere*), coordinated by C105, C224. The active site is located at the bottom of the structure, in a cleft between N-terminal and C-terminal domains of each monomer, here represented as a *grey*-coloured cavity. (C and D) Electrostatic surface potential representation of the model structure and its positively charged substrate access channel, marked with *black arrows*, accommodating the L-malate.

iron uptake systems under copper stress, and demonstrated that this induction was directly correlated with cellular copper content (though importantly, iron content was essentially unchanged under copper stress) (Chillappagari et al., 2010). They also observed similar patterns of transcriptional induction of genes whose products are involved in biosynthesis of the amino acids cysteine, arginine and, critically, the branched-chain amino acids. Conversely, few of the known ROS detoxification genes were induced by copper stress, again supporting a model of toxicity in which direct generation of ROS is of negligible importance.

The microarray data also showed that expression of almost half of the genes encoding known Fe-S cluster-dependent proteins in *B. subtilis* had been upregulated under copper stress (Chillappagari et al., 2010), including *leuC* and *ilvD* encoding the dehydratases in the branched-chain amino acid biosynthetic pathway that were shown to be targets of copper toxicity in *E. coli* (Macomber & Imlay, 2009). Unfortunately, these researchers did not specifically test whether these two Fe-S cluster-dependent enzymes were functionally inactivated under copper stress in *B. subtilis* in an analogous manner to that observed in *E. coli* (Chillappagari et al., 2010); that conclusion remains an assumption.

Instead, the authors focused on another aspect of the perturbation of iron homeostasis by copper stress by studying a different set of genes that were induced under copper stress in *B. subtilis*. Transcripts of four genes of the SUF operon, responsible for Fe-S cluster biogenesis in *B. subtilis* (Albrecht et al., 2010), were induced by copper stress and thus the authors hypothesised that copper may be capable of damaging these newly formed clusters, which would be expected to be sterically vulnerable (Fig. 7), as well as those that are solvent exposed within their mature holo-enzyme structure (Chillappagari et al., 2010). Indeed, the ability of Cu(I) ions to disintegrate SufU-assembled Fe-S clusters was demonstrated in vitro (Chillappagari et al., 2010). The researchers then constructed a conditional mutant strain of *B. subtilis* in which the expression of one of the genes of the SUF operon, *sufU*, encoding the main Fe-S cluster scaffold protein on which the clusters are assembled and which transfers the clusters to target proteins (Albrecht et al., 2010), could be reduced in a controlled manner. This strain was cultured under conditions in which low-level expression of *sufU* was achieved, which resulted in these cells being more susceptible to copper toxicity (Chillappagari et al., 2010). Induction of Fur-regulated genes was shown to be correlated with expression of *sufU*, implying that the increased SUF expression observed in copper-stressed *B. subtilis* cells, resulting in an increased rate of Fe-S cluster biogenesis, may be the cause of the cellular iron starvation response. They concluded that the copper-dependent induction of the Fe-S cluster biogenesis system, combined with the copper-dependent reduction in cluster biogenesis efficiency, the increased cluster demand due

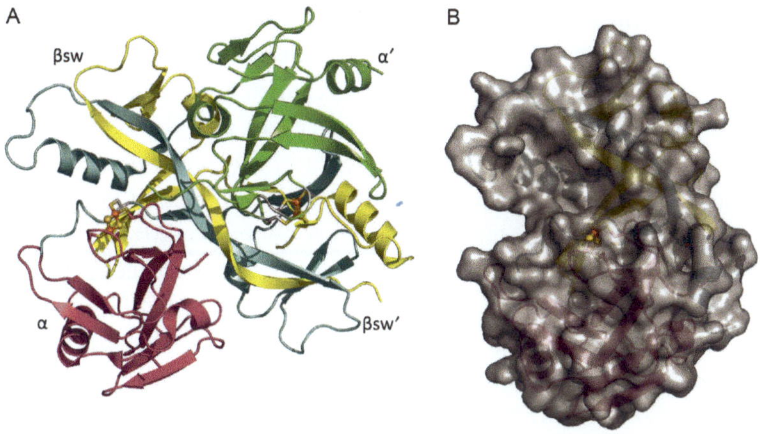


**Fig. 7** The Fe-S cluster of the cluster biogenesis scaffold proteins IscA and SufA(U) are solvent exposed and vulnerable to attack by excess Cu(I) or Co(II) ions. (A) Cartoon representation of the crystal structure of the Fe-S cluster scaffold protein IscA from *Thermosynechococcus elongatus* (PDB ID: 1X0G; Morimoto et al., 2006) containing a solvent-exposed [2Fe-2S] cluster, with β domains (βsw) represented in *blue* and *yellow* and α domains depicted in *magenta* and *green*. (B) A semitransparent surface model of the proposed structure of a monomer assembly of holo-IscA in solution. The partially exposed [2Fe-2S] cluster is coordinated by asymmetric cysteinyl ligation with C37, C101, C103 from the α domain and C103 from the β domain. The [2Fe-2S] cluster is represented as a rhombus with iron (*orange sphere*) and sulphur (yellow sphere), and the coordinating residues represented as *sticks*. The *E. coli* IscA and SufA homologues show 30.7% and 29.6% sequence identity with *T. elongatus* IscA, with all ligand residues conserved.

to destabilisation of dehydratase clusters, and the induction of other cluster-requiring proteins, creates an intracellular iron sink that leads to consumption of the primary iron pool and subsequent induction of the Fur regulon (Chillappagari et al., 2010).

The interaction between copper toxicity and iron homeostasis has been further expanded by studies of two different Gram-negative beta-proteobacteria, the purple nonsulphur photoheterotrophic bacterium *Rubrivivax gelatinosus* (Azzouzi et al., 2013) and the mammalian pathogen *N. gonorrhoeae* (Djoko & McEwan, 2013). In both cases, mutant strains that were defective in their ability to efflux excess copper were studied, and in the case of *N. gonorrhoeae*, the cells were cultured in the presence of branched-chain amino acids to circumvent the effects of copper on the dehydratase enzymes. These strains were found to exhibit a reduced ability to synthesise the iron-containing porphyrin prosthetic group, haem. In both cases, this haem deficiency manifested itself as a marked reduction in aerobic respiration via their *cbb3*-type terminal respiratory oxidase enzymes, which require multiple haem groups. In *R. gelatinosus* additional effects were observed on the photosynthetic machinery, with a reduction in the abundance of both the photosystem reaction centre and the light-harvesting complexes (Azzouzi et al., 2013), whereas in *N. gonorrhoeae* the most pronounced effect was a diminution of the activity of the $H_2O_2$-detoxifying, haem-dependent enzyme catalase (Djoko & McEwan, 2013).

Both studies identified the same molecular target of copper toxicity, namely the coproporphyrinogen III oxidase enzyme HemN, through the identification by chemical analysis of a pigment that was hyperaccumulated under copper stress as coproporphyrin III, the substrate of HemN (Azzouzi et al., 2013; Djoko & McEwan, 2013). In *N. gonorrhoeae* this was also confirmed through feeding experiments using haem biosynthetic intermediates (Djoko & McEwan, 2013). Consistent with previous studies described above, HemN is an Fe-S cluster-dependent enzyme, with the cluster residing in an active site pocket in which it is likely to be solvent accessible (Layer, Moser, Heinz, Jahn, & Schubert, 2003; Fig. 8). Although in vitro studies are awaited to confirm the prediction that Cu(I) ions are able to disintegrate the HemN [4Fe-4S] cluster in a manner analogous to that shown for the dehydratases (Macomber & Imlay, 2009), it seems likely that these observations represent another manifestation of the same phenomenon.

The effect of copper on the haem supply to catalase in the pathogen *N. gonorrhoeae* has important implications, as *N. gonorrhoeae* cells that were pretreated with copper were subsequently demonstrated to show increased sensitivity to $H_2O_2$ (Djoko & McEwan, 2013). At the host–pathogen interface, including within the macrophage phagolysosomal compartment, pathogenic bacteria face a combined onslaught from ROS (including $H_2O_2$) and reactive nitrogen species derived from the host oxidative and nitrosative bursts, respectively, as well as copper and zinc excess (Botella et al., 2011; Ong et al., 2014; White et al., 2009) and deficiency in iron and manganese (Kehl-Fie & Skaar, 2010). Bacteria have evolved a complex and interconnected network of cellular defences to each of these stresses in the form of ROS-detoxification enzymes, copper and zinc efflux machineries and high-affinity iron and manganese acquisition systems, all of which provide a selective advantage to pathogens during infection (Corbett et al., 2012; Dintilhac et al., 1997; Horsburgh et al., 2001; Wolschendorf et al., 2011). But several of the key ROS-detoxifying enzymes, such as catalases and peroxidases, require a haem cofactor. Thus, the host response to infection involves an attack with copper that is designed to 'blunt' the pathogen's first

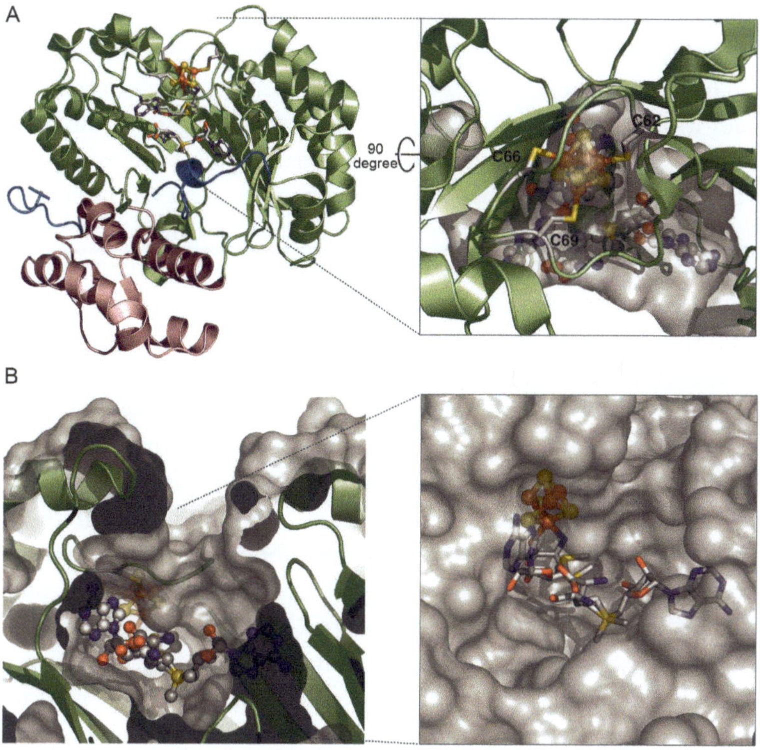


**Fig. 8** The Fe-S cluster of the haem biosynthetic enzyme HemN is solvent exposed and vulnerable to attack by excess Cu(I) ions. (A) Cartoon representation of a monomeric assembly of the crystal structure of the radical SAM-dependent *E. coli* enzyme coproporphyrinogen III oxidase HemN (PDB ID: 1OLT; Layer et al., 2003). HemN consists of two distinct domains (*green* and *pink*) and an elongated N-terminal region termed a trip-wire (*blue*). The catalytic domain is located in the core of the N-terminal β-barrel domain. The [4Fe-4S] cluster and two SAM molecules are bound in close proximity (expanded view, *right*), inside a solvent accessible pocket, where the Fe-S cluster is coordinated by C62, C66, C69 and the methionine group of SAM1. The N-terminal trip-wire and the C-terminal domain are proposed to participate in substrate binding. (B) Surface model of the solvent accessible protein exterior, indicating the active site pocket with two molecules of SAM leading towards the [4Fe-4S] cluster. The Fe-S cluster is represented as a rhombus model with iron (*orange sphere*) and sulphur (*yellow sphere*).

line of defence to one of the host's primary weapons, ROS (Djoko & McEwan, 2013).

Again, it is noteworthy that this effect of copper on haem biosynthesis cannot fully explain copper toxicity; supplementation of *N. gonorrhoeae* cultures with hemin in the presence of copper completely restored haem levels, but only partially restored bacterial growth, suggesting further copper toxicity mechanisms operate under these conditions (Djoko & McEwan, 2013). Nonetheless, the emerging evidence suggests that copper-mediated damage to Fe-S clusters, either inside certain target enzymes in which they are solvent exposed or during Fe-S biogenesis, is a conserved molecular mechanism of bacterial copper toxicity.

Interestingly, a remarkably similar mechanism occurs under conditions of cobalt toxicity. Research in the Fontecave group using *E. coli* as a model system clearly demonstrated that Fe-S cluster-containing enzymes are also a target for excess cytosolic Co(II) ions accumulated under growth conditions containing high exogenous concentrations of cobalt (Ranquet et al., 2007). They demonstrated that the activities of two Fe-S cluster-dependent enzymes, the TCA cycle enzyme aconitase and the tRNA maturation methylthiotransferase MiaB, were significantly reduced in *E. coli* cells that were cultured in media containing elevated cobalt concentrations (Ranquet et al., 2007). Furthermore, they demonstrated that the [2Fe-2S]-containing ferrichrome reductase enzyme FhuF purified from these cells showed reduced iron content relative to control cells, and instead contained trace quantities of cobalt. However, in vitro assays showed that none of these three Fe-S cluster-containing enzymes is susceptible to inactivation by Co(II). Genetic experiments implicated the Fe-S cluster biosynthetic machinery in the mechanism of cobalt toxicity. It was shown that both the constitutive Fe-S cluster biosynthetic scaffold protein IscU and the stress-induced paralogue SufA (Fig. 7), could incorporate cobalt ions in vitro, displacing bound iron, and that IscU could even transfer the resulting putative mixed-metal clusters to target Fe-S cluster-containing proteins (Ranquet et al., 2007). While a role for cobalt-catalysed production of ROS has not yet been conclusively ruled out in this system, it seems likely that this inhibition of Fe-S cluster biogenesis and/or transfer to target Fe-S cluster-requiring enzymes is mediated through direct attack by cytosolic Co(II) ions on the scaffold-bound clusters.

A different study made similar observations in *Salmonella enterica* cells exposed to cobalt stress (Thorgersen & Downs, 2007). It was found that two Fe-S cluster-dependent enzymes in the TCA cycle, aconitase B and succinate dehydrogenase, were strongly inhibited in *S. enterica* cells that were cultured in the presence of elevated cobalt concentrations, whereas the enzyme malate dehydrogenase, which does not require a cluster, was unaffected.

These authors also presented data suggesting a second possible toxic effect of cobalt on *S. enterica*, namely by affecting the sulphur assimilation pathway, due to the observed dependence of cobalt toxicity on the type of sulphur source provided in the culture (Thorgersen & Downs, 2007). The authors suggested that this was a result of direct competition between iron and cobalt at the active site of the sirohaem synthase CysG (Thorgersen & Downs, 2007), which plays a dual role of inserting either iron or cobalt into the porphyrin factor II in the biosynthesis of sirohaem or vitamin B12, respectively (Fazzio & Roth, 1996). Consistent with such a model, cobalt stress in the *E. coli* laboratory overexpression strain BL21 was shown to lead to misincorporation of cobalt into protoporphyrin IX by ferrochelatase (Majtan, Frerman, & Kraus, 2011). Unfortunately, however, no data were presented to demonstrate altered cellular ratios of sirohaem/B12 under copper stress, which would be expected from such a mechanism, nor to conclusively disprove that this effect was simply caused by differential extracellular chelation of cobalt ions by the sulphur-containing compounds added to the growth medium (Thorgersen & Downs, 2007). Such extracellular cobalt chelation, combined with the known effects of cobalt on Fe-S cluster biogenesis and their own observations of the effect of sulphur sources on the biosynthesis of cysteine and glutathione, could potentially explain these results, and thus further conclusions on this interesting potential mechanism of cobalt toxicity await further experimental testing.

A very different example of intermetal competition in the cytosol caused by metal toxicity was provided by the Hausinger group in their study of nickel toxicity in *E. coli* (Macomber et al., 2011). Several studies have shown that nickel excess leads to a dysregulation of bacterial iron homeostasis, but whether this is a direct effect of Ni(II) ions on the cytosolic iron pools, is caused through binding of nickel to Fur, or is an indirect result of oxidative stress or metal displacement cascades is unclear (Gault, Effantin, & Rodrigue, 2016; Wang, Wu, & Outten, 2011). Instead, it was shown that a very specific bacteriostatic effect of nickel excess in *E. coli* is a result of a block in glycolysis, caused by binding of Ni(II) ions to the enzyme fructose-1,6-bisphosphate-aldolase, FbaA, which inactivates the enzyme catalytically (Macomber et al., 2011). Exposure of *E. coli* cells to high concentrations of exogenous nickel for just 20 min was shown to reduce the FbaA activity to ~30% of that of control cells through a reversible process that did not involve new protein synthesis, whereas overexpression of FbaA from a plasmid could restore nickel tolerance to cells lacking the primary nickel exporter, RcnA.

Crucially, in vitro studies demonstrated that brief treatment of FbaA with excess Ni(II) led to replacement of one of the two FbaA-bound Zn(II) ions by nickel (Fig. 9). Surprisingly, subsequent in vitro analysis of mutant forms of the FbaA protein (Macomber et al., 2011) that lacked some of the known zinc-coordinating residues (Hall et al., 1999) implied that the Zn(II) ion that is displaced by Ni(II) is not the catalytic zinc ion, but a second, structural Zn(II) ion (Macomber et al., 2011). This zinc ion, located in the vicinity of the active site, has not been ascribed a specific role in catalysis but may play a structural role in stabilising the active site through hydrogen bonding of one of its ligands, Asp144, to His110, which is a ligand for the catalytic Zn(II) ion (Hall et al., 1999). It was proposed that displacement of this second, structural zinc ion leads to an 'allosteric' effect on the catalytic site (Macomber et al., 2011), perhaps through nickel's preference to adopt a square planar geometry rather than the tetrahedral arrangement observed for this Zn(II) site (Hall et al., 1999). However, it should be noted that the catalytic site itself was not actually mutagenised and tested within this study (Macomber et al., 2011); although mutations to the catalytic site will inevitably abrogate catalytic activity, which was used to assess the effect of the mutations, in vitro assays for nickel binding and concomitant zinc loss would be illuminating to confirm whether the catalytic zinc is actually resistant to Ni(II) exchange, as was proposed. This is important future work, as crystal structures of FbaA are ambiguous with respect to the precise structure and extent of solvent exposure of the two zinc-binding sites (Blom, Tetreault, Coulombe, & Sygusch, 1996; Hall et al., 1999). Both appear to be relatively solvent accessible, especially considering the coordination of an inhibitor molecule to the catalytic zinc ion in one FbaA structure, the absence of which would presumably make this zinc ion highly accessible (Hall et al., 1999; Fig. 9).

It is also worth noting that metal toxicity can lead to mis-metalation and metal-dependent inactivation of cytosolic transcriptional regulators. For example, the nickel-responsive transcriptional regulator InrS from *Synechocystis* PCC 6803 was found to respond initially to an extreme metal shock, demonstrating derepression of its target gene (encoding a nickel-efflux system) 1 h after the cells experienced very high concentrations of either zinc or copper (Foster, Pernil, Patterson, & Robinson, 2014). Yet when assayed 48 h later, this mis-metalation had corrected itself, and no zinc- or copper-dependent induction of the target gene was observed. This suggests that the sudden inflow of excess metal ions may have enabled these metal ions to wrongly associate with this nickel sensor, giving rise to incorrect gene expression, but that the copper and zinc adaptive response

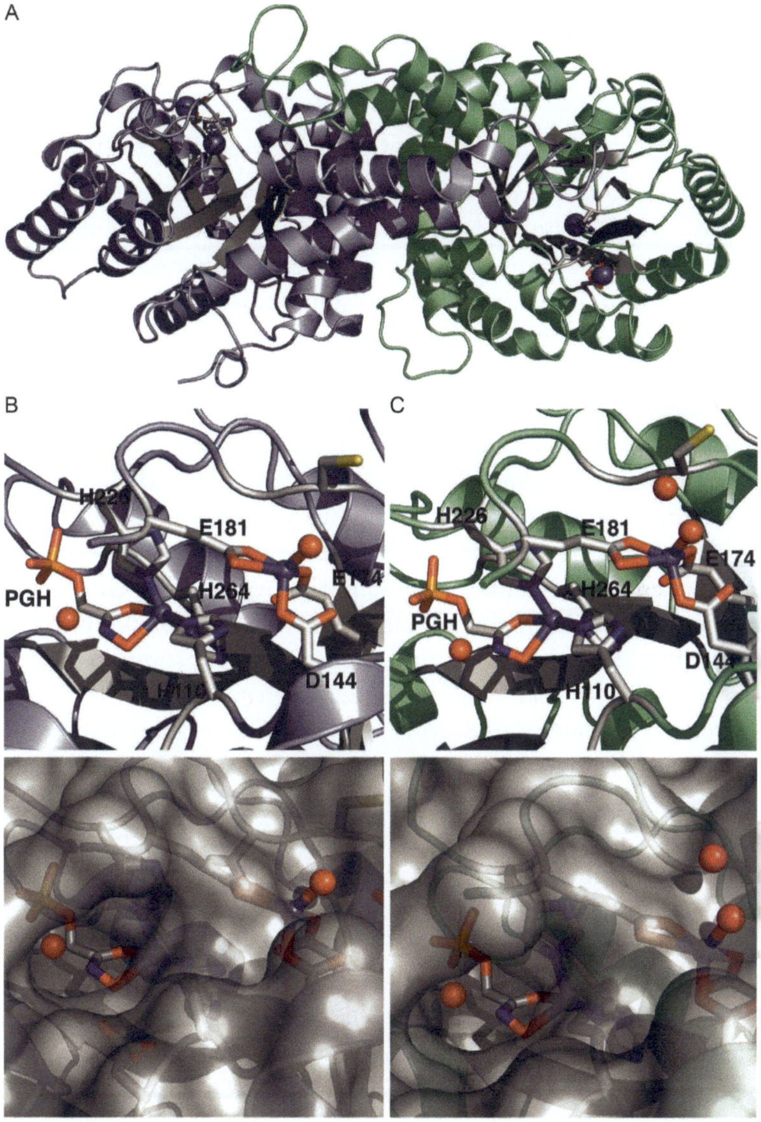


**Fig. 9** The two bound zinc ions in the structure of *E. coli* FbaA are potentially solvent exposed, one of which is vulnerable to attack by excess Ni(II) ions. (A) Cartoon representation of the crystal structure of a homodimer of the class II fructose-1,6-bisphosphate aldolase (FbaA) from *E. coli* (PDB ID: 1B57; Hall et al., 1999) in complex with the substrate analogue and inhibitor phosphoglycolohydroxamate (PGH). $ZnCl_2$ is necessary for protein crystallisation, and several zinc-binding sites were identified in the structure apart from the two critical ones (*blue spheres*). (B and C) The enzyme possesses two Zn-binding sites, one catalytic and the second of unknown function. Both appear solvent exposed. The catalytic zinc is coordinated in trigonal bipyramidal geometry by

*(Continued)*

(involving induction of their own specific export systems) was able to resolve this problem over time.

## 3.4 Extracellular Metal Competition in Metal Toxicity

An obvious, but surprisingly understudied mechanism of excess metal toxicity is caused through direct intermetal competition at the point of import. The precise nature of the first metal-binding site for a metal ion on its specific importer is generally unknown, due to the difficulty of determining such sites in insoluble transporters. One clear exception is the ATP-binding cassette (ABC) family, a ubiquitous class of transporters, bacterial members of which are known to transport a diverse range of substrates including both metal ions and metal complexes (Berntsson, Smits, Schmitt, Slotboom, & Poolman, 2010). ABC transporters are generally composed of three protein components; a membrane permease, an ATPase, and an extracellular solute binding protein (SBP). The SBPs are generally either soluble in the periplasm of Gram-negative bacteria or associated with the membrane via a lipid-modification in Gram-positive bacteria. They bind their respective substrate with high affinity at the cell surface, and then transfer it to their respective transporter complex, thereby delivering the substrate for import.

A number of studies, primarily by McDevitt and coworkers, have investigated the molecular and cellular properties of the ABC transporter PsaABC from *S. pneumoniae*, which was previously demonstrated to be a manganese importer (Dintilhac et al., 1997) and is essential for virulence (Berry & Paton, 1996). They observed that manganese import via this transport system was inhibited by elevated exogenous concentrations of zinc, leading to growth inhibition at Zn(II):Mn(II) ratios greater than 30:1 (Eijkelkamp et al., 2014; McDevitt et al., 2011). This inhibition could be reversed by the addition of extra manganese to the growth medium and was dependent on the presence of the intact PsaABC transporter. Growth of *S. pneumoniae* at high Zn(II):Mn(II) ratios led to a significant decrease in accumulated

**Fig. 9—Cont'd** H110, H226 and H264 (represented as *sticks*), and the hydroxyl and enolate oxygen atoms of PGH bound deep in a polar and solvent-accessible active site. The second zinc is coordinated by D144, E174, E181 and a water molecule (*red sphere*). D144, E174 and E172 are involved in stabilising the positions of two histidine residues (H110, H226 and H264, respectively) that are ligands to the catalytic zinc, possibly stabilising the catalytic site. It is currently unclear which zinc ion is displaced by nickel (Macomber et al., 2011).

cellular manganese, but had no significant effect on total cellular zinc levels, consistent with the mechanism of growth inhibition being caused by competitive inhibition of the manganese importer by excess zinc, rather than PsaABC-dependent zinc import leading to zinc-dependent toxicity. These conditions also led to the induction of manganese import genes (McDevitt et al., 2011), and actually produced a similar overall transcriptional response to that observed in the Δ*psaA* strain under standard growth conditions (Jacobsen et al., 2011). Notably, the PsaABC system seems incapable of importing zinc (Couñago et al., 2014; Jacobsen et al., 2011; McDevitt et al., 2011).

In vitro studies have investigated the molecular basis of this competition between zinc and manganese for the manganese transporter SBP PsaA. Evidence has been presented to show that PsaA binds a number of divalent ions (Couñago et al., 2014), but that zinc, unlike manganese, binds in an essentially irreversible manner, becoming refractory to extraction even with strong metal chelators (Couñago et al., 2014). Conversely, binding of Mn(II) is kinetically faster than binding of Zn(II) ions (Li et al., 2014). Interestingly Cu(II) appears to bind in a similar irreversible manner to Zn(II) (Couñago et al., 2014), but data were not presented from growth under high Cu(II):Mn(II) ratios (McDevitt et al., 2011). It would be interesting to test whether this inhibitory mechanism of manganese uptake is also replicated in equivalent experiments with Cu(II).

Extensive crystallographic data have clarified the atomic structures of PsaA in multiple forms (Couñago et al., 2014; Lawrence et al., 1998; McDevitt et al., 2011), allowing an unprecedented molecular view of the mechanism of zinc competition with manganese for binding to this SBP. The structures of the Mn(II)- and Zn(II)-loaded forms are essentially identical in crystallo (Lawrence et al., 1998; McDevitt et al., 2011), but both demonstrate a 'closed' structure relative to that in the metal-free form (Couñago et al., 2014), in which the two domains of the protein are separated by a movement around a 'hinge' region to yield a deep groove in which the metal-binding ligands are solvent exposed, structural properties that are common to the SBPs. Remarkably, a structure was also obtained from crystals of a metal-free PsaA mutant protein, D280N, which had been soaked with Mn(II) and displayed the 'open', apo-like form but had a Mn(II) ion bound (Couñago et al., 2014). It has been proposed that this transition is important for protein function and that binding of Zn(II) makes this opening transition more difficult and thus hinders PsaA function (Couñago et al., 2014), a model that is consistent with data comparing the stabilisation of the PsaA structure caused by binding of Zn(II) compared with Mn(II) (Couñago et al., 2014; McDevitt et al., 2011).

Importantly, *S. pneumoniae* cells that were cultured under these high Zn(II):Mn(II) ratio subsequently showed decreased survival under conditions of oxidative stress, including by human immune cells, akin to the decreased survival observed with the Δ*psaA* strain (McDevitt et al., 2011). This result is consistent with the downregulation of transcription under these growth conditions of the SOD gene, *sodA* (Eijkelkamp et al., 2014). Evidence was also provided that suggested that infection with *S. pneumoniae* led to an increase in the Zn(II):Mn(II) ratio in several physiological niches in infected mice and that the detected ratio was sufficiently high postinfection to suggest that zinc-mediated inhibition of streptococcal manganese import could play a role within the infected host (McDevitt et al., 2011). Interestingly, it has been reported that host zinc deficiency can exacerbate streptococcal infections in mice (Strand et al., 2003).

It seems highly likely that this effect of excess exogenous zinc on manganese import will be to some extent conserved across other manganese SBPs, which themselves are highly conserved. For example, the structure of the equivalent proteins from the manganese importing ABC transporters from Gram-positive bacteria exhibit a similar overall structure, using conserved ligand residues to coordinate their substrate manganese ion (Gribenko et al., 2013), and show similar metal-binding preferences and kinetics (Vigonsky et al., 2015). It has also been shown that excess zinc can inhibit transport by a related ABC transporter, whose metal specificity is unclear but appears to transport manganese or iron (but not zinc), in another streptococcus, *S. pyogenes* (Janulczyk, Ricci, & Björck, 2003). Importantly, an analogous effect was observed when *S. pneumoniae* was cultured under elevated cadmium concentrations, due to competition between Cd(II) ions and Mn(II) for the PsaA protein (Begg et al., 2015). It remains to be studied whether this is a more general phenomenon, and whether the SBPs of other metal-specific ABC transporters such as those for zinc or iron import are also susceptible to inhibition by other exogenous metals.

An important point to note here is that this toxicity mechanism, whereby excess zinc accumulates at sites of infection and thereby retards bacterial manganese uptake, will synergise with other mechanisms of nutritional immunity whereby the mammalian immune system attempts to restrict access of the pathogenic microbe to essential manganese ions (Corbin et al., 2008; Garcia et al., 2017).

## 4. CONCLUSIONS

The last decade has seen a resurgence in interest in the practical utilisation of metal toxicity towards microorganisms, in light of the worrying spread of resistance to traditional antibiotics amongst human and animal pathogens and the difficulties experienced both in academia and the pharmaceutical industry in developing new antibiotics. Bacterial metal toxicity has useful potential applications in medicine, biotechnology and commerce.

This increase in interest in the application of metal toxicity has resulted in a surge in research aimed at determining the molecular mechanisms of bacterial metal toxicity. These studies have been aided by our increased understanding of bacterial metal homeostasis systems, the development of advanced methods for precise genetic manipulation of model bacteria, and by new chemical, biochemical and biophysical techniques for the study of metal homeostasis and metalloproteins. Although much remains to be uncovered, a common theme is emerging from these studies so far, namely that one of the main effects of an excess of one essential metal ion within a bacterial cell is the disruption of the homeostasis of other essential metals.

Iron homeostasis in particular seems to be especially vulnerable to attack by metals that are more competitive, i.e. that are higher in the Irving–Williams series. Toxic concentrations of both copper and cobalt have been shown to affect the stability of enzymes that contain solvent-exposed Fe-S clusters, either by directly attacking the clusters in target Fe-S cluster-dependent enzymes or by interfering with the Fe-S cluster biosynthetic machinery. This has been shown to alter numerous metabolic pathways, including key aspects of carbon metabolism such as the TCA cycle and both haem biosynthesis and branched-chain amino acid biosynthesis, but it is anticipated that future studies will identify other Fe-S-dependent pathways that are vulnerable to this mechanism of metal toxicity. A side effect of this damage to iron homeostasis seems to be the release of protein-bound iron, increasing the size of the labile iron pool, which in turn gives rise to real or perceived increase in oxidative stress.

Other examples presented also involve metal–metal competition. Nickel stress in *E. coli* has been shown to involve displacement of zinc ions from a critical enzyme in glycolysis, whereas zinc stress in *S. pneumoniae* derives from direct competition with manganese for the binding site on the cell surface protein that provides manganese for import, leading to zinc-dependent manganese starvation. This latter example is of additional importance given the accumulated evidence that the mammalian immune system uses manganese starvation as a key weapon to fight invading pathogens and to prevent establishment of infection.

Importantly, in all cases studied thus far and presented herein, the molecular mechanisms of metal toxicity that have been uncovered, while important, are insufficient to fully explain the observed toxicity of the metal towards bacterial growth. This shows that further mechanisms are at play, providing fruitful avenues for future research. Applications of new experimental approaches, including metalloproteomics and metabolomics, promise to expand our knowledge of these molecular mechanisms, allowing us to find new ways to combat important pathogenic bacteria and to assess the risk of resistance occurring to future metal-based antibacterial products.

## ACKNOWLEDGEMENTS

K.J.W. is supported by a Wellcome Trust/Royal Society funded Sir Henry Dale fellowship (098375/Z/12/Z). Additional support by a BBSRC DTP studentship (to A.B.-S.) is also gratefully acknowledged.

## REFERENCES

Aguirre, J. D., Clark, H. M., McIlvin, M., Vazquez, C., Palmere, S. L., Grab, D. J., et al. (2013). A manganese-rich environment supports superoxide dismutase activity in a Lyme disease pathogen, *Borrelia burgdorferi*. *Journal of Biological Chemistry*, *288*, 8468–8478.

Albrecht, A. G., Netz, D. J., Miethke, M., Pierik, A. J., Burghaus, O., Peuckert, F., et al. (2010). SufU is an essential iron-sulfur cluster scaffold protein in *Bacillus subtilis*. *Journal of Bacteriology*, *192*, 1643–1651.

Almirón, M., Link, A. J., Furlong, D., & Kolter, R. (1992). A novel DNA-binding protein with regulatory and protective roles in starved *Escherichia coli*. *Genes and Development*, *6*, 2646–2654.

Amachawadi, R. G., Shelton, N. W., Shi, X., Vinasco, J., Dritz, S. S., Tokach, M. D., et al. (2011). Selection of fecal enterococci exhibiting *tcrB*-mediated copper resistance in pigs fed diets supplemented with copper. *Applied and Environmental Microbiology*, *77*, 5597–5603.

Amano, A., Shizukuishi, S., Tsunemitsu, A., & Tsunasawa, S. (1992). Identity of amino acid sequences of superoxide dismutase purified from both anaerobically maintained and aerated *Porphyromonas gingivalis*. *Oral Microbiology and Immunology*, 7, 368–371.

Andreesen, J. R., & Makdessi, K. (2008). Tungsten, the surprisingly positively acting heavy element for prokaryotes. *Annals of the New York Academy of Sciences*, *1125*, 215–229.

Andreini, C., Bertini, I., Cavallaro, G., Holliday, G. L., & Thornton, J. M. (2008). Metal ions in biological catalysis: From enzyme databases to general principles. *Journal of Biological Inorganic Chemistry*, *13*, 1205–1218.

Anjem, A., & Imlay, J. A. (2012). Mononuclear iron enzymes are primary targets of hydrogen peroxide stress. *Journal of Biological Chemistry*, *287*, 15544–15556.

Anjem, A., Varghese, S., & Imlay, J. A. (2009). Manganese import is a key element of the OxyR response to hydrogen peroxide in *Escherichia coli*. *Molecular Microbiology*, *72*, 844–858.

Ayres, P. G. (2004). Alexis Millardet: France's forgotten mycologist. *Mycologist*, *18*, 23–26.

Azzouzi, A., Steunou, A. S., Durand, A., Khalfaoui-Hassani, B., Bourbon, M. L., Astier, C., et al. (2013). Coproporphyrin III excretion identifies the anaerobic coproporphyrinogen III oxidase HemN as a copper target in the $Cu^{+}$-ATPase mutant $copA^{-}$ of *Rubrivivax gelatinosus*. *Molecular Microbiology*, *88*, 339–351.

Baker, J., Sengupta, M., Jayaswal, R. K., & Morrissey, J. A. (2011). The *Staphylococcus aureus* CsoR regulates both chromosomal and plasmid-encoded copper resistance mechanisms. *Environmental Microbiology*, *13*, 2495–2507.

Baker, J., Sitthisak, S., Sengupta, M., Johnson, M., Jayaswal, R. K., & Morrissey, J. A. (2010). Copper stress induces a global stress response in *Staphylococcus aureus* and represses *sae* and *agr* expression and biofilm formation. *Applied and Environmental Microbiology*, *76*, 150–160.

Baker-Austin, C., Wright, M. S., Stepanauskas, R., & McArthur, J. V. (2006). Co-selection of antibiotic and metal resistance. *Trends in Microbiology*, *14*, 176–182.

Balasubramanian, R., Kenney, G. E., & Rosenzweig, A. C. (2011). Dual pathways for copper uptake by methanotrophic bacteria. *Journal of Biological Chemistry*, *286*, 37313–37319.

Balasubramanian, R., Smith, S. M., Rawat, S., Yatsunyk, L. A., Stemmler, T. L., & Rosenzweig, A. C. (2010). Oxidation of methane by a biological dicopper centre. *Nature*, *465*, 115–119.

Banci, L., Bertini, I., Ciofi-Baffoni, S., Kandias, N. G., Robinson, N. J., Spyroulias, G. A., et al. (2006). The delivery of copper for thylakoid import observed by NMR. *Proceedings of the National Academy of Sciences of the United States of America*, *103*, 8320–8325.

Banci, L., Bertini, I., Ciofi-Baffoni, S., Kozyreva, T., Zovo, K., & Palumaa, P. (2010a). Affinity gradients drive copper to cellular destinations. *Nature*, *465*, 645–648.

Banci, L., Bertini, I., Ciofi-Baffoni, S., Poggi, L., Vanarotti, M., Tottey, S., et al. (2010b). NMR structural analysis of the soluble domain of ZiaA-ATPase and the basis of selective interactions with copper metallochaperone Atx1. *Journal of Biological Inorganic Chemistry*, *15*, 87–98.

Barnett, J. P., Scanlan, D. J., & Blindauer, C. A. (2012). Fractionation and identification of metalloproteins from a marine cyanobacterium. *Analytical and Bioanalytical Chemistry*, *402*, 3371–3377.

Begg, S. L., Eijkelkamp, B. A., Luo, Z., Couñago, R. M., Morey, J. R., Maher, M. J., et al. (2015). Dysregulation of transition metal ion homeostasis is the molecular basis for cadmium toxicity in *Streptococcus pneumoniae*. *Nature Communications*, *6*, 6418.

Berntsson, R. P.-A., Smits, S. H. J., Schmitt, L., Slotboom, D.-J., & Poolman, B. (2010). A structural classification of substrate-binding proteins. *FEBS Letters*, *584*, 2606–2617.

Berridge, M. J., Bootman, M. D., & Roderick, H. L. (2003). Calcium signalling: Dynamics, homeostasis and remodelling. *Nature Reviews. Molecular Cell Biology*, *4*, 517–529.

Berry, A. M., & Paton, J. C. (1996). Sequence heterogeneity of PsaA, a 37-kilodalton putative adhesin essential for virulence of *Streptococcus pneumoniae*. *Infection and Immunity*, *64*, 5255–5262.

Beyer, W. F., & Fridovich, I. (1991). In vivo competition between iron and manganese for occupancy of the active site region of the manganese-superoxide dismutase of Escherichia coli. *Journal of Biological Chemistry*, *266*(1), 303–308.

Biasini, M., Bienert, S., Waterhouse, A., Arnold, K., Studer, G., Schmidt, T., et al. (2014). SWISS-MODEL: Modelling protein tertiary and quaternary structure using evolutionary information. *Nucleic Acids Research*, *42*, W252–W258.

Binet, M. R., & Poole, R. K. (2000). Cd(II), Pb(II) and Zn(II) ions regulate expression of the metal-transporting P-type ATPase ZntA in *Escherichia coli*. *FEBS Letters*, *473*, 67–70.

Blair, J. M., Webber, M. A., Baylay, A. J., Ogbolu, D. O., & Piddock, L. J. (2015). Molecular mechanisms of antibiotic resistance. *Nature Reviews. Microbiology*, *13*, 42–51.

Blindauer, C. A., Harrison, M. D., Parkinson, J. A., Robinson, A. K., Cavet, J. S., Robinson, N. J., et al. (2001). A metallothionein containing a zinc finger within a four-metal cluster protects a bacterium from zinc toxicity. *Proceedings of the National Academy of Sciences of the United States of America, 98*, 9593–9598.

Blindauer, C. A., Harrison, M. D., Robinson, A. K., Parkinson, J. A., Bowness, P. W., Sadler, P. J., et al. (2002). Multiple bacteria encode metallothioneins and SmtA-like zinc fingers. *Molecular Microbiology, 45*, 1421–1432.

Blom, N., Tetreault, S., Coulombe, R., & Sygusch, J. (1996). Novel active site structure in *Escherichia coli* fructose 1,6-bisphosphate aldolase. *Nature Structural Biology, 3*, 856–862.

Booth, S. C., Weljie, A. M., & Turner, R. J. (2015). Metabolomics reveals differences of metal toxicity in cultures of *Pseudomonas pseudoalcaligenes* KF707 grown on different carbon sources. *Frontiers in Microbiology, 6*, 827.

Booth, S. C., Workentine, M. L., Weljie, A. M., & Turner, R. J. (2011). Metabolomics and its application to studying metal toxicity. *Metallomics, 3*, 1142–1152.

Botella, H., Peyron, P., Levillain, F., Poincloux, R., Poquet, Y., Brandli, I., et al. (2011). Mycobacterial $P_1$-type ATPases mediate resistance to zinc poisoning in human macrophages. *Cell Host & Microbe, 10*, 248–259.

Boyd, P. W., Jickells, T., Law, C. S., Blain, S., Boyle, E. A., Buesseler, K. O., et al. (2007). Mesoscale iron enrichment experiments 1993–2005: Synthesis and future directions. *Science, 315*, 612–617.

Boyer, E., Bergevin, I., Malo, D., Gros, P., & Cellier, M. F. M. (2002). Acquisition of Mn (II) in addition to Fe (II) is required for full virulence of Salmonella enterica serovar Typhimurium. *Infection and Immunity, 70*, 6032–6042.

Braun, V., & Hantke, K. (2011). Recent insights into iron import by bacteria. *Current Opinion in Chemical Biology, 15*, 328–334.

Brown, E. D., & Wright, G. D. (2016). Antibacterial drug discovery in the resistance era. *Nature, 529*, 336–343.

Bsat, N., Herbig, A., Casillas-Martinez, L., Setlow, P., & Helmann, J. D. (1998). *Bacillus subtilis* contains multiple Fur homologues: Identification of the iron uptake (Fur) and peroxide regulon (PerR) repressors. *Molecular Microbiology, 29*, 189–198.

Busenlehner, L. S., Pennella, M. A., & Giedroc, D. P. (2003). The SmtB/ArsR family of metalloregulatory transcriptional repressors: Structural insights into prokaryotic metal resistance. *FEMS Microbiology Reviews, 27*, 131–143.

Cameron, V., House, C. H., & Brantley, S. L. (2012). A first analysis of metallome biosignatures of hyperthermophilic archaea. *Archaea, 2012*, 789278.

Canfield, D. E. (1998). A new model for Proterozoic ocean chemistry. *Nature, 396*, 450–453.

Carrondo, M. A. (2003). Ferritins, iron uptake and storage from the bacterioferritin viewpoint. *EMBO Journal, 22*, 1959–1968.

Casey, A. L., Adams, D., Karpanen, T. J., Lambert, P. A., Cookson, B. D., Nightingale, P., et al. (2010). Role of copper in reducing hospital environment contamination. *Journal of Hospital Infection, 74*(1), 72–77.

Catling, D. C., Zahnle, K. J., & McKay, C. P. (2001). Biogenic methane, hydrogen escape, and the irreversible oxidation of early Earth. *Science, 293*, 839–843.

Cavaco, L. M., Hasman, H., & Aarestrup, F. M. (2011). Zinc resistance of *Staphylococcus aureus* of animal origin is strongly associated with methicillin resistance. *Veterinary Microbiology, 150*, 344–348.

Cavet, J. S., Graham, A. I., Meng, W., & Robinson, N. J. (2003). A cadmium-lead-sensing ArsR-SmtB repressor with novel sensory sites complementary metal discrimination by NmtR and CmtR in a common cytosol. *Journal of Biological Chemistry, 278*, 44560–44566.

Cavet, J. S., Meng, W., Pennella, M. A., Appelhoff, R. J., Giedroc, D. P., & Robinson, N. J. (2002). A nickel-cobalt-sensing ArsR-SmtB family repressor: Contributions of cytosol and effector binding sites to metal selectivity. *Journal of Biological Chemistry, 277*, 38441–38448.

Changela, A., Chen, K., Xue, Y., Holschen, J., Outten, C. E., O'Halloran, T. V., et al. (2003). Molecular basis of metal-ion selectivity and zeptomolar sensitivity by CueR. *Science*, *301*, 1383–1387.

Chillappagari, S., Miethke, M., Trip, H., Kuipers, O. P., & Marahiel, M. A. (2009). Copper acquisition is mediated by YcnJ and regulated by YcnK in *Bacillus subtilis*. *Journal of Bacteriology*, *191*, 2362–2370.

Chillappagari, S., Seubert, A., Trip, H., Kuipers, O. P., Marahiel, M. A., & Miethke, M. (2010). Copper stress affects iron homeostasis by destabilizing iron-sulfur cluster formation in *Bacillus subtilis*. *Journal of Bacteriology*, *192*, 2512–2524.

Chivers, P. T., Benanti, E. L., Heil-Chapdelaine, V., Iwig, J. S., & Rowe, J. L. (2012). Identification of Ni-(L-His)$_2$ as a substrate for NikABCDE-dependent nickel uptake in *Escherichia coli*. *Metallomics*, *4*, 1043–1050.

Chivers, P. T., & Sauer, R. T. (2000). Regulation of high affinity nickel uptake in bacteria $Ni^{2+}$-dependent interaction of NikR with wild-type and mutant operator sites. *Journal of Biological Chemistry*, *275*(26), 19735–19741.

Christman, M. F., Morgan, R. W., Jacobson, F. S., & Ames, B. N. (1985). Positive control of a regulon for defenses against oxidative stress and some heat-shock proteins in *Salmonella typhimurium*. *Cell*, *41*, 753–762.

Cooper, S. R., McArdle, J. V., & Raymond, K. N. (1978). Siderophore electrochemistry: Relation to intracellular iron release mechanism. *Proceedings of the National Academy of Sciences of the United States of America*, *75*, 3551–3554.

Corbett, D., Wang, J., Schuler, S., Lopez-Castejon, G., Glenn, S., Brough, D., et al. (2012). Two zinc uptake systems contribute to the full virulence of Listeria monocytogenes during growth *in vitro* and *in vivo*. *Infection and Immunity*, *80*, 14–21.

Corbin, B. D., Seeley, E. H., Raab, A., Feldmann, J., Miller, M. R., Torres, V. J., et al. (2008). Metal chelation and inhibition of bacterial growth in tissue abscesses. *Science*, *319*, 962–965.

Cotruvo, J. A., Jr., & Stubbe, J. (2012). Metallation and mismetallation of iron and manganese proteins *in vitro* and *in vivo*: The class I ribonucleotide reductases as a case study. *Metallomics*, *4*(10), 1020–1036.

Couñago, R. M., Ween, M. P., Begg, S. L., Bajaj, M., Zuegg, J., O'Mara, M. L., et al. (2014). Imperfect coordination chemistry facilitates metal ion release in the Psa permease. *Nature Chemical Biology*, *10*, 35–41.

Cvetkovic, A., Lal Menon, A., Thorgersen, M. P., Scott, J. W., Poole, F. L., II, Jenney, F. E., Jr., et al. (2010). Microbial metalloproteomes are largely uncharacterized. *Nature*, *466*, 779–782.

Damo, S. M., Kehl-Fie, T. E., Sugitani, N., Holt, M. E., Rathi, S., Murphy, W. J., et al. (2013). Molecular basis for manganese sequestration by calprotectin and roles in the innate immune response to invading bacterial pathogens. *Proceedings of the National Academy of Sciences of the United States of America*, *110*, 3841–3846.

Di Pilato, V., Arena, F., Giani, T., Conte, V., Cresti, S., & Rossolini, G. M. (2014). Characterization of pFOX-7a, a conjugative IncL/M plasmid encoding the FOX-7 AmpC-type β-lactamase, involved in a large outbreak in a neonatal intensive care unit. *Journal of Antimicrobial Chemotherapy*, *69*, 2620–2624.

Dintilhac, A., Alloing, G., Granadel, C., & Claverys, J.-P. (1997). Competence and virulence of *Streptococcus pneumoniae*: Adc and PsaA mutants exhibit a requirement for Zn and Mn resulting from inactivation of putative ABC metal permeases. *Molecular Microbiology*, *25*, 727–739.

Djoko, K. Y., Donnelly, P. S., & McEwan, A. G. (2014). Inhibition of respiratory Complex I by copper(II)-bis(thiosemicarbazonato) complexes. *Metallomics*, *6*, 2250–2259.

Djoko, K. Y., & McEwan, A. G. (2013). Antimicrobial action of copper is amplified via inhibition of heme biosynthesis. *ACS Chemical Biology*, *8*, 2217–2223.

Dollwet, H. H. A., & Sorenson, J. R. J. (1985). Historic uses of copper compounds in medicine. *Trace Elements in Medicine*, *2*, 80–87.

Domínguez, D. C., Guragain, M., & Patrauchan, M. (2015). Calcium binding proteins and calcium signalling in prokaryotes. *Cell Calcium*, *57*, 151–165.

Drakesmith, H., & Prentice, A. M. (2012). Hepcidin and the iron-infection axis. *Science*, *338*, 768–772.

Dupont, C. L., Butcher, A., Valas, R. E., Bourne, P. E., & Caetano-Anollés, G. (2010). History of biological metal utilization inferred through phylogenomic analysis of protein structures. *Proceedings of the National Academy of Sciences of the United States of America*, *107*, 10567–10572.

Eijkelkamp, B. A., Morey, J. R., Ween, M. P., Ong, C. Y., McEwan, A. G., Paton, J. C., et al. (2014). Extracellular zinc competitively inhibits manganese uptake and compromises oxidative stress management in *Streptococcus pneumoniae*. *PloS One*, *9*, e89427.

El Ghazouni, A., Baslé, A., Gray, J., Graham, D. W., Firbank, S. J., & Dennison, C. (2012). Variations in methanobactin structure influences copper utilization by methane-oxidizing bacteria. *Proceedings of the National Academy of Sciences of the United States of America*, *109*, 8400–8404.

Elguindi, J., Wagner, J., & Rensing, C. (2009). Genes involved in copper resistance influence survival of *Pseudomonas aeruginosa* on copper surfaces. *Journal of Applied Microbiology*, *106*, 1448–1455.

Espitia, P. J. P., Soares, N. D. F. F., dos Reis Coimbra, J. S., de Andrade, N. J., Cruz, R. S., & Medeiros, E. A. A. (2012). Zinc oxide nanoparticles: Synthesis, antimicrobial activity and food packaging applications. *Food and Bioprocess Technology*, *5*, 1447–1464.

Faúndez, G., Troncoso, M., Navarrete, P., & Figueroa, G. (2004). Antimicrobial activity of copper surfaces against suspensions of *Salmonella enterica* and *Campylobacter jejuni*. *BMC Microbiology*, *4*, 19.

Fazzio, T. G., & Roth, J. R. (1996). Evidence that the CysG protein catalyzes the first reaction specific to B12 synthesis in *Salmonella typhimurium*, insertion of cobalt. *Journal of Bacteriology*, *178*, 6952–6959.

Feeney, K. A., Hansen, L. L., Putker, M., Olivares-Yañez, C., Day, J., Eades, L. J., et al. (2016). Daily magnesium fluxes regulate cellular timekeeping and energy balance. *Nature*, *532*, 375–379.

Feliciano, P. R., Drennan, C. L., & Nonato, M. C. (2016). Crystal structure of an Fe-S cluster-containing fumarate hydratase enzyme from *Leishmania* major reveals a unique protein fold. *Proceedings of the National Academy of Sciences of the United States of America*, *113*, 9804–9809.

Ferguson, A. D., Hofmann, E., Coulton, J. W., Diederichs, K., & Welte, W. (1998). Siderophore-mediated iron transport: Crystal structure of FhuA with bound lipopolysaccharide. *Science*, *282*, 2215–2220.

Ferrer, M., Golyshina, O. V., Beloqui, A., Golyshin, P. N., & Timmis, K. N. (2007). The cellular machinery of *Ferroplasma acidiphilum* is iron-protein-dominated. *Nature*, *445*, 91–94.

Festa, R. A., Jones, M. B., Butler-Wu, S., Sinsimer, D., Gerads, R., Bishai, W. R., et al. (2011). A novel copper-responsive regulon in *Mycobacterium tuberculosis*. *Molecular Microbiology*, *79*, 133–148.

Foster, A. W., Osman, D., & Robinson, N. J. (2014). Metal preferences and metallation. *Journal of Biological Chemistry*, *289*, 28095–28103.

Foster, A. W., Patterson, C. J., Pernil, R., Hess, C. R., & Robinson, N. J. (2012). Cytosolic Ni(II) detection in cyanobacterium: Nickel detection follows nickel affinity across four families of metal sensors. *Journal of Biological Chemistry*, *287*, 12142–12151.

Foster, A. W., Pernil, R., Patterson, C. J., & Robinson, N. J. (2014). Metal specificity of cyanobacterial nickel-responsive repressor InrS: Cells maintain zinc and copper below the detection threshold for InrS. *Molecular Microbiology*, *92*, 797–812.
Fraústo da Silva, J. J. R., & Williams, R. J. P. (2001). *The biological chemistry of the elements: The inorganic chemistry of life* (2nd ed.). UK: Oxford University Press.
Freeman, J. L., Persans, M. W., Nieman, K., & Salt, D. E. (2005). Nickel and cobalt resistance engineered in *Escherichia coli* by overexpression of serine acetyltransferase from the nickel hyperaccumulator plant *Thlaspi goesingense*. *Applied and Environmental Microbiology*, *71*, 8627–8633.
Gabriel, S. E., & Helmann, J. D. (2009). Contributions of Zur-controlled ribosomal proteins to growth under zinc starvation conditions. *Journal of Bacteriology*, *191*, 6116–6122.
Garcia, Y. M., Barwinska-Sendra, A., Tarrant, E., Skaar, E. P., Waldron, K. J., & Kehl-Fie, T. E. (2017). A superoxide dismutase capable of functioning with iron or manganese promotes the resistance of *Staphylococcus aureus* to calprotectin and nutritional immunity. *PLoS Pathogens*, *13*, e1006125.
Gault, M., Effantin, G., & Rodrigue, A. (2016). Ni exposure impacts the pool of free Fe and modifies DNA supercoiling via metal-induced oxidative stress in *Escherichia coli* K12. *Free Radical Biology and Medicine*, *97*, 351–361.
Gladyshev, V. N., & Zhang, Y. (2013). Comparative genomics analysis of the metallomes. In *Metallomics and the cell* (pp. 529–580). Netherlands: Springer.
Golynskiy, M. V., Gunderson, W. A., Hendrich, M. P., & Cohen, S. M. (2006). Metal binding studies and EPR spectroscopy of the manganese transport regulator MntR. *Biochemistry*, *45*, 15359–15372.
González-Flecha, B., & Demple, B. (1995). Metabolic sources of hydrogen peroxide in aerobically growing *Escherichia coli*. *Journal of Biological Chemistry*, *270*, 13681–13687.
Graham, A. I., Hunt, S., Stokes, S. L., Bramall, N., Bunch, J., Cox, A. G., et al. (2009). Severe zinc depletion of *Escherichia coli*: Roles for high affinity zinc binding by ZinT, zinc transport and zinc-independent proteins. *Journal of Biological Chemistry*, *284*, 18377–18389.
Gribenko, A., Mosyak, L., Ghosh, S., Parris, K., Svenson, K., Moran, J., et al. (2013). Three-dimensional structure and biophysical characterization of *Staphylococcus aureus* cell surface antigen-manganese transporter MntC. *Journal of Molecular Biology*, *425*, 3429–3445.
Grifantini, R., Sebastian, S., Frigimelica, E., Draghi, M., Bartolini, E., Muzzi, A., et al. (2003). Identification of iron-activated and -repressed Fur-dependent genes by transcriptome analysis of *Neisseria meningitidis* group B. *Proceedings of the National Academy of Sciences of the United States of America*, *100*, 9542–9547.
Griffith, J. S., & Orgel, L. E. (1957). Ligand-field theory. *Quarterly Reviews, Chemical Society*, *11*, 381–393.
Groisman, E. A., Hollands, K., Kriner, M. A., Lee, E.-J., Park, S.-Y., & Poyes, M. H. (2013). Bacterial $Mg^{2+}$ homeostasis. *Annual Review of Genetics*, *47*, 625–646.
Grossoehme, N., Kehl-Fie, T. E., Ma, Z., Adams, K. W., Cowart, D. M., Scott, R. A., et al. (2011). Control of copper resistance and inorganic sulfur metabolism by paralogous regulators in *Staphylococcus aureus*. *Journal of Biological Chemistry*, *286*, 13522–13531.
Gu, M., & Imlay, J. A. (2013). Superoxide poisons mononuclear iron enzymes by causing mismetallation. *Molecular Microbiology*, *89*, 123–134.
Guerra, A. J., Dann, C. E., 3rd, & Giedroc, D. P. (2011). Crystal structure of the zinc-dependent MarR family transcriptional regulator AdcR in the Zn(II)-bound state. *Journal of the American Chemical Society*, *133*, 19614–19617.
Gunther, M. R., Hanna, P. M., Mason, R. P., & Cohen, M. S. (1995). Hydroxyl radical formation from cuprous ion and hydrogen peroxide: A spin-trapping study. *Archives of Biochemistry and Biophysics*, *316*, 515–522.

Hall, D. R., Leonard, G. A., Reed, C. D., Watt, C. I., Berry, A., & Hunter, W. N. (1999). The crystal structure of *Escherichia coli* class II fructose-1,6-bisphosphate aldolase in complex with phosphoglycolohydroxamate reveals details of mechanism and specificity. *Journal of Molecular Biology, 287*, 383–394.

Hammer, N. D., & Skaar, E. P. (2011). Molecular mechanisms of Staphylococcus aureus iron acquisition. *Annual Review of Microbiology, 65*, 129–147.

Hantke, K. (2005). Bacterial zinc uptake and regulators. *Current Opinion in Microbiology, 8*, 196–202.

Hao, Z., Lou, H., Zhu, R., Zhu, J., Zhang, D., Zhao, B. S., et al. (2014). The multiple antibiotic resistance regulator MarR is a copper sensor in *Escherichia coli*. *Nature Chemical Biology, 10*, 21–28.

Hao, X., Lüthje, F., Rønn, R., German, N. A., Li, X., Huang, F., et al. (2016). A role for copper in protozoan grazing—Two billion years selecting for bacterial copper resistance. *Molecular Microbiology, 102*, 628–641. http://dx.doi.org/10.1111/mmi.13483.

Harrison, A., Ray, W. C., Baker, B. D., Armbruster, D. W., Bakaletz, L. O., & Munson, R. S., Jr. (2007). The OxyR regulon in nontypeable *Haemophilus influenzae*. *Journal of Bacteriology, 189*, 1004–1012.

Hattori, M., Iwase, N., Furuya, N., Tanaka, Y., Tsukazaki, T., Ishitani, R., et al. (2009). $Mg^{2+}$-dependent gating of bacterial MgtE channel underlies $Mg^{2+}$ homeostasis. *EMBO Journal, 28*, 3602–3612.

Helmann, J. D., Soonsanga, S., & Gabriel, S. (2007). Metalloregulators: Arbiters of metal sufficiency. In D. H. Nies & S. Silver (Eds.), *Molecular microbiology of heavy metals* (pp. 37–71). Germany: Springer Berlin/Heidelberg.

Hensley, M. P., Tierney, D. L., & Crowder, M. W. (2011). Zn(II) binding to *Escherichia coli* 70S ribosomes. *Biochemistry, 50*, 9937–9939.

Herbig, A. F., & Helmann, J. D. (2001). Roles of metal ions and hydrogen peroxide in modulating the interaction of the *Bacillus subtilis* PerR peroxide regulon repressor with operator DNA. *Molecular Microbiology, 41*, 849–859.

Hinsa-Leasure, S. M., Nartey, Q., Vaverka, J., & Schmidt, M. G. (2016). Copper alloy surfaces sustain terminal cleaning levels in a rural hospital. *American Journal of Infection Control, 44*, e195–e203. http://dx.doi.org/10.1016/j.ajic.

Hohle, T. H., & O'Brian, M. R. (2014). Magnesium-dependent processes are targets of bacterial manganese toxicity. *Molecular Microbiology, 93*, 736–747.

Hohle, T. H., & O'Brian, M. R. (2016). Metal-specific control of gene expression mediated by *Bradyrhizobium japonicum* Mur and *Escherichia coli* Fur is determined by the cellular context. *Molecular Microbiology, 101*, 152–166.

Horsburgh, M. J., Clements, M. O., Crossley, H., Ingham, E., & Foster, S. J. (2001). PerR controls oxidative stress resistance and iron storage proteins and is required for virulence in Staphylococcus aureus. *Infection and Immunity, 69*, 3744–3754.

Horsburgh, M. J., Wharton, S. J., Cox, A. G., Ingham, E., Peacock, S., & Foster, S. J. (2002). MntR modulates expression of the PerR regulon and superoxide resistance in *Staphylococcus aureus* through control of manganese uptake. *Molecular Microbiology, 44*, 1269–1286.

Hu, P., Brodie, E. L., Suzuki, Y., McAdams, H. H., & Andersen, G. L. (2005). Whole-genome transcriptional analysis of heavy metal stresses in *Caulobacter crescentus*. *Journal of Bacteriology, 187*, 8437–8449.

Imlay, J. A. (2015). Diagnosing oxidative stress in bacteria: Not as easy as you think. *Current Opinion in Microbiology, 24*, 124–131.

Imlay, J. A., Chin, S. M., & Linn, S. (1988). Toxic DNA damage by hydrogen peroxide through the Fenton reaction *in vivo* and *in vitro*. *Science, 240*, 640–642.

Inkinen, J., Makinen, R., Keinanen-Toivola, M. M., Nordstrom, K., & Ahonen, M. (2016). Copper as an antibacterial material in different facilities. *Letters in Applied Microbiology*, *64*, 19–26. http://dx.doi.org/10.1111/lam.12680.

Irving, H., & Williams, R. J. P. (1953). The stability of transition-metal complexes. *Journal of the Chemical Society*, 3192–3210.

Jacobsen, F. E., Kazmierczak, K. M., Lisher, J. P., Winkler, M. E., & Giedroc, D. P. (2011). Interplay between manganese and zinc homeostasis in the human pathogen *Streptococcus pneumoniae*. *Metallomics*, *3*, 38–41.

Jang, S., & Imlay, J. A. (2007). Micromolar intracellular hydrogen peroxide disrupts metabolism by damaging iron-sulfur enzymes. *Journal of Biological Chemistry*, *282*, 929–937.

Janulczyk, R., Ricci, S., & Björck, L. (2003). MtsABC is important for manganese and iron transport, oxidative stress resistance, and virulence of *Streptococcus pyogenes*. *Infection and Immunity*, *71*, 2656–2664.

Jen, F. E., Djoko, K. Y., Bent, S. J., Day, C. J., McEwan, A. G., & Jennings, M. P. (2015). A genetic screen reveals a periplasmic copper chaperone required for nitrite reductase activity in pathogenic *Neisseria*. *FASEB Journal*, *29*, 3828–3838.

Jeschek, M., Reuter, R., Heinisch, T., Trindler, C., Klehr, J., Panke, S., et al. (2016). Directed evolution of artificial metalloenzymes for in vivo metathesis. *Nature*, *537*, 661–665.

Jickells, T. D., An, Z. S., Andersen, K. K., Baker, A. R., Bergametti, G., Brooks, N., et al. (2005). Global iron connection between desert dust, ocean biogeochemistry, and climate. *Science*, *308*, 67–71.

Jin, B., Newton, S., Shao, Y., Jiang, X., Charbit, A., & Klebba, P. E. (2006). Iron acquisition systems for ferric hydroxamates, haemin and haemoglobin in *Listeria monocytogenes*. *Molecular Microbiology*, *59*, 1185–1198.

Jo, I., Chung, I.-Y., Bae, H.-W., Kim, J.-S., Song, S., Cho, Y.-H., et al. (2015). Structural details of the OxyR peroxide-sensing mechanism. *Proceedings of the National Academy of Sciences of the United States of America*, *112*, 6443–6448.

Jordi, A., Basterretxea, G., Tovar-Sánchez, A., Alastuey, A., & Querol, X. (2012). Copper aerosols inhibit phytoplankton growth in the Mediterranean Sea. *Proceedings of the National Academy of Sciences of the United States of America*, *109*, 21246–21249.

Kehl-Fie, T. E., & Skaar, E. P. (2010). Nutritional immunity beyond iron: A role for manganese and zinc. *Current Opinion in Chemical Biology*, *14*, 218–224.

Kehl-Fie, T. E., Zhang, Y., Moore, J. L., Farrand, A. J., Hood, M. I., Rathi, S., et al. (2013). MntABC and MntH contribute to systemic *Staphylococcus aureus* infection by competing with calprotectin for nutrient manganese. *Infection and Immunity*, *81*, 3395–3405.

Kenney, G. E., & Rosenzweig, A. C. (2013). Genome mining for methanobactins. *BMC Biology*, *11*, 17.

Key, H. M., Dydio, P., Clark, D. S., & Hartwig, J. F. (2016). Abiological catalysis by artificial haem proteins containing noble metals in place of iron. *Nature*, *534*, 534–537.

Kim, H. J., Graham, D. W., DiSpirito, A. A., Alterman, M. A., Galeva, N., Larive, C. K., et al. (2004). Methanobactin, a copper-acquisition compound from methane-oxidizing bacteria. *Science*, *305*, 1612–1615.

Kimura, T., & Nishioka, H. (1997). Intracellular generation of superoxide by copper sulphate in *Escherichia coli*. *Mutation Research/Genetic Toxicology and Environmental Mutagenesis*, *389*, 237–242.

Kirsten, A., Herzberg, M., Voigt, A., Seravalli, J., Grass, G., Scherer, J., et al. (2011). Contributions of five secondary metal uptake systems to metal homeostasis of *Cupriavidus metallidurans* CH34. *Journal of Bacteriology*, *193*, 4652–4663.

Kleiner, D. (1978). Inhibition of the respiratory system in *Azotobacter vinelandii* by divalent transition metals. *FEBS Letters*, *96*, 364–366.

Kung, F. C., Raymond, J., & Glaser, D. A. (1976). Metal ion content of Escherichia coli versus cell age. *Journal of Bacteriology*, *126*, 1089–1095.

Lane, T. W., & Morel, F. M. M. (2000). A biological function for cadmium in marine diatoms. *Proceedings of the National Academy of Sciences of the United States of America*, *97*, 4627–4631.

Lane, T. W., Saito, M. A., George, G. N., Pickering, I. J., Prince, R. C., & Morel, F. M. (2005). Biochemistry: A cadmium enzyme from a marine diatom. *Nature*, *435*, 42.

Lawrence, M. C., Pilling, P. A., Epa, V. C., Berry, A. M., Ogunniyi, A. D., & Paton, J. C. (1998). The crystal structure of pneumococcal surface antigen PsaA reveals a metal-binding site and a novel structure for a putative ABC-type binding protein. *Structure*, *6*, 1553–1561.

Layer, G., Moser, J., Heinz, D. W., Jahn, D., & Schubert, W. D. (2003). Crystal structure of coproporphyrinogen III oxidase reveals cofactor geometry of radical SAM enzymes. *The EMBO Journal*, *22*, 6214–6224.

Le Brun, N. E., Crow, A., Murphy, M. E., Mauk, A. G., & Moore, G. R. (2010). Iron core mineralisation in prokaryotic ferritins. *Biochimica Biophysica Acta*, *1800*, 732–744.

Lee, J.-W., & Helmann, J. D. (2006). The PerR transcription factor senses $H_2O_2$ by metal-catalysed histidine oxidation. *Nature*, *440*, 363–367.

Lee, J. W., & Helmann, J. D. (2007). Functional specialization within the Fur family of metalloregulators. *Biometals*, *20*, 485–499.

Lee, S. J., McCormick, M. S., Lippard, S. J., & Cho, U.-S. (2013). Control of substrate access to the active site in methane monooxygenase. *Nature*, *494*, 380–384.

Li, X., Pal, U., Ramamoorthi, N., Liu, X., Desrosiers, D. C., Eggers, C. H., et al. (2007). The Lyme disease agent *Borrelia burgdorferi* requires BB0690, a Dps homologue, to persist within ticks. *Molecular Microbiology*, *63*, 694–710.

Li, N., Yang, X. Y., Guo, Z., Zhang, J., Cao, K., Han, J., et al. (2014). Varied metal-binding properties of lipoprotein PsaA in *Streptococcus pneumoniae*. *Journal of Biological Inorganic Chemistry*, *19*, 829–838.

Liu, T., Ramesh, A., Ma, Z., Ward, S. K., Zhang, L., George, G. N., et al. (2007). CsoR is a novel *Mycobacterium tuberculosis* copper-sensing transcriptional regulator. *Nature Chemical Biology*, *3*, 60–68.

Ma, Z., Chandrangsu, P., Helmann, T. C., Romsang, A., Gaballa, A., & Helmann, J. D. (2014). Bacillithiol is a major buffer of the labile zinc pool in Bacillus subtilis. *Molecular Microbiology*, *94*, 756–770.

Ma, Z., Cowart, D. M., Scott, R. A., & Giedroc, D. P. (2009). Molecular insights into the metal selectivity of the copper(I)-sensing repressor CsoR from *Bacillus subtilis*. *Biochemistry*, *48*, 3325–3334.

Macomber, L., Elsey, S. P., & Hausinger, R. P. (2011). Fructose-1,6-bisphosphate aldolase (class II) is the primary site of nickel toxicity in *Escherichia coli*. *Molecular Microbiology*, *82*, 1291–1300.

Macomber, L., & Imlay, J. A. (2009). The iron-sulfur clusters of dehydratases are primary intracellular targets of copper toxicity. *Proceedings of the National Academy of Sciences of the United States of America*, *106*, 8344–8349.

Macomber, L., Rensing, C., & Imlay, J. A. (2007). Intracellular copper does not catalyse the formation of oxidative DNA damage in *Escherichia coli*. *Journal of Bacteriology*, *189*, 1616–1626.

Majtan, T., Frerman, F. E., & Kraus, J. P. (2011). Effect of cobalt on *Escherichia coli* metabolism and metalloporphyrin formation. *Biometals*, *24*, 335–347.

Martin, M. E., Byers, B. R., Olson, M. O., Salin, M. L., Arceneaux, J. E., & Tolbert, C. (1986). A Streptococcus mutans superoxide dismutase that is active with either manganese or iron as a cofactor. *Journal of Biological Chemistry*, *261*, 9361–9367.

Martin, J. H., & Fitzwater, S. E. (1988). Iron deficiency limits phytoplankton growth in north-east Pacific subarctic. *Nature, 331*, 341–343.

Martin, J. E., Waters, L. S., Storz, G., & Imlay, J. A. (2015). The Escherichia coli small protein MntS and exporter MntP optimize the intracellular concentration of manganese. *PLoS Genetics, 11*, e1004977.

Massé, E., & Gottesman, S. (2002). A small RNA regulates the expression of genes involved in iron metabolism in *Escherichia coli*. *Proceedings of the National Academy of Sciences of the United States of America, 99*, 4620–4625.

Massé, E., Vanderpool, C. K., & Gottesman, S. (2005). Effect of RyhB small RNA on global iron use in *Escherichia coli*. *Journal of Bacteriology, 187*, 6962–6971.

McDevitt, C. A., Ogunniyi, A. D., Valkov, E., Lawrence, M. C., Kobe, B., McEwan, A. G., et al. (2011). A molecular mechanism for bacterial susceptibility to zinc. *PLoS Pathogens, 7*, e1002357.

Mijnendonckx, K., Leys, N., Mahillon, J., Silver, S., & Van Houdt, R. (2013). Antimicrobial silver: Uses, toxicity and potential for resistance. *Biometals, 26*, 609–621.

Molteni, C., Abicht, H. K., & Solioz, M. (2010). Killing of bacteria by copper surfaces involves dissolved copper. *Applied and Environmental Microbiology, 76*, 4099–4101.

Moore, C. M., Gaballa, A., Hui, M., Ye, R. W., & Helmann, J. D. (2005). Genetic and physiological responses of *Bacillus subtilis* to metal ion stress. *Molecular Microbiology, 57*, 27–40.

Morby, A. P., Turner, J. S., Huckle, J. W., & Robinson, N. J. (1993). SmtB is a metal-dependent repressor of the cyanobacterial metallothionein gene *smtA*: Identification of a Zn inhibited DNA-protein complex. *Nucleic Acids Research, 21*, 921–925.

Morimoto, K., Yamashita, E., Kondou, Y., Lee, S. J., Arisaka, F., Tsukihara, T., et al. (2006). The asymmetric IscA homodimer with an exposed [2Fe-2S] cluster suggests the structural basis of the Fe-S cluster biosynthetic scaffold. *Journal of Molecular Biology, 360*, 117–132.

Morrissey, J. A., Cockayne, A., Hill, P. J., & Williams, P. (2000). Molecular cloning and analysis of a putative siderophore ABC transporter from Staphylococcus aureus. *Infection and Immunity, 68*, 6281–6288.

Nakashige, T. G., Zhang, B., Krebs, C., & Nolan, E. M. (2015). Human calprotectin is an iron-sequestering host-defense protein. *Nature Chemical Biology, 10*, 765–771.

Nandakumar, R., Espirito Santo, C., Madayiputhiya, N., & Grass, G. (2011). Quantitative proteomic profiling of the *Escherichia coli* response to metallic copper surfaces. *Biometals, 24*, 429–444.

Niederhoffer, E. C., Naranjo, C. M., Bradley, K. L., & Fee, J. A. (1990). Control of *Escherichia coli* superoxide dismutase (*sodA* and *sodB*) genes by the ferric uptake regulation (*fur*) locus. *Journal of Bacteriology, 172*, 1930–1938.

Noyce, J. O., Michels, H., & Keevil, C. W. (2006). Potential use of copper surfaces to reduce survival of epidemic methicillin-resistant *Staphylococcus aureus* in the healthcare environment. *Journal of Hospital Infection, 63*, 289–297.

Noyce, J. O., Michels, H., & Keevil, C. W. (2007). Inactivation of influenza A virus on copper versus stainless steel surfaces. *Applied and Environmental Microbiology, 73*, 2748–2750.

Olson, J. W., Mehta, N. S., & Maier, R. J. (2001). Requirement of nickel metabolism proteins HypA and HypB for full activity of both hydrogenase and urease in *Helicobacter pylori*. *Molecular Microbiology, 39*, 176–182.

Ong, C. L. Y., Gillen, C. M., Barnett, T. C., Walker, M. J., & McEwan, A. G. (2014). An antimicrobial role for zinc in innate immune defense against Group A *Streptococcus pneumoniae*. *Journal of Infectious Disease, 209*, 1500–1508.

Osman, D., Waldron, K. J., Denton, H., Taylor, C. M., Grant, A. J., Mastroeni, P., et al. (2010). Copper homeostasis in *Salmonella* is atypical and copper-CueP is a major periplasmic metal complex. *Journal of Biological Chemistry, 285*, 25259–25268.

Outten, C. E., & O'Halloran, T. V. (2001). Femtomolar sensitivity of metalloregulatory proteins controlling zinc homeostasis. *Science*, *292*, 2488–2492.

Panina, E. M., Mironov, A. A., & Gelfand, M. S. (2003). Comparative genomics of bacterial zinc regulons: Enhanced ion transport, pathogenesis, and rearrangement of ribosomal proteins. *Proceedings of the National Academy of Sciences of the United States of America*, *100*, 9912–9917.

Park, S., & Imlay, J. A. (2003). High levels of intracellular cysteine promote oxidative DNA damage by driving the Fenton reaction. *Journal of Bacteriology*, *185*, 1942–1950.

Parr, R. G., & Pearson, R. G. (1983). Absolute hardness: Companion parameter to absolute electronegativity. *Journal of the American Chemical Society*, *105*, 7512–7516.

Patterson, C. J., Pernil, R., Dainty, S. J., Chakrabarti, B., Henry, C. E., Money, V. A., et al. (2013). Co(II)-detection does not follow $K_{Co(II)}$ gradient: Channeling in Co(II) sensing. *Metallomics*, *5*, 352–362.

Pearson, R. G. (1963). Hard and soft acids and bases. *Journal of the American Chemical Society*, *85*, 3533–3539.

Philips, S. J., Canalizo-Hernandez, M., Yildirim, I., Schatz, G. C., Mondragón, A., & O'Halloran, T. V. (2015). Allosteric transcriptional regulation via changes in the overall topology of the core promoter. *Science*, *349*, 877–881.

Pi, H., Patel, S. J., Argüello, J. M., & Helmann, J. D. (2016). The Listeria monocytogenes Fur-regulated virulence protein FrvA is an Fe (II) efflux P1B4-type ATPase. *Molecular Microbiology*, *100*, 1066–1079.

Posey, J. E., & Gherardini, F. C. (2000). Lack of a role for iron in the Lyme disease pathogen. *Science*, *288*, 1651–1653.

Ranquet, C., Ollagnier-de-Choudens, S., Loiseau, L., Barras, F., & Fontecave, M. (2007). Cobalt stress in *Escherichia coli*: The effect on the iron-sulfur proteins. *Journal of Biological Chemistry*, *282*, 30442–30451.

Rathnayake, I. V. N., Megharaj, M., Krishnamurti, G. S. R., Bolan, N. S., & Naidu, R. (2012). Heavy metal toxicity to bacteria—Are the existing growth media accurate enough to determine heavy metal toxicity? *Chemosphere*, *90*, 1195–1200.

Ray, P., Girard, V., Gault, M., Job, C., Bonneu, M., Mandrand-Berthelot, M.-A., et al. (2013). *Pseudomonas putida* KT2440 response to nickel or cobalt induced stress by quantitative proteomics. *Metallomics*, *5*, 68–79.

Reeder, N. L., Kaplan, J., Xu, J., Youngquist, R. S., Wallace, J., Hu, P., et al. (2011). Zinc pyrithione inhibits yeast growth through copper influx and inactivation of iron-sulfur proteins. *Antimicrobial Agents and Chemotherapy*, *55*, 5753–5760.

Rehder, D. (2015). The role of vanadium in biology. *Metallomics*, *7*, 730–742.

Remy, L., Carrière, M., Derré-Bobillot, A., Martini, C., Sanguinettit, M., & Borezée-Durant, E. (2013). The *Staphylococcus aureus* Opp1 ABC transporter imports nickel and cobalt in zinc-depleted conditions and contributes to virulence. *Molecular Microbiology*, *87*, 730–743.

Riether, K., Dollard, M. A., & Billard, P. (2001). Assessment of heavy metal bioavailability using Escherichia coli zntAp::lux and copAp::lux-based biosensors. *Applied Microbiology and Biotechnology*, *57*(5-6), 712–716.

Rodriguez, G. M., Voskuil, M. I., Gold, B., Schoolnik, G. K., & Smith, I. (2002). *ideR*, an essential gene in *Mycobacterium tuberculosis*: Role of IdeR in iron-dependent gene expression, iron metabolism, and oxidative stress response. *Infection and Immunity*, *70*, 3371–3381.

Rosch, J. W., Gao, G., Ridout, G., Wang, Y.-D., & Tuomanen, E. I. (2009). Role of the manganese efflux system *mntE* for signalling and pathogenesis of *Streptococcus pneumoniae*. *Molecular Microbiology*, *72*, 12–25.

Rutherford, J. C., Cavet, J. S., & Robinson, N. J. (1999). Cobalt-dependent transcriptional switching by a dual-effector MerR-like protein regulates a cobalt-exporting variant CPx-type ATPase. *Journal of Biological Chemistry, 274,* 25827–25832.

Saito, M. A., Bertrand, E. M., Dutkiewicz, S., Bulygin, V. V., Moran, D. W., Monteiro, F. M., et al. (2011). Iron conservation by reduction of metalloenzyme inventories in the marine diazotroph *Crocosphaera watsoni*. *Proceedings of the National Academy of Sciences of the United States of America, 108,* 2184–2189.

Saito, M. A., Sigman, D. M., & Morel, F. M. M. (2003). The bioinorganic chemistry of the ancient ocean: The co-evolution of cyanobacterial metal requirements and biogeochemical cycles at the Archean-Proterozoic boundary? *Inorganica Chimica Acta, 356,* 308–318.

Salgado, C. D., Sepkowitz, K. A., John, J. F., Cantey, J. R., Attaway, H. H., Freeman, K. D., et al. (2013). Copper surfaces reduce the rate of healthcare-acquired infections in the intensive care unit. *Infection Control & Hospital Epidemiology, 34,* 479–486.

Schäfer, G., & Kardinahl, S. (2003). Iron superoxide dismutases: Structure and function of an archaic enzyme. *Biochemical Society Transactions, 31,* 1330–1334.

Schirrmeister, B. E., de Vos, J. M., Antonelli, A., & Bagheri, H. C. (2013). Evolution of multicellularity coincided with increased diversification of cyanobacteria and the Great Oxidation Event. *Proceedings of the National Academy of Sciences of the United States of America, 110,* 1791–1796.

Schreiter, E. R., Wang, S. C., Zamble, D. B., & Drennan, C. L. (2006). NikR-operator complex structure and the mechanism of repressor activation by metal ions. *Proceedings of the National Academy of Sciences of the United States of America, 103,* 13676–13681.

Schwarz, G., & Mendel, R. R. (2006). Molybdenum cofactor biosynthesis and molybdenum enzymes. *The Annual Review of Plant Biology, 57,* 623–647.

Semrau, J. D., Jagadevan, S., DiSpirito, A. A., Khalifa, A., Scanlan, J., Bergman, B. H., et al. (2013). Methanobactin and MmoD work in concert to act as the 'copper-switch' in methanotrophs. *Environmental Microbiology, 15,* 3077–3086.

Sevcenco, A.-M., Krijger, G. C., Pinkse, M. W. H., Verhaert, P. D. E. M., Hagen, W. R., & Hagedoorn, P.-L. (2009). Development of a generic approach to native metalloproteomics: Application to the quantitative identification of soluble copper proteins in *Escherichia coli*. *Journal of Biological Inorganic Chemistry, 14,* 631–640.

Silver, S., Phung, L. T., & Silver, G. (2006). Silver as biocides in burn and wound dressings and bacterial resistance to silver compounds. *Journal of Industrial Microbiology and Biotechnology, 33,* 627–634.

Smith, C. L., Stauber, J. L., Wilson, M. R., & Jolley, D. F. (2014). The use of immobilised metal affinity chromatography (IMAC) to compare expression of copper-binding proteins in control and copper-exposed marine microalgae. *Analytical and Bioanalytical Chemistry, 406,* 305–315.

Stähler, F. N., Odenbreit, S., Haas, R., Wilrich, J., Van Vliet, A. H., Kusters, J. G., et al. (2006). The novel *Helicobacter pylori* CznABC metal efflux pump is required for cadmium, zinc, and nickel resistance, urease modulation, and gastric colonization. *Infection and Immunity, 74,* 3845–3852.

Strand, T. A., Hollingshead, S. K., Julshamn, K., Briles, D. E., Blomberg, B., & Sommerfelt, H. (2003). Effects of zinc deficiency and pneumococcal surface protein a immunization on zinc status and the risk of severe infection in mice. *Infection and Immunity, 71,* 2009–2013.

Sun, X., Xiao, C.-L., Ge, R., Yin, X., Li, H., Li, N., et al. (2011). Putative copper- and zinc-binding motifs in *Streptococcus pneumoniae* identified by immobilized metal affinity chromatography and mass spectrometry. *Proteomics, 11,* 3288–3298.

Taudte, N., German, N., Zhu, Y. G., Grass, G., & Rensing, C. (2016). Restoration of growth by manganese in a mutant strain of *Escherichia coli* lacking most known iron and manganese uptake systems. *Biometals, 29*, 433–450.

Thelwell, C., Robinson, N. J., & Turner-Cavet, J. S. (1998). An SmtB-like repressor from *Synechocystis* PCC 6803 regulates a zinc exporter. *Proceedings of the National Academy of Sciences of the United States of America, 95*, 10728–10733.

Thieme, D., Neubauer, P., Nies, D. H., & Grass, G. (2008). Sandwich hybridization assay for sensitive detection of dynamic changes in mRNA transcript levels in crude *Escherichia coli* cell extracts in response to copper ions. *Applied and Environmental Microbiology, 74*, 7463–7470.

Thompson, D. K., Beliaev, A. S., Giometti, C. S., Tollaksen, S. L., Khare, T., Lies, D. P., et al. (2002). Transcriptional and proteomic analysis of a ferric uptake regulator (Fur) mutant of *Shewanella oneidensis*: Possible involvement of Fur in energy metabolism, transcriptional regulation, and oxidative stress. *Applied and Environmental Microbiology, 68*, 881–892.

Thorgersen, M. P., & Downs, D. M. (2007). Cobalt targets multiple metabolic processes in *Salmonella enterica*. *Journal of Bacteriology, 189*, 7774–7781.

Tottey, S., Harvie, D. R., & Robinson, N. J. (2007). Understanding how cells allocate metals. In D. H. Nies & S. Silver (Eds.), *Molecular microbiology of heavy metals* (pp. 1–33). Germany: Springer Berlin/Heidelberg.

Tottey, S., Patterson, C. J., Banci, L., Bertini, I., Felli, I. C., Pavelkova, A., et al. (2012). Cyanobacterial metallochaperone inhibits deleterious side-reactions of copper. *Proceedings of the National Academy of Sciences of the United States of America, 109*, 95–100.

Tottey, S., Waldron, K. J., Firbank, S. J., Reale, B., Bessant, C., Sato, K., et al. (2008). Protein-folding location can regulate manganese-binding versus copper- or zinc-binding. *Nature, 455*, 1138–1142.

Touati, D., Jacques, M., Tardat, B., Bouchard, L., & Despied, S. (1995). Lethal oxidative damage and mutagenesis are generated by iron in Δ*fur* mutants of *Escherichia coli*: Protective role of superoxide dismutase. *Journal of Bacteriology, 177*, 2305–2314.

Van der Heijden, J., Vogt, S. L., Reynolds, L. A., Peña-Díaz, J., Tupin, A., Aussel, L., et al. (2016). Exploring the redox balance inside gram-negative bacteria with redox-sensitive GFP. *Free Radical Biology and Medicine, 91*, 34–44.

Vigonsky, E., Fish, I., Livnat-Levanon, N., Ovcharenko, E., Ben-Tal, N., & Lewinson, O. (2015). Metal binding spectrum and model structure of the *Bacillus anthracis* virulence determinant MntA. *Metallomics, 7*, 1407–1419.

Vita, N., Landolfi, G., Baslé, A., Platsaki, S., Lee, J., Waldron, K. J., et al. (2016). Bacterial cytosolic proteins with a high capacity for Cu(I) that protect against copper toxicity. *Scientific Reports, 6*, 39065.

Vita, N., Platsaki, S., Baslé, A., Allen, S. J., Paterson, N. G., Crombie, A. T., et al. (2015). A four-helix bundle stores copper for methane oxidation. *Nature, 525*, 140–143.

Waldron, K. J., & Robinson, N. J. (2009). How do bacterial cells ensure that metalloproteins get the correct metal? *Nature Reviews. Microbiology, 6*, 25–35.

Waldron, K. J., Rutherford, J. C., Ford, D., & Robinson, N. J. (2009). Metalloproteins and metal sensing. *Nature, 460*, 823–830.

Wang, S., Wu, Y., & Outten, F. W. (2011). Fur and the novel regulator YqjI control transcription of the ferric reductase gene *yqjH* in *Escherichia coli*. *Journal of Bacteriology, 193*, 563–574.

Watanabe, S., Kita, A., Kobayashi, K., & Miki, K. (2008). Crystal structure of the [2Fe-2S] oxidative-stress sensor SoxR bound to DNA. *Proceedings of the National Academy of Sciences of the United States of America, 105*, 4121–4126.

Watson, A. J., Bakker, D. C., Ridgwell, A. J., Boyd, P. W., & Law, C. S. (2000). Effect of iron supply on Southern Ocean $CO_2$ uptake and implications for glacial atmospheric $CO_2$. *Nature, 407*, 730–733.

Weaver, L., Michels, H. T., & Keevil, C. W. (2008). Survival of *Clostridium difficile* on copper and steel: Futuristic options for hospital hygiene. *Journal of Hospital Infection, 68*, 145–151.

Weinberg, E. D. (1984). Iron withholding: A defense against infection and neoplasia. *Physiological Reviews, 64*, 65–102.

White, C., Lee, J., Kambe, T., Fritsche, K., & Petris, M. J. (2009). A role for the ATP7A copper-transporting ATPase in macrophage bactericidal activity. *Journal of Biological Chemistry, 284*, 33949–33956.

Wolfram, L., & Bauerfeind, P. (2009). Activities of urease and nickel uptake of *Helicobacter pylori* proteins are media- and host-dependent. *Helicobacter, 14*, 264–270.

Wolschendorf, F., Ackart, D., Shrestha, T. B., Hascall-Dove, L., Nolan, S., Lamichhane, G., et al. (2011). Copper resistance is essential for virulence of Mycobacterium tuberculosis. *Proceedings of the National Academy of Sciences of the United States of America, 108*, 1621–1626.

Wu, J., & Weiss, B. (1991). Two divergently transcribed genes, *soxR* and *soxS*, control a superoxide response regulon of *Escherichia coli*. *Journal of Bacteriology, 173*, 2864–2871.

Yazdankhah, S., Rudi, K., & Bernhoff, A. (2014). Zinc and copper in animal feed—Development of resistance and co-resistance to antimicrobial agents in bacteria of animal origin. *Microbial Ecology in Health and Disease, 25*, 25862.

Zhang, L., McSpadden, B., Pakrasi, H. B., & Whitmarsh, J. (1992). Copper-mediated regulation of cytochrome c553 and plastocyanin in the cyanobacterium Synechocystis 6803. *Journal of Biological Chemistry, 267*, 19054–19059.

MIX
Papier | Fördert gute Waldnutzung
FSC® C083411

Kolibri 360 GmbH
Ferdinand-Jühlke-Straße 7
99095 Erfurt, Deutschland
produktsicherheit@kolibri360.de